TRIZ For Engineers: Enabling Inventive Problem Solving

TRIZ For Engineers: Enabling Inventive Problem Solving

Karen Gadd
Oxford Creativity

A John Wiley & Sons, Ltd., Publication

This edition first published 2011

Registered office
John Wiley & Sons Ltd, The Atrium, Southern Gate, Chichester, West Sussex, PO19 8SQ, United Kingdom

For details of our global editorial offices, for customer services and for information about how to apply for permission to reuse the copyright material in this book please see our website at www.wiley.com.

Library of Congress Cataloguing-in-Publication Data

Gadd, Karen.
TRIZ for engineers : enabling inventive problem solving / Karen Gadd.
p. cm.
Includes index.
ISBN 978-0-470-74188-7 (cloth)
1. TRIZ theory. 2. Inventions. 3. Engineering–Methodology. 4. Problem solving–Methodology. I. Title.
T212.G33 2011
620.0028–dc22

2010034198

A catalogue record for this book is available from the British Library.
ISBN pbk: 9780470741887
ISBN ePDF: 9780470684337
ISBN oBook: 9780470684320
ISBN epub: 9780470975435

Set in 9/11 pt Times by Toppan Best-set Premedia Limited

Contents

About the Author

Karen Gadd has been teaching TRIZ and problem solving with engineering teams from major companies for over 13 years. Her mission is to make TRIZ learning straightforward and the TRIZ Tools easy to use. She has worked on nothing but TRIZ since discovering and learning its power to give us all the routes, to all the solutions, to all engineering problems.

In 1998 Karen started Oxford Creativity to concentrate on developing simple and practical TRIZ problem solving for the European market. Karen has taken TRIZ to major companies including Rolls-Royce, British Nuclear Group, Bentley Motors, BAE Systems, Nissan, Pilkington, Borealis and Sanofi Aventis. Oxford Creativity is now well established as one of the world's top TRIZ companies and has helped to make TRIZ well known and widely used throughout Europe and encouraged top companies to create expert TRIZ teams for innovative problem solving.

Karen studied Mechanical Engineering at Imperial College, and has an MBA from London Business School. After working in strategy and corporate planning in the City of London she returned to live in Oxford and was a tutor at Oxford's Business School the European School of Management ESCP-EAP (based in Paris, Oxford, Madrid and Berlin). From 1995–2002 she was a Governor of Coventry University. Karen's career has been dedicated to creating new enterprises which make a difference – she founded both MUSIC at OXFORD and the European Union Baroque Orchestra and ran both for over ten years and raised millions in corporate sponsorship to make their activities possible. These successful music organisations still flourish. MUSIC at OXFORD transformed Oxford's music scene and is now approaching its 30th season of top professional classical concerts. EUBO has celebrated 25 years of launching the careers of talented young musicians and has been so successful in its mission, that there are now former EUBO students in every major professional baroque ensemble in the world. Karen launched Oxford Creativity to make TRIZ accessible to everyone and transform and launch careers of TRIZ enthusiasts and champions. There are now thousands of engineers who have learned TRIZ from Karen and who intelligently daily apply TRIZ to solve difficult technical and scientific problems.

Karen is long married, has four children and three grandchildren and lives happily in Oxford and the Lake District. Karen has recently become a director of the Orchestra of St.John's. Concerts and singing are still her interests and part of her activities, as well as speaking at conferences throughout the world on the success and power of TRIZ.

Acknowledgements

I would like to thank the following people for their support and help.

My Managing Director and daughter, Lilly Haines-Gadd, for her unfailing enthusiasm, support, patience and encouragement.

My colleagues who have worked with me to create the simple approaches to TRIZ in this book – we have all learned from each other and I am grateful for all they have taught me – most especially Henry Strickland, Andrew Martin, Andrea Mica.

All those who have helped me put my ideas and TRIZ solutions into a form to communicate it to others. Merryn Haines-Gadd, our graphic designer, who turns my thoughts into pictures, and Eric Willner and Nicky Skinner of Wiley for their help, optimism and common sense advice and Caroline Davies of Oxford Creativity.

Our whole TRIZ community and all those who learned TRIZ from me and my colleagues, and who have joined us in our quest to make TRIZ accessible within their organisations and beyond, especially:

Frédéric Mathis (Mars), Ric Parker (Rolls-Royce), Dave Knott (Rolls-Royce), Pauline Marsh (BAE Systems), Simon Brodie (RAF) Mike West (Babcock), and Professor Derek Sheldon (Institution of Mechanical Engineers).

Those who have worked to bring TRIZ to the world and first introduced me to the power of its toolkit, most especially, Ellen Domb and Sergei Ikovenko.

My thanks to all the engineers who have taught me, inspired me and worked with me to solve problems with TRIZ. To all the TRIZ teams of engineers we have worked with (but we cannot name for security reasons) whose clever ideas during TRIZ sessions helped us to extend and develop our TRIZ thinking.

My own family of engineers: my father Kenneth Gadd who claims that as an old engineer (contemporary, but unknowing of Altshuller) he represents a generation whose engineering prowess was unmatched by any previous or succeeding generations. All my fellow Imperial College engineers, including my husband Geoff Haines (a middle-aged engineer) and to my son Jonathan Haines-Gadd (a young engineer), and those few members of my family who don't yet work with TRIZ but who have supported me in my championship for such challenging causes – my mother Kathleen Gadd, my daughter Rebecca Haines-Gadd and my TRIZniks of the future, my grandchildren Isobelle, Livia and Freddie.

Most especially, the great Genrich Altshuller – my gratitude and respect for his extraordinary vision to uncover and summarise the world's engineering genius grows with all the TRIZ work I undertake. I have returned to his source material at every stage of my learning, and I strive to merely interpret the power and logic of his TRIZ.

This is what I offer with this book, faithfulness to Altshuller's TRIZ tools and I hope simple, clear innovative ways for understanding and using them. TRIZ requires both right (creative) and left (logical and systematic) sides of the brain to make it work. The only known way to join these two is with laugher and humour and the TRIZ cartoons attempt to achieve this. I have been assisted in this by the wonderful cartoonist Clive Goddard, who has worked with me to create TRIZ cartoons to help show just how much fun TRIZ offers. I hope we have succeeded.

Foreword

Since discovering TRIZ, Karen Gadd has been an enthusiastic, near-evangelical, promoter of the methodology. She has introduced this exciting modern approach to innovation to many companies, including my own, as well as countless individuals. I am very pleased that she has now captured the essence of TRIZ in this well-written and very readable book, with its colourful and amusing illustrations, making TRIZ accessible to an even wider audience.

TRIZ is a contradiction in terms: it is free thought by numbers; it is a top-down approach to lateral thinking; it is a structured approach to brainstorming. Emerging from post-cold-war Russia, TRIZ is based on an intense and systematic study of all the world's great inventions. All of these can be broken down to be rendered as a combination of simple physical phenomena. There are actually relatively few conceptual solutions to problems (about 100) and 2,500 or so scientific principles. By presenting the designer, or inventor with novel (at times implausible) combinations of principles, new ways of solving problems can be discovered.

Look around you. Mix static electricity and photographic imaging and you have the instant, dry printing of the photo-copier. Mix suction and vortices with centrifugal force and you have the bag-less simplicity of the new generation of vacuum cleaners. The list is endless, as are the possibilities for new inventions.

We don't realise how bounded are our conceptual spaces and how linear our thought processes. We are locked in by past experience, by 'rules' passed down by our teachers, by the 'it'll never work' attitudes of those around us. It is far easier to suppress innovation than to stimulate it. If, like the Queen in Alice in Wonderland, you sometimes believe as many as six impossible things before breakfast, you may be able to get by without TRIZ. Many seemingly impossible things are indeed impossible, but some are not. TRIZ will set you free to explore the world of the apparently impossible.

TRIZ is an essential part of the modern engineer's or inventor's tool kit. This book will not make you an expert overnight, but it will hopefully stimulate your interest and make you want to explore more deeply the world of TRIZ.

Professor Ric Parker, FREng
Director of Research and Technology
Rolls-Royce Group

Introduction

Teoriya Resheniya Izobretatelskikh Zadatch = Theory of Inventive Problem Solving

Теория Решения Изобретательских Задач

TRIZ is an engineering problem solving toolkit which successfully summarizes past solutions and successes to show us how to systematically solve future problems. TRIZ comes from Russia, initially and primarily the work of Genrich Altshuller, a great engineer and inventor, perhaps one of the greatest engineers of the twentieth century, whose work helps all other engineers. All good engineers live with both uncertainty and certainty – uncertainty about where to find the solution to the next problem and certainty that a solution will be found. TRIZ enhances and speeds up this process by directing us to the places full of good solutions to our particular problems. TRIZ focuses our problem understanding to the particular, relevant problem model and then offers conceptual solutions to that model. Good engineers reduce wasted time with TRIZ as they head straight for the valid solutions and use their valuable time to define their problem accurately, find all the solutions to that problem and then develop those solutions. TRIZ and other toolkits help in all these various stages of problem understanding but only TRIZ helps in the solution locating stage. The best engineers enjoy complex problem solving and finding new, better innovative solutions – TRIZ enhances their abilities as innovators and just trims out the wasted empty trials and dead ends. TRIZ keeps engineers doing what they do best – solving problems – and takes away nothing but time wasting, brain deadening, complex and irrelevant detail. TRIZ helps engineers power forward to useful and practical answers.

TRIZ is a toolkit – each tool is simple to use, and between them they cover all aspects of problem understanding and solving. The only challenge with TRIZ is learning which tool to use when, and this comes with practice and familiarity. Complete TRIZ algorithms are hard to master, as they set out to cover all problem situations, and are about as useful as an algorithm to help you complete 18 holes in golf. This book takes you through each TRIZ tool in turn and suggests when and where to use them, and offers some simple problem solving flowcharts for each tool. Once each TRIZ tool is mastered, it will become part of your problem solving tool; some you will use everyday and some just occasionally, but they are all useful for engineering problems, and are great thinking tools to help good engineers become great engineers.

In Russian TRIZ is written as above. TRIZ became known and available outside Russia after 1993, although there were some occasional TRIZ activities before that in some Western companies. In the USA there was some resistance to a Russian technique with a strange acronym name, which to most people was incomprehensible and unpronounceable. Some attempts were made to anglicise and rename it as TIPS (Theory of Inventive Problem Solving) – this was not widely adopted and in the USA TRIZ is now pronounced as TREES and in Europe TRIZZZZ.

At my first Altshuller conference in California I was surprised that there seemed to be three TRIZ camps: the Russian, the American and the European, and they didn't seem to be mixing much. (There are now other powerful camps from Korea and Japan). I called my paper TIPSY TRIZ and talked about the western TIPS acronym representing the resistance to Russian thoroughness, the temptation to oversimplify TRIZ in order to gain acceptance and to break down the initial resistance to this rigorous toolkit. I talked about the UK TRIZ serious successes with major engineering companies, and the TRIZ impact on UK engineering, but at the end I made jokes about trying to problem-solve whilst under the influence of vodka – only the Europeans laughed, the Americans viewed me with disapproval, and the Russians with incomprehension. The paper was mostly about overcoming the difficulties of selling an unknown but brilliant Russian process to Western companies who initially view it with weariness and suspicion, but once familiar with TRIZ are almost always smitten, embrace it with enthusiasm, and how company experts in other toolkits always argue that TRIZ is like ….. and name their favourites. Persuading them that TRIZ is unlike any other toolkit, but complimentary to the others, first involves getting them to use TRIZ to successfully solve problems. This demonstrates that TRIZ covers the parts that no other toolkits even attempt – and that TRIZ is a toolkit for moving from vague problem to defined problem, and then to locating relevant conceptual solutions distilled from all of science and engineering – and that no other toolkit comes close.

I was telling dispiriting stories of how despite its unique power, how hard it can be to sell TRIZ to European companies, and then (I hoped) the more cheering tale of how Oxford Creativity used TRIZ to overcome the problems of getting engineers and engineering communities to adopt TRIZ in the UK. In particular I described our spectacular success in taking TRIZ to Rolls-Royce and the effect of our TRIZ training and problem solving with many hundreds of their engineers. Getting Rolls-Royce to adopt TRIZ took over three years and its introduction was due to TRIZ being initially championed by their R&T director Ric Parker assisted by Dave Knott – a rare prior TRIZ convert earlier, as he had heard a TRIZ lecture some years before and written his most successful patent the next day. With Ric and Dave's help and enthusiasm we overcame the inertia and hostility and TRIZ has been one of their core competence tools since 2000. Since then our other successes have included BAE Systems where even the most experienced and curmudgeonly engineers can be turned around to energetic enthusiasm.

I have always been amazed that, until some small understanding is established, how strong the resistance and inertia to TRIZ can be at a corporate and personal level. This is despite its huge value to engineers, its documented successes, and once accepted and learned, its transforming power on even the most plodding of engineers to think clearly and problem-solve innovatively, quickly and effectively.

Recently whilst teaching a seminar at a college in Oxford University to six technical directors of an international engineering company, we asked them why they had been considering TRIZ with us since 1999 yet ten years later were cautiously allowing themselves a one-day seminar. "We thought TRIZ was either too trivial or too complex for us" came the answer from these clever engineers: all were in desperate need of new solutions, new products and greater understanding of future technologies. After the one day they said TRIZ switched the lights on for them in an area of new technology, where before they had been groping around each equipped with the relatively small torch of their own knowledge and ideas. They have since all learned TRIZ and are working to establish it throughout the company.

The TRIZ tools were developed in Russia by engineers for engineers with thousands of man-years of work (and many women-years). Russia 30-50 years ago was a very different culture to our own and time was not of the essence for them and to learn TRIZ the Russian way was, and is, rigorous, requires great application of thought, with lots of worked examples and at least three months is recommended. This is not practical in Europe today and together with other TRIZ users I have endeavoured to create TRIZ Workshops which do not compromise the thoroughness or rigour of TRIZ but will give an understanding of the best TRIZ tools in days, rather than months.

We have also used TRIZ to solve the essential contradiction of learning TRIZ – how to teach a powerful, set of problem solving tools to engineers and managers who are hard pressed to spare the time. We explain the TRIZ tools and show how to use them after each workshop on their own problems; back at work there needs to be sufficient time and practical application for TRIZ to become useful and effective for problem solving for individuals and teams. Enough experience, and even some small successes in the real world, as well as solving big problems that matter, and it will become second nature. TRIZ is very much about analogous thinking, so learning TRIZ is a bit like learning to swim or to drive – once we have been taught and have some confidence, to master it and really do it on our own we need to practise and build the skills and confidence to succeed. Then we know we are able to do it again and better – success and improvement will depend entirely on actually doing it … as often as possible. Nobody will commit to TRIZ until they understand and experience the power and speed of TRIZ to solve problems which will help ensure their company's future – but once committed to using TRIZ every hour invested in TRIZ will repay many dividends for you for ever.

It has been my privilege to teach (with my colleagues in Oxford Creativity) many thousands of really good and clever engineers in the last thirteen years. Almost all engineers seem to me to be very nice and trustworthy people (mostly men in the UK), with many virtues, including hard work, an appetite for understanding everything about the problem from the big picture to the relevant detail, responsible attitudes, a passion for good solutions, a genuine mistrust of trivia or flash quick answers, good humour and a genuine sense of humility.

All of us at OC teach TRIZ in our very different styles and with as much fun as possible. Despite a reputation for full days, hard work and light heartedness, TRIZ classes are acknowledged as enjoyable, useful if exhausting experiences and we hope delegates leave with a great deal to think about and practice. We teach TRIZ in two bites; two days learning the essential TRIZ tools, followed by two days on the TRIZ problem solving process. Over 95% of delegates sign up for the second course, which is encouraging.

This book is a result of thirteen years problem solving, teaching, (and learning) TRIZ. I offer this book in the same spirit as an apprentice had to offer a 'masterpiece' in the hope that he / she could now enter the ranks of their trade accepted by their masters. I hope I offer it in a spirit of humility and I do not offer my way of making TRIZ because I have left TRIZ mostly unchanged, but worked to reveal its simple and powerful logic, assisted by flow charts, pictures cartoons and even jokes.

However I have made one significant new approach and offer two routes for System Analysis – the traditional TRIZ Substance Field Analysis and the five classes of the 76 Standard Solutions and a simpler alternative used by elements of the TRIZ community of Function Analysis and three simple categories of 76 Standard Solutions – how to deal with harm, how to deal with insufficiency and how to detect/ measure something. I have taught both in major companies but have only used the second, easier system for many years.

I feel passionately that TRIZ should now be communicated clearly and simply without losing any of its rigour; made simple and straightforward but never, I hope, made trivial. I have always encouraged jokes, fun and laughter when learning and using TRIZ as it seems to make everyone more creative. I was pleased to see that scientific research has shown that the only way to become truly creative in scientific and engineering problem solving is by joining the left side of the brain (alleged to be systematic) with the right side of the brain (alleged to be creative). To make the right side of our brain join up with our left side we have to laugh and see the fun in situations.*

As in our classes I offer TRIZ tools with jokes and (I hope) humorous stories – not to make light of TRIZ but because I believe this research, which claims the importance and power of humour. I am a great believer in keeping as much fun and enjoyment in life as possible. I hope the cartoons I have commissioned from the wonderful Clive Goddard and the jokes I use here do not offend – I have a very English sense of humour.

Accompanying this book is a website (www.triz4engineers.com) that contains additional material and case studies. There are expanded versions of the 40 Principles and the Oxford Standard Solutions and an Effects database and links to other versions of the TRIZ effects. This website invites TRIZ engineers to contribute their own successes with TRIZ – including case studies – and an opportunity to share problem solving with other engineers.

*W. Wayt Gibbs, Side Splitting, *Scientific American*, January 2001.

Part One
TRIZ Logic and the Tools for Innovation and Clarity of Thought

TRIZ Tools for Creativity and Clever Solutions

What is TRIZ?

TRIZ is a unique, rigorous and powerful toolkit which guides engineers to understand and solve their problems by accessing the immense treasure of past engineering and scientific knowledge. TRIZ helps us find the surprisingly few relevant and practical answers to our real problems. This is made simple by the TRIZ summary of all the conceptual answers to engineering and scientific problems.

TRIZ is the only solution toolkit which exists so far in the world that offers engineers help beyond brainstorming at the actual *concept–solution locating* and *problem solving* moments. There are wonderful toolkits for understanding problems, with analysis processes for capturing the requirements, analysing the systems, looking at processes and pinpointing actual causes of problems. There are also many rigorous and useful toolkits for the time after problem solving has occurred, including processes for selecting solutions and developing them, with useful ways of evaluating and predicting costs etc.; but for the actual moment of problem solving – the search and capture of the right solutions or new concepts – there is only TRIZ. Until TRIZ the assumption has been that clever engineers and scientists would somehow find the right answers either individually or collectively by brainstorming and using their experience and knowledge.

TRIZ for Engineers: Enabling Inventive Problem Solving, First Edition. Karen Gadd.

BEFORE – Preparation for Problem Solving	DURING – Problem Solving	AFTER – Solution Selection and Development
TRIZ and other tool-kits for Requirements / needs capture System analysis Find any root causes of problems etc.	Brainwave ideas Brainstorming Creativity Tools and/or **TRIZ** **TRIZ Conceptual Solutions** 40 Principles 76 Standard Solutions 8 Trends all used with relevant **World's Knowledge** and **TRIZ Creativity Tools** including – Ideal Solutions Smart Little People Size-Time-Cost etc.	**TRIZ and other Tool-kits** for Concept development Concept selection process (such as Pugh Matrix) Successful innovation and new technologies etc.

Use all the available tools for each stage of problem solving

Before TRIZ no one had looked to systematically summarize all the published solutions to scientific and engineering problems and seek similarities, overlaps and patterns. When the founder of TRIZ, Altshuller, examined patent databases he found only 40 solutions to solving contradictions and a total of about 100 conceptual solutions for improving systems and solving problems which can be used together with the 2500 or so scientific and engineering theories (called effects in TRIZ) to solve problems. Together with his TRIZ community Altshuller developed TRIZ as the 'science of creativity'derived from all scientific and engineering success and offered a practical problem solving toolkit for engineering systems. The principal TRIZ tools direct us to find all the ways for improving and solving problems in existing systems and processes. They also help us find relevant concepts to develop the next-generation systems, and they offer systematic processes for inventing and developing new products. TRIZ has simple general concept lists to help us solve any problem; these TRIZ solution triggers are distilled from analysing all known published successful, scientific and engineering solutions. TRIZ tools are unique in their power for problem solving but TRIZ also offers simple tools for problem understanding, for system analysis, for understanding what we want, and for stimulating new ideas, creative thought and innovative solutions.

Who Uses TRIZ and Why?

For good engineers TRIZ offers the best of all worlds: the individual tools are straightforward, the problem-solving process is systematic and repeatable, and when we move fast with TRIZ we can uncover all the possible solutions and maintain and use our brains at their most creative. Engineers understand TRIZ better than anyone else because it comes from engineering success. It was developed by engineers for engineers, and although TRIZ works well on anything, it seems to work best on engineering problems. When first hearing of TRIZ, however, most engineers resist it, perhaps because it sounds too good to be true; but also maybe because it is new to them, not-invented-here and counter-intuitive to the way they approach problems. Unfortunately TRIZ is also seen as just another part of the overcrowded market in corporate toolkits (which range from excellent to trivial). Taking on a new toolkit makes demands on their brain power, company time and money and other hard-pressed resources, and this requires belief in its efficacy and power. Six Sigma and Lean are both promoted to improve efficiency and deliver cost savings and therefore widely adopted – TRIZ also delivers efficiency and similar cost savings plus innovation, fast problem solving and clear thinking. TRIZ is slowly becoming recognized as a unique addition to all the other company toolkits, as it supplements and complements them all and fills in their missing essentials – the big hole they all possess of no directed problem solving to relevant solution concepts.

One reason cited for the more hesitant uptake of TRIZ is its perceived complexity. This is a misapprehension, as each TRIZ tool is fairly simple and straightforward and learning the entire toolkit should offer no problems for engineers. TRIZ is quite a large and rigorous toolkit but many of the tools overlap, as TRIZ is designed to suit all types of problem solvers. Part of the genius of TRIZ is that it was developed to allow individuals to build their own, personal TRIZ toolkit which suits their problem solving style best. This is a bit like having a well-equipped gym with a large range of equipment – individuals choose and use only part of that equipment depending on what suits them and which fitness problem they are tackling. The TRIZ Toolkit is much the same – there are tools which have specific purposes and tools which an individual finds suit them well, and which they will use extensively.

TRIZ and Other Problem-Solving Toolkits

It is big claim to make that TRIZ is the only toolkit which helps engineers locate all the useful solutions to their problems, and understandably this can create some tension with 'problem-solving experts' who use the many other wonderful engineering toolkits. These other toolkits have immensely useful guidance to help us find inefficiencies in processes, the root causes of problems, help us understand all requirements, all stakeholders' perspectives, what everyone wants, and analyse products, processes and systems, and once we have solutions (for solution finding we're only offered brainstorming) they help us rank solutions. However, none have anything practical to offer for solving problems, and perhaps should not be labelled as 'problem-solving toolkits'.

Many good toolkits but none for problem solving

TRIZ is a comprehensive toolkit with simple tools for understanding 'what we want' and detailed tools for 'system analysis', which are helpful for everything from invention of new systems to improving old ones. BUT unlike any other Toolkits, TRIZ has tools for finding solutions. TRIZ will complement and fill in the gaps (hitherto ignored) of how to actually find concepts to solve problems. The many engineering quality and production experts who come to learn TRIZ, enthusiastically and thoroughly apply it, and find it is a very valuable addition to whatever toolkit they also use – whether it is Total Quality Management, Quality Function Deployment, Taguchi, Lean, Pugh Matrix, Kepner-Tregoe, 8 Ds, Six Sigma, Value Engineering etc. – especially when they need innovation and want to find powerful solutions.

Other toolkits may offer different and additional benefits such as a more thorough analysis of requirements, or solution selection, but whatever toolkits we normally use TRIZ offers something extra, and powerful and is a good additional toolkit for any engineer.

Only TRIZ helps engineers solve problems

One TRIZ delegate, Jon Theuerkauf, Six Sigma expert and Head of Best Practice for HSBC (Jon is now Credit Suisse's Head of Centre of Excellence), explained 'At the beginning of the TRIZ Workshop I said, "There is nothing new under the sun" as I thought I knew most problem-solving toolkits and was sure that TRIZ would just contain the same tools as all the others. TRIZ showed me something "new under the sun" with its unique problem-solving tools of the 8 Trends, 40 Principles and Standard Solutions. I am impressed and I am not easily impressed.'

Innovation – Fool's Gold or TRIZ?

Innovation has been in fashion for a long time – and heads up company value and mission statements – and unlike other trendy corporate initiatives has not peaked and faded. Probably because like searching for the philosopher's stone, most companies have never found how to systematically locate useful innovation to achieve its promise of success and hopefully endless future riches; practical, sustained innovation remains elusive to most companies. Corporate culture innovation initiatives often establish the need for innovation, identify innovation gaps and stimulate the desire to close those gaps. There are many innovation departments and consultancies with smart presentations, articles and conference talks telling us that innovation is needed more than ever, especially in difficult times and as the pace of new technologies increases etc. Non-TRIZ innovation toolkits are mostly brain-provocation exercises – helping to create new viewpoints of problems, clearer understanding and provoking ideas by stirring up or helping recall what we already know – and offer innovation by suggesting different combinations of known entities. Companies looking for innovation often keep asking the same questions over and over again ... hoping that it will somehow get different answers (close to Einstein's definition of insanity). Although 'innovation' is blazoned across corporate values, mission statements, slogans and encouraging, motivational material, it is hard to achieve, apart from those companies who have embedded the TRIZ toolkit to deliver technical and business innovation.

Einstein said that the definition of insanity is to do the same thing over and over again and hope for different results. He also said 'If at first, the idea is not absurd, then there is no hope for it' which is very aligned to TRIZ thinking using the **Ideal Solution** which helps to shoot us away from our comfortable problem definition and solution places and suggest new definitions for the problem and different solutions far from our previous positions. If a solution is wild and wacky we should not instantly reject it – but look for the good and the benefits it delivers. Negativity or psychological inertia will help us reject new ideas and suppress innovation – TRIZ offers tools to combat these.

'Einstein develops his theory of negativity.'

Psychological Inertia can happen to anyone

What does TRIZ Offer?

TRIZ Delivers Systematic, Guaranteed Innovation and Creativity

By helping us to find all the solutions to problems and new concepts TRIZ delivers systematic innovation. By learning TRIZ and following its rules many companies have found that they can accelerate creative problem solving for both individuals and project teams. Such companies who successfully encourage and apply TRIZ can demonstrate that the spectacular results they obtain reflect that they are using the success and knowledge of the whole world, and not merely the spontaneous, random and occasional creativity of individuals or groups of engineers within their organization.

TRIZ – Helps Us Understand the Problem and All its Solutions

TRIZ will help us ask the right questions and locate most, if not all, of the solutions including those we have within our knowledge and experience and others beyond that which we didn't previously know about (delivering innovation). Most companies find TRIZ will give them the many new answers they are seeking, and feel confident that they have covered all the possibilities for new solutions. In a wide range of industries from nuclear clean-up to new medical devices TRIZ has been helping teams locate many good solutions, delivering high patent rates, intelligent choices and confidence of a good innovative strategy for future business.

TRIZ Simplifies Systems to Maximize Benefits and Minimize Costs and Harms

This is fundamental to the TRIZ approach of seeing how to achieve the greatest benefits for the least costs and harms. One evolutionary trend in engineering is to complicate systems as they develop and offer more functions and then later simplify them without losing any functionality; TRIZ suggest ways of moving straight to the high functionality delivered by a simple system. This is almost an underlying philosophy of TRIZ and offers many tools to assist this process. Called TRIMMING it is the TRIZ approach to systematically simplify systems while retaining all functionality.

TRIZ Helps Us Overcome Psychological Inertia

Engineers love finding good solutions and TRIZ offers systematic routes towards the solution places and in addition offers ways of breaking any mental habits which prevent innovation, clarity of understanding and thought, or keep us stuck in the same old solution space. In TRIZ this is called 'breaking psychological inertia' and the TRIZ tools to help us include Thinking in Time and Scale, Ideal Outcomes, Smart Little People and Size-Time-Cost. TRIZ also gives a confidence that confusing, more difficult, complex problems can be systematically understood and solved. The TRIZ problem understanding and solving processes build habits of asking the right questions at the right levels and subduing complexity, whilst the TRIZ creativity tools shake our brains, attitudes, prejudices and help us seek new areas, new ideas, new solutions and helps make us creative and innovative.

When all seems lost – try TRIZ

How TRIZ Works

TRIZ offers us clarity of understanding of problem situations and solution triggers to help us solve problems. TRIZ has a set of simple but powerful tools that take us from a problem situation to a problem model – which is focussed only on the relevant part of the problem – exactly where and when the problem is occurring. This drilling down to the exact problem area to define the problem requires a very exact understanding of the problem and its context. Once the problem is accurately defined TRIZ offers matches of problem solutions to problem types. For example if the problem contains something that is harmful then TRIZ offers all the recorded solutions to dealing with harm at a very conceptual level such as stop, eliminate, correct or transform into good.

The Golden Rule of TRIZ

There is one essential TRIZ tool which is fundamental to all approaches to problem solving. This is the IDEALITY Equation which is the starting and end point of all problem solving. Ideality is like the

Golden Rule of TRIZ – and improving Ideality is the aim of all problem solving, i.e. achieve more benefits, less costs, less harms.

$$\textbf{Ideality} = \frac{\textbf{All Benefits (Primary + Secondary Benefits i.e. all outputs we want)}}{\textbf{Costs (all inputs) + Harms (all outputs we don't want)}}$$

All the other TRIZ tools are there to improve Ideality, in this book Ideality terms are used such as Ideality Balance – a positive Ideality Balance means a product is viable and achieving market acceptance (its benefits exceed its costs and harms). An Ideality Audit is a complete check of all inputs and outputs of the system we've got compared to the system we want, and Ideality Tactics describe the various processes/sequences for combining relevant TRIZ tools to solve particular problems.

The great TRIZ Toolkit – every player is needed

The TRIZ Toolkit

The TRIZ Toolkit is straightforward and although rigorous and powerful is still fairly easy for engineers to learn and apply fairly quickly. The main tools include:

- **40 Principles** for solving contradictions accessed through the **Contradiction Matrix** and **Separation Principles**
- **8 Trends of Evolution** for perfecting systems – used for future system development
- **Effects** – engineering and scientific concepts arranged for easy use. A simple list of questions and answers to access all the relevant technical and scientific conceptual answers – a list to deliver all the ways to solve problems without technical language or jargon. (see www.TRIZ4engineers.com)
- **Thinking in Time and Scale** for problem context, understanding and solving (9 boxes)
- **Ideal – Ideality, the Ideal Outcome, Ideal Solution, Ideal System and Ideal Resources** for understanding requirements and visualizing solutions

- **Resources and Trimming** – for clever and low cost solutions
- **Function Analysis and Substance Field Analysis** – system analysis for understanding the interrelationship of functions
- **Standard Solutions** – for solving any system problems. Creating and completing systems, simplifying systems, overcoming insufficiency, dealing with harms, future development and smart solutions for technical problems.
- **Creativity Triggers** for overcoming psychological inertia and for understanding systems and visualizing solutions including Size-Time-Cost and Smart Little People

The unique parts of the toolkit are the solution tools (40 Principles, 8 Trends, Standard Solutions and Effects) and the tools to access these solutions which include the separation principles, contradiction matrix, function analysis and substance field analysis. These are used to solve engineering problems – especially difficult ones which have resisted the usual brainstorming methods. In addition are the amazing TRIZ thinking tools which once learned are used daily and deliver clarity of thought and effective approaches to both problem understanding and solving. The most popular is 'Thinking in Time and Scale (9-Boxes)' a powerful tool which delivers great mental clarity. Imagining Ideal Solutions or Systems is very much part of the TRIZ thinking for both capturing all benefits and seeing how to deliver them for the least cost and harm. Fundamental to this approach is reducing costs by delivering essential functions using available resources. Good TRIZ thinkers conjure up such Ideal Systems – which is also the natural territory of great inventors. One example of an Ideal System was created by Garrett Morgan – the Safety Fire Hood which he invented in 1912 to help fire-fighters. The hood is designed to supply cool air in a smoke-filled room and cleverly uses available resources because it has a long inlet tube at ground level (where the air tends to be cool as smoke and fumes rise in a fire).

Cleverly matching available resources to deliver good solutions is fundamental to TRIZ

One great strength and problem of learning TRIZ is that we each like different tools and each have slightly different TRIZ toolkits. The TRIZ Tools are all different but some overlap and duplicate each other because they are designed to suit different styles of problem solving. Most of us will only need and will use about 80% of the tools; we will reject the ones which don't suit us but still have a complete TRIZ toolkit. Which 20% we reject depends on each person. TRIZ offers a complete toolkit for everyone, no matter what their learning styles, preferences or experience. TRIZ achieves this by having a wide-ranging toolkit which contains tools to suit all situations and approaches. The genius of the toolkit is that it has tools for all problem-solving types, and allows each of us to build the toolkit which suits us best. However it is important to learn and use all the tools so that when working in TRIZ teams we can work together effectively, and not reject too many tools. There is a danger of becoming familiar with a small number of the TRIZ Tools and only using them – and some TRIZ gatekeepers insist that everyone else in their company limits themselves to their particular choices – which is like playing only three players in football and keeping all your best talent on the reserve bench.

Gatekeepers limit their corporate TRIZ to their favourite tools & keep much huge talent on the bench

Many of the TRIZ Thinking Tools are very simple brain prompts and can be learned in a few minutes such as Ideal Outcome, Smart Little People and Size-Time-Cost. Some of the analysis tools take longer to master such as the Separation Principles and the Contradiction Matrix but reward any investment of effort as they offer routes to locate all the known ways to create systems which deliver contradictory solutions.

The Toolkit Chart

Table 1.1 is a complete chart of the TRIZ Toolkit. The horizontal axis of the chart shows the TRIZ Tools and the vertical axis shows all that TRIZ delivers.

Which Boxes should be Ticked?

I could put marks in the boxes based on what I have seen work in problem-solving sessions for ten years. Each of my TRIZ colleagues would mark a slightly different set. In a sense every box could be ticked as all the tools can be used in many ways. The ticking of the boxes of which tool when would vary from person to person, and when I fill the chart in for myself it is based only on my experience. Academic study is needed about where and how the TRIZ Tools deliver the best solutions and how to comprehensively complete the chart below.

TRIZ Toolkit

TRIZ guides us to achieve the following benefits

TRIZ guides us to achieve the following benefits	No.	Benefit	Ideal Outcomes, Benefits, System, Functions & Resources	Resources	Thinking in Time and Scale (9boxes)	BAD Solutions Locate and develope solutions (which have GOOD & BAD within them)	Solution Park Capture everyone's ideas for transformation with TRIZ to better ideas	Prism of TRIZ Concept solutions to concept problems	Analogy 'Life and death' solutions from others
Understand & Solve Problems	1	Solve / eliminate Problems							
	2	Get the functions right & deliver all the right functions							
	3	Deliver opposite benefits (solving physical contradictions)							
	4	Reduce costs							
	5	Simplify systems							
	6	Improve systems							
	7	Improve / make something better without downsides							
	8	More benefits							
	9	Less inputs							
	10	Less harms							
	11	Understand and map essential requirements							
	12	Define and locate problems							
	13	Detect problems							
	14	Understand and map the context for problem situations							
	15	Choose the right directions for problem solving							
	16	Locate and recognise available resources							
Invent, Locate Future & Next Generation Systems	17	Forecast future products							
	18	Invent systems							
	19	Next stage and next generation of all or parts of technical systems							
	20	Match new technologies to unfulfilled needs							
Team Building, Innovation & Creativity	21	Audit Innovation							
	22	Understand the relevance / context of our own and other's work							
	23	Work well in teams							
	24	Co-ordinate work and ideas							
	25	Share knowledge / solutions & respect each other's solutions							
	26	Access relevant, good, innovtive solutions to problems							
	27	Generate solutions from within our own brains							
	28	Locate solutions previously unknown to us							
	29	Access relevant world knowledge							
	30	Sort / Map solutions							

Table 1.1 TRIZ Toolkit chart

40 Inventive Principles	Physical Contradiction	Separation Principles	Technical Contradiction	Contradiction Matrix	8 Trends of Technical evolution	Function Analysis	HARMS TRIZ list of all the ways the world knows to deal with anything harmful	INSUFFICIENCY TRIZ list of all the ways the world knows to deal with anything Insufficient	Substance Field analysis	76 Standard Solutions	Effects Database	Size Time, Cost	Smart Little People Nano Crowd	Subversion	Combine Solutions (potato / lettuce)	ARIZ Algorithm of Inventive Problem Solving

TRIZ Creativity Tools

TRIZ wakes up your brain

The TRIZ Creativity Triggers complement all the other TRIZ tools and can be used on their own or kept as part of the waiting and available toolkit to be used as and when appropriate.

TRIZ Creativity Triggers (to be applied at any/all stages as required)

1. Define and seek an *Ideal Outcome* which solves the problem itself.
2. *X-Factor* – Imagine something which solves the problem.
3. *Thinking in Time and Scale* – Nine Boxes Maps (*Time* steps Before, During, After. *Scale* – zoom in and out – look at the details and the whole context to understand the problem situation and see all possible solutions).
4. *BAD Solution Park* – Capture all IDEAS – everyone's BAD solutions (even if unusual or unattainable etc.).
5. *Subversion* – Try the solution the other way round – Invert it – try the opposite.
6. *Prism of TRIZ* (understand essential problem and see if someone else has already solved this problem).
7. Model the problem and solutions with *Smart Little People.*
8. Apply *Size-Time-Cost* (exaggeration thinking tool).
9. *Simple language* (no technical jargon or acronyms to obscure simple truths).
10. Look for *Life and Death analogies* (Has someone got solutions which are to them critical and therefore very good?).
11. Distil solution *IDEAS* to the *CONCEPTS* behind them, multiply solutions with more Ideas for each concept.
12. *Ask Why?* To identify benefits instead of features or functions.
13. *Combine* all the *good* from various *solutions* (e.g. create a carrot–cabbage with edible roots *and* leaves).

The TRIZ Creativity Triggers are simple and powerful and help us to:

- think fast and powerfully – all are good brain prompts;
- understand and define what we want;
- reveal requirements – everyone's needs are made clear;
- understand the system and its problems;
- prompt our brains to find lots of solutions;
- overcome Psychological Inertia;
- deliver clarity of thought;
- locate different solutions;
- see others' viewpoints.

Creativity Prompts – Smart Little People and Size-Time-Cost

How to model problems and solutions with Smart Little People and applying Size-Time-Cost to problems are shown below. The rest of these Creativity tools are dealt with in subsequent chapters.

Smart Little People (SLP)

This powerful and simple TRIZ Creativity tool can be mastered in a few minutes and is thereafter forever useful. Smart Little People are imaginary tiny beings who represent the different elements of the problem we are trying to understand and solve. It works as a mental trick because it is based on empathy, or creating some personal analogy with the problem. Empathy means becoming the object/problem and looking to see what can be done from its position and viewpoint. If we imagine ourselves becoming so tiny that we are in the problem area and seeing the problem in great detail then this can be useful and harmful. This is useful for problem understanding but harmful because may we resist solving a problem if the solution means ourself, as a tiny being, is going to be destroyed, dissolved in acid, mashed up, dissected, boiled etc. This is overcome by using a crowd/multitude of disposable Smart Little People, for which we feel no responsibility, based on the premise that 'A single death is a tragedy – a thousand a statistic.'

Smart Little People works by modelling the different aspects of the problem (causes and solutions) with different groups of rival or complementary Smart Little People. They are *Smart* because they have the ability and insight to create/solve problems and be anywhere, doing anything. *Little* means they are as tiny as necessary – molecular level if required. Rival teams of smart little people can be created and some can cause the problem and others solve it; they do whatever is required even if this means they get destroyed. Figure 1.1 uses SLP to illustrate a composite material.

Model the Problem with Smart Little People

We draw the Problem Zone by using SLP to model the problem and the causes and conditions and solve it. When this is used in engineering problem solving it can be met with a certain sneering incredulity that such a simplistic tool can possibly help with a serious technical problem. I remember working with a team of fairly curmudgeonly, oldish, experienced engineers on a very serious and public problem of the leaking of engine oil in passenger planes contaminating cabin air. When the oil appeared in the cabin air of the aircraft it caused sickness and fainting and the problem was threatening safety and generating lawsuits.

It was a three-week session (working on two very different problems simultaneously) and about halfway through I suggested that we model the leaking oil molecules as Smart Little People and track and draw their progress from the engine to the cabin. This was greeted with some hilarity by the engineers who

Figure 1.1 Using Smart Little People to illustrate a composite material

grudgingly undertook the task but with typical engineering thoroughness and accuracy. This was very effective and productive and mapped at least seven good places where we could deal with the problem; the models we produced delivered such clarity of understanding that progress became very rapid. For safety the problem was solved in a number of places but the one cause of the problem was the use of synthetic oils, which are much more toxic in cabin air. The solution suggested was less use of synthetic oils. (I was also working on a problem with another company about reducing problems of carbonization of oils. Again great clarity of understanding was achieved by modelling with SLP and for this the solution was to use only synthetic oils in aero-engines.) SLP helps show the causes of simple and complex problems.

Altshuller's Famous Use of Smart Little People

In much of the TRIZ literature is the original famous Altshuller example of how he designed a perfect marine cable to prevent tethered mines in the sea from being detected and removed.

As shown in Figure 1.2, minesweepers are used to destroy mines at sea by dragging a loop of cable which ensnares the mine retaining cable. The mine then detonates or floats to the surface. Altshuller's challenge was to design a cable which would tether the mine to the seabed but also allow the minesweeper cable to pass through it. Altshuller drew the zone of conflict as if with populated the smart little people, and by imagining a tiny person holding the feet of the tiny person above he saw the answer (Figure 1.3).

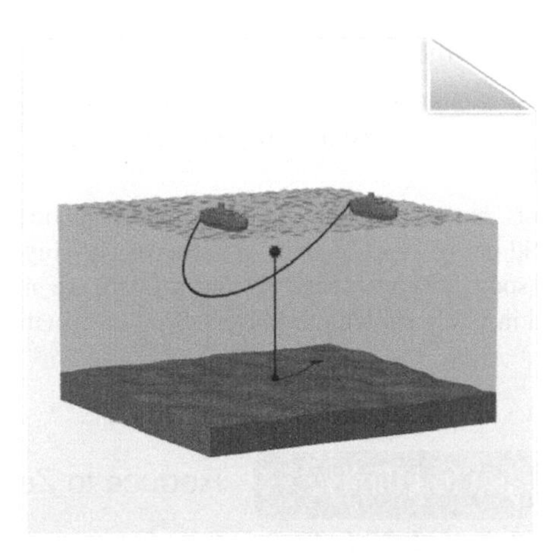

Figure 1.2 Minesweepers

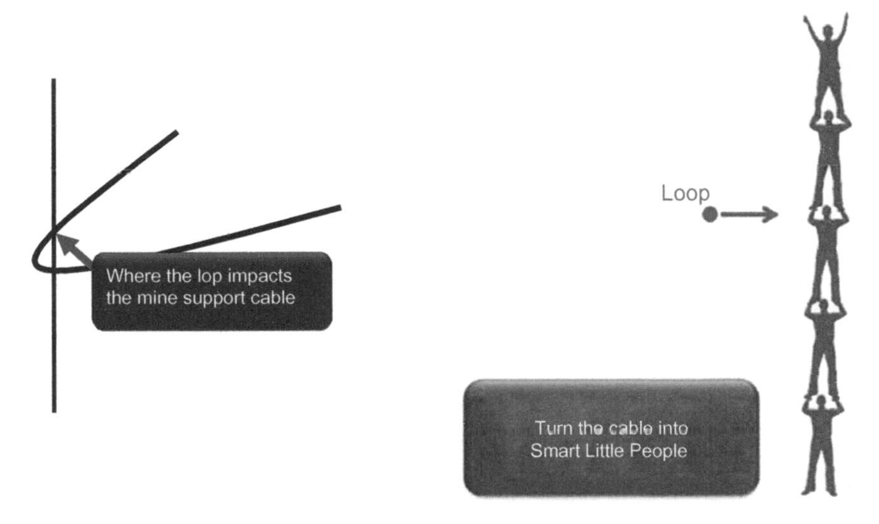

Figure 1.3 Altshuller's challenge

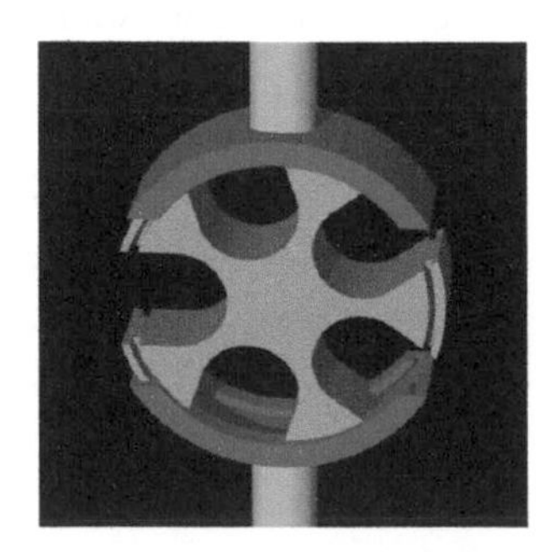

Figure 1.4 Altshuller's answer

The device which was developed and is now widely used works like a rotating door (Figure 1.4). It is based on the principle of the smart little person letting go with one hand to let the cable pass through while still hanging on with the other hand. Then rejoining the first hand and letting go with the second hand so the cable passes through but the vertical link is always maintained.

Size-Time-Cost for Visualizing Solutions

This is a very powerful tool for stimulating new solutions and overcoming Psychological Inertia as it shoots us away from our current solutions and narrow image of the problem, and re-constructs it in an extreme and different way.

Size-Time-Cost simply suggests that we take our problem and exaggerate all possible options and then consider the prompts we would get if we exaggerate *size* (make very big and very small), *time* (very fast and very slow) and *cost* (spend lots and spend nothing). These are six extreme places to look for solutions; they effectively and instantly stretch our imaginations to visualize new solutions.

Six simple steps:

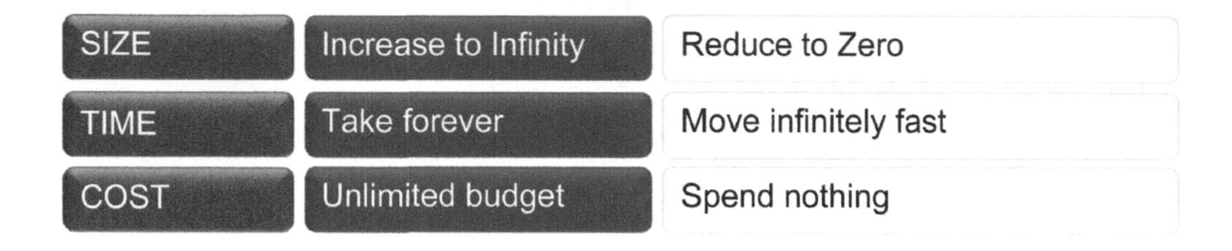

This could be applied to anything. Imagine designing a new system for self-propelled travel (instead of a bicycle, scooter, roller blades, skate board or a wheelchair). Apply Size-Time-Cost to this problem and visualize some innovative ideas and systems which could solve this problem.

Problem Challenge

Apply Size-Time-Cost and imagine solving the problem of self-propelled travel by trying six solution types, making the device:

Very Large	Very Tiny
Very Fast	Very Slow
Very expensive	Very inexpensive

Write in all ideas which occur including 'wacky and wonderful ideas' for all six of the extreme solution places and then see if those extreme ideas suggest any practical and real solutions which could be explored. This is just one of the TRIZ Creativity Triggers. When thinking of solutions it may be worth trying all the other solution prompts.

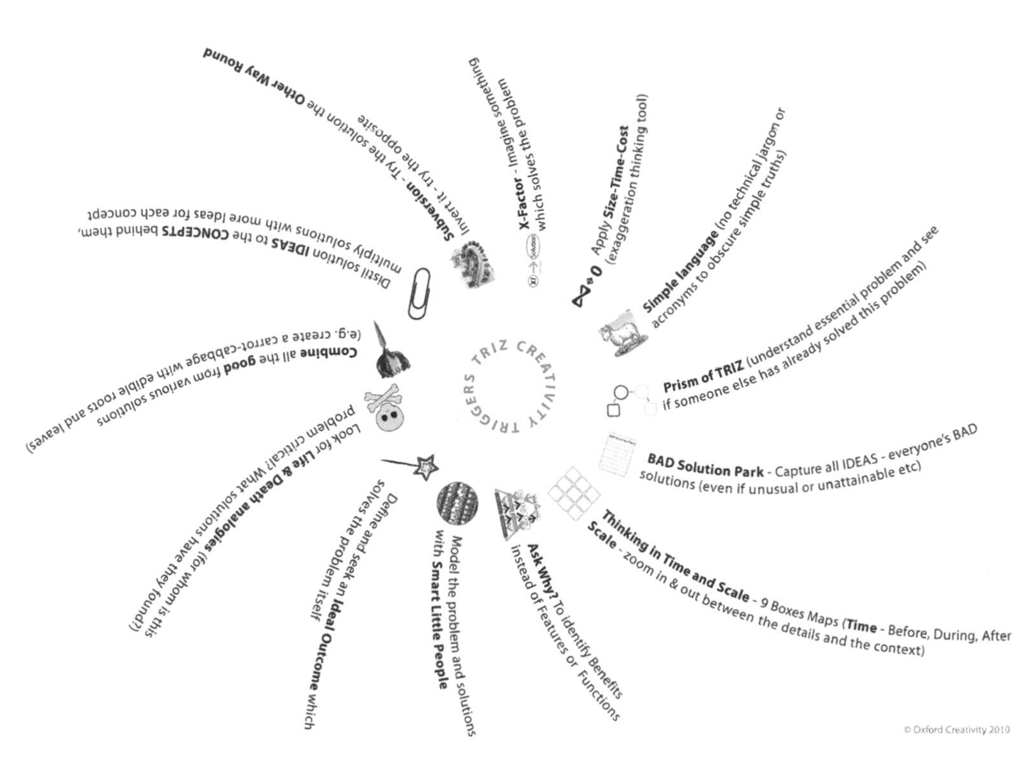

The TRIZ Creativity Triggers should not be in a list which implies priority but in a circle or mind map

TRIZ for Everyone – No Matter What Your Creativity

The simple tools of Smart Little People and Size-Time-Cost are easy to learn and apply and stimulate our understanding of problems and help us visualize and imagine solutions. They are both powerful and useful and just two tools of the thirteen or so in the TRIZ Creativity Triggers. They aid our creativity and help us to think quickly and clearly.

Altshuller developed these tools while working with his engineers, training them in TRIZ and in response to what was happening in live problem-solving sessions – he 'bottled' the simple but practical ways to stimulate clever ideas.

Altshuller once said that TRIZ made the very creative people three times more creative but the non-creative people were made ten times more creative. Altshuller had studied the differences between different creative abilities and developed TRIZ as a universal tool-kit and process for many different types of problem solvers.

Clever and motivated engineers approaching TRIZ will eventually use about 50–80% of the TRIZ tools to help them solve problems – but everyone builds their own individual TRIZ toolkit and there are many variations. TRIZ was designed to do this and developed to cover many different thinking styles, with tools to help everyone quickly and effectively move from problem to the best solutions. Some people don't need so many of the tools because they are instinctive TRIZ thinkers. For those of us who lack this marvellous ability Altshuller uncovered the instinctive processes used in quick and clever thinking. He built TRIZ tools to help anyone do the clever thinking by stepping through it slowly, consciously and systematically to cover every stage of the problem-understanding and problem solving process.

We are now benefiting as our engineering communities can learn to quickly and effectively TRIZ problem-solve, using less resources and more ingenuity – theirs and other people's. Although TRIZ initially may seem complex because there are many tools, and these overlap significantly, each tool is simple to understand and some tools can be applied immediately after only a few hours of learning and practice. The basis of TRIZ is to encourage everyone in a team to come up with as many good ideas and solutions as possible, combining the best of each as appropriate to eventually find good solutions to any problem.

TRIZ can help us:

1. solve problems
2. think clearly, powerfully and see the wood for the trees when confronted with a complex problem situation
3. be creative (invent new systems, find next generation systems, come up with lots of new ideas etc.)
4. be innovative (find new ways of using existing systems, existing technologies etc.)
5. get teams to work together to pool their brain power and experience, enhanced by the distilled world's knowledge of the TRIZ tools
6. improve existing systems and increase the ideality of systems by lowering costs, removing harms or increasing benefits
7. use our resources: we can often find quick, cheap solutions to problems
8. get quick solutions
9. get a structure to brainstorm around difficult problems – even with those unfamiliar with TRIZ
10. structure and use our thinking time effectively – we know that we'll be looking in the right direction and places
11. turn harm into good

And can't ...

Solve every problem – although it can solve most

Replace our brain power – there is still a creative leap required – it's just much smaller than the leap we'd have without TRIZ

Tell us which TRIZ tools to use when – we need to learn the tools and the process and intelligently use both

Churn out solutions – it will only give us triggers – we still have to work and think hard, but we know that we'll be going in the right direction to the right places.

Reasons to use TRIZ

TRIZ is systematic, auditable and repeatable

During problem-solving sessions, we may uncover some great solutions without using TRIZ. Using TRIZ will not only uncover all the good and bad solutions we would have found anyway, but many more, and will ensure that no solutions are missed. We will always come out of a problem-solving session with several workable and good solutions.

TRIZ is based on proven successful patents

So we know they work – there is less risk involved in using an existing technology or technique in a new field than developing our own custom-made solution.

TRIZ uses the world's knowledge

Brainstorming and other techniques randomly help unlock some of the knowledge in the room; TRIZ helps us systematically unlock our own knowledge and helps us intelligently access the relevant world's knowledge.

We can build our own toolkit from the TRIZ tools

We show the principal TRIZ tools, but we don't have to use them all. Everyone finds tools that they prefer, and then sticks to the tools that work best for them: we don't have to bend our way of thinking and working around an inflexible toolkit.

TRIZ is quick

Like any new skill we have to practice, and at first we may go slowly. However, once we've become familiar with the tools, we will start using them automatically and problem solving will become quicker and more effective.

It can be used in groups

Having one or two TRIZ-trained people in a large group can facilitate incredibly effective, structured brainstorming sessions, because they can focus the group's attention on a few key areas that they know will lead to good solutions.

It's not just for engineers

The principles of TRIZ apply to any problems, or situations that need more innovation and creativity, and have been used to great success to solve management problems.

It doesn't often need software – just your brain power and the TRIZ processes

Learn TRIZ properly and separately from the software companies before trying any TRIZ software. Software can impede good TRIZ thinking and hinders working well in groups to solve problems. I know less than five enthusiasts who once having learned TRIZ, use the software rather than the thinking skills of TRIZ. I know many more people who have the software on their laptops and rarely use it when problem solving with TRIZ. There are a number of software packages on the market. Mostly they distort TRIZ to make it fit with some software vision. The software may well get in the way and hinder rather than help. Getting started with TRIZ problem solving is the hardest part – learning and using the software and then entering things into a software package can give the illusion that we are doing something useful. If we need handholds because we are uncertain (or feeling sluggish) then use the excellent free TRIZ problem processes and packs.

TRIZ Processes are explained in this book

All the TRIZ problem-solving routes are shown in this book. They are easy to use, also there are Ideality algorithms which offer thorough understanding of what we want (all requirements), TRIZ system analysis, problem definition (the gaps between our requirements and what the system delivers, i.e. the problems) and finally to problem solving using the excellent TRIZ problem-solving tools, which software packages may claim to have enhanced, but in reality expand the detail and blur the simplicity of TRIZ.

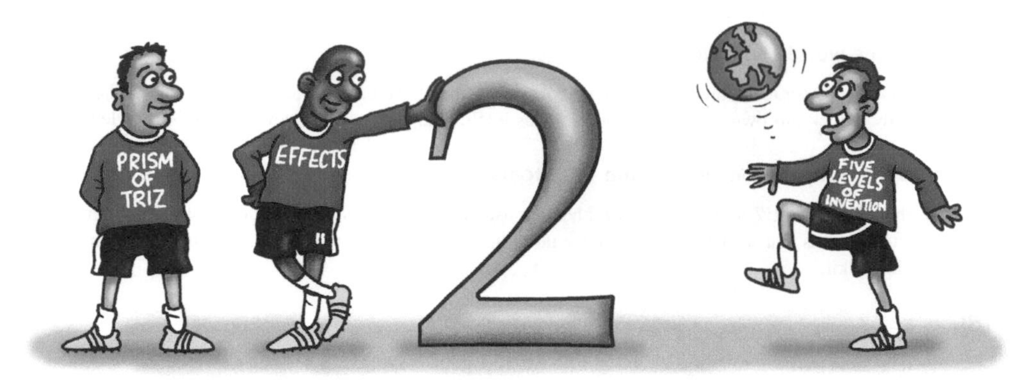

TRIZ Knowledge Revolution to Access All the World's Known Solutions

Problem Solving – Resolving Defined Problems

Problem solving is both complex and varied. It defies most simple attempts at categorization as it is needed at all stages of creating and using systems – invention, design, manufacture, use, maintenance and disposal. Problems can range from simple (easy to solve) to very difficult (require innovative solutions) and furthermore different problems need different experience and knowledge to match them to the right solutions. TRIZ helps us systematize and audit the problem-solving process so that the many different problem types can be selected for different approaches and be solved with various and appropriate TRIZ tools. TRIZ helps us find the appropriate routes from innovative problems to innovative solutions and offers a classification of both problem type and problem difficulty (in five levels).

Limited Time for Understanding and Solving Important Problems

In many leading engineering companies minimum time is given to the actual problem-solving stage, because without TRIZ it is hard to measure and assess. Dave Knott of Rolls-Royce presented a paper about this at the 2001 Bath Etria TRIZ conference and talked about the relative importance of elements of the design cycle based on designer attributes, based on his research as Head of Design Technology. Looking at the Rolls-Royce Design Process and its various stages he was concerned that anything that didn't have a formal process, which was difficult to audit and measure, would be starved of resources and time and its importance overlooked. In his analysis of the design process he mapped the most relevant stages and their relevant importance.

TRIZ for Engineers: Enabling Inventive Problem Solving, First Edition. Karen Gadd.

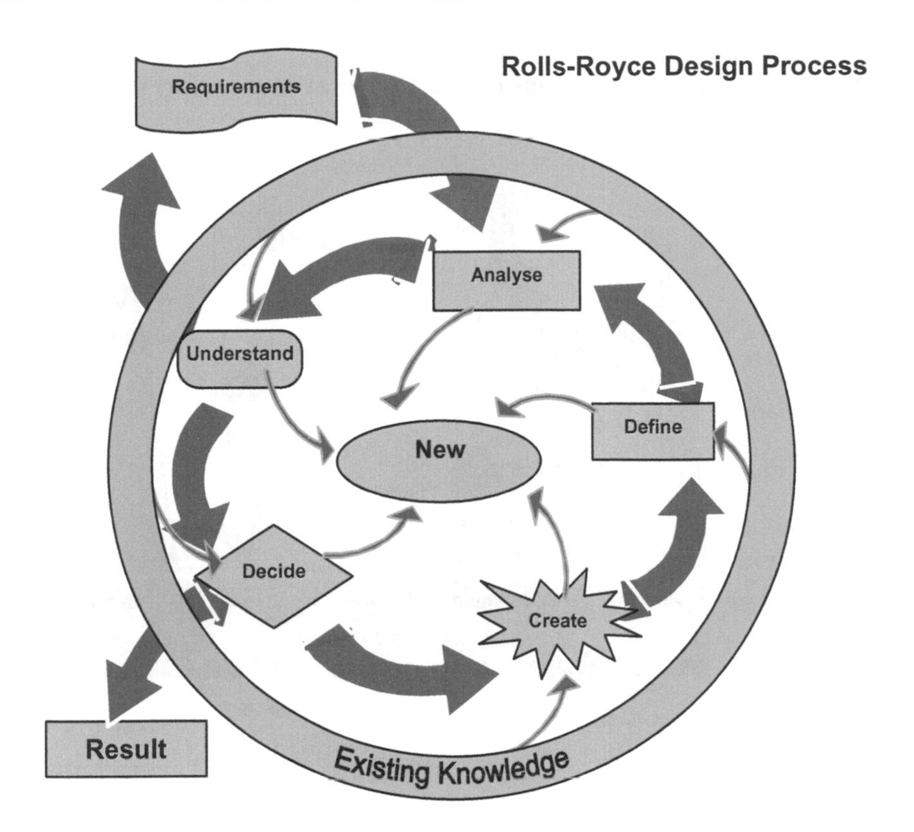

He was concerned that although in-house research showed that the stages of create and understand were the two most important attributes, they were the easiest to neglect because they are not mandatory.

Adding TRIZ to the Rolls-Royce Toolkit for these two most important stages helped overcome this problem, and give structure and an audit trail for the create and understand stages of the process, as shown in the diagram below.

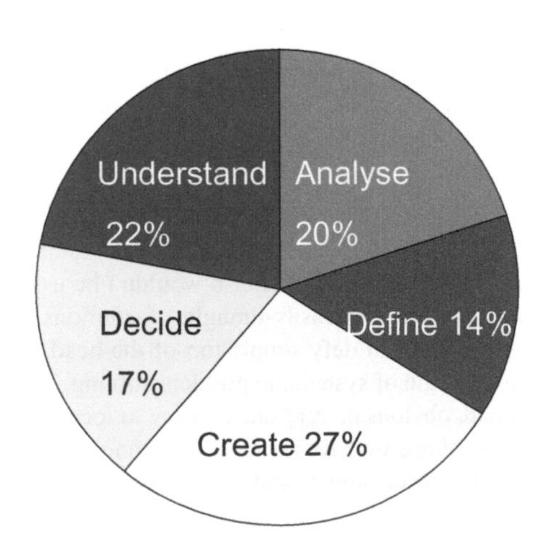

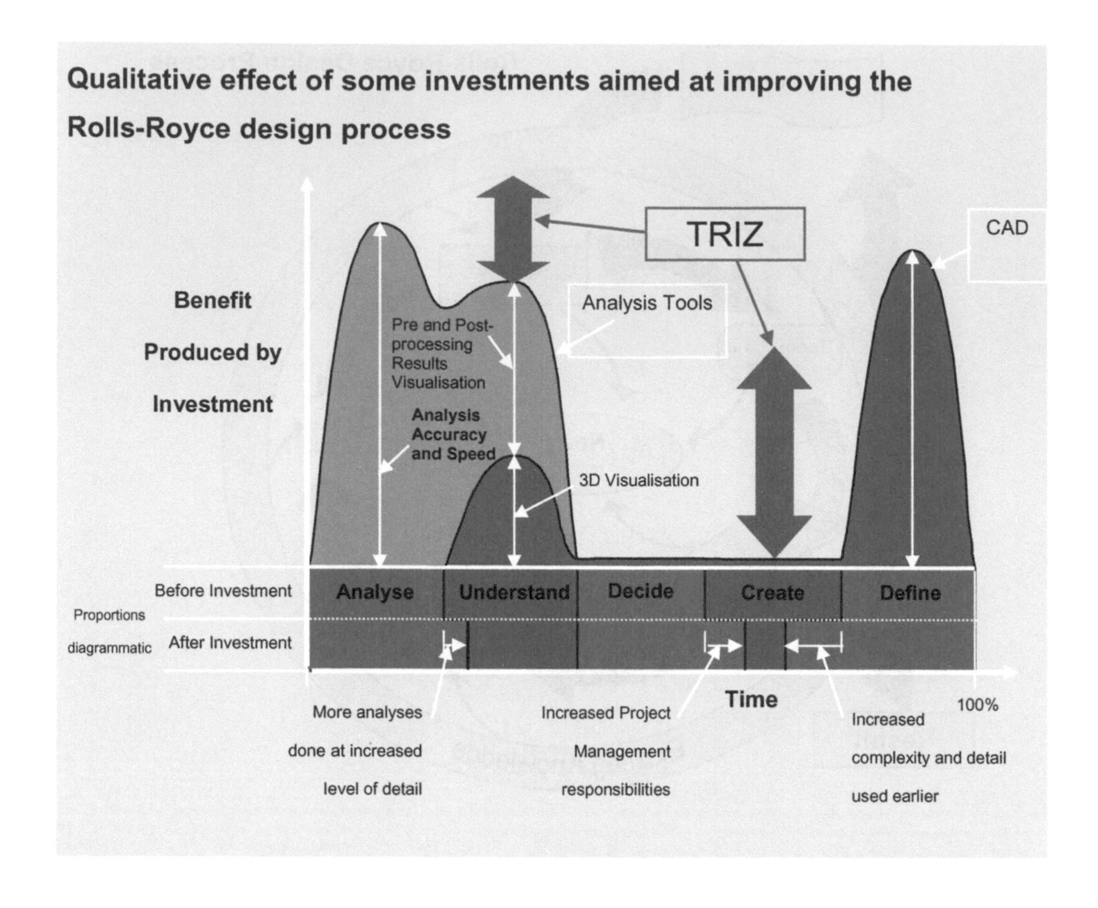

Some illustration of Dave Knott's success with TRIZ on Rolls-Royce problems is given at the end of this chapter in a case study entitled 'TRIZ Application in Rolls-Royce' and an article entitled 'Post TRIZ article on Rolls-Royce'.

From Random to Systematic Problem Solving

One particular challenge is persuading people of the importance of processes when creativity and innovation are needed for problem solving and finding new concepts. As long as some results, sometimes good, are achieved there will be resistance to the idea that we can systematize and use processes for that wonderful Eureka moment we all love when we suddenly make the right connections and solve a problem. Some people argue that it is enough to get results – it doesn't matter if the method is random. Engineers however normally like processes such as TRIZ for problem solving.

Brainstorming is straightforward and obviously works or it wouldn't be used so widely. There are some very simple problems which are susceptible to easily-thought-of solutions, and for these brainstorming will help. But there are many problems that defy simple top-of-the-head problem solving and need to be tackled systematically, using the logic of systematic problem solving. The solutions to easy and hard problems can also be good and bad, obvious or very clever, easy to locate or only found after extensive study. But as we have systematic methods we have no need to confine critical stages of invention, design etc. to random ideas, guesswork, brainstorming or sudden brainwaves as the principal route to finding a wide range of important solutions.

'We can learn the repeatable processes to achieve success or just be Paddling Pirates'

Problems Vary – Some Are Easy, Some Are Difficult

Altshuller showed us that when solving a problem we should be open to a wide range of solutions. He also demonstrated that the depth of approach was equally important – some problems are so straightforward that an average group of engineers with the relevant knowledge can solve them by brainstorming. However, some are much harder and we need to apply the relevant processes and techniques for technical problem solving to deal with them.

Tools for Problem solving-from Abacus to TRIZ

TRIZ systematizes and offers short cuts to help anyone solve these hard problems; it uses clever brain power and hard work in a directed, efficient way. This is achieved in TRIZ by removing much of the random searching, which previously needed many trials, along with many dead-ends. TRIZ directs us to the useful and fruitful places where good, relevant solutions have been found before, by using the TRIZ Concept Solutions. TRIZ thus shows us how to solve any problem, no matter how hard or complicated. TRIZ could help move technology forward much faster by helping clever engineers and scientists use their precious problem-solving time much more effectively.

TRIZ Five Levels of Inventiveness/Creativity

Altshuller researched engineering problems and their solutions and categorized them into five levels of difficulty. When he examined published patents he established that the problems solved in Levels One and Two, about 77% of problems, offer fairly simple, obvious, predictable answers, and only need company and industry knowledge to find solutions. He said that these patents were almost trivial and could be written by any engineer, with the help of the very simple and accessible required knowledge. The other patents, in Levels Three, Four and Five initially do not have obvious and easily found solutions, and have therefore mobilized much inventive creativity and accessed a wide range of knowledge. TRIZ tools help us with all problem types from the trivial to the seemingly impossible, and TRIZ will help us solve problems at all five levels. Techniques such as brainstorming nearly always help with the simpler problem solving of the first two levels.

The five levels are important to understand, because in many companies all problems with different levels of difficulty of problem solving are just lumped together; it is argued that because brainstorming has solved simple problems, it will solve all problems. Altshuller offers TRIZ as a better technique for solving the harder problems. Indeed TRIZ is shown again and again to help solve difficult and seemingly 'impossible' problems.

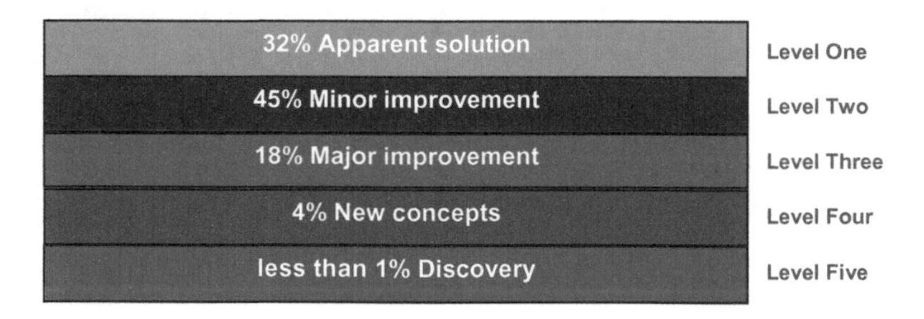

The higher levels of inventiveness are about going deeper and wider for more solutions, deeper by identifying the relevant functions and then seeing how to deliver those functions by using more inventiveness, looking at more options, going beyond the obvious solutions. Looking wider for solutions is to go beyond your own knowledge, and that of your colleagues and your company. Level One is about using knowledge easily available, and solving a simple problem in an obvious way. Level Two problems require knowledge or solutions outside your company but still easily available within your industry – aerospace companies use aerospace knowledge, automotive companies traditionally look to other automotive engineers for relevant solutions. Level Three solutions require a search outside your industry but still within a particular discipline – mechanical engineers would still be using known mechanical engineering solutions at Level Three. One example might be James Dyson's invention of his bagless vacuum cleaner, which used cyclone separation – a new technology for vacuum cleaners but a well-established technology in other industries such as saw mills, where it had been used since the nineteenth century. This level is about clever analogous thinking – looking for proven, tested solutions from other industries. This gives us low-risk innovation.

In Level Four new technical systems are created by bringing together solutions from wide boundaries of knowledge (a mechanical problem being solved by chemistry would be an example). Level Four successes come from new combinations of ground-breaking technologies to produce new solutions and new materials. These Level Four solutions are a result of looking much wider than one discipline at new technologies and new ways of using them.

One example of Level Four success might include the step change development of rechargeable lithium batteries which are an important technology for new energy sources. Searches for smart and efficient energy storage is seen as key to harvesting and using energy more effectively. Research into the development of rechargeable lithium batteries is seeking to deliver significant increases in energy and power density. Such batteries store electricity from renewable sources to power portable devices such as laptop computers, mobile phones, electric cars and bikes etc. One requirement is to greatly improve both the recharge and discharge rates especially for applications such as electric cars where fast recharge is a significant benefit that is being sought.

Global research is being undertaken in new materials to synthesize new nanomaterials by trying combinations of properties to develop materials for lithium batteries. Combining disciplines including solid-state chemistry and electrochemistry has created new useful materials. In Korea Professor Cho Jae-Phil and his team have replaced the graphite in lithium batteries with a certain kind of silicon to store eight times the power.

While at St Andrews University in Scotland, Professor Peter Bruce and his team in Chemistry have created rechargeable lithium batteries using oxygen as a readily available infinite and renewable resource. This represents a step change in the whole technology as LiCoO2 electrodes are replaced by a porous carbon electrode which allows Li+ and e- in the cell to react with oxygen in the air. This reduces cost while giving a ten-fold increase in energy storage capacity. In the USA at MIT, Byoungwoo Kang and Gerbrand Ceder have also developed materials for a high-power battery and created a lithium-based 'super battery' which releases its charge 100 times faster than previous batteries.

Level Five – Discovery – is the exciting, sometimes unexpected breakthroughs made by science, producing new systems and new materials which offer exciting new solutions that then have to be realised by matching them to unfulfilled needs. Often Level Five inventions have increased the scope of modern science by accidental breakthrough. Although Altshuller estimated Level Five inventions as representing less than 0.4% of registered patents, their impact can be very high.

The five levels of innovation/inventiveness were researched and uncovered by Altshuller to help us understand that within invention and problem solving are various degrees of both difficulty and scope. By understanding the different levels we can appreciate when we need more systematic and creative approaches to solving a problem. We can also appreciate when we should look wider beyond our own expertise and company knowledge for solutions. Uncovering the wonders of Level 4 and Level 5 cannot probably be timetabled or planned but will occur with a sufficient investment in research.

An example of Level Five could be the exciting announcement of Black Silicon. This was discovered by accident by physicist Eric Mazur and his team at Harvard USA. This amazing new material hit the scientific headlines as it is 100 to 500 times more sensitive to light than ordinary silicon. This powerful new solution matches many important unfulfilled needs. Black Silicon gives us:

- extremely fine computer and other electronic displays;
- better telecommunication by light along fibre-optic cables – Black silicon absorbs light at the critical wavelengths for optical communication (1–2 microns) unlike ordinary silicon;
- more efficient ways of converting sunlight to electricity – for much higher solar energy capture;
- vastly improved monitoring of the environment for global warming – Black Silicon offers excellent detection of clouds, pollution and dirt etc. that change the quality of the air;
- new medical devices – the spiky surfaces are much thinner than the sharpest needles and offer powerful options to deliver continuous small doses of essential drugs through the skin.

'Being able to use a material that is cheap and abundant and that industry has learned how to process offers a tremendous opportunity in a multibillion-dollar market,' Mazur comments.

It was accidentally discovered by shining a super powerful laser onto a silicon wafer. The laser's output briefly matches all the energy produced by the sun falling onto the Earth's entire surface at a given moment in time. After the silicon wafer had been subjected to over 500 pulses, the silicon turned black. Examination of the surface showed it wasn't burned but had an etched surface – a minute forest of billions of regularly spaced, tiny sharp tipped spikes all about two-thousandths of an inch high. When light is shone onto the spikes, it bounces about endlessly and hardly ever bounces out again. This means that the Black Silicon absorbs nearly all the light and this has many practical applications of which only some have been matched so far.

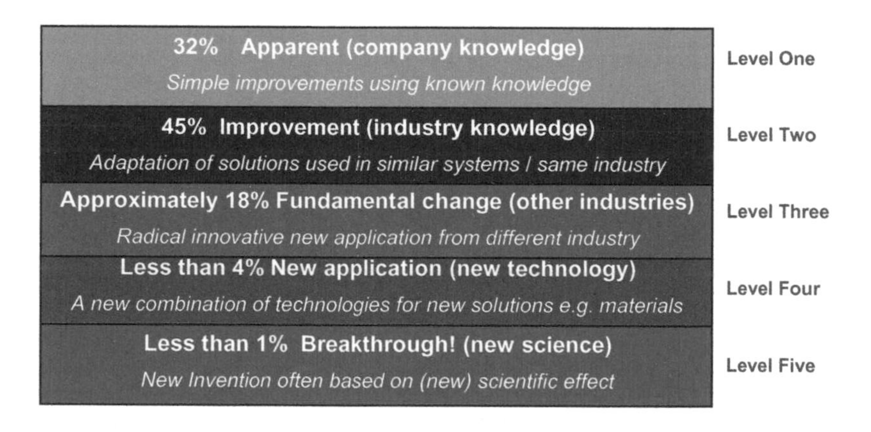

32% Apparent (company knowledge)
Simple improvements using known knowledge
Level One
45% Improvement (industry knowledge)
Adaptation of solutions used in similar systems / same industry
Level Two
Approximately 18% Fundamental change (other industries)
Radical innovative new application from different industry
Level Three
Less than 4% New application (new technology)
A new combination of technologies for new solutions e.g. materials
Level Four
Less than 1% Breakthrough! (new science)
New Invention often based on (new) scientific effect
Level Five

TRIZ helps at all the first three levels but is particularly useful at Level Three. TRIZ helps us look beyond our own area and gives us confidence to look for more interesting, stronger solutions that overcome contradictions and give us everything we want. TRIZ helps us raise our game from Levels One and Two and seeks to join the great and interesting inventors at Level Three and above. TRIZ logic is geared towards systematically locating good, clever answers, particularly those which simplify rather than complicate, and which minimize the inputs and resources needed to produce elegant, simple but powerful design solutions. Quick access of established good solutions through the Internet, together with TRIZ tools for understanding all known concepts, is now part of today's TRIZ process. We are fortunate to have free and immediate access to the patent databases which means that for the first time ever, the world's knowledge is available to us without leaving our desks. In this way TRIZ's moment is here. It teaches us to look for existing, proven good solutions by intelligently interrogating these powerful sources of all the world's answers rather than trying to conjure up an answer out of our own heads either individually or by brainstorming with our colleagues.

Systematic TRIZ or random brain storming?

How to Access Our Own and the World's Knowledge

The TRIZ tools and problem-solving process work with other established toolkits to help us to identify and define the real problem but once we have identified our problems, then applying the TRIZ toolkit will help us locate solutions – both those we know and those we don't know. Like many creativity systems one of the benefits TRIZ offers is to help us access solutions, which we already knew, but were eluding us. There are TRIZ tools to help us use all our education and experience to efficiently locate and adapt good solutions (solutions which are within our capability and knowledge). But a greater benefit of TRIZ is that it gives us routes to find solutions from the knowledge not yet familiar to us but proven elsewhere in the world.

To solve any problem with TRIZ, we need to put it through the Prism of TRIZ. This involves focusing it down (converging) from the real, factual problems to a simple conceptual problem. TRIZ then has processes to locate the conceptual answers to this problem, which then need to be expanded back by the Prism (diverging) from a handful of conceptual solutions to all the possible practical, real-life, factual solutions. The Prism of TRIZ is the model we use to access the particular and relevant TRIZ answers to our problem.

The Prism of TRIZ takes us from problem to solution using both our own and the world's knowledge through analogous thinking and is a simple process to follow and helps us systematically access solutions.

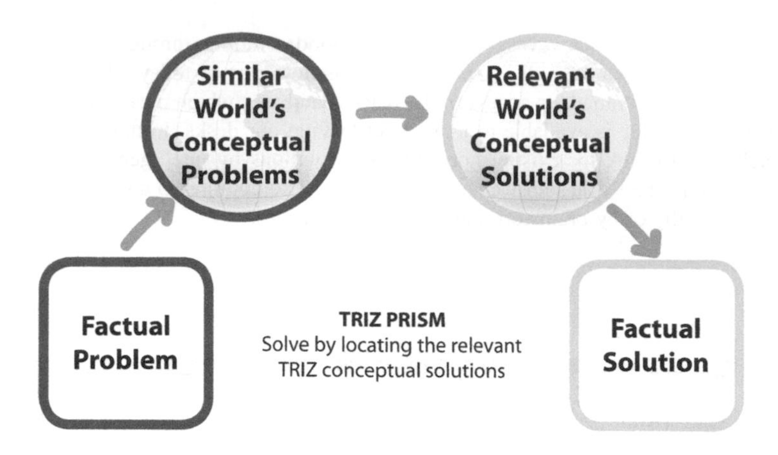

The Prism of TRIZ is the route we follow, after we have identified and defined problems, and it shows us how to find solutions. This simple TRIZ problem-solving process is systematic, and the TRIZ Solution Concepts are codified answers – all the important problem-solving power and logic contained in patents and recorded knowledge is reduced into simple TRIZ lists. The power of these lists is that they offer all the proven answers to problems, and when matched to the right problem, these relevant TRIZ conceptual answers become practical answers by engineers and scientists extending and applying their relevant knowledge. These conceptual solutions trigger both the factual answers which we know about (or could have invented with our knowledge and experience). They also show us routes to all other solutions which already exist and are documented in the patent database but may be outside our experience or knowledge. Unfamiliar technical concepts are easy to research and locate now and, with the right technical bent to understand how they will offer new solutions, can be added to our relevant and practical solution list.

TRIZ 'Dictionary' of the 100 World's Conceptual Solutions to Any Engineering Problem

By analysing the world's patent database Altshuller identified all known solutions, sorted them and then distilled them to four simple generic lists which apply to all areas of science and technology. For example TRIZ research showed that most problems contain contradictions and that there are only forty ways to solve them known as the 40 Inventive Principles. Similar research also revealed that technology evolution is predictable and can be mapped in 8 Trends. Altshuller also researched and categorized 76 Standard Solutions. These solutions occur in patents but don't necessarily solve a contradiction; these general solutions overlap with the 8 Trends and the 40 Principles, giving us a total of about 100 conceptual answers to any problem (depending on how the problem categories are divided). The 76 Standard Solutions are very general solutions, and for some, are easier to apply to problem solving than the 40 Principles or the 8 Trends but as the answers can be very general, they require more technical input to get the concept answers back to a useable solution.

Using the Prism of TRIZ Together with 40 Ways of Solving Contradictions

A good solution often contains drawbacks – or contradictions. When faced with a problem any good engineer nearly always has solution ideas, and wants the satisfaction of locating good solutions. The joy of TRIZ is that it can work with those first top-of-the-head solutions to start the problem-solving process, particularly if the first ideas have both good things and bad things (contradictions). Indeed TRIZ will help improve those solutions a little or a lot, or even use the first instant ideas to take us to relevant solutions which may be a long way from the first ideas. For example when we change or adapt engineering systems to meet new needs, our solutions – the proposed improvements – suggest changes to give us some new advantage and often mean that something else get worse – a contradiction may emerge. TRIZ offers just 40 ways of solving a contradiction, and has created two simple routes to locate which of the 40 Principles will help us solve our problem.

Get everything you want with TRIZ

The problem can be solved by using the relevant 40 Inventive Principles (one of the four TRIZ lists). Solving contradictions is explained in Chapter 5 – and each time a contradiction is solved with TRIZ one or more of the 40 Inventive Principles are used.

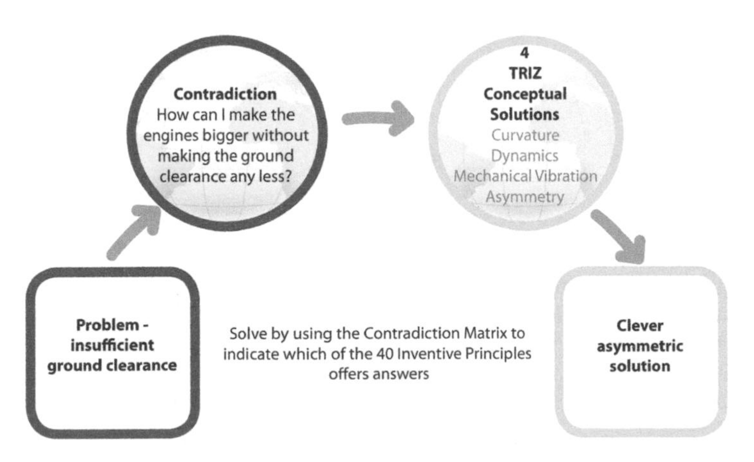

Using the Prism of TRIZ together with the TRIZ Function Database

TRIZ Function Database = over 2,500 Engineering and Scientific Effects

This lists all the engineering and scientific concepts which Altshuller found in the patent databases and arranged to make them simply understood, locatable and easy to use. They are not offered to us in complicated lists with highly technical titles: they are just categorized by what they do – the function they provide. To use the effects we must define the function we want to achieve for our product/system, and then simply match from the effects a function which is closest to our defined desired function. This process is simply following the Prism of TRIZ.

If our problem is to find out 'HOW TO?' do something, find the right function, then we can use the TRIZ effects as a first filter for answers before delving further into the world's knowledge. The effects database is the result of the initial analysis of the patents in the last century. This produced the identification of approximately 2,500 relevant scientific and engineering theories used in patents (called effects in TRIZ). These are arranged in a way that triggers ideas – their usefulness and application is put in context first rather than the name of the concept, or how it works. Altshuller in his book *And Suddenly the Inventor Appeared* said:

> In textbooks of physics the effects and phenomenon described are very *neutral* ... But what if the same *effects* could be described in an *inventive manner*? For instance: *The substance will expand when heat is applied:* therefore this phenomenon is not just about HEAT but about EXPANSION and could be used in all cases when we need to expand something; for example when we need to control very small and precise movements. If we rewrite all the textbooks of physics we will get a very powerful tool, a catalogue of effects and phenomenon.

This is precisely what Altshuller did and left for us all to use.

HOW to Find All the Ways of Removing the Water from the Glass Using Our Knowledge and the World's Knowledge

This is a simple creativity exercise used in Oxford Creativity TRIZ workshops.

A glass of water is placed on the table and delegates are given 2 minutes to write down all the ways they can think of to remove the water without moving the table or the glass.

When delegates are thinking of all the ways they can think of to get the water out of the glass, they have gone straight into a powerful and creative mindset – they understand the problem and they can immediately think of solutions. Normally in a couple of minutes each delegate will think of about eight ways of solving the problem. In a room of about a dozen people when everyone's solutions are gathered, and there is brainstorming for a few minutes we will get about 15–25 solutions. This shows how we can get more solution ideas by combining our brain power and knowledge – a good argument for brainstorming.

What Sort of Solutions Do We Think of First?

If there are a lot of chemists present they will think of changing the water. If they are big engine people they will think about pressure and big forces. If they are biologists they will think of plants and animals. The 15–25 immediate solution ideas will vary according to the mix of experience of people in the room, and if we analyse the all the ideas we will find only about 5-8 very general concepts such as displacement, chemical change, phase change etc. will have been used to produce 15-25 solution ideas. For example if someone has suggested put stones in the glass, and someone else has suggested filling it with mercury and another idea is to put another glass inside the original on – all these solution ideas are based on the one concept of displacement.

How Do We Get Beyond Just the Education and the Brain Power in the Room to Use All the Solutions Known in the World and Access the World's Knowledge?

We use the TRIZ Effects database. Altshuller's ambition was to catalogue all useful concepts into the effects list, so that it will be accessible, simple and logical. Thus the usefulness of effects is understood immediately, without needing any prior knowledge of their technical complexity or any huge knowledge of physics, chemistry etc. (see www.TRIZ4engineers.com for effects database).

We have to phrase our question in a way that defines the problem in simple terms and describes the function we want.

PROBLEM/SIMPLE QUESTION	ANSWER
FUNCTION WE WANT	Particular Concepts

This means that we have to ask our question or state our problem in very simple terms as the function we want. The original problem is stated as 'how to remove water from a glass'. To make this question simpler and more general involves a basic translation to more general non-specific descriptions:

water = liquid

remove = moving

This helps us define ... FUNCTION WE WANT = **MOVE A LIQUID** or

Simple Question = How to move a liquid?

The original question 'How to remove water from a glass' has now become **'HOW CAN WE MOVE A LIQUID?'**. We need to do this simple process to access and use the effects database easily and effectively.

PROBLEM/SIMPLE QUESTION	ANSWER
FUNCTION WE WANT (move liquid)	Particular Concepts
HOW CAN WE?	(How can we **move a liquid**?)

All we need is to ask which concepts deliver the **FUNCTION** we want? **Or answer** our Simple Question

We look up in the effects database which concepts deliver the function we want of **moving a liquid**?

The TRIZ effects database will give us all the answers found in patents – this is like brainstorming with everyone in the world. We are asking the TRIZ database for how many concept solutions there are to provide the function **move a liquid?** We are offered 99 answers in concept form.These are all in the public domain and there are many free versions available to everyone, but they all need practice to use quickly and effectively.

How can we move a liquid? *The TRIZ Effects databases suggests that there are 99 Ways ...*

Absorption (physical), Acoustic Cavitation, Acoustic Vibration, Adsorption, Aerosol, Anti-bubble, Archimedes Screw, Archimedes' Principle (Buoyancy), Barus Effect, Bernoulli Effect, Boiling

Brownian Motion, Brownian Motor, Capillary Action, Capillary Condensation. Capillary Evaporation, Capillary Porous Material, Capillary Pressure, Capillary Wave Effect, Cavitation

Centrifugal Force, Chromatography, Coanda Effect, Condensation, Converse Piezoelectric Effect, Coriolis Force, Coulomb's Law, Cyclone Separation, Desiccant Material, Desiccation

Diamagnetism, Diffusion, Displacement, Distillation, Elasticity, Electric Field, Electro-hydrodynamics, Electrolysis, Electro-osmosis, Electrophoresis, Electrostatic Induction, Electro-wetting, Evaporation

Explosion, Ferro-fluid, Ferromagnetism, Fluid Hammer, Foam, Foil (fluid mechanics), Forced Convection, Free Convection, Funnel Effect, Gravitation, Gravitational Convection (non heat)

Hydraulic Jump, Hydraulic Ram, Inertia, Injector, Ionic Exchange, Jet Flow, Kaye Effect, Leidenfrost Effect, Lorentz Force, Magneto-striction, Marangoni Effect, Mechanocaloric Effect

Micro-electro-mechanical Systems, Mixed Convection, Nuclear Fission, Onnes Effect, Osmosis, Pascal's Law, Permeation, Pump, Ranque Effect, Rayleigh-Bénard Convection

Resonance, Screw, Shock Wave, Solvation, Sorption, Super Thermal Conductivity, Super-cavitation, Super-fluidity, Surface Tension, Thermal Expansion, Thermo-capillary Effect

Thermo-mechanical Effect, Thermo-phoresis, Turbulence, Ultrasonic Capillary Effect, Ultrasonic Vibration, Weissenberg Effect, Wetting, Wind

These 99 conceptual solutions greatly expand the knowledge used in the room originally.

Simple Brainstorm without TRIZ: 5–8 concepts = 15–25 solution ideas

Systematic TRIZ Effects: 99 concepts = Every solution idea to this problem known to the world.

We have just used the Prism of TRIZ to remove the trial and error and accessed all known answers. We are able to easily locate all the concepts which can help us deliver, collate and list all the known solutions to this problem which exist anywhere. Someone else has done all this work for us and only very recently has it become freely available to everyone. All we need to do is to get into the habit of using these free databases on all problems – simple or complicated, easy or difficult.

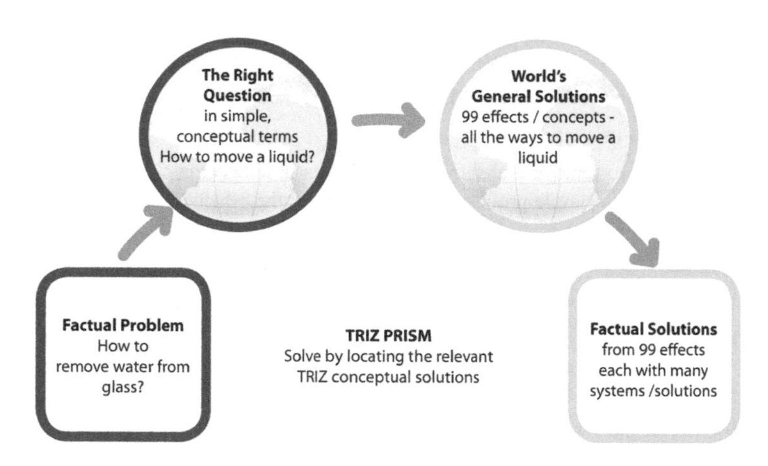

In a very general form we have now got all the solutions. Within these 99 scientific effects some are within our own knowledge, and the others include existing solutions which are proven and well documented but are outside and beyond our own education and experience. **To make these into useful solutions we have to use our brains.**

We have used the Prism by reducing our problem to a very simple question. The TRIZ problem-solving framework thus allows engineers and scientists working in any one field to access a very general form of all the good solutions and practices of all other industries and all fields of science and engineering.

Back to First Principles

TRIZ pulls together engineering and scientific theory into one framework by reducing all the solutions and concepts in the patent database to simple lists of what they can do in very simple terms, rather than what they are – their very technical names or how they do it. Keeping it simple and very general is the key to searching the entire world's knowledge to find the right functions, the ways of solving contradictions, to see and apply the evolutionary trends all products follow towards perfection, and all useful solutions for our problems. All this enables us to access scientific and engineering experience and solutions, which may be essential to solve our problem but unfamiliar to us.

Using these lists and becoming competent in the TRIZ process helps each of us realize our full potential. Part of the power of the simple TRIZ lists (the TRIZ triggers) is that the answers are in very simple general forms and by returning us to first principles – reminds us of why we wanted to be an engineer or scientist in the first place. TRIZ achieves this in different ways – the principal way is by breaking down specialism – it teaches us to look confidently outside our own knowledge, experience, company expertise and even our whole industry to find proven, but to us and our company, unusual solutions to our problem. This process is made possible by reducing all problems to their most basic descriptions – avoiding unnecessary detail.

To Share Knowledge KEEP IT SIMPLE – Avoid Too Much Detail Initially or We May Lose Our Way

All engineers are aware of a tendency when describing problems to slip into technical jargon, use confusing acronyms, and elaborate with detail both relevant and irrelevant. Good use of TRIZ requires a discipline and an ability to describe everything in simple terms and rigorously restrict detail – think in concepts not detailed facts. This helps break psychological inertia as we move away from solutions and think in general terms about the problem and solution and see more possibilities.

There are many stories of getting lost in the detail. One of my favourites is of a young teacher who had accepted a temporary job teaching a class of four-year-olds out in one of the most isolated, rural parts of north Wales. One of her first lessons involved teaching the letter S so she held up a big colour picture of a sheep and said: 'Now, who can tell me what this is?' No answer. Twenty blank and wordless faces looked back at her. 'Come on, who can tell me what this is?' she exclaimed, tapping the picture determinedly, unable to believe that the children were quite so ignorant.

The 20 faces became apprehensive and even fearful as she continued to question them with mounting frustration. Eventually, one brave soul put up a tiny, reluctant hand. 'Yes!' she cried, waving the picture aloft. 'Tell me what you think this is!' 'Please, Miss,' said the boy warily. 'Is it a three-year-old Border Leicester?' There all knew it was a sheep but they were assuming the required answer was much more complex. (Copyright Guardian Newspapers – Wednesday 2 November 2005)

Keeping it simple helps us achieve a complete understanding of the essential problem we are tackling, but it takes thought and understanding to explain something in its simplest form.

Keeping it is simple is no longer fashionable

Many industries have identical general problems but stay within their own industry when looking for solutions. TRIZ helps break these barriers down and shares solutions of apparently dissimilar problems at first glance. At a TRIZ meeting in Manchester some engineers from a glass company talked about a problem of identifying damaged rollers which scratch the glass they are supporting. There are many hundreds of rollers and identifying which one was scratched seemed troublesome and complicated. One TRIZ mantra is to get the problem component to tell you itself – that here is a problem. Another company which manufactured photographic paper had a problem of scratched paper rollers supporting photographic paper that scratched the paper. As they routinely used TRIZ they had solved this problem using the TRIZ rule of getting the damaged roller to tell them that it was damaged. A simple hunt through the characteristics and features (resources) of the rollers suggested that the damaged roller could be identified with ultrasonic detectors – a solution which the glass company then investigated and adopted (see Chapter 7 on resources).

TRIZ = Sharing of Solutions

Like many creativity tools TRIZ often leads us to access solutions which we already knew, but were eluding us. More importantly it also guides us to all the world's recorded solutions so we can locate solutions and use other companies' and other industries' solutions which we don't know but may solve our problem and are tried and tested in their own areas. This takes engineering experience, courage and creativity to successfully climb back up the Prism of TRIZ from concepts to useful solutions.

TRIZ offers a raft of creativity tools to help us for such tasks and see how to apply concepts in ways relevant to our problem. We need to think differently and use our brains to their full extent. The TRIZ tools to help us do this include Time and Scale, Smart Little People, and Size-Time-Cost – all are very powerful creativity tools which help jolt our brains and translate concepts into useful solution ideas.

Creative people (especially engineers) enjoy imagining new solutions

TRIZ also takes us to all the solutions to a problem – if we already knew all the answers (very rare) it does at least it reassure us that we have thought of everything. It also helps us become efficient in locating and sharing solutions. The TRIZ approaches to simple solutions are of huge importance today for knowledge sharing. In a world of downsizing, fragmentation and new corporate cultures, knowledge management, based on TRIZ, can help make both in-house and global solutions readily available to anyone appropriate, anywhere in the company at any time. Ideally if the person storing the information and the person accessing it have an understanding of the TRIZ process it will be simple. If everyone has a knowledge of TRIZ then the use and sharing of knowledge in the world would become very effective indeed.

Conclusion: TRIZ Access to the World's Knowledge

By applying the TRIZ Solution Concepts we are accessing a distillation of all known problem solutions in a systematic way – and proving the TRIZ concept that nearly all problems have been solved before. Much time is wasted without TRIZ as we often re-invent the wheel instead of accessing the smart and proven solutions. TRIZ guides us to the solutions which we need, which exist, which are innovative but which without TRIZ we often don't know how to find.

The Prism of TRIZ is a simple but powerful process for locating solutions and helping us to achieve innovation and creativity in problem solving by leading us to all recorded solutions. It helps us with any kind of problem, of any kind of difficulty and at any stage including how to invent, evolve, fix/mend, design, develop, improve, simplify, adapt systems intelligently and helps us find the next generation systems, products and processes. TRIZ thinking leads us to the most cost-effective routes for achieving these by using all the resources we've got together with the world's patent/knowledge base for proven (analogous) solutions.

There are simple lists in TRIZ which distil the world's knowledge:

- 2500 Effects/Concepts arranged in a HOW TO list (that will help us achieve a function)
- 40 Inventive Principles for solving contradictions
- 8 Trends of Evolution(which all systems follow as they develop)
- 76 Standard Solutions.

TRIZ has thus distilled the world's knowledge into these four simple lists which we can use to access all the conceptual solutions in the world. When we apply them to our problems, it is rather like brainstorming with every inventor and engineer who has ever published solutions. We use the TRIZ concepts list to trigger good ideas for solutions both to get the most out of our brains, to remind us of all the solutions we know, and also to go beyond our own experience to find new solutions (new to us). TRIZ will thus suggest all the solutions to our problem, which includes all those we know about, and all those we don't know.

TRIZ and TRIZ Software – a personal view

Beyond the simple effects database there seems to be no better and comprehensive English version of Altshuller's original catalogue of effects. There are expanded versions available through the interesting but often complex software packages developed by Russian TRIZ experts in the 1990s in the USA – the two principle ones being Invention Machine and Ideation. Initially I sold TRIZ software for Desktop Engineering in the UK. The TRIZ expanded effects database was very useful and interesting to engineers but I was surprised that even when companies bought the software and were trained in it – how little use it got. Eventually I realised that thorough understanding of TRIZ needs to come first. The software packages needed good TRIZ grounding to use effectively and once you had good TRIZ skills many people did not usually need software for most aspects of problem solving.

The effects database is now available free in several forms on the Internet and also available in some paper forms (less complete than the software versions). It is useful in some circumstances but with such intelligent search engines available on the patent databases – anyone with good TRIZ approaches can learn to find the relevant information fairly quickly. With TRIZ training, the simple TRIZ effects and familiarity with the patent sites we can access all the world's published solutions.

Case Study: TRIZ in Rolls-Royce

The following is taken from two articles written by Dave Knott, CEng FIMechE published in *CIPA Journal* (Chartered Institute of Patent Agents) article 'TRIZ in Practice: Better Ideas, Faster' and 'Applying TRIZ in a large organisation' from *The Post*. Dave Knott recommends certain TRIZ tools here, which do not include the Oxford TRIZ Standard Solutions (of similar importance to the 40 Principles) as they were not taught at Rolls-Royce.

Get Hooked on Faster, More Creative Thinking

As part of its continuing drive to devise better ideas faster, Rolls-Royce has been applying the TRIZ technique to innovative problem-solving.

As much a philosophy as a physical system, TRIZ came from Russia and is the result of identifying common factors from more than two million successful inventions. It is a way of harnessing the creativity of the mind and steering it along proven paths of success.

Dave Knott, Company Specialist in Design Technology, explained, 'Russian patents agent Genrich Altschuller discovered that one or more of just 40 inventive principles lay behind all the patents analysed over decades after the Second World War. This knowledge has been refined into a set of tools to help us understand a problem, identify the resources available to solve it and apply TRIZ principles to generate ideas and provide solutions.'

The new Trent 900 Fuel Manifold derived from a TRIZ workshop to solve a technical issue resulted in a very welcome spin-off: a 95 per cent reduction in build time.

Dave first used TRIZ principles after attending an event in Loughborough back in 1996. He applied it to a particular problem he was wrestling with at the time – and it ultimately resulted in a patent. 'I suppose I was hooked after that,' Dave said. 'It promotes creative thinking in finding engineering and manufacturing solutions. TRIZ changes the way you approach problems, and tends to encourage the simple, low-cost solution.' Rolls-Royce used Oxford Creativity to deliver TRIZ training for over five years but in time, as capability and understanding grew, a network of facilitators was established and the company launched its own training, centred on proven engineering challenges.

A TRIZ website and 'community of practice' have been established to ensure the global network in this creative approach to technical creativity can grow and prosper. Dave Hopkins, Assistant Chief Design Engineer for the F136 engine for JSF, says: 'TRIZ was very effective in helping us to identify areas for improvement.' Jerry Goodwin, Chief of Advanced Propulsion Systems Design, recommends that all his designers receive TRIZ training: 'We've used this methodology a number of times and it has not just helped generate new concept avenues but also resulted in the generation of a number of patent applications.'

Business Benefit of TRIZ

Innovation and Engineering Design have been key factors in enabling Rolls-Royce to establish leading positions in highly competitive global markets, providing power for land, sea and air. TRIZ is proving a valuable tool in generating better ideas, when compared to conventional brainstorming techniques. Better ideas translate into better products that are cheaper and faster to produce, resulting in better business performance.

One quantitative example of the benefit of TRIZ is shown in Figure 1. This shows the results of an initial evaluation conducted on the ideas generated to solve a design problem. The ideas were generated in two parts: the first a conventional brainstorm; the second a TRIZ-facilitated session. The same number of ideas were generated in each session, but the results clearly showed that the quality of the ideas generated by TRIZ were higher than in the brainstorm. About 75% of the ideas judged to be of high benefit and high feasibility came from the TRIZ session. Information like this is very helpful in encouraging the use of TRIZ within a large organisation.

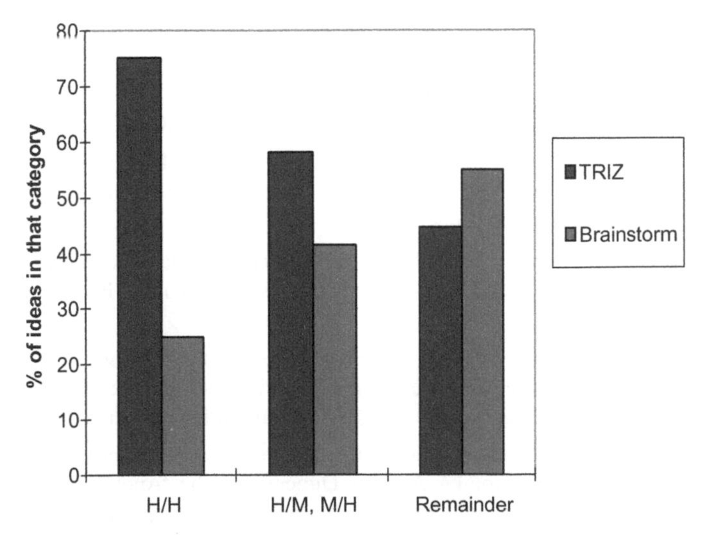

Figure 1 Distribution of ideas evaluated against benefit/feasibility using a high, medium and low (h, m and l) scale

TRIZ Tool	Benefit
Functional Analysis	Encourages functional thinking, and captures all the interactions within a system, even the harmful effects. Particularly useful for cost-reduction problems.
Resources	Identifies all the potential resources available to solve a problem, even harmful/waste effects and products. The cheapest solution is likely to use an existing resource.
Contradiction	Helps clarify and generalize the key problem to be solved.
Ideal Outcome	Helps to overcome self imposed constraints.
Smart Little People	Engages the imagination, and helps break 'Psychological Inertia'.
40 Inventive Principles	All technical problems are solved by applying one of 40 Principles. Most important and useful TRIZ tool (in my experience) see below.
8 Evolutionary Trends	Helps predict likely technological directions. Useful when looking to improve an existing system.

Figure 2 The key TRIZ tools being used within Rolls-Royce

Of all the TRIZ tools one of the easiest to use, and the one proving most popular within Rolls-Royce is the 40 Inventive Principles. Part of the TRIZ research showed that all inventive problems are solved by applying one or more of these 40 Principles. The benefits of this approach are:

- Good ideas are generated faster than with conventional brainstorming, because the Principles direct thinking along profitable lines.
- A broader range of ideas is generated, because the Principles take you down paths you would not otherwise have travelled. In Rolls-Royce high-quality, new ideas have been generated even on old problems.
- The 40 Principles are largely intuitive and one of the easiest tools to use, and therefore there is no steep learning curve when they are being used infrequently. Once they are familiar to you, they become self-reinforcing, because you see them at work in products all around you.

For example Figure 3 shows how one innovation is illustrative of a common principle at work in Rolls-Royce products, namely Principle No. 17: Another Dimension. The 'Another Dimension' Principle suggests that there is likely to be benefit from thinking around how the dimensionality of a feature can be increased or more use be made of another dimension, e.g. making a two-dimensional feature three-dimensional. In the example in Figure 3, an improvement has been made to the end walls of the

Principle 17. Another Dimension

- If an object or feature is 1 or 2D, consider making it 2 or 3D

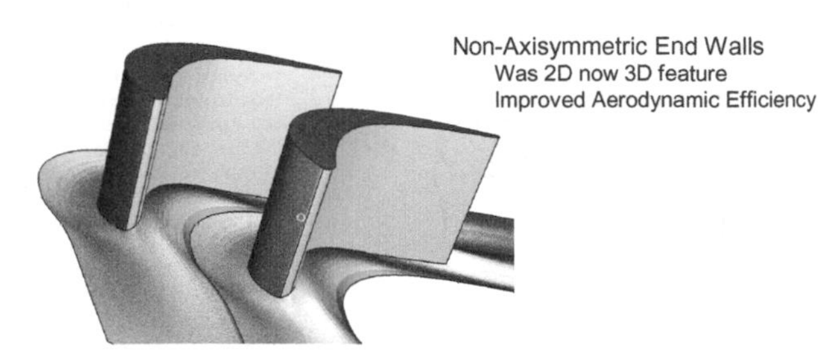

Figure 3 Example of one of the 40 Inventive Principles at work in Rolls-Royce products

main gas path in a gas turbine engine. These end walls have been two-dimensional surfaces of revolution for decades. It has now been found that putting three-dimensional features onto these walls (i.e. making them non-axisymmetric) improves their aerodynamic efficiency significantly.

Another inventive principle that is generally very useful is Principle No. 13: The Other Way Round, which encourages you to think about doing the opposite of what is currently done. This principle was applied to a problematic automated method for calculating turbine throat areas in which the software to to calculate a complex aerodynamic parameter often failed as the complex geometry involved changed. The problem was still not fixed after 6–8 months of development and then TRIZ was tried.

A new approach required and the 40 Principles were used to generate some ideas. Principle 13: The Other Way Round yielded the idea of modelling air in passage rather than the metal itself, i.e. producing a solid model of the air in the passage rather than a model of the metal that formed the passage. In other words use the 'negative' of what was being used. The result: problem solved.

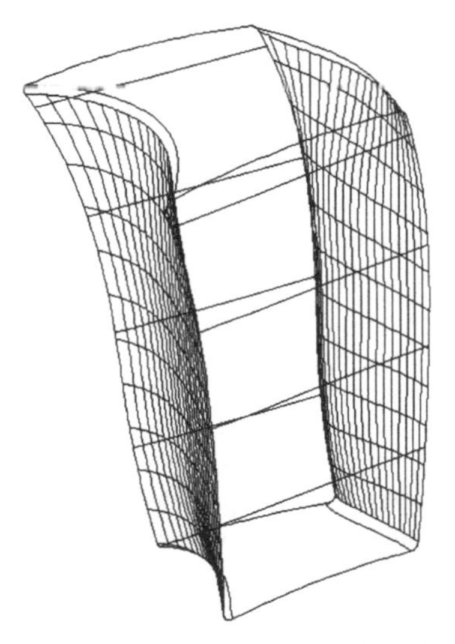

Figure 4 Produce a solid model of the air in the passage rather than of the metal that formed the passage

TRIZ Illustrated in a Rolls-Royce Patent

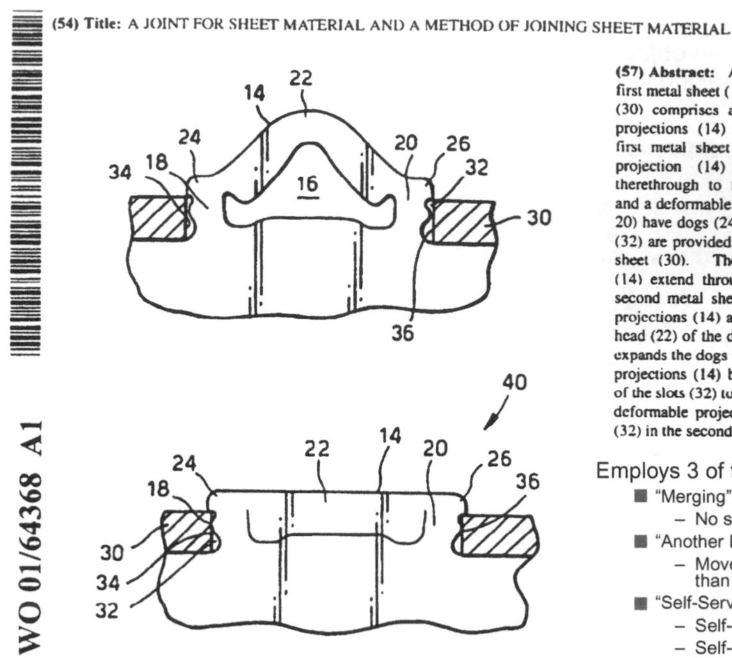

(57) Abstract: A joint (40) for joining a first metal sheet (10) to a second metal sheet (30) comprises a plurality of deformable projections (14) on an edge (12) of the first metal sheet (10). Each deformable projection (14) has an aperture (16) therethrough to form two sides (18, 20) and a deformable head (22). The sides (18, 20) have dogs (24, 26). A plurality of slots (32) are provided through the second metal sheet (30). The deformable projections (14) extend through the slots (32) in the second metal sheet (30). The deformable projections (14) are deformed such that the head (22) of the deformable projection (14) expands the dogs (18, 20) of the deformable projections (14) beyond the sided (34, 36) of the slots (32) to prevent withdrawal of the deformable projections (14) from the slots (32) in the second ductile sheet (30).

Employs 3 of the 40 Inventive Principles

- "Merging"
  - No separate rivet
- "Another Dimension"
  - Moves in the plane of the sheet rather than normal to it.
- "Self-Service"
  - Self-Locating
  - Self-Locking

Figure 5 A Rolls-Royce patent employing 3 of the 40 Principles

Many innovations employ more than one of the 40 Principles at the same time. RR Figure 5 shows a Rolls-Royce patent for a novel method of joining sheet metal together that does not require expensive fixturing or welding. Three of the 40 Principles can be seen in this invention, any one of which could have stimulated the idea. This also illustrates the fact that TRIZ can also be very useful in the solving of manufacturing problems.

In common with many other companies, Rolls-Royce is increasingly seeking to maximize the return on its intellectual property, through the licensing of patents, technology and know-how, as a means of exploiting its technology in markets that are outside its core business. This feature of the 'Knowledge Economy' is fully in line with the TRIZ ethos of re-using inventions and knowledge from other fields.

Effective change requires people to be provided with an improved capability AND have the motivation to use it AND also the opportunity to use it. It is shown as a multiplying relationship in the equation because, no matter how big the improvement, if you only have one or even two out of the three the benefit will still be zero. Therefore it has to be recognized that providing engineers and designers with training in TRIZ alone is not enough. The training certainly gives them an improved capability, and a degree of initial motivation, but unless they are provided with the opportunity to use it, and an environment that nurtures innovation and creativity, the benefits in the long term will be limited. Now motivation and opportunity are very much leadership, organizational and cultural issues, and therefore TRIZ needs to be implemented as part of a holistic approach to creativity and innovation.

Finally, in common with many other tools, one of the critical factors in deriving maximum business benefit from TRIZ is applying it to the **right problem at the right time**. In the case of TRIZ the right time is usually as early as possible.

Fundamentals of TRIZ Problem Solving

What is Problem Solving?

Problems are gaps between requirements and their fulfilment by systems, and problem solving evolves systems to both work better and minimize those gaps. There are problems at all stages of creating and using systems, and we need solutions for system improvement both for faultless operation and to fulfil more of the requirements. Therefore problem solving can involve fixing problems within systems and adapting them (or finding a more appropriate system) to better match the requirements.

To start problem solving we need to understand gaps between the requirements and the system (plus any shortfall in its performance). Ideally we would like a perfectly formulated problem statement and then locate and implement its most relevant, practical and clever answers. Problem solving is rarely that straightforward – defining problems takes careful work and solving them needs access to relevant solutions (both in our heads and beyond our knowledge) and difficult problems need either good luck or systematic TRIZ processes. Problem solving can be about finding a good range of solutions to create or improve systems, or can be about solving the 'mystery problems' (or hidden causes).

TRIZ for Engineers: Enabling Inventive Problem Solving, First Edition. Karen Gadd.

Most problems have more than one solution.

Mystery problems – solved by finding the cause or culprit – involve identifying the relevant information

Finding Solutions – Systematic or Eureka Moments?

When we find answers by ourselves or with others in a team, it is satisfying, fulfilling and often fun. We all remember with pride those moments when we reach the most pleasing moment of finally knowing how to solve difficult and puzzling problems. This is one aspect of the thrill of creativity, and because our best solutions often seem to come to us in a flash of inspiration (after a period of confusion or aridity of ideas), we may be led to believe that all good solutions are to be found in this way: that real creativity is an uncontrolled, lucky and sudden mental leap from problem to solution.

Moving across the gap from problem to solution can be a big leap (non-TRIZ) or it can be an ordered progress of small steps in the right direction towards the relevant answers (TRIZ). Moving towards the

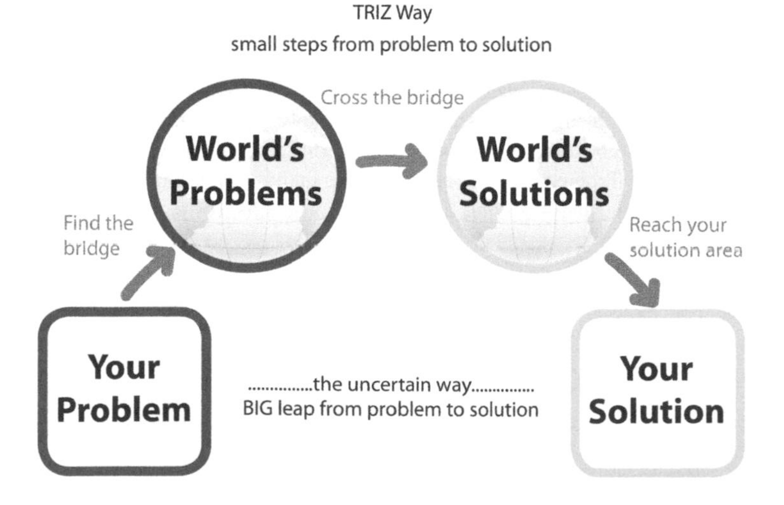

right solutions can be like crossing a bridge which has been built in the right place from the blueprints of previous successful solutions from similar problems. This bridge is the Prism of TRIZ, explained in Chapter 2.

The role of TRIZ in problem solving is to make this journey from problem to solution as predictable as possible, to point us in the right direction and to enable us to problem solve whenever we want – to travel certainly to the relevant solution areas and be able to solve problems on demand. This needs the conversion of TRIZ conceptual answers to real, practical answers by adding relevant knowledge. (See the TRIZ case studies for successful conversion of TRIZ concept to practical solution). By following the steps of the Prism of TRIZ we are jettisoning all irrelevant, confusing information and focusing on the real problem to locate the small number of relevant solutions. It is a disciplined few steps which make us concentrate to define the exact nature of the problem before we develop real, practical solutions by adding back our knowledge to transform the few relevant concept solutions to a useable answer.

I'm a Genius – I Don't Need TRIZ Thinking

Systematic problem solving on demand can be a very unattractive thought to some who believe themselves to be 'instinctive problem solvers' and assume they are innately better than most others at making the right connections due to their sheer brain power. This implies that problem solving has no logic, no map, no steps which can be mapped and recreated on demand – and it is only certain kinds of clever people with their unique brains who can solve difficult and puzzling problems. Our most famous heroes of science often tell of sudden breakthroughs which seem to emphasize their unique cleverness but don't mention years of relevant education and meticulous work. There are stories going back hundreds of years of famous artists who tried to destroy all evidence of their apprenticeship years, to suggest that they simply emerged as ready-made geniuses, and that learning from the past and others played no part in their achievements. Even childhood prodigies such as Mozart were strictly schooled in the tools of their trade before their creativity emerged.

Such myths of inherent scientific or problem-solving genius are reinforced by problem-solving tales about both real and fictional characters. Fictional heroes such as the doctor in the TV series 'House' or detectives such as Sherlock Holmes are often assumed not just to have superior intellects, but special 'innate' problem solving abilities which help them focus on the relevant facts and solve mysteries. This means that when they make that jump to solution it is often assumed that they have spontaneously found the answer from sheer brain power. However closer analysis shows that they are usually portrayed as meticulously and systematically distinguishing between the relevant and irrelevant information and their innate ability is to be systematic. This enables them to make the right connections, make sense of right knowledge and instil the self-discipline to concentrate on the problem and not keep leaping to possible solutions. Altshuller said we all have this innate ability (but many of us have to re-awaken it) and that solving problems can be like solving a murder mystery. He emphasized that the Sherlock Holmes approach was a very good one (not just because Sherlock Holmes is clever) but because he doesn't wrongly jump to premature solutions. Holmes solved problems with careful, methodical, systematic searches to help isolate, recognize and concentrate on the relevant facts. The Sherlock Holmes approach is not about some magical, innate ability but one of rigorous logic (like 'TRIZ Thinking') which is valuable for anyone to reawaken and adopt. TRIZ teaches us to think like Holmes.

When engineering teams learn TRIZ together, and radically improve their problem-solving abilities, if there is a champion problem solver (someone who seems to naturally come up with more ideas than everyone else put together) – then they often remain ahead of the others. This is because TRIZ is a multiplier of problem-solving ability. For those who seem innately good at it, TRIZ makes them better and faster; for those who appear normally slower at solving hard problems, it provides a proven method of always getting to the right answers. But what is important is that although the 'hero problem solver' is valuable, with TRIZ-trained engineers we don't need a hero to get us there faster – all engineers become good problem solvers, and with team work all problems can be solved.

Working together effectively and consistently out-performs individual efforts – even of self professed heroes

TRIZ Conceptual Solutions

The essence of nearly all the useful discovered solutions is contained in the TRIZ solution concepts which offer very general prompts of how to solve problems. Locating the small number of the right TRIZ solution concepts (from these lists) to a particular problem shows us all the particular places where useful good solutions can be found, and made practical and useable by adding in relevant experience and knowledge. Therefore if we have a system which has something 'harmful' then we can step through the 24 ways of dealing with that harm. For example if we define a problem as 'a toothbrush which damages gums', we have a harm. To deal with this we would simply try each of the 24 solution concepts for HARMS on our design of the toothbrush and the process of using it to clean

HARMFUL ACTIONS – 24 ways

1. **Eliminate** – Trim out the Harm – 6 ways
2. **Stop** – Block the Harm – 11 Ways
3. **Transform** – Turn Harm into Good – 4 ways
4. **Correct** – Put right the harm – 3 ways

teeth. We would know then that we had explored all possible theoretical solutions to solve the problem, and that our challenge would be to create practical answers.

The power of dealing with HARM with the TRIZ solution concepts does not just apply to technical systems. These powerful lists can be used for anything which has harm including management, business or social problems. They suggest all the conceptual ways known in the world of removing, blocking, using and transforming harm as shown in the real life example illustrated below.

Many TRIZ solutions to problems – counteract harm is just one

Thinking in Time and Scale

At the beginning of problem solving we use two of the simplest and most powerful TRIZ tools Ideal Outcome to help us understand what we want and Thinking in Time and Scale to map the context of problems (also often called the 'System Operator' or '9 Boxes'). Thinking in Time and Scale helps us understand the many big issues related to a problem, and see that there may be many places and times to find a range of relevant solutions. Some problems are not difficult, just complicated, and mapping them in the 9-Boxes helps us understand initially confusing issues; then we can deal with the problems at source (not symptoms). When we better understand a complex situation we can sort out the many solution possibilities, and sometimes all that is needed is our common sense to solve the problems.

Thinking in Scale highlights all the *places* to solve a problem. There may be many good solutions at all levels – at the system level, or in the detail at component level (sub-systems), or in the big picture at the super-system level. Thinking in Time helps us understand all the many different *times* for solutions in a process or when using a system. This includes the **before** (particularly when prevention is better than cure), **during** (options for reaction to situations) and **after** (we always need solutions for fixing or dealing with something which we cannot always prevent). For example when teaching 9-Boxes Solution Maps (see Chapter 4) I ask teams of engineers to plot in time and scale (using all 9 boxes) the many solutions they know (or can imagine) which would prevent death or serious injury of a Formula One racing driver by filling in the 9-box sheet shown on the next page.

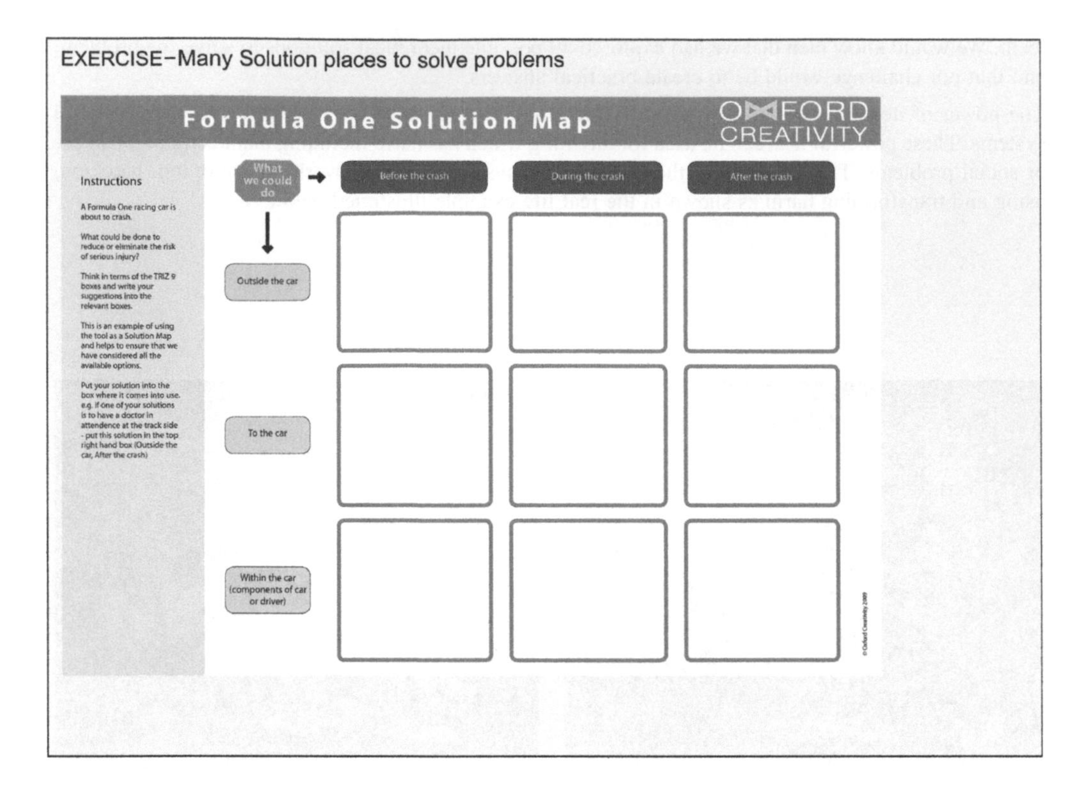

Many Solutions to Any Problem

Thinking of all possible solutions (with or without TRIZ solution triggers) and sorting them in *Time and Scale* helps prevent an inappropriate narrow selection of solutions. When we look at problems solved by TRIZ this is often a characteristic – a wider trawl was made for the best solutions. For example an industrial baking problem in a global food company was solved by TRIZ and resulted in a very inexpensive and simple solution. The engineer who owned the problem told us a story about using 9 Boxes and sharing knowledge in her company to find this alternative, cheaper and better solution.

The problem was that when making biscuits on an industrial scale the edges of large batches were burnt. The engineers involved in the process were at the final stages of approval for an edge-trimming machine (to remove the burnt bits very neatly). Before approving such an expensive and imperfect solution a TRIZ session helped assess alternative solutions. This encouraged looking beyond the first obvious solutions to check if someone else knew of other ways of solving this problem. As in this case better solutions are often found by looking beyond the knowledge of the original team, which meant the engineers never got their large, expensive, complex trimming machine. This is because one of the female chemists from another department delivered a better solution. She explained that all we need to do is to change the type of flour, to one which won't burn at the baking temperature, even at the edges. Most good bakers (home and industrial) could have solved this with their knowledge. The engineers had a solution, but it was far from ideal. Any amount of problem-solving time and brainstorming from the original team of engineers for solutions would probably not have produced this good cost-effective solution as they were only using their own knowledge, and engineering approaches for solutions.

TRIZ always helps us look outside our area of expertise and our own ideas, even though we might want to stick to our imperfect solutions – because they are ours. With TRIZ we can locate different and often better solutions which probably already exist and are documented, but in other disciplines and industries outside our experience.

When problem solving, TRIZ will fairly quickly identify most (if not all) good alternative solutions and not leave such important decisions to chance or confined to one area of expertise. One problem I worked on involved the emptying of cake moulds without damaging the cake or leaving any behind. Two days of TRIZ function analysis showed the problem to be an insufficient action when removing the cake, which was solved using the Oxford TRIZ Standard Solutions which suggested adding a field – they chose vibration. (I wonder if any of them had ever noticed their mothers gently tapping the cake tins at home to get their cakes out intact.) A similar problem was solved in the nuclear industry when a long stick used to remotely move nuclear contaminated materials was working insufficiently – a check through the TRIZ 35 solutions for insufficiency suggested adding vibration – which very simply solved the problem.

Pulsing power boosts many kinds of systems

TRIZ for Sharing Solutions

Once when problem solving in an automotive company we were working on the failure of a small cover, made of cloth which kept ripping in one place. In the problem-solving session I asked when they cut out the cover which direction was the cloth arranged; I suggested that if they cut it 'on the cross' (on the diagonal), then they would get more flexibility when stretched and it would probably not fail and rip. This solved the problem and anyone with basic dressmaking skills would have known how to do this; however, such skills were not possessed by any of the engineers who had worked on it. But the same team admitted that one of their female engineers had recently solved another problem involving cloth – which provided an inexpensive boot cover for occasional use. The problem was that material would not fold where required but kept folding in random places. To get the cover to fold in the designated places their female engineer suggested applying a simple design trick used in window blinds – this simply required sewing just one seam where each fold was required, which then solved the problem.

Even with TRIZ when problem solving it is wise to involve a great mix of experience and knowledge, and whenever possible also use people who are not experts in the problem area. We have used TRIZ in many companies for design and process problems, and we always request a good range of experience, as many different areas of expertise and some naivety is also useful. (New graduates are good for this as they still have a can-do attitude because they haven't learned to think in the many constraints of a problem.) With a broad base of experience and knowledge in a team the TRIZ solution triggers are applied in many different ways, and this creates a good range of practical solutions. One way we achieve this is by solving two different problems in the same session each with their own expert teams. The teams are asked to solve the problem which they own *and* the other team's problem, which they often know nothing about. This proves successful every time and breaks the psychological inertia of believing that the problem you own is very difficult to solve (because we are familiar with the many constraints) but the other team's problem always seems fairy trivial. We are able to offer many (naive) answers with confidence to others' problems – some of which form the starting point for good solutions.

Such a wide range of experience from various team members or from different teams is always helpful. At one of our public TRIZ classes, an engineer who was working on a small domestic product with a small engine had the problem of reducing its noise (due to a change of regulations). She was the only one from her company but the class included several aerospace engineers. When she drew up her problem the aerospace engineers laughed and said the answer is easy, just look at the airflows – they had the relevant expertise, she did not. She was delighted with their solutions which they worked with her to develop. Quite rightly she was not embarrassed that they knew the answer, which was obvious in their industry, but not in hers. The TRIZ Creativity Triggers, such as the Prism of TRIZ, encourage us to look outside our own industry and expertise because someone else probably has the answer. Locating proven solutions from other industries has never been easier with the Internet and all the on-line patent databases which are an invaluable treasure trove to any engineer. Nowadays there is no excuse for not finding the best answer; inappropriate and therefore needlessly costly solutions should become a thing of the past.

In problem-solving sessions with mixed teams once everyone understands the problem, what is wanted, and where and when the system does not deliver this, then the solutions are often a simple mix of:

TRIZ solution triggers
AND
The right, relevant knowledge
AND
Common sense

Logic of TRIZ Problem Solving

It was only after using TRIZ for some time that I appreciated the simple power of its problem solving and its logic. It was so simple, and yet it had not been stated very clearly in any of the TRIZ literature or teaching or I had somehow missed it. It is simply and obviously that problems are gaps between SYSTEMS and NEEDS / requirements / benefits we want. When I suddenly understood it, I drew the simple picture below.

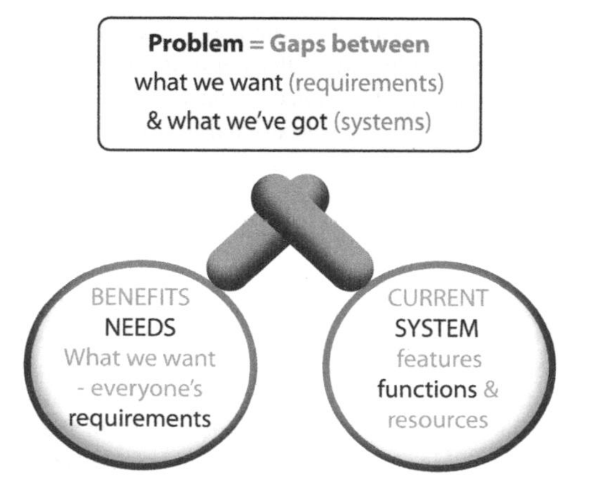

Systems are created (using resources) to provide functions which deliver benefits to fulfil **needs**. Whenever there is a mismatch between needs and the system's outputs then there are problems. This mismatch includes any gaps between the price (real inputs) and what we want to pay or invest, and includes any deficiencies and harms and contradictions.

To problem solve we have to understand the system and how it works, how well it fulfils defined needs and also understand any requirements or benefits we are additionally seeking. We can use simple TRIZ tools to define both the 'system we have' and the 'system we want' and use TRIZ to solve any problems by closing the gaps between them. Analysing our system requires the TRIZ Thinking in Time and Scale Tool (for its context) and TRIZ Function Analysis to understand all its functions (good, bad, insufficient and missing) and their relevant detail. Understanding our needs is achieved with a TRIZ Ideality Audit, and unless we undertake a proper analysis, we can be mismatching needs, or overlook important requirements. There are many situations when we loosely define requirements/needs and provide a totally inappropriate system to meet only some of our requirements.

Make sure you define everything you want

Problem understanding is therefore simply defining the gaps between NEEDS (what we want) and the SYSTEM's outputs, and TRIZ problem solving involves closing them, and getting the system to work as it should. The TRIZ problem-solving toolkit is a useful addition to any of the other engineering toolkits within companies and the application of TRIZ enhances such work. TRIZ additionally and uniquely then shows how problems can be solved. This is by applying one or more of the TRIZ solution concept lists and using relevant experience together with an intelligent and systematic application of resources. After achieving good problem understanding we should have focused to a succinct problem list which we can tackle one at a time. We can then follow the TRIZ problem-solving process for each problem and find the small number of relevant TRIZ solution concepts which will help us solve each problem.

Understand the Problem – Where's the Fun in That? We Like Solutions

When we start problem solving – whether beginning with system analysis, or requirements/needs analysis or while defining our Ideal (and hence roughly defining what we want and what our system may develop towards) – we need to capture solutions as they occur to us. The many worthy and detailed problem-understanding, system-analysis, root-cause and requirements-capture toolkits (non-TRIZ) generally encourage a thorough understanding of the problem – both requirements and the system – before matching them by problem solving. Nearly always any capture of solutions/ideas is discouraged until the prescribed point when 'solutioneering' is allowed. Problem solving is either given a fairly brief time and expected to happen during a brainstorming session, or given a very extended time in research departments (in-house and university). Outsourcing or timetabling the problem solving stage can lose many good ideas because solutions are spontaneously occurring to everyone involved from the first moment of problem understanding. Whenever we are beginning to grasp a problem our brain is working hard and effectively to solve it. We can't help conjuring up answers (systems to match the requirements) even when the formal problem-solving processes we are subjected to may forbid us to do so yet (a bit like 'no pudding until all the vegetables are eaten' – or no fun of problem solving is allowed until the problem is defined). The spontaneous solutions which pour out of clever engineers in such circumstances may be quashed and ignored. This is a great loss as many solutions to partially understood problems are valuable starting points for problem solving (not end points or even chosen directions but they may contain some good but problematic solution ideas which the TRIZ processes can help transform from half-formed solutions to real solutions).

Spontaneous 'BAD' Solutions

More than anyone else engineers are more inclined to think about solutions rather than problems! Start describing a problem to engineers and after a few minutes they are listening with only a part of their brain, because they have thought of answers and are thinking hard about their own clever ideas and how to develop them. This is normal. Sometimes even the person describing the problem presents the problem in terms of their own solution with something like 'We need to do this' or 'The problem is this and we could ...' or 'How can we achieve this?'. The person tasked with describing the problems has couched the problem together with their solution and can't seem to help jumping into solution mode even during problem description. Always check that a problem statement doesn't contain solutions.

BAD Solutions or Ugly Babies

We all love our own solutions – and for problem-solving sessions we tell everyone to call them 'my ugly baby' or 'my bad solution'. This is because we always love our own solutions as they always look good to us (as we created them); but everyone else can see what is wrong with them. Once we think up *our own solution* we find it hard to concentrate on anything else. Our solutions came from *our* creativity and *our* experience and *our* genius. All solutions have something good and bad about them.

As engineers are trained to react to problems by thinking up solutions (unlike scientists, who are apparently better at analysing problems), it is curious that so many engineering 'problem-solving' kits forbid such natural behaviour. This may be because it is seen as harmful that engineers tend not to ask for enough information about the problem, once they are into solution mode. There is a very simple way to overcome this problem – we just call all spontaneous solution ideas 'BAD SOLUTIONS' and encourage their capture and then park them in a BAD SOLUTION PARK. This helps ensure that they can be temporarily forgotten, but not lost, while problem understanding is properly completed. Solutions are what engineers do best, and finding them feels more fulfilling. Curiously this is the way that engineers enjoy themselves and have fun. (I asked a number of my TRIZ colleagues what sort of party I should have to celebrate the ten years of Oxford Creativity's TRIZ work – they all asked for a problem-solving party.)

Bad Solution Parks

We call them Bad Solutions to encourage a false modesty in the person posting the idea and because every solution idea which occurs (particularly before the problem is understood) almost certainly will have problems. It will also have some good things – some analysis of how a solution system could match some of the requirements – but there will be bad things, such as insufficiencies, harms or not meeting some important needs. Bad Solutions are the starting point of the TRIZ contradiction problem-solving process which is designed to keep the good in solutions but deal with anything bad. TRIZ Thinking for example suggests that all harm should be regarded as resource to turn into good if possible, and that Bad Solutions are a very valuable resource to be captured and harvested in all problem-solving sessions.

At Oxford Creativity the introduction of Bad Solution Parks into our TRIZ problem-solving sessions transformed everything we did and linked it to the fundamental logic of TRIZ problem solving. This simple trick or device encourages the quick capture of all powerful ideas from everyone in the room. The rule is everyone quickly and briefly writes down any and every solution idea that occurs to them (on extra sticky notes) and parks them on the Bad Solution Park. After quickly jotting down their ideas they are able to mentally return to analysing and understanding the problem. This enables the capture of everyone's powerful insights but does not impede thorough problem understanding. When a solution is parked on a sticky, then the idea can be temporarily forgotten (let go) and can be returned to later. One way to discourage immediate but inappropriate solution development is to make it clear that all solutions will be considered later and that no ideas will be lost. By labelling these ideas as Bad Solutions (even the very good ones) we are asking people to capture their ideas, but not judge or assess them in any way; all idea prompts no matter how bad, good, trivial, wacky or impractical are allowed.

Bad Solution Parks can take many forms such as a simple sheet where all ideas written on stickies are simply stuck on randomly to be analysed later – or it can allow some simple subsequent analysis of the solution to show how it is bad and good.

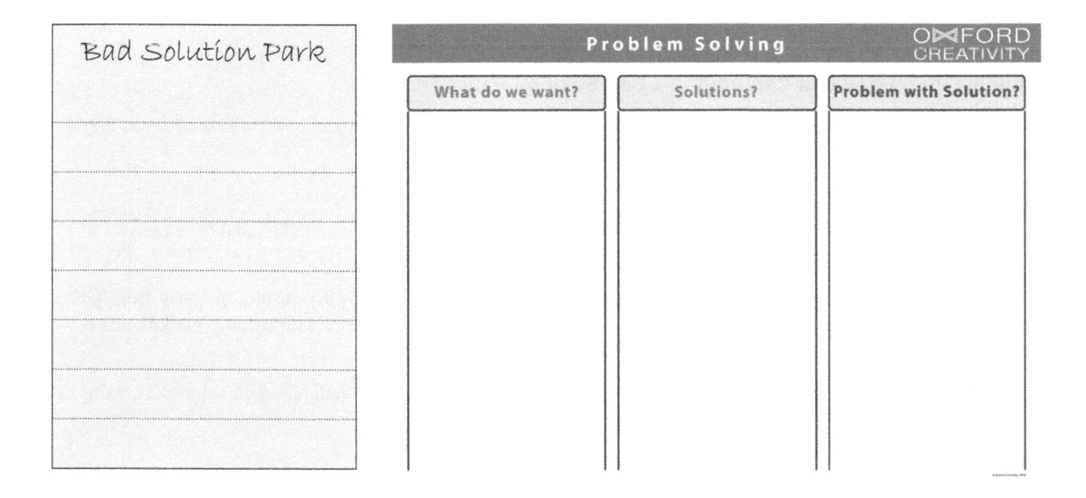

Different Ideas from Different Team Members

Transforming, realizing and delivering top-of-the-head solutions means we can then use the TRIZ toolkit for solving problems with Bad Solutions as a starting point and which are transformed to better solutions using TRIZ concepts plus resources. Teams are important in this because any Bad Solution will gain new strengths and applications from the various insights of others. Team thinking is one of biggest resources for transforming, combining and improving each others' solutions and seeing new ways of using solution concepts. Every team member will see different solutions to each problem.

When a team starts problem solving together everyone has their own individual solutions and ideas which are very valuable but can lead to a battle of solutions.

We may all have different valid and good solutions to a problem

Engineering Genius is Captured in a Bad Solution Park

As soon as a problem session has started everyone should use stickies to capture and park *all solutions* as they occur.

This is very valuable because:

- Every solution from the each team member contains individual, clever understanding, and then analysis of both needs and systems which could provide them.
- Clever solutions have been created by everyone present (this process always results in ideas from all members of the team) – this fills up the Bad Solution Park. Unlike other problem-solving toolkits this is the beginning (not the end) of the solution finding process.
- A full solution park is a repository of everyone's creative and engineering genius – each solution contains the knowledge from each individual's unique experience.
- The best solutions often come from combining everyone's solutions.

We can use all Bad Solutions as starting points for the TRIZ problem-solving process. It is always worth sticking with this process and trusting the process (it is fast and it works).

TRIZ Innovation Audit Trails – Importance of Hindsight in Problem Solving

One way of ensuring that most (if not all) of the good solutions have been uncovered is always to complete a TRIZ problem-solving audit trail: record each of the problem-understanding stages, and list all solutions thought of, tried and rejected. This is important as rejected solutions may be useful for slightly different

Innovation is not often audited or measured and in big companies if it is not mandatory or measured it will be neglected – no matter how important it is – and this is what happens to the most important stage of problem solving.

problems later on. Also it helps acknowledge all the team's work, particularly when the people with the decision-making power reject many good solutions but casually enquire later why the rejected solutions were not adopted. It may be wrongly implied that the engineering team had failed to locate all possible good solutions – this particularly applies when designing new-generation systems. An Innovation Audit shows exactly which solutions the engineering team have uncovered, and demonstrates the TRIZ processes and tools which have been used, and records all the systematic innovation steps which have been followed. Also an Innovation Audit ensures that everyone understands that there is a systematic and complete problem-understanding and problem-solving process (which includes an Ideality Audit, Bad Solution Park and 9-Box TRIZ Solution Park) and that all the essential elements of this process are recorded not just the solutions. It shows that systematic innovation has been undertaken and provided results.

TRIZ problem-solving audit trails also help when previously elusive but very good and simple answers are uncovered. Sometimes after guiding a team to use TRIZ to find a really good answer to their problem they say 'Ah! but this is obvious', and are almost embarrassed to tell the story of how long this solution has eluded them. When a really good solution seems simple and obvious, the telling of the problem story can make the problems look trivial in retrospect. However, when there is an innovation audit trail everyone can appreciate that, before it was solved, the problem looked very perplexing and was too difficult to solve with random problem-solving methods, such as brainstorming. Part of the fun of problem solving, especially difficult problems or inventive problem solving (when we don't know how to get to the best answer without TRIZ), is the challenge of uncovering these really good but initially elusive solutions.

TRIZ Basic logic – Improving Ideality

Problem solving has its many stages from invention through design, manufacture, use to disposal and each stage has its own timeline and context. Fundamentally problem solving needs the accurate completion of four essential and overlapping activities:

1. **Recognizing and meeting needs** – the benefits everyone involved wants from a system.
2. **Improving or providing systems** – to deliver systems with the right functions to meet needs. This requires an understanding of how and what the system actually delivers – all benefits (if they already exist), what it should be delivering and any downsides.
3. **Recognizing and mobilizing resources** – access available inputs (and define acceptable costs – not just money but time, space, skills, materials etc.). Minimize inputs when appropriate.
4. **Dealing with problems, insufficiencies and harms** – solving contradictions, overcoming insufficient functions, simplifying to reduce cost and complexity, dealing with harms (unacceptable outputs) and other problems are the gaps between the system we want and the defined needs delivered by the system we want.

All these elements are summarized in the simple TRIZ golden rule – the Ideality Equation – essential to TRIZ problem understanding and problem solving is a real understanding of the Ideality of the situation.

What we want from any system can be defined by the relationship between our acceptable inputs/costs which provides the system with functions/outputs which can be both good (benefits) and bad (harms). The Ideality Equation informs us about the balance between inputs and outputs, both good and bad. Very simply when we problem solve we are trying to increase ideality. One important element is the definition of a benefit – it is only a benefit if it fulfils a need, any outputs which are not needed are harms. So if I am choosing a washing machine and I want four wash-cycle programmes and the machine

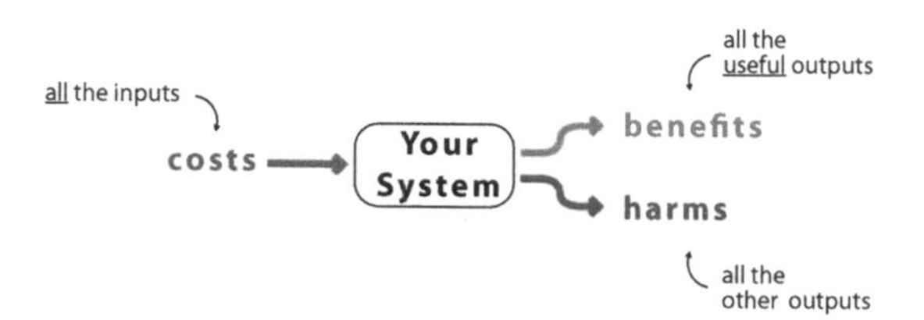

has twelve, then only the four I want are benefits – the other eight are harms and are probably adding cost and complexity.

In reality we normally start with the *Ideality we want* – the benefits we want, the inputs we are prepared to make (costs) and the unfavourable or unnecessary outputs (harms) we will tolerate. This is a very simple good starting point for any problem solving no matter how complex or straightforward, difficult or simple.

Choosing Systems to Meet All Needs

We are familiar with defining *Ideality we want* by matching systems to needs, or outputs to available inputs. So when choosing a car we have a set of needs (including a budget) and some idea of the system available to meet them.

Exercise: Roughly define the *Ideality we want*

Assume you are buying a car today. Begin by mapping all outputs/everything you want before you try and select cars (systems) that might fulfil them. Start by understanding the needs/what you want before looking at the cars available (systems) and trying to decide which one you want.

When first recording your needs do not sort them or judge them, allow opposite and conflicting benefits or features (such as big and small) or allow needs which you think you can't afford. This is an exercise in establishing what you want – not how you are going to get it.

(This can be applied to any item we want to buy and this approach has certainly revolutionised my approach to shopping – particularly for household appliances.)

Many systems available – we need to select one which really meets our real benefits

Systems Provide Functions Which Provide Benefits

Choosing the right system takes careful thought. For engineers creating and improving systems this is a major part of their work. We create systems to give us the right functions which deliver benefits to provide our needs. Benefits are provided by functions. Any gaps or problems with the system are shortcomings or missing, excessive, insufficient, conflicting or harmful functions. Problem solving is therefore about getting all the functions right and the right functions until we perfect the Ideality and the system gives us exactly what we want (no more, no less); the gaps have disappeared and we get what we want for the right cost. In TRIZ we can reduce or close those gaps each time we problem solve by applying the TRIZ Solution Concepts.

Problems – the mismatch between systems and needs can occur because:

- we initially chose the wrong system;
- the system has never been right in the first place;
- it has changed (the system stops performing as it should in some way);
- the needs have changed so that the system's previously acceptable performance no longer meets changing conditions (such as new regulations on noise or pollution, harsher conditions, changing markets).

Problem solving can also be the search for systems to meet needs (invention) or ways of providing systems for minimum costs or inputs. This applies to all problems – technical, personal, family, social, financial, management etc. TRIZ problem solving helps us create and improve systems to better meet needs, with the least costs or inputs and the least harm or unwanted or unnecessary outputs. Before we invent anything we should first check to see if the system already exists. The philosophy of TRIZ is very much about building on existing technologies (evolution) rather than going over old ground. We can systematically locate existing systems by identifying the functions we want and then search for all the systems which deliver our required functions. We are used to this connection being made by accident in famous inventions. For example James Dyson said that the breakthrough which led to his famous bagless vacuum cleaner occurred when he noticed an existing system which delivered the essential function he was seeking – how to separate dust from air – when he was in a sawmill. This uses cyclone separation to remove dust from air by rotation and gravity to separate the mixtures of solids and fluids – vortex separation.

When we start problem solving we are usually hemmed in and concerned with constraints and problem boundaries, and this simple *Ideality definition of what we want* helps us define the desired shift from the *system we've got* to the *system we want.*

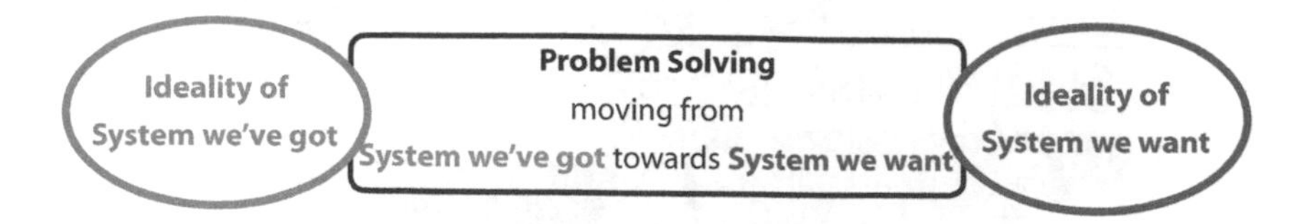

In corporate problem solving we ask people to define these two system states, and this works on both technical and management situations to help define these problem gaps.

Ideality applies to everything in life and defining both the 'ideality we want' and the 'ideality we've got' is a very useful and quick way of defining our problem gaps. There are several culture change tools which illustrate the differences between the existing situation and the desired situation and get staff to identify the gaps – this gives them a clear understanding of what is wrong and often a desire to close the gaps. The problem gap between the required ideality and the real-life ideality for the pirate in the cartoon below defines the simple and common problem of too much time being needed to get management approval, which slows them down, wastes their time and makes them less effective. This is part of a pirate-themed ideality audit exercise to get engineers to identify what they want to change in their working processes.

Many organisational traits can prevent your engineers from being effective

Normal problem solving of identifying and closing these gaps involves getting the right functions (and getting the functions right) to deliver this acceptable ideality. Problem solving with TRIZ often involves getting beyond even this acceptable ideality – for better or higher ideality – which means even more benefits *and* fewer costs *and* fewer harms. Taken to extremes the ultimate Ideality is the TRIZ tool known as the Ideal System which delivers *only* benefits with no costs and harms, and by defining it we narrow our thinking to accurately define all the benefits we want.

Using the Bad Solution Park at All Stages of Problem Solving

At all stages of problem understanding and solving we need Solution Parks while analysing the problem, the system, the requirements and Ideal System.

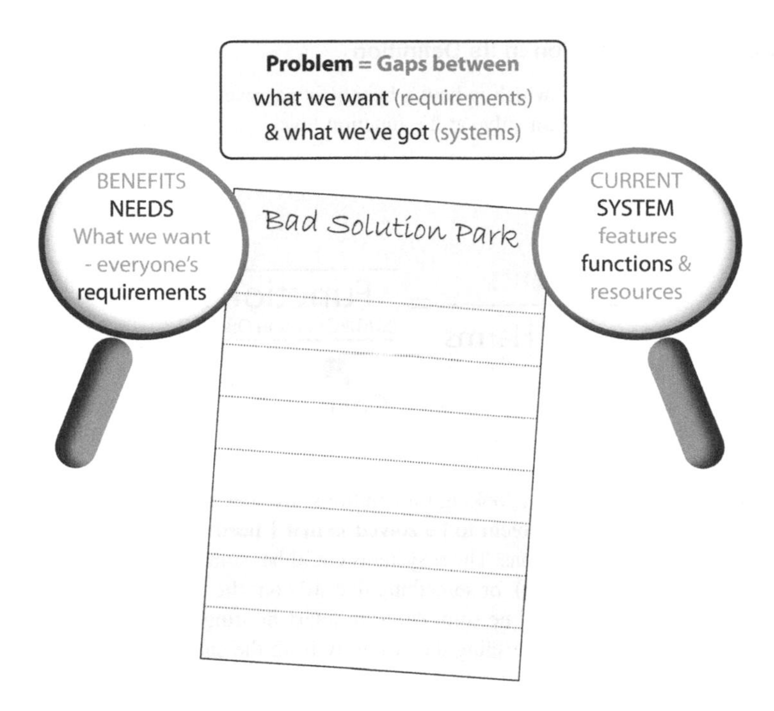

Once the benefits we want are recognized and defined, then our brains want to match them to systems to deliver them and we often erupt with spontaneous solutions. In TRIZ when we begin requirement capture we ask teams to define their IDEAL OUTCOME (as this helps everyone define the benefits/needs) and inevitably takes everyone involved straight to thinking about solution systems and then to the next step of resource capture (how to use what is available so we get everything we want). This is a very fast form of problem solving which can short cut all the usual formal steps, and throw up some very good ideas very quickly.

Functions or Benefits? Functions Imply How We Get Something but Benefits Contain No Solutions in their Descriptions

TRIZ offers tools to help us create and improve systems, which give us functions which provide benefits. A good function is a solution which provides a benefit which fulfils a need. (In TRIZ it is not a benefit unless it is needed.) So if a benefit I want is a clean carpet, my benefit of clean carpet does not contain in its definition any solution or function of how cleanness is to be achieved – simply what I want.

In problem solving requirements, needs and benefits have specific meanings.

NEED/REQUIREMENT = lack or want of something

BENEFIT = a good output which fulfils a need only describes what we want – offers no solutions

FUNCTION & FEATURES = ways benefits are provided

Problem solving = providing the right **benefits** to more exactly meet **needs** by providing or improving **functions**

A Benefit Contains no Solution in its Definition

The solutions are in the functions which is how a benefit is achieved. Functions deliver benefits and are provided by systems, we problem-solve at the function level.

$$\text{Ideality} = \frac{\text{Benefits}}{\text{Costs \& Harms}} = \text{Functions (SUBJECT action OBJECT)}$$

Benefits
Good functions
Insufficient Function
Missing Function
Harmful Function
Harms
Costs

When faced with problems and we are looking for solutions, we seek systems to provide the functions which provide benefits. So if my problem to be solved is that I need to achieve a clean carpet. I am looking for systems to give me solutions. These systems could be: some cover that prevents dirt reaching the carpet (some kind of protector); or something that lifts out the dirt (like a vacuum cleaner or a carpet shampoo machine); or it could be some kind of smart flooring that does not get dirty or self cleans in some way by vibrating and pushing the dirt away from the carpet. Each system provides different functions: the cover provides the function of 'protection from dirt'; the vacuum cleaning provides the function of 'removing the dirt and dust'. Functions contain solutions in their descriptions (unlike benefits). When we have a benefit, then asking *how* we can get this benefit will lead us to functions.

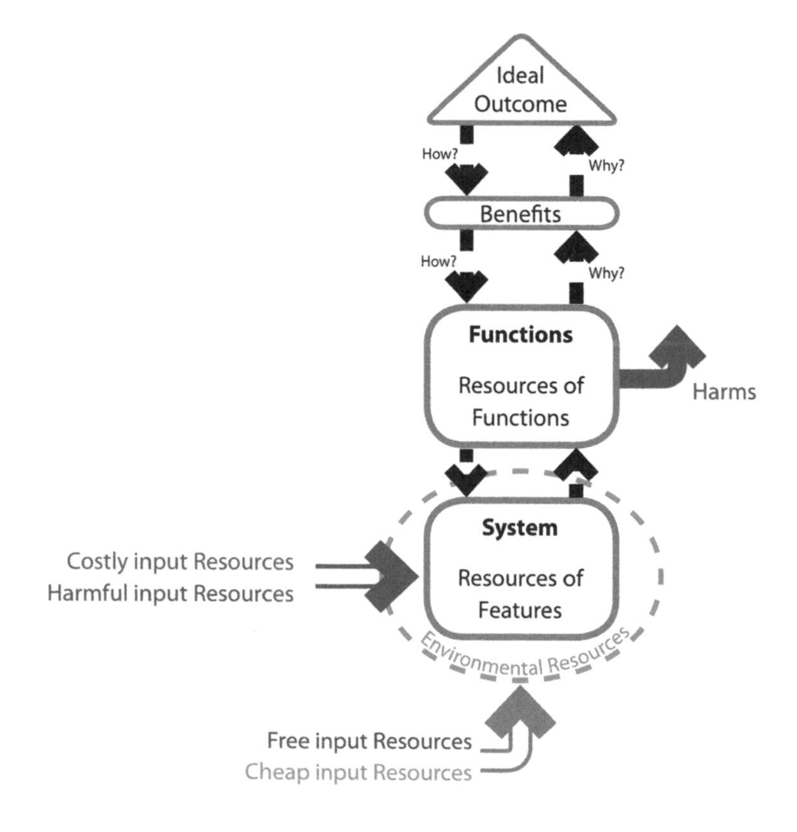

Asking *why* we have a function leads us to benefits as shown in the diagram above. Functions of course are provided by systems. So for the carpet we have a *benefit* we want (clean carpet) which if we ask '*How* can we achieve it?', might get the suggestion of the *function* of removing dirt. If in turn we ask '*How* can we achieve this function?', one answer would be with a system such as a vacuum cleaner. This may look a little tortuous but the logic of this is essential to understand and apply when defining and solving problems. We can start with what we want from our Ideal Outcome and all the benefits; asking how? and why? links everything we want to how we get it all with resources.

Avoiding Premature Solutions – Ask WHY?

Problem-solving sessions are sometimes organized to improve terrible, inadequate, complicated, expensive systems (sometimes still at the design phase) which are some senior engineer's Bad Solution (ugly baby) and often they will fight to keep their design idea – no matter how imperfect. The only times I have seen bad systems superseded by much better ones is when we ask WHY enough times to check that this terrible design actually meets the requirements it is designed to satisfy. Only when we establish that it doesn't even fulfil its original brief do we have the opportunity to persuade the parent of the ugly baby to work with their team to design a more appropriate system.

Boss imposes his solution which misses the primary benefit but delivers a secondary one

The simple trick of asking HOW? to see ways of solving the problem, and asking WHY? repeatedly to understand benefits/wants or causes is always useful. Asking such questions helps to establish both solutions and understand why we have a system (asking why tells us what benefits it provides and repeatedly asking 'Why do we have this problem?' often helps us funnel down to the essential problem causes as well). Asking why, also reveals when some poor choices of function providers have been made to deliver certain benefits. It can also get us to check that we are delivering the benefits we really want, even if that seems far away from the initial problem and its immediate solutions.

The links between the system from its highest level, the Ideal Outcome, and its lowliest inputs (its Resources) are linked by asking why? and how? as shown below. Many root cause toolkits are based on this fundamental logic. The ***WHY? and HOW? triangle*** below illustrates the simple logic and links between the Ideal Outcome, benefits, functions, systems and resources.

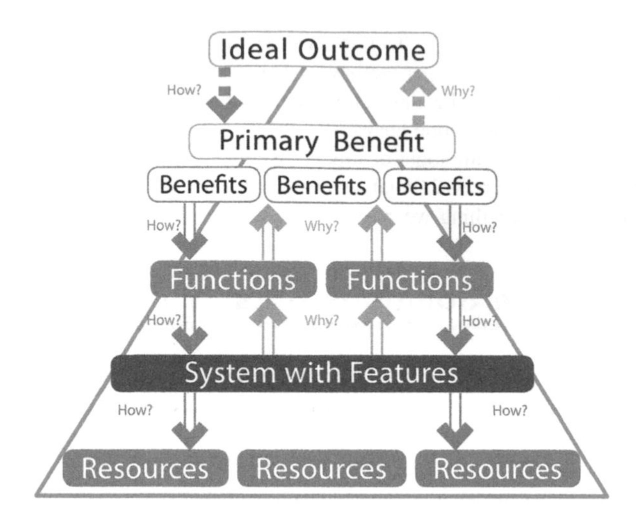

In one problem-solving session a new design of a military armoured vehicle was too heavy for its air transport (it was going to fall through the floor of the plane). It looked like an old-fashioned military tank – it was in many ways just an old-fashioned tank with some extra functionality. The designers were very proud of it as it was deemed safe being so heavily armoured. Initial analysis revealed a simple contradiction it needed to be *heavy* (armoured) and it needed to be *light* (for air transport and fast and efficient travel under its own power, and to avoid enemy attack).

Asking why it was designed as it was, revealed that it needed to be fast moving under its own power, stable in the desert, safe when under enemy attack and hard to see. Asking why it had the features it did, revealed the functions they provided, and asking why again revealed the benefits required. Delivery of these benefits could have been in other ways, by many different functions, and alternative ways of delivering the main functions were revealed. These were more sophisticated, better, cheaper and safer options than an adapted old-fashioned military tank. What the exercise revealed (in the first half hour) was what was really wanted – a military vehicle that is invisible, protects the crew, moves fast under its own power and survives attack, and many possible and better ways of achieving it than something very heavy and tank-like.

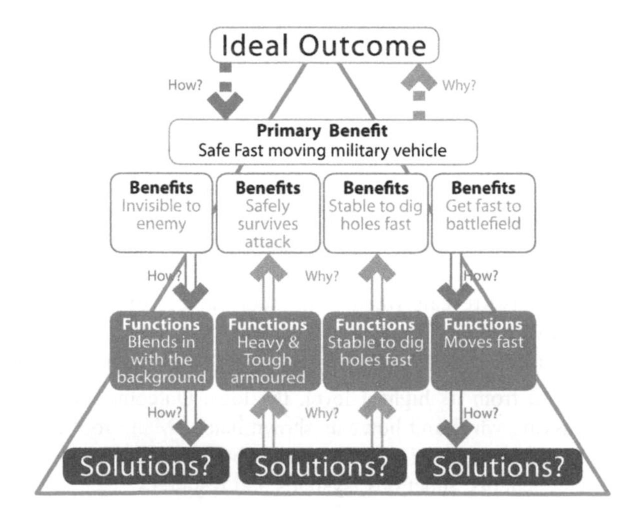

Ask Why?

This is to move us back to the benefits/requirements definition and reveal hidden irrelevant solutions to minor problems. Major problems are concerned with the primary benefit. The description of the problem should be closer to the ultimate goal and primary benefit. Then the more important it is likely to be, and its solution more relevant. Hence we ask why to access the important problems and solutions as they both have their own hierarchy and asking why moves us upwards to the important issues.

Asking WHY and HOW as Practical Problem Tools

Linking benefits and solutions by asking why is a very simple and practical tool in many problem situations. For example in a TRIZ session in a food manufacturing plant, there was a problem of puddles of water around machinery after cleaning with water hoses.

What is the problem?

Puddles of water after cleaning the machine

WHY are there puddles of water?

Because the machinery cleaned by being hosed down with water and there is insufficient drainage

WHY is this a problem?

Because the water may pose a health risk to people, and is unhygienic for the foodstuffs being processed

WHY is the machinery hosed down?

Because the machinery needs to be clean for food production

What do we really want? What is our IDEAL?

At this point we can establish that the *benefit* we want – amongst the outcomes we need to seek are:

clean machinery with some system that doesn't need good drainage

or

clean the machinery with water without leaving a residue of water

or

ways of removing surplus water?

How?

What functions deliver benefits? What resources are available to deliver those functions? What is available in and around our system to help us get what we want? When there are time steps each with different solutions it is always worth plotting the problem in the time steps and also in scale (in 9 boxes, see next page). This gives us the opportunity to understand the context and also plot available resources in each box which can offer *how* we solve the problem.

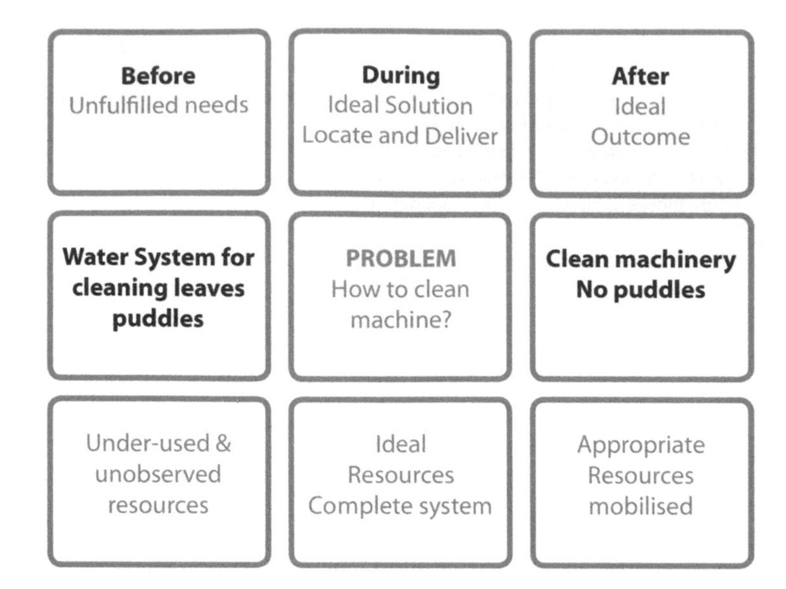

In the under-used and unobserved resources box we filled in *steam cleaners* used on other adjacent machines and this somewhat nonsensical but real and annoying problem was solved, as they could clean this machine with steam in the same way as the machines nearby.

Before we asked why, they were talking about installing new drains etc. and without asking why, we would have been dealing with some ancillary, rather irrelevant problem with its own unnecessary and inappropriate solutions, instead of using the resources they already had.

Simple Questions to Ask in Problem Solving

WHAT is the problem (or what is the problem with our solution)?

WHY is this a problem? (Ask why as many times as necessary until we reach a clearly defined requirement/ benefits we want.)

WHAT is our *ideal*? What benefits/solutions are we seeking?

HOW? That is, which functions would deliver those benefits?

HOW? What available solutions/ resources could provide the functions and solve the problem?

Defining the *ideal* and all benefits then repeatedly asking how again shows us ways of solving the problem. When we are starting with some bad solution then asking why? takes us back to the heart of the problem which is always close to the *primary benefit* – and asking why? again takes us from solutions back to the benefits, and asking why? again takes us to the *ultimate goal* and to the *ideal*.

For example to warn motorists of impassable flooded roads we need to have some FLOODS AHEAD signs in place during stormy weather on flooded roads. **What is the difficulty with the solution of the simple sign?** It is very important to warn motorists, but it is difficult to get such signs in place at exactly the right time on flooded roads, and remove them as soon as the floods have subsided.

Why is this a problem? Motorists need accurate warnings to avoid flooded roads as they are very dangerous and it is difficult to send maintenance vehicles to flooded areas to put the signs in place. It is also difficult to remove the signs at exactly the right moment, but we need to re-open he road as soon as floods subside. If we put this problem though our ***WHY? and HOW? triangle***. We can get the following sequence leading us to solutions from resources.

What is our IDEAL? What benefits? Our Ideal solution would be a warning to drivers that is there when the road floods and not there after the floods subside.

How? What functions/systems/solutions are we seeking? A road sign that is clearly visible and appears as the floods occur and disappears when the floods subsides and the road is passable again.

What available solutions/ resources could solve the problem? We have heavy rain, flood water, motorist and anything in the area just before the floods etc. Ideally we would like to use only these to create a warning to motorists when the road has flooded which removes itself as soon as the floods have gone.

Ideal Solutions use available resources to deliver exactly what we want at the right time and place

Whatever the problem, it is worth stating the Ideal, and then looking for its answers in resources. If there are no obvious solutions, then look hard and wide for system solutions in all available resources. If there is already a system but with problems then ask repeatedly WHY? of each of those problems until we reach a point of seeing what we really want (what benefits are missing / or any harms) and then search for solutions to solve these problems preferably using everything that is already there.

Stakeholder Needs and the Ideal

One important contribution of the Ideal is that it helps us to completely understand and accurately define the essence of what we really want. Everyone has different needs and therefore a different Ideal – getting everyone to define their Ideal is a very good way of highlighting the differences between different people – defining all the different stakeholder needs.

There is no one right solution for everyone – we need different solutions

One simple way of visualizing this is to suggest to everyone involved that they can have a magic wand to wave over their problem – and magically this would give them the main or primary things they want (and any other benefits which relate to the delivery of this one main thing). The Ideal System really gives them what they want and the idea of a magic wand helps them understand this quickly and with some accuracy. This is powerful and simple because even clearly stating the main thing we want is often difficult and takes some thought, and there are many jokes about getting this definition wrong. Such jokes usually involve magic wands, or fairies, or genies coming out of bottles and are often about carelessly or ill defining the main or primary benefit we want, and its essential attributes. The cartoon shows the ridiculous inadequacy of achieving a partial solution, getting the wrong solution (through sloppy language or words with double meanings) or getting the right solution but in the wrong time or context.

The dangers of offering solutions rather than defining what your really want

Start by Imagining an Ideal System

To start any problem-solving session with a group of engineers by defining the Ideal System creates a good (and often quick) consensus about the important needs, (the primary and other benefits – what we really, really want) and what everyone wants. Defining the Ideal also helps everyone to see the big picture, and also prevents losing sight of any important big issues by diving straightaway into too much technical detail about the real system, or dwelling on problems (assumed and real) of conflicting needs between different interested parties. Defining everyone's (each stakeholder) Ideal System is just a quick statement of needs and even if this produces a list of very different and contradictory ideal systems, which might have opposite or conflicting needs, TRIZ offers solutions to contradictions and these will be resolved when we get to problem solving.

Problem Solving at the Right Price with TRIZ – Use Trimming and Resources

The TRIZ logic of improving Ideality has important assumptions about and suggestions for minimizing inputs. This is done by being very clever with resources. In TRIZ we look to minimize costly input resources and use free or cheap input resources as well as any resources within the system itself. One important element of TRIZ is to look at the harmful input resources and if they are inevitable and unavoidable then step through the simple TRIZ suggestion list of how to turn harms into benefits.

TRIZ Trimming

TRIZ also has a very powerful tool called TRIMMING which is part of the TRIZ system and function analysis process – this is a very lean approach of examining every component to see if it can be removed or trimmed without adversely affecting the necessary outputs of the system. It aims at maintaining all benefits while decreasing costs, harms and complication. This can apply to both technical and management problems. Recently we have been working with companies to trim out unnecessary process steps in management systems. One simple success reduced many barriers to communication, by simply taking away or trimming out the barriers to teams communicating directly with each other.

Don't keep your teams in separate silos-Removing unnecessary complex systems may deliver huge benefits

Define Problem	Solve Problem	Solution Selection and Development
Map Context Define Ideal Capture Needs Analyse System and Problems Locate causes of problems Map Process etc. **Preparation for Problem Solving**	Fill BAD Solution Park Apply TRIZ tools to transform bad solutions and achieve opposites and correct system problems etc. **40 Principles** for Contradictions **76 Standard Solutions** for reducing harms, maximizing benefits **8 Trends** for future products Relevant world's knowledge Plus Resource application Enough time and space for all the above activities etc.	**TRIZ and other Tools for helping with** Concept development Concept Selection Process (use other tools such as Pugh Matrix) Successful Innovation and New Technologies etc.

Problem-Solving Steps – Before, During and After

In engineering communities successful problem solving relies on the proper attention being given to all three stages – yet within many large and small companies there is a consensus that the middle box 'Solve Problem' is the most important and yet that is the stage given the least time, resources and processes (or only quite trivial processes). One reason for this might be that whereas the 'Define Problem' and 'Solution Selection and Development' boxes are well populated by other engineering toolkits, the middle box only has TRIZ; until TRIZ is well known, generally adopted, and well used then this critical Solve Problem stage will continue to be left to the methods only appropriate to solve simple problems, often involving a random search for solutions produced by engineering teams brainstorming together for a very limited time.

Thinking in Time and Scale

Altshuller observed that about one in ten engineers are very creative and see lots more solution possibilities than most people when faced with a problem. He observed that these creative engineers look at problems and, unlike the rest of us, they often instinctively activate three powerful approaches which encourage lots of ideas. He called it TALENTED THINKING and uncovered and mapped the simple processes which help to achieve it.

Talented Thinking

The three strategies to ensure that we all achieve the same level of very creative thinking are:

1. **Develop IDEAL OUTCOMES / SOLUTIONS to problems** – which inspire real solutions with clever use of available resources.
2. **View our own ideas / solutions without constraints or judgement** – don't judge and therefore eliminate any emerging ideas as impracticable, too difficult or wacky. (The TRIZ tool for achieving these first two is Ideal Outcome or Ideal Final Result.)
3. **Expand views of problems in the context of Time and Scale** – simultaneously view a subject in nine simple views or windows seeing its context in both time and scale.

The TRIZ tool for this is called **Thinking in Time and Scale** (also called 9-Windows and 9-Boxes), and helps us view all aspects of problems and solutions and use our brains and knowledge more effectively to both understand and solve problems. By requiring us to think in **Time and Scale** in 9-Boxes, as shown below, we can use the vertical axis to focus in and look at the detail, or pull back and see the whole context, and the horizontal axis for time steps of past, present and future.

TRIZ for Engineers: Enabling Inventive Problem Solving, First Edition. Karen Gadd.

How to Become a More Inventive, More Creative Engineer

This particular TRIZ thinking technique often separates the creative and the less creative. In order to become creative we simply have to learn, adopt and use this TRIZ thinking tool **Thinking in Time and Scale** or **9-Boxes.** It is a simple tool and is often used instinctively by children and many creative people. The rest of us may have been educated in a way that teaches us to focus in on the problem, and we have to unlearn this habit of concentrating in one area. We can do this by adopting and applying the TRIZ 9-Boxes to all our problems as this helps us enhance both our problem understanding and our ability to see many solutions.

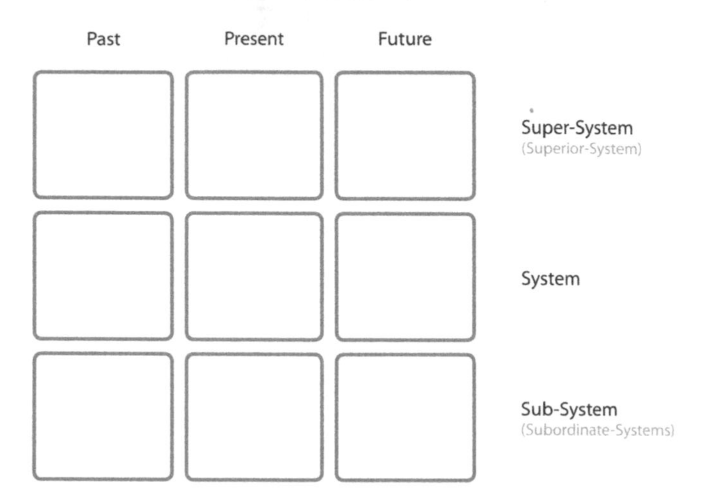

Using the above 9-Boxes enables the mapping of any situation, problem or system in TIME and SCALE. **In TIME** means thinking about our system as it is now (present) as well how it was in the past, and how it may be in the future. **In SCALE** means understanding the context of our system – up into the big picture and down into the details. When first looking at a problem or situation mapping it in a 9 Box context map is an excellent way of summarizing the arrays of facts and data which may be relevant and will help us understand the situation. This helps us take in complicated information and present it in a way that we can understand and communicate it to others.

Time & Scale – Helps you Organise, Summarise and Communicate complex data

Thinking in Time helps us think about the trends from past to future. We can see when significant changes have occurred and have brought us to our present situation, and how this is relevant to the future and where we are going and where we would like to be. Mapping the past helps us understand the present; mapping the future gives us insight to what is needed now to get there, what we would like and where we would like to be.

Thinking in Scale helps us place the answers to the relevant questions such as the context of our situation or problem. When looking at a technical system the questions we might ask could be: Is it part of another system? Which industry does it belong to? What is the physical context? What is the environment surrounding it? What market does it operate in? Who and where are its customers and competition? Thinking in Scale is also the thinking about the details: What are the components? Which details are relevant? Thinking in Scale is like looking at our situation on Google Earth – we can pull right back and look at the system in context and see everything around the system – and also reverse this and focus in and look at all the details and sub-systems.

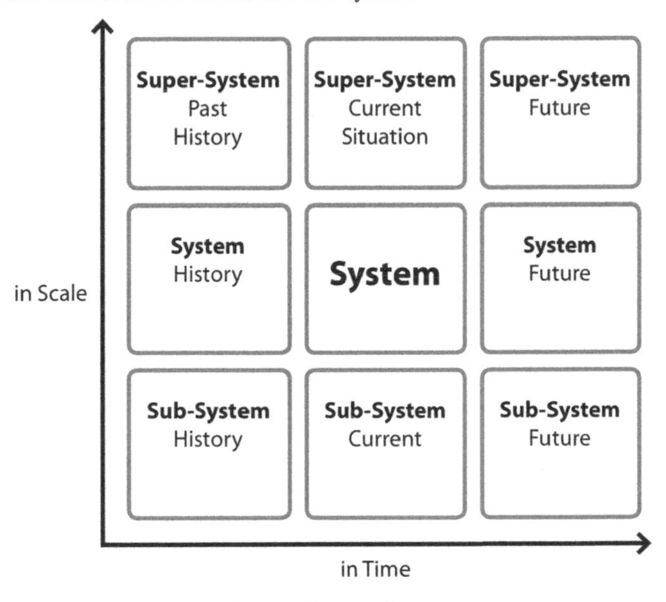

Time and Scale – Helps with All Problems Types

Once we have the relevant level and understanding, we can solve problems and be creative at all levels. Altshuller's name for this as 'talented thinking' is very apt, as it enables us to comprehend the many views of the situation simultaneously, the problems or system we are examining – the relevance of time and context – the big picture and the details. We are both going into more detail but also keeping any complication firmly under control, staying at the point that we can judge and assess all the relevant knowledge, keeping it together at the front of our minds. Nine concepts are apparently the maximum number we can intelligently deal with simultaneously (when we are stressed it falls to five) – so this is stretching our understanding of something to its maximum but more. In over ten years of teaching and using this tool I have found that most people struggle initially with this as it is so simple, but once through that first barrier, it is a tool many people use daily for ever.

Inventive Engineers – Thinking in Time and Scale for System Context and All Requirements

When Altshuller observed the 'naturally' creative individuals who saw the many possibilities when faced with a problem, he concluded that they had retained creative abilities that for many are somehow educated out of us. He showed that most children are good at thinking in Time and Scale and he wanted to help all of us think like the children (or the most creative engineers). He built the TRIZ methods to help us both appropriately focus and widen our thinking, and break any psychological inertia which encourages us to draw a box around the problem area, stay within the box, and only look for solutions within that box. To illustrate the differences in approach between the two groups, Altshuller used the following example.

If we ask most people to think about a tree – they will think about a tree.

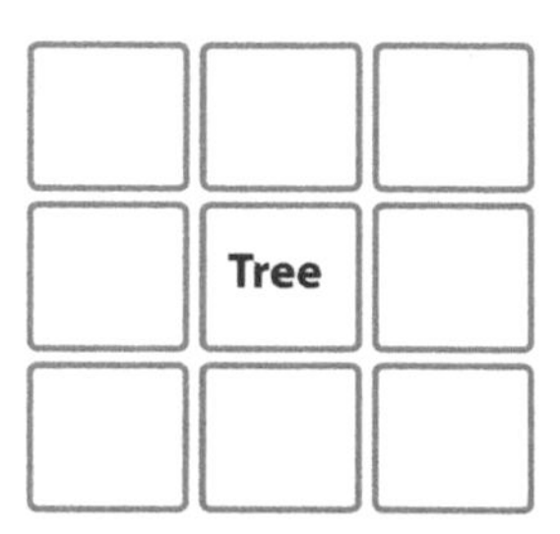

If we ask the very creative engineers (and most young children) to think about a tree, they use their talented thinking and would also think about the context and the history and future of that tree; they immediately and automatically think in Time and Scale.

Thinking in Scale means looking beyond the system of the tree to the big picture and thinking about the context such as the forest (super-system) **and** the details, the leaves, the twigs, roots etc. (sub-systems).

Thinking in Time means thinking about the tree in the past, the present and the future; the time steps in the process of growing and then using a tree. Time means understanding what Altshuller called 'the lines of development' looking to the future to see how we would like our system to develop – defining the Ideal System in a Future box is very helpful. Alternatively we can also use Time to think about one incident (or step in a process) and think about before and after the incident occurs.

So if we have a problem we may be able to see the possibilities of preventing it (past) or deal with it as it occurs (present) or let it happen and then see all the ways of putting it right (future) or minimizing the harm.

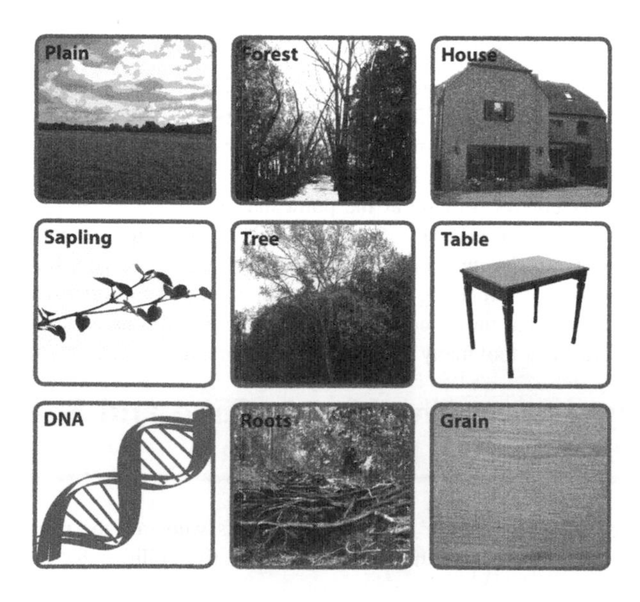

Altshuller set a simple problem for using the 9-Boxes to find solutions. Imagine that our **Wood (table)** in the 9-Boxes above is painted red and is used in a kindergarten. After a short time our red table is very scratched – and this is the problem to solve – we don't want unsightly tables because of scratches – so what can we do? (The problem is in the bottom right hand box – **Wood Details.**)

How can we produce red tables which after heavy use, still have no unsightly scratches?

Most engineers suggest repainting the table and solve the problem locally in both Time and Scale.

If we seek solutions in all 9-Boxes we will come up with many more solutions than just by looking at solutions in the Wood Details box which is where many of us would start and finish the problem (such as just repaint the table). If we paint out the scratches all we are doing here is dealing with the problem after it has occurred. The 9-Boxes suggest what we could do to prevent it, or deal with it as it happens and what we could do at the system or super-system level.

If we explore the other eight windows for solutions we would come up with lots of solutions including stopping the table getting scratched (by covering the tables?) or making the table 'un-scratchable' (by more radically changing the material of the table to a harder wood or by making it out of red plastic?) and not just dealing with them after they occur. By searching for solutions in all nine boxes we are building a solution map which is one of the very useful applications of this tool.

The famous TRIZ solution to this problem is one that most children will come up with – a solution that means that you let the tables get scratched but it doesn't show. A solution in the Past windows – **Grow red trees**. The solution of growing red trees may have popped into many minds when thinking about the problem but would have been immediately mentally dismissed or 'splatted' as being impractical, too expensive and taking too long. Applying constraints to solutions as they form in our minds is a very effective way of constricting creativity and innovation. It is very common for unusual solutions to be immediately dismissed as impracticable both in our own minds and in problem-solving sessions.

9 boxes offers systematic creativity for finding new ideas

Why Use Time and Scale?

For some it is daunting to start to use a tool like 9-Boxes, which can be used in so many ways, but as it is such a useful aid to many steps in problem understanding and solving, it is well worth mastering.

Thinking in Time and Scale helps us:

- widen our field of thinking;
- see the big picture and the detail and the relationship between them;
- map the significance of time in terms of the history, the process steps, conflicting benefits etc.;
- realize that there are many valid approaches to solving a problem;
- remove many constraints in our thinking;
- communicate the history.

There are few tools which are immediately so useful for most people in most situations. The only difficulty with this tool is that you can use it in so many situations, on its own or in conjunction with other tools and this makes it initially a little confusing. I use it for every stage of problem solving and with new and long-standing problems.

Time and Scale Can Be Used in at Least Four Ways

- **Context Map** – History/context of a problem.
- **Solution Map** – For initial Bad Solutions and again for TRIZ improved solutions after applying the relevant TRIZ problem-solving tools.
- **Needs Map** – What we want our system to do (all requirements and their contradictions).
- **Causes and Effects and Hazards Maps** – Locating the relationship between causes and effects and hazards of problems at different moments in time and different places in scale in a system and its environment.

In addition there is the 9-Box Resources Map which is useful at all stages of problem solving and can be built up alongside all the others. TRIZ thinkers look for resources everywhere and in all steps of a process, and adding in relevant resources in each box is helpful, particularly when looking for solutions. The 9 box diagram below reflects the simplicity of this tool – at the Super-system level we can see the whole picture while at the sub-system level we are focussing in on details.

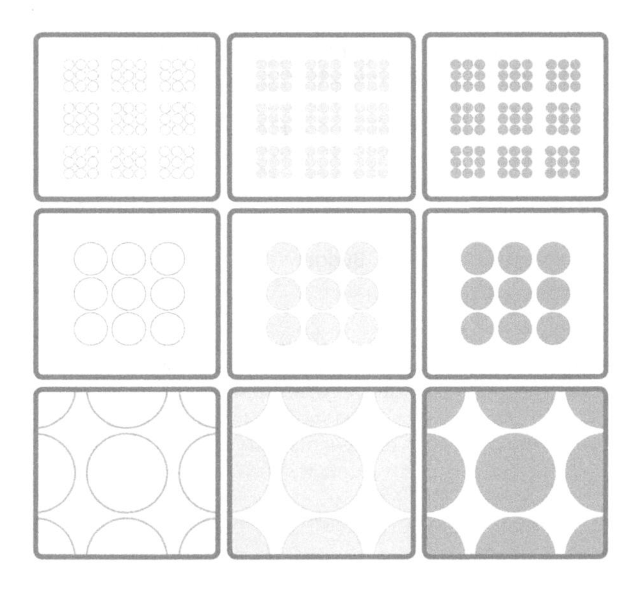

Without 9 boxes it can be hard to see the wood for the trees or the water for the sea

Context Map

History and context of the problem

Mapping the context of a problem can be a very quick and simple map or very detailed 9-Box map with much work needed to complete it, depending on what is required. The time scales are chosen to whatever is appropriate to the particular problem. The time steps do not have to be consistent: the map could show10–20 years in the past and only 5 years into the future, depending on what is most helpful for problem understanding.

Super System History Markets? Customers? Global Use? Industry?	**Super System** Which industry is it part of? Name the bigger system to which it belongs	**Super System Future** Markets? Customers? Global Use?
History of System Previous Systems Design requirements	**System** What does it deliver? Budget	**Systems Future** Next generation of system
Sub-Systems History Suppliers? Skills needed? Buy-in % etc.	**Current Sub-Systems** What does the system consist of? Components and details	**Sub-Systems Future** Staff Skills Buy-in %

A Context Map can be used for anything: an important personal problem (such as which university to choose) or for a simple product such as a small hinge or a turbine blade, or could be for a large problem such as nuclear clean-up of contaminated materials or the future of global use of oil. A general Context Map could be something as shown below – but its contents can be adapted for what is appropriate – a more specific map is shown for the very general context of UK nuclear clean up – briefly showing some of the reasons for the problem in the Past and History boxes and the position we need to reach in the Future boxes.

Super System History U.K. Energy Requirements	**Super System** U.K. Concerns on Nuclear Waste Contamination	**Super System Future** U.K. Highly Regulated
History of System Produce Energy UK Nuclear Power Stations	**System** UK Clean up Nuclear Waste	**Systems Future** Safe/secure storage of Nuclear Contaminated Materials
Sub-Systems History Spent Fuel Rods etc. (Stored in Ponds)	**Current Sub-Systems** Empty / Clean Ponds	**Sub-Systems Future** Effective Very Long term Safe systems & Places for storage

Solution Maps

When looking at any system or problem there are many ways to solve a problem in scale and time. The Formula One Safety solutions and red kindergarten tables examples were about finding all the ways of dealing with the problems, and mapping many solutions in 9-Boxes. Mapping all possible solutions in a 9-Box Solution Map organizes, inspires and stretches our thinking. The solutions are also all the contributions, all the elements to ensure success or avoid disaster. Therefore if I want to deliver a successful seminar on TRIZ, there are things I can do before (prepare) and after the meeting (feedback); the environment is important, a good room – and the details are important and I can fill in a 9-Box Solution Map to remind myself of all the elements needed to ensure the outcome I want. One element can wreck everything. I remember being at a large IT conference in Alabama where no expense had been spared, and several thousand people were in a marquee to hear the chairman's address which was inaudible because someone outside was mowing the lawn (and refused to stop).

When facing a complicated problem situation with many possible solutions a Solution Map can be a very quick way of understanding the many possible options and helping the search for more solutions. For example in the UK the nuclear clean-up programme has a particular problem of emptying the 50-year-old nuclear ponds which contain spent fuel rods and other nuclear rubbish such as swarf and anything contaminated which needs storing safely. A longer-term solution is being sought and there are many possible ways of emptying the ponds and dealing with the contaminated materials, the water and the sludge which has formed and covers the bottom of the ponds several feet deep.

One teaching exercise for the Solution Maps is to create a Time and Scale Solution Map for emptying the ponds and dealing with the materials including the contaminated sludge. The Context Map might be as follows:

Air 10 Ft of water Inches of sludge	Pond Air above pond Surroundings	No change in Radioactivity Pond Air above pond Surroundings
Need to move items From pond to safe storage	Nuclear Contaminated Items covered In sludge	Items safely removed from pond
Radioactive particles within the sludge exist and are mobile	Sludge water Pond bottom Pond sides Surface of water	No change in radioactivity Sludge water Pond bottom Pond sides

The problem has been defined as how to remove contaminated items from the pond without the outside area becoming contaminated, which could be caused by the disturbed nuclear sludge floating to the surface. The solutions below are ways dealing with or preventing this problem. There are many other ways the ponds could be safely emptied, which could be plotted in another Solution Map.

The challenge is to empty the ponds without bringing the sludge to the surface, and many solution ideas are shown in the Solution Map. This problem could have many other solutions as well as those shown below – these are just ideas and not tested.

Before Prevent!	**During** Deal with it!	**After** Put Right-Correct
Super-System Stop sludge moving? Solidify water? Layering with oils Gel Ice / Cool down water	**Super-System** Empty quickly Segment with shutters or sleeves	**Super-System** Barrier to stop radiation escaping Lead foam Seal off pond for two years
System Solidify sludge Encapsulate sludge Remove sludge	**System** Local vacuum system while moving items	**System** Move sludge back down quickly flocculate?
Sub-System Break surface tension of items so no particles adhere Standing waves to arrange sludge	**Sub-System** Remove one sleeved item or move very slowly	**Sub-System** Sludge moves down Push back down Microsieve - coffee?

When looking for solutions to problems it is always worth trying to find as many as possible in the 9-Boxes of a Time and Scale map. For example one problem I was facing ten years ago concerned my young son regularly hurting himself while trying daring jumping tricks while roller-blading and skate-boarding, which resulted in him breaking his arm on more than one occasion. I soon realized there were many solutions which my son and I could implement together (this needed cooperation).

Before	**During**	**After**
Super System Culture of safety & reasonable targets	**Super System** Only use proper facilities with supervision	**Super System** Ambulance and medical help available & phones to ring them
System Learn how to do it Safety lessons Instill common sense Learn your limits	**System** Minimise injury while skateboarding and rollerblading Concentrate - do it well	**System** Lessons learned Parents/other people there to react and help in case of an accident
Sub-Systems Buy good equipment Maintain it well Add safety kit helmet and pads etc.	**Sub-Systems** Good blades Good skate board Good wheels etc. Protection in place Use them properly	**Sub-Systems** First aid kits Comfort Right knowledge about moving injured children

We mapped the possibility of an accident in Time and Scale and found a great number of solutions from banning him from anything dangerous, to arranging for some lessons, and lots more practice to make him better at the dangerous manoeuvres (always try TRIZ Principle 13 – try opposite solutions). Note for an event the time boxes should be labelled BEFORE–DURING–AFTER. There are many solutions but assuming that children will always want to try dangerous activities like jumping on skate-boards, we created the Time and Scale Map which gave my son some simple understanding of the issues and actions needed. Others have applied this before skiing.

Cushion in Advance helps us avoid future, likely hazards

Needs Map

What we want our system to do at all stages and in all places (all requirements and their contradictions).

Needs/requirements often conflict and create contradictions. Needs Maps are simple ways to complement the other TRIZ tools for requirements (the Ideal and Ideality Audit). They help us uncover both physical (achieving opposite benefits) and technical (solutions without downsides) contradictions in time and scale between different stages of use of a device, system or process and also highlight conflicts between different stakeholders such as suppliers and consumers (should a product last forever or regularly need replacing).

In everyday life we often need different solutions at different times (absent/present umbrella)

Contradictions such as in the umbrella above illustrate how many systems, both industrial and domestic, would ideally only be present when used and then disappear (or become very small) when not required. Anyone who has tried to store a lawn mower or mops, vacuum cleaners, ironing boards etc. in a tiny living space will be familiar with this problem. Many devices we would like to be there and big when in use and as small or unobtrusive as possible when not in use, such as landing wheels on an aeroplane, are systems with contradictions which can be solved with TRIZ.

Altshuller in CAAES described how to use talented thought to capture and understand contradictory requirements and illustrates this by applying this to the design of the fastest speed boat in the world. He said we must break psychological inertia and *smash* any pictures of known, familiar speedboats in our minds and not begin by looking at the current fastest speed boat and look for marginal improvements such as:

- lengthen hull
- streamline hull
- more powerful engine.

Instead we start by mapping what we want – the requirements in 9-Boxes to see the contradictions and lines of development. In practical terms this means plotting the different elements and required features for each stage of its use in a 9-box map. Therefore if we take the elements/subsystems of the speedboat and think about their contributions to the main thing (Prime Output) we are trying to achieve *increased speed* which means:

The IDEAL SPEED BOAT needs a tiny hull and huge engine which gives us clashing ideas/contradictions.

Engine	more power	bigger engine = more speed
Hull	decrease resistance	smaller hull = smaller resistance = greater speed

Finding clashing contradictions / opposite requirements is an essential element of using TRIZ to solve 'unsolvable problems' and are just physical contradictions which can be solved with the TRIZ 40 Inventive Principles. Whenever faced with clashing ideas it is worth trying the tools in the TRIZ Creativity Prompts which include **Size-Time-Cost** and '**Other way round** – Invert it – try the opposite'. Altshuller describes using both these tools as turning our imagination into anti-imagination and seek the anti-speedboat.

Size-Time-Cost on anti-speedboat = Huge engine needs huge hull = Expand it to … infinity

Shrink it to … nothing = No resistance = greater speed

Ideal hull = no hull

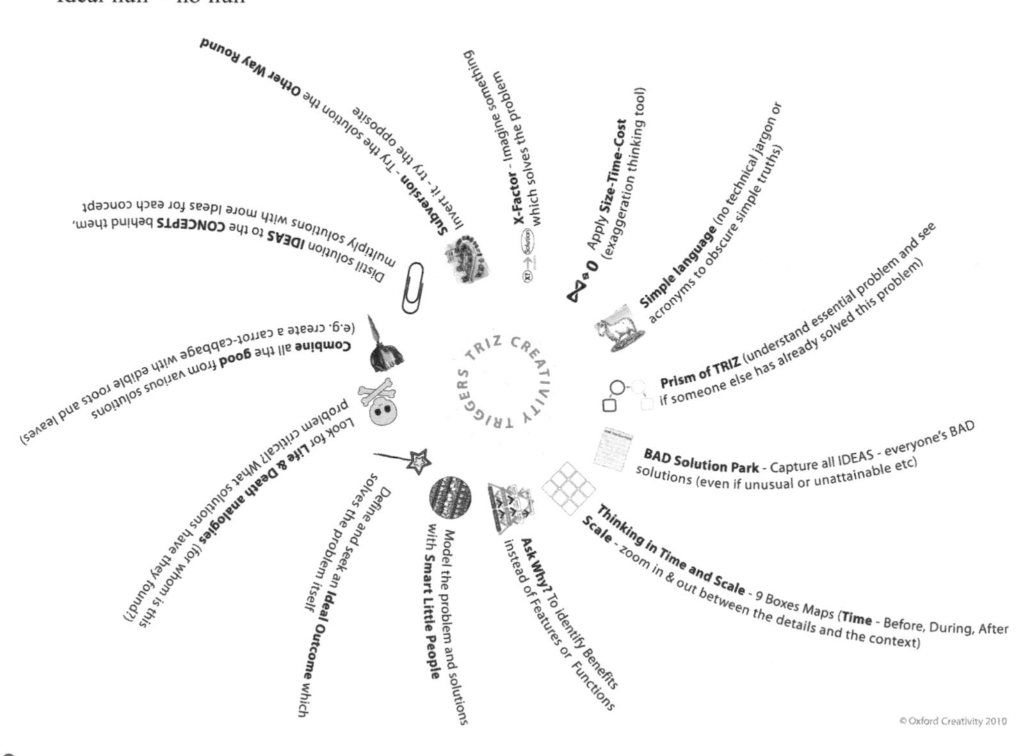

Ask why, to identify benefits not features.

Why do we have the feature of a large hull? So the boat will stay afloat

Why and when does it need to stay afloat? When moving and in harbour?

When moving could it be kept up by other lifting forces? When in harbour it could have other solutions to keep it afloat (floats, harbour supports etc.) So why is the hull mostly empty when moving?

Anti-Speedboat does not need stay above water when moving therefore needs no flotation space and the whole hull can be filled with engine. Using 9-Boxes we can immediately see the contradictions in the requirements of a system such as a boat.

In Harbour	Moving	In Harbour
Engine Small & unobtusive	**Engine** Big & Powerful	**Engine** Small & unobtrusive
Hull Wide to float	**Hull** Narrow to move fast	**Hull** Wide to float

Altshuller argues that in its development towards perfection the speedboat's Ideal Hull would become 'no hull' or at least a very narrow hull which offers no resistance in the water. Similarly the route to perfection for an Ideal Engine would be a very powerful (and hence very large) engine.

The Ideal Speedboat would be an 'anti-speedboat' as big an engine as possible, and as small a hull as possible, which being an Ideal Speedboat would move very fast. Mapping in 9-Boxes reveals the contradictions of needs and solutions between different system levels and at different points in time then the speedboat needs to be 'ideally' designed to move fast, but when in harbour it needs to be able to float. At different times our speedboat needs to be wide (to float) and narrow (to move fast) and needs to be both a speedboat and an anti-speedboat. He describes each of the 9 boxes as a mental screen to view the challenge/problem and he says on page 119,

> Great Passions are constantly clashing on the mental screen of the talented thinker. Contradictory tendencies collide, different steps in time. Mapping how we would like a system to be say big at one stage and really small at another is a very simple use of this tool. If we were to plot the Ideal speedboat, conflicts arise and peak, opposites fight it out … In the heat of this struggle the imagination sometimes changes into anti-imagination.

There are many systems such as cars, which face similar contradictions and the desire for opposite benefits. Plotting opposite benefits in time steps in the 9 Boxes helps us understand and map all needs. We can also plot needs in scale – for example a good watch face needs to be hard at the supersystem level but soft (self-repairing) at the subsystem level.

Causes and Effects and Hazards Maps

Locating the relationship between causes and effects and consequences at different points in Time and Scale.

The Time and Scale Map is a very powerful and simple way of locating, understanding and communicating the causal links between elements and events and their impact on systems and their super-systems. For example one tiny invention was essential to the development of skyscrapers and created city skylines dominated by high-rise buildings: this was a small brake device to make elevators/lifts safe, created and publicly demonstrated by Mr Otis in 1854 and gave confidence in their safety. That confidence was well placed because all lifts in public building are required to have an Otis safety brake and it has never failed. The only elevator/lift failures have been in private buildings not fitted with the Otis Safety Brake.

This story can be plotted in 9-Boxes which shows the big super-system result (cities changed) by one tiny engineering device.

Cities need multi-storey buildings - 12 levels achieved	Over 12 storey skyscrapers begin in USA and Europe	Cities changed HUGE HIGH RISE Buildings
Crude freight hoist between 2 floors put in NY building	Lifts accepted as safe	Lifts used everywhere
People distrust lifts/ elevators Otis invents lift brake	Otis - Lift safety brake Proved in public	Same invention in use (& has never failed)
1830-50	**1852**	**Now**

Links between causes and effects are needed to show how to deliver benefits and detect problems and harms. Traditional root-cause toolkits offer rigorous searches for the causes of problems to locate any causal changes which need to be pinpointed and understood. The TRIZ 9-Box Hazard Map is very complimentary and useful to be used in conjunction with other root cause toolkits. In TRIZ, tracing causes of problems is helped by the simple problem-solving logic of asking repeatedly WHY? and also using the 9-Box Hazard Map to see the relationship between causes and effects and to locate the causal links between consequences (good and bad) and events in Time and Scale.

Consequences, Connectivity, Interrelationships

Sometimes one tiny change or new element in a system can have phenomenal consequences which can be very good or bad. Small events can fundamentally change how or whether whole systems are created, or whether they are safe or will fail, or can cause huge benefits or harms and expenses. Famous catastrophes were often caused by some small element failing, or being wrongly designed or wrongly used such as in many bridge collapses, the Comet aircraft, the BP Gulf of Mexico oil spill or the Piper Alpha disasters, which are well documented. Mapping any famous disasters in 9-Boxes delivers interesting insights.

Some great changes in the world have come about by some small and apparently fairly insignificant inventions such as saddle bags for camels – which after the year 500 led to changes in the world's trading routes as camels were used instead of donkeys and could cross deserts. Mecca became a very wealthy trading post, as a result of the camel saddle bag, as it was on the new camel routes. Mecca is in a rocky valley with no significant agriculture but good water sources which made it ideal to develop as a successful commercial city and later of course as the great focus for pilgrimage and homage.

We can plot events (including great inventions) in 9-Boxes to understand their significance over time and space. We can plot their interfaces over time with its super-systems and sub-systems and put in context systems' structures and their impact either locally or globally. We can also use Time and Scale Maps to understand links between system levels and the effects that changes at one level will have on other levels over time and place. This includes good things such as camel saddlebags and the Otis Safety Brake, and hazards and harms such as fatigue failures in aircraft, or bad design choices of some small element costing huge amounts in larger systems.

Social Harms in Time and Scale

Any problem with complex causes can be mapped in 9-Boxes as can their solutions. Problems may have super-system causes and sub-system solutions. One delegate plotted personal safety in their suburb of London in 9-Boxes after a discussion of how to change cities for the better, and make them better – should it be changed by super-system factors such as legislation or sub-system factors such as the local residents?

Past	Now	Future
No drug culture No gambling	Government wish to provide safety in streets? Violence, drugs & gun culture in some areas	**Ideal Outcome Culture of safety everywhere**
Reputation as safe (dull) area Lots of churches etc. No massage parlours, nightclubs etc.	**Safety in city streets? Adequate police force?**	Places where people can meet socially for sport, arts, fun etc.
No-one is afraid Lots of people go out Places and activities for young people (sports facilities)	People feel unsafe? Insufficient lighting? Family units changing No-one knows neighbours	Young people grow up to be responsible citizens. Safe streets where everyone knows each other their neighbours

Mothers work together to use small influences to solve fear on streets problem

History - or what made this possible and wanted	**Big picture context**	**General outcomes Likely, worst or best?**
History or causes of problem (if known)	**General problem description (for us)**	**Outcomes Likely, worst or best?**
History of details	**Details of problem components (Suppliers Customers etc.)**	**Outcomes for components & details**

We can map the future boxes in a number of ways – these can include what is likely to happen, what we dread happening (worst outcomes) or what we want to happen (best outcomes). The trends over time from the past to the future give us the clarity to see what may happen, and connect the links between past actions and future consequences. Looking at the 9-Boxes should suggest some actions (what we could do now) to create the outcomes we want for the future.

Using Time and Scale to Map Hazards

What is a hazard? The unexpected or the unforeseen? Many situations are only hazardous if a complex combination of conditions exist – we can map the hazards in Time and Scale to understand and anticipate them, and see where and when we might prevent, deal with them or put them right afterwards.

Hazards may be caused by a combination of circumstances:

- Links mean that a change in one part influences the others either positively or negatively.
- A breakdown in one part causes problems in other parts and/or in the system as a whole.
- A problem in one area may be removed by changing a different part of the system.

Remember: There are many solutions and many options in Time and Scale to choose between to resolve any problem – lots of ways to tackle a given problem.

Tying into a Pipe Containing Hazardous Fluids

If we were facing a potentially hazardous situation of tying into a pipe containing a harmful fluid – before we breach the pipe we could map potential hazards in Time and in Scale steps and highlight the links, all the causes that must be in place to create a hazard.

Map Hazards in Scale - Before we breach

To create a Hazard there has to be Hazardous chemical/content in pipe

AND

Reasons for the chemical to exit pipe (pressure difference, gravity etc)

AND

Condition of pipe contents that allows it to flow out

Before	During	After
	Pressure Difference	
	Breach Pipe	
	Pipe contents able to flow out	

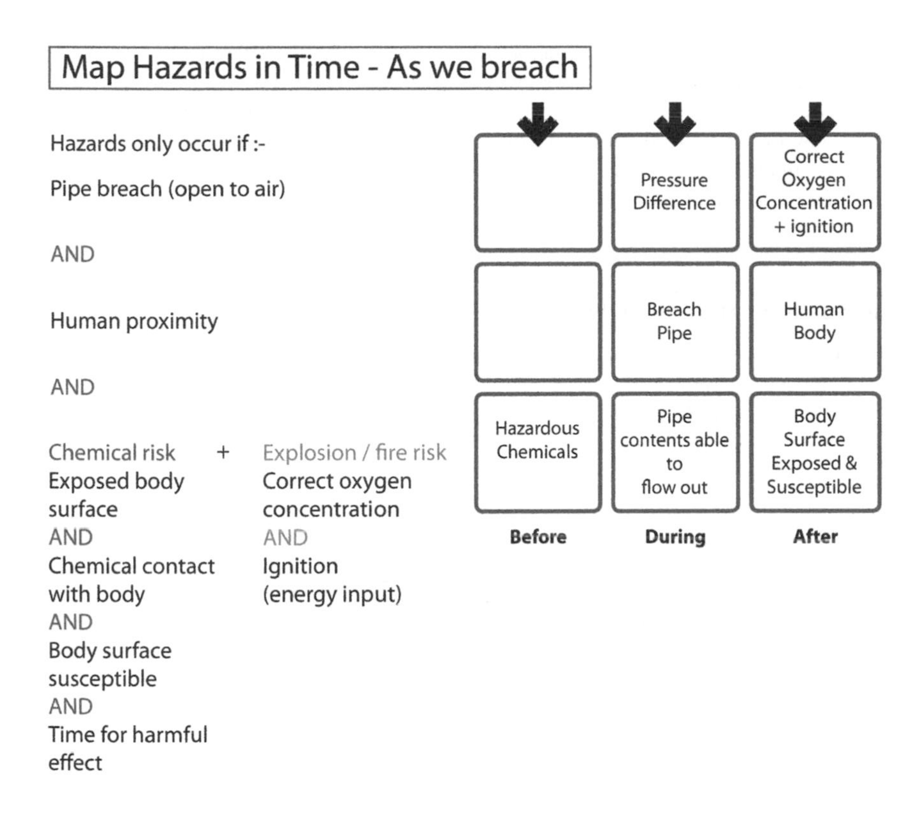

Thinking about all the **time** steps helps us remember what can be done *before*, *during* and what can be put right *after* the hazard has occurred. Thinking what can we do afterwards includes can we counteract or neutralize any harmful effects? For example, if the chemical is an acid can we add an alkali/base.

Thinking about what we can do *before* – preventing the hazards includes blocking or preventing any one of the conditions – ensures that we do not create a hazard.

Thinking about all the **scale** steps reminds us that we can do this at any level of the system and some possible solutions include:

Hazard Example	**Pipe breach (open to air)**
Solutions can be found at all levels in scale and time - for example:-	
Super-system	Do it in an inert environment (no oxygen)
System	Use a self sealing system
Sub-system	Solidify pipe contents

Preventing Hazards in your Life

Thinking in Time and Scale – using it in your own life to face your worst downsides.

Preventing problems – prevention is better than cure – a stitch in time saves nine.

We have all had situations where a little care before (or during) would have saved so much time to correct something or put it right. Prevention of things going wrong such as stopping pipes bursting in cold weather, careful preparation in cooking, levelling a floor properly before laying a wood covering, running out of petrol etc. are all lessons we learn in life. These lessons have a lot to do with common sense (which we can lose sight of in the complexities of business or technical problems). But even with a clear view of our problems it is sometimes hard to see the way forward because we are overwhelmed by the many options and all the details – we can't see the wood for the trees. This is where drawing a simple 9-Box map helps us see both the many causes of hazards and the many solutions to prevent harms.

Time and Scale for Hazardous Situations

Try this simple exercise – what is one of the worst things that could happen to you? For many of us it could be our house catching fire and our family being trapped and in danger. Draw a 9-Box action plan of all the things we could do to prevent, stop, reduce and deal with the danger and harm of such a situation.

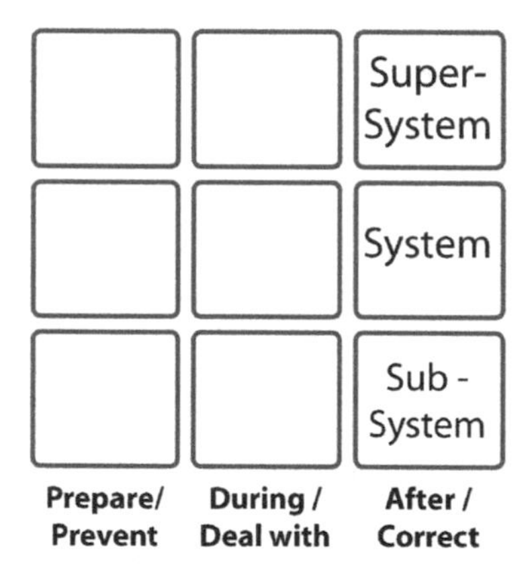

Fire in Home

In the **Prepare** and **Prevent** Column could be insurance, fireproof materials, fire-doors, fire alarms in place, no open-flame heating or cooking, no playing with matches etc. It is worth adding very small details such as new battery in fire alarm, and teaching the children a fire drill of where to go, covering mouth and noses and to stay low on the floor if there is smoke. Investing in simple smoke hoods which have a long pipe to the ground to daw up cooler air.

In the **During / Deal with** column could be taking the right action at time of fire, react to fire alarm, use fire blankets and extinguishers, inform fire brigade and neighbours, escape etc.

In the **After / Correct** column could be hospital treatment, rebuild house, insurance claims, counselling etc.

Sometimes we don't know which box to put something like having a fire brigade and hospital (by being available should they be in the **Before** column?) The answer is generally to put them in the box where they will come into action / be used – so in this case the **After** column.

After this exercise one delegate on a course said he had gone home and had a fire drill. To his horror, his wife and children couldn't get out because of burglar proof window locks. (Conflicting requirements – solve the contradiction of keeping your house secure which makes it hard to get into but easy to get out of – are your windows and doors designed to solve this contradiction?

Unidentified Manufacturing Problem – Scrap Rate Rises Dramatically

Some years ago I was called into a major company to look at a number of manufacturing problems. It was a three-day workshop with two different teams of engineers, each with their own problem, which they had been unable to solve over a period of months. The drawing of the Time and Scale Maps produced immediate solution directions for the first problem and eventually with the right information and a hazard map identified the cause of the second problem.

The first problem was a new piece of equipment which performed very poorly and was an upgrade of a previously excellent version, promoted by the machine manufacturer, and redesigned to give less than half the performance. The team reverted to the original machine and had no further problems.

The second problem was caused more by accountancy policy than by manufacturing. The problem was a ten-fold rise in scrap (from 1% to 10%) of a very expensive, complex, cast shape. The manufacturing process had many complicated steps and we mapped these backwards to the beginning – the first step – which was the purchase of a complex shaped ceramic core. When we asked what was changed before and after the scrap rate worsened, there was much examining of data until it was remembered that the whole company had been asked to reduce all costs by 20% by their accountants. The external supplier, the company that provided the cores had been asked to reduce the cost of each ceramic core from $5 to $4 – this they had achieved but with some loss in quality – this saving of $1 at the beginning of the process was costing $5000 plus for each piece scrapped. Mapped in 9-Boxes below:

NEW cost cuttingculture!	**Tenfold rise in scrap Costing $5,000** Huge rise in overall costs despite 20% savings on purchases big & small	**Time wasted Big losses Manufacturing slowed down**
20% Cost Savings required on all purchases	**PROBLEM** Something has gone wrong in one of the many manufacturing steps	Return to low scrap rates and previous accurate and reliable manufacturing processes
$5 ceramic core saved 20% = $1 now only $4!	Supplier slightly reduces specification to achieve requested price reduction	Return to original specification & price **right ceramic core costing $5 again**

What neglect or small savings now could cost the earth later? (Savings of £s now will cost £millions later? Jam today, disaster tomorrow.)

Recently I was given an interesting example by someone from the UK Ministry of Defence of a small cost saving (of a few pounds) some years ago of taking some simple clips away from a design. The consequence of this is now costing the nation millions each year.

The clips were in a nuclear reactor in a UK submarine and the clips were originally there to hold and restrain the moderator rods – keep them in place should the submarine turn upside down. If the submarine is the right way up then gravity will keep the rods in place. It was believed that the submarine would never go upside down and therefore the clips were superfluous and they were removed from the design and manufacture.

The purpose of these moderator rods is to control (or moderate) the nuclear reaction (a nuclear reaction needs fission-generating neutrons) and they do this by absorbing neutrons. While the rods are fully in place they absorb the neutrons and prevent nuclear reaction. Removing them completely, on the other hand, could allow the reaction to become uncontrolled.

A simple Time and Scale Diagram of the situation shows the relationship between the really small cost saving and the subsequent huge costs to the maintenance facilities. The decision to remove the clips was probably on cost savings, so I have divided the super-system into two separate categories of the Commercial environment and the Natural environment. It was the possibilities of a disaster in the natural environment which were insufficiently analysed when thinking about why the clips were needed. When seismic studies were undertaken it was seen as a possibility (very remote) that an earthquake could happen. If during an earthquake the submarine turned upside down then the rods would come out and the reactor would go super critical.

The only possible solution was to build a restraint for the whole submarine which cost millions as it was not possible to modify the submarine design and put the clips back.

	Build	In dock / Maintenance	Consequence (assess risk / effect)
Commercial Environment	**Cost Pressures Safety**	Re-assessment of need to keep submarine upright	**Huge costs of building restraints to keep whole Sub upright**
Natural Environment	**Insufficiently Analysed**	**Earthquake could cause Sub to roll over**	**Potential major disaster on shore**
Submarine		**Safe Submarine**	
Reactor	**Omit rod restraints for small cost savings**		**Definite minor cost savings on clips**

Simply by mapping our problem situation in 9-Boxes we can check:

- An understanding of harmful up-front cost savings which may be inappropriate.
- Is everything in place? The power of careful preparation (often cheaper and easier in the long run).
- The costs of not integrating a good, available technology in a system (see Trends in Chapter 9 – Uneven development of parts).
- All stakeholders interests are mapped and covered and we define the problem and requirements from all viewpoints.

Use 9-Boxes to Understand History/Context of a Problem

Looking at the bigger picture can help us identify the likely causes, and hence solution directions. The dangers of blanket directives (such as save 20% on cost everywhere) are very dangerous (law of unintended consequences).

Any directive must be clear about the desired outcome. For example:

- **NOT** lower cost per manufactured item produced;
- **BUT** lower cost per GOOD manufactured item produced.

Lower cost can imply lower quality – do we really want this?

	Big picture Context	**Expensive Outcomes**
Cheapest Quote	General Problem	**Costly Outcomes**
Small cost Savings		

Problem example E-Bay has created a global shopping place / market stalls

Problem - I am an unknown landscape artist (with no agent or outlets to sell my paintings) how does E-Bay help me? (based on real life success story)

Internet develops credit cards for safe and easy global purchasing Global delivery / postal services available & widely used E-Bay launched	E-Bay becomes known and popular	Global market established Global prices reduce experts' profits
People selling & People buying	General problem description - for us **How to buy & sell in global markets**	Local unknown painter starts to sell paintings in global market Finds out what is popular and acceptable prices from the bidding process - market sets the prices
Articles available in various places but mostly with only local marketing. Experts travel the world to buy in one place and sell elsewhere Individuals buy from experts	Prices are set by big market Lots of small sellers emerge many (especially men) start to buy things on E-Bay they previously bought elsewhere or didn't buy at all	Antique, art & other professional dealers gain and lose markets Artists & painters etc. gain many customers in global markets

Conclusion: TRIZ Aim is to Increase Ideality and Subdue Complexity

TRIZ is a system of straightforward steps and rules you follow, to first understand the problem (all the benefits you want and all the costs and harms) and then solve it. Complexity is subdued by understanding the context of your problem – thinking in 9-Boxes and by reducing your problem to the essentials.

Time and Scale – an Important TRIZ Tool

Altshuller describes the power of this most useful TRIZ tool:

> The screens of talented thought. Three stages and nine screens, depiction and anti-depiction – this is after all an elegant simplification of the pattern. A genuinely talented thought has many stages above the system (the super-system, the super-super system …). Behind the tree one should see not only the wood but also the biosphere in general and not only the leaf but the cell of the leaf. Many screens should be to the left of the system (the recent past, the distant past) and to the right (the future, the far distant future). The depiction of the screen becomes now large, now small, now the actions slows down, now speeds up … Complicated? Yes, indeed. The world in which we live is constructed in a complicated way. And if we want to learn about it and transform it, our thinking must reflect this world correctly. The complex, dynamic, dialectically developing world should find in our consciousness its full model which is complex, dynamic and dialectically developing. A mirror reflecting an image of the world should be large and many sided. (*CAESS* p. 121)

Altshuller is of course right – there are many more than 9-Boxes in a system and over its lifetime as shown below.

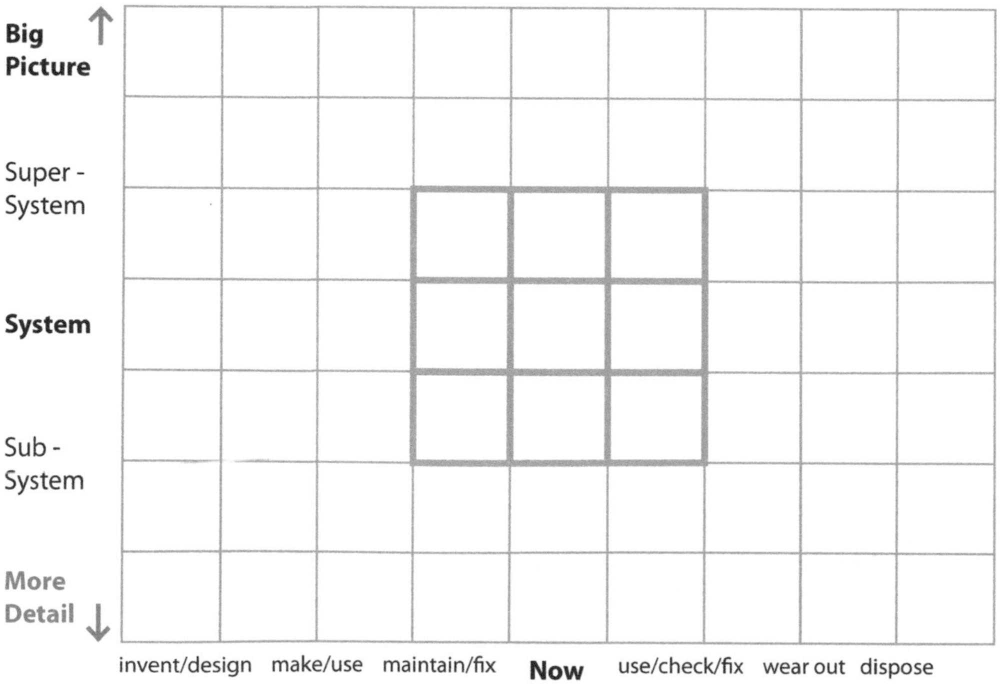

Big Picture ↑

Super - System

System

Sub - System

More Detail ↓

invent/design make/use maintain/fix **Now** use/check/fix wear out dispose

When using the 9-Boxes it is helpful to reduce any situation to as near to 9 as possible, either we can combine as many of these boxes to just 9 or we can move the central 9 boxes around the system – like shining a search light on the relevant areas. In problem solving sessions we try to stick to 9 boxes, but when absolutely necessary to go wider, try not to go much higher than 12 or 15 boxes to describe a situation.

This TRIZ tool of Time and Scale is useful to everyone not just engineers. Useful for a simple understanding of situations but invaluable when the complication is confusing us, and we can cannot understand and think clearly (see the wood for the trees). For simple but confusing situations it is unique and powerful. More importantly for technical problems it helps our comprehension and gives us clarity of thought for solutions. Engineering problems are rarely simple and subduing complication to reduce a situation to several simple problems is helpful to everyone both to understand and then to see what to do.

Beyond talented thinking is the highest levels of problem solving. Genius is often seeing clearly everything that is there at all levels and Altshuller contended that 9-Boxes was an often applied tool of not just talented thought but genius.

> Even among geniuses such thinking is by no means encountered every day. In reality the 'full screen pattern' (9-Boxes) shows the thinking of a genius in his starry hours, which are extremely rare even in the lives of great thinkers and artists. 'The full picture' is the IFR and approaching this idea is ASIP (Algorithm for the solution of Inventive Problems).

Altshuller argued that as we can predict the likely direction of our system by defining the Ideal Outcome (Ideal Final Result or IFR) and using the TRIZ tools such as the Trends of Evolution (Chapter 9) we can define our future boxes with some confidence and move forwards.

Nine boxes then is not just the stuff of talented thought but genius. If we look at our great scientists and engineers it is often seen that they could think in 9 boxes naturally, and retain and master all the detail of their specialist areas but simultaneously see it in context – see the big picture and see the clashes between different boxes and see the significance of time. Great physicists and mathematicians such as Richard Feynman and Paul Dirac had the kind of minds that recognized patterns, connections between system levels, symmetries, contradictions between systems and anti-systems (such as anti-matter), considered how time does not always go forward at all system levels. They could picture all the possibilities, and could see and explain the significance of time and scale from the big picture down to the molecular level'. Genius does 9-Boxes instinctively – Altshuller wanted us all to recognize how a genius thinks and have the choice to emulate.

Case Study: Applying Time and Scale to Nuclear Decommissioning Research Sites Restoration Limited – an Estimating Workshop

Requirement

Research Sites Restoration Limited (RSRL) is responsible for the closure of Harwell, a nuclear research site whose history dates back to the dawn of the UK nuclear industry in the 1940s. There are many buildings with varying levels of nuclear contamination that must be cleaned up before the buildings can be demolished, and RSRL must maintain up-to-date estimates for the nuclear decommissioning of each building

The TRIZ workshop was held because RSRL needed to revise the decommissioning estimate for the clean-up of radioactive contamination within one specific building to enable it to be demolished safely. This workshop was part of a larger project to revise the process for estimating nuclear decommissioning.

Problem

There is a large building with radioactive contamination and a management team with operational and maintenance experience, an in-depth knowledge of the building, but limited decommissioning experience. A new annual funding limit has placed a ceiling on how much work can be completed in a single financial year. Historical estimating practices lack focus and are risk adverse.

TRIZ was selected by the new estimating manager as a facilitation tool, to help shape behaviour during a workshop. The workshop was arranged to scope out the decommissioning project in line with the new requirements to inform the production of the plan and cost estimate.

Tackling the Problem

The two tools that were selected for use in the workshop were:

- Ideal Outcome:
  - Looking at the situation from a 'can do' rather than a 'can't do' perspective.
  - Allowing a creative mindset by focussing on what we want going forward rather than what we've had to accept in the past.
- Thinking in Time and Scale (9-Boxes):
  - Looking at how the structure had been contaminated through time and how that contamination had been managed.
  - Looking at how funding changed over time.
  - Looking at where the building was in relation to the site, the organisation and the industry.
  - Looking at where rooms and areas were in relation to the building.

1	2	3
4	5	6
7	8	9

Thinking in Time and Scale was the first tool introduced to the team and was used in conjunction with floor plans of the building. It was decided that individual rooms constituted the sub-system, the system was the building and the super-system the site/organisation.

The first task involved establishing a 'best' room and a 'worst' room (Cell 8) and then establish why they were the best and worst rooms (Cell 7) and how many similar rooms existed. Next was to understand the relationship between the rooms and the building and whether the contamination was confined to a room or covered a wider area (Cell 5). After this the way the building had been constructed was analysed (Cell 4) to see if there were any hazards or opportunities. It was established that the building had been constructed in three parts, that it could be naturally sectioned but that there were a number of shared services such as ventilation and drains that crossed these boundaries. Next the team looked at the problem from a site/organisation perspective (Cell 2) in terms of where the building is with regards to other structures, what access is like and any services it provides or shares with the site. Finally they looked at what was there before the building was constructed (Cell 1); was it green field or was it brown field?

Before looking at the 'Future' cells (Cells 3, 6, 9) it was decided to introduce Ideal Outcome to further help shape the thought process and to control the psychological 'headwind' that safety can create within the industry. Different types of contamination require different safety/containment management solutions – the need to protect people and also uncontaminated areas within the building. Our building has had many uses throughout its history and safety management has evolved from very few to today's strict requirements.

Results

Completing Cells 3, 6 and 9 was probably the purpose of the session, to establish the objectives of the project. Ideal Outcome was used in conjunction with our risk, uncertainty and opportunity process. The team articulated what they thought would be an ideal outcome for a particular activity or package of work and then identified the assumptions and exclusions required to facilitate a successful outcome. The ideal outcomes were recorded in Cells 3, 6 and 9; and the assumptions and exclusions reviewed in terms of impact and probability of occurrence and recorded on the Probability Impact Grid (PIG).

Ideal Outcomes included:

- overall project duration
- individual work-package durations
- year-on-year liability / risk reduction
- process cost reduction
- new waste disposal routes
- no remote-handled intermediate-level waste
- no site dependencies.

Outcome

The workshop was successful on two fronts:

- It acted as launch event for the project and ensured that all stakeholders were involved in developing an understanding of what the project requirements were.
- It produced a comprehensive scope document and ensured that the problem was considered from multiple perspectives with risk, uncertainty and opportunities documented.

The information was written up into a scope document and was used to develop the project plan using Primavera and the base cost estimate.

Next Steps

The approach has been adopted for other projects that required estimate revisions completed.

Part Two
The Contradiction Toolkit

Uncovering and Solving Contradictions

Contradictions – Solve or Compromise?

TRIZ engineers are comfortable with paradox, contradiction and ambiguity – as was Leonardo da Vinci and many great inventors. Every engineering problem contains at least one contradiction but for many engineers contradictions remain unnecessarily associated with uncertainty, difficult choices and compromise: unnecessary because there are a small and finite number of practical answers to solving any contradiction (40), which are all simple basic principles, and help every engineer solve any problem. Unnecessary also because this invaluable and powerful list of answers, the TRIZ 40 Inventive Principles, is public domain and has been openly used in Russia for over 50 years and more recently has been freely accessible to almost everyone. Contradictions are solved by applying the relevant TRIZ Inventive Principles which are simple ways of achieving clever solutions. They are straightforward and easy to understand. Some examples are shown below.

Principle No 4 – Asymmetry

Change the shape or properties of an object from symmetrical to asymmetrical

TRIZ for Engineers: Enabling Inventive Problem Solving, First Edition. Karen Gadd.

Principle No 17 – Another Dimension

Move into an additional dimension – from one to two – from two to three

Principle No 13 – The Other Way Round

Invert the action used to solve the problem, (e.g. instead of cooling an object, heat it)

What is a Contradiction?

A contradiction is a simple clash of solutions. Either we want opposite solutions, or by introducing a new solution, i.e. an improving change to one feature in a system, another feature in our system has got worse. Engineers recognize contradictions as familiar situations, such as when we improve strength by adding more material we find that this solution often makes weight get worse. Contradictions can also be the need for opposite benefits which are achieved with opposite features or functions – an everyday item such as an umbrella has the benefits of being both large and small. There are many situations of wanting opposites such as white and black – TRIZ shows us all the way to achieve such opposite benefits.

Principle No 33 – Colour Change

Achieving opposite benefits / opposite features within a system may initially seem impracticable and silly – even though every polar bear has solved how to be both black and white (it has black skin for warmth but its transparent fur allows it to appear white to merge into the snowy landscape). The clever polar bear solution and all other clever answers to contradictions are in the TRIZ 40 Inventive Principles to help engineers systematically find simple and useful solution triggers to questions that at first glance may seem impossible.

Spotting Contradictions – But and And

When we say our solution idea gives us this *but*, we then have a problem. That is when striving to make something better we then find we are consequently making other things worse … there is usually a contradiction (called a Technical Contradiction in TRIZ). I make my window much bigger for more light *but* then I have less privacy and it is noisier, and it is colder in winter and hotter in summer.

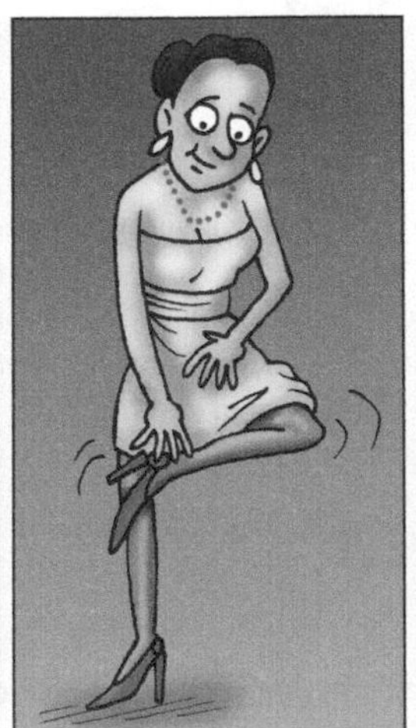

Principle No 15 – Dynamics – can deliver opposite benefits

Additionally when we are trying to choose between two opposite solutions wanting black *and* white, sharp *and* blunt (a knife is sharp to cut *and* blunt to hold in our hands), or high heels for glamour *and* low heels for comfort, in shoes we have again uncovered a contradiction; these opposite solutions are called Physical Contradictions in TRIZ.

We solve both these types of contradictions by using the TRIZ 40 Inventive Principles. This is one of the three TRIZ solution lists which show us how to maximize the good things and minimize the bad things in engineering problem solving. TRIZ research uncovered these three fundamental but simple solution lists, the 40 Inventive Principles, the 8 TRIZ Trends of Technical Evolution and the 76 TRIZ Standard Solutions.

Systems Meet Needs

In real life when problem solving we are trying to match a system to our needs. Contradictions occur when we have opposite needs (I want a small car for parking *and* a big car for safety) or when we focus on one need and it conflicts with another – I want a powerful car *but* I also want greater fuel efficiency.

When choosing something like a car we are faced with many such contradictions – I would like my car small when I am by myself *but* big when I have my family with me, or I would like it to offer many sophisticated features *but* I would like it to be simple and economic to maintain. How to get everything we want from a system like a car involves recognizing the contradictions between all our needs, and then understanding and using all the TRIZ ways of solving them – whenever possible. In these situations we are familiar with contradictions and must look for systems which give us all our requirements and not compromise or choose between them.

Compromise or Solve?

When faced with a contradiction we can compromise (optimize) or solve it. Contradictions are at the heart of many problems. TRIZ thinking helps us first to recognize (uncover) contradictions and then shows us how to solve contradictions by using the 40 Inventive Principles systematically.

Making mobile phones smaller has both advantages and disadvantages

40 Inventive Principles

Fact or Fiction

Altshuller studied thousands of patents to extract solutions and concepts from the clever ones. He defined a clever patent as one that offered solutions to contradictions and/or cleverly applied knowledge from another industry. He found that about 20% of patents were 'clever' and he noted all the successful concepts they employed. After 35,000 patents he had allegedly found just 37 concepts for solving contradictions and after 50,000 patents had found only 40 concepts – known as the TRIZ 40 Inventive Principles.

Today after millions of patents have been analysed we still stand at these 40 concepts/principles for solving a contradiction. These 40 Inventive Principles are simple solution triggers ('tricks' Altshuller called them in his book he wrote for children *And Suddenly the Inventor Appeared*) to show us all the ways the world knows to solve particular contradictions. So once we uncover contradictions TRIZ directs us to the relevant 40 Principles which will help us solve our contradictions. These powerful solution triggers then need to be turned into practical ideas; this process needs relevant knowledge combined with brain power and experience to produce practical and relevant solutions to contradictions.

Altshuller never earned enough to live on from TRIZ – so he wrote science fiction under the name of H. Altov. His most famous TRIZ book in the West is *And Suddenly the Inventor Appeared*. It offers one of the few easily readable TRIZ books and is also fun and interesting but gives very few clues of how to actually use TRIZ.

The book is based on some of the published work Altshuller produced for children and the examples such as use of magnetic materials to help ploughing are not supposed to offer practical examples for most engineers today. He was a funny and charismatic man with a phenomenal memory and almost total recall of everything he had ever read. This ability probably helped him uncover the very few solution concepts in the patent database when working as a patent officer in the Russian Navy in 1946–9.

As well as the three TRIZ fundamental solution lists Altshuller found there are about 2,500 scientific and engineering concepts in the patent database. This is remarkable if we remember how many geniuses and great scientists have worked in this field. No other patent experts had seen the simple and underlying patterns and tried to measure the very few conceptual solutions and scientific concepts in all patents. Indeed very few great scientists even saw this pattern in their own work. Edison produced over a 1023 patents using only 23 concepts while Tupolev filed 1010 patents using only 35 scientific concepts.

What is a Contradiction?

Two types of contradiction – Technical and Physical.

Technical Contradiction

I get something good *but* I then also get something bad.

We think of a solution to improve something *but* something else gets worse.

Often we assume that when faced with this dilemma that we must choose one solution, technical parameter or feature at the expense of the other – without TRIZ we initially assume we can't have both. We believe that the two features are inevitably linked and that when we improve one, then the other will inevitably get worse in some way.

"It appears we've accidentally laid off all the people who 'do' things..."

Improving one thing can make something else get worse

The cartoon on page 31 uses TRIZ Inventive Principle 4 – Asymmetry – and can be seen in use in airplanes throughout the world: it was first adopted by Boeing, who have made TRIZ available to their engineers for many years.

We need practice to recognize and uncover technical contradictions – looking for situations when our solution gives us what we want (i.e. something good, useful, or beneficial) *but* it comes with some bad consequence for some previously good feature or function. Something is made worse (i.e. harmful or costly) by the solution.

Simple examples of Technical Contradictions occur everywhere:

- Digital camera – we want small pixels (better resolution) but this gives us increased noise.
- Cooling fan – how can we get good airflow, *but* without noise?
- A larger heat sink dissipates more heat *but* is bigger.
- I enjoy eating cream cakes *but* they are bad for me.

It is always worth thinking about domestic and work situations to see if we have solved a contradiction or accepted a compromise. The height of our chairs, desks, kitchen worktops are often all set at the same height for people of different shapes and sizes – a compromise and probably unsuitable for most individual needs – but one we accept often without noticing – as with many everyday objects.

Technical contradictions are situations when there are two conflicting benefits which irretrievably seem linked in some way – and conventional thinking makes us assume we must choose one and forgo the other; accept that when we get a good thing it comes at the expense or loss of another benefit. TRIZ shows us how to avoid compromise and have both benefits.

Principle No 22 – Blessing in Disguise

Don't Compromise – or Choose Between Two Conflicting Solutions – Have Both

When we want to achieve two apparently incompatible outcomes/solutions TRIZ has gathered together all the answers to this dilemma in the 40 Inventive Principles, and offers us all the ways to successfully have both. The basis for TRIZ is not to compromise, or optimize contradictions, but fundamentally solve them – not try and choose between two good things but systematically locate solutions that will give us both. The TRIZ Toolkit contains these powerful solution lists of all the ways the world knows to help us get everything we want: all the recorded ways of uncovering, embracing and then solving contradictions.

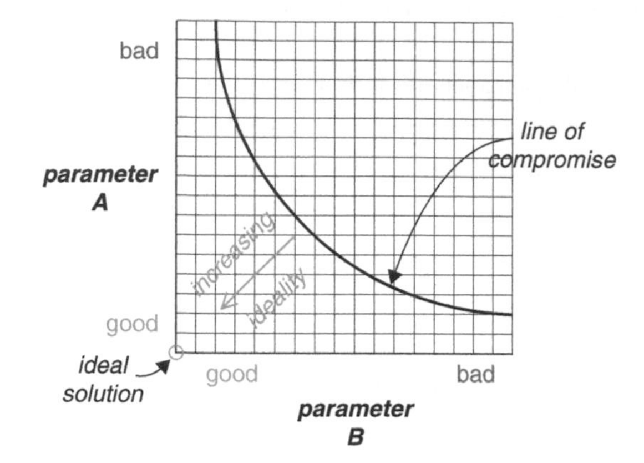

For engineers the line of compromise is familiar territory – but with TRIZ we are aiming to get off this curve and draw a new curve, where we can improve one parameter without changing the other parameter. This means moving to a new line closer to the good/good point of the graph above – this is achieved by applying the relevant 40 Principles to our problem.

The essential power of TRIZ in problem solving is to focus/concentrate on what we really want, and not be distracted by the important but distracting details of the problem situation. By focussing in on delivering the primary (essential) benefits and solving the resulting contradiction – often solving a single contradiction – we have found the place where the powerful, good solutions can be found. This often leads to answers where all the other benefits can also be accommodated and the system is matched to all our needs. The random chaos of problem solving is reduced – the innumerable solution options are set aside – we are directed to model the one essential, uncluttered problem and to use TRIZ to locate all its possible solutions – normally distilled to no more than about a dozen principle concepts. We move out of this simple, focussed problem solving as we add back the relevant detail for our particular problem.

Many engineers ask what about the many real situations when there are more than two variables – with TRIZ we solve one problem at a time by reducing that problem to one simple model. We must concentrate first on the solution which gives us the main benefits we are really seeking and the other distracting variables will be dealt with in turn. Most problems have at their heart, one essential problem with a finite number of answers. Converge to that place, locate the answers TRIZ suggests, then diverge back to the reality and the additional detail of the problem situation.

Physical Contradiction

We want opposite solutions – for example, high and low.

We want a high garden fence for our garden boundary – but we also want a low fence so it doesn't get blown down in high winds. Clever solutions to such contradictions exist in the world – we have to find a fence that is *there* for privacy but *not there* for high winds – and such fences exists and is called a Hit and Miss.

THE ADVANTAGE OF THIS PANEL IS THAT IT KEEPS YOUR PRIVACY BUT LETS THE WIND THROUGH...

Physical Contradictions need careful thought but once we can recognize that we want opposite solutions or benefits, then TRIZ can show us all the ways to have both. Human ingenuity and design skill has always been used to deliver opposite solutions and there are often obvious ways of getting both.

CAN SOMEONE PLEASE EXPLAIN TO ME THE EXPRESSION: 'THE ELEPHANT IN THE ROOM'?

Physical contradictions include when something is 'there' and 'not there'

When we can't spontaneously solve the problem of how to achieve opposite benefits then we use the TRIZ solution lists to prompt clever ideas – to direct us to the small number of simple but powerful ways known in the world to have opposite solutions.

Physical Contradictions mean that we want opposite benefits/solutions/features. We improve something, but the *same* technical parameter gets worse – I want my coffee cup *hot* next to my coffee but *cold* next to my hand and lip, and when on the table so it doesn't leave burn marks.

Another simple domestic example can be seen from using glue – if we make it easy to spread (less viscous or runnier) then we can get more even coverage but it is harder to control and place exactly where we want it. We want it runny and not runny (viscous).

A coffee cup has to be hot AND cold – at the same time but in different places

Physical contradictions are solved by separating the solutions in different ways which gives both solutions. We can separate in time, in space, on condition or in scale.

Separate in time means having one solution at one time (long board pointer) and the opposite solution at another time (short board pointer).

Separate in space means to have one solution in one place (cold outside of coffee cup) and the opposite in another place (hot inside cup).

We use the separation principles – separate in time and space etc. – to see how to get systems to give us opposite requirements but at different times and in different places.

Principle No 7 – Nested Doll

Separate in time examples are everywhere as there are many situations and occasions when we want systems *there* at one time – when we want them but *not there* at another time when we don't. We want different things at different times. For example, I would like a tiny umbrella in my bag and a sufficiently big umbrella when it is raining and small again when not raining. I would like no protection systems round me when driving my car in normal conditions but would like airbags etc. to appear when I need them in an accident. I would like the centre of town suitably equipped for drunken clubbers when they emerge from pubs and nightclubs – but nothing unnecessary there in the daytime.

Principle No 19 – Periodic Action

TRIZ helps us think of solutions to understand how to create or locate systems to give us what we want. All the ways of overcoming Physical Contradictions are in the 40 Principles – the list of the Inventive Principles help us create systems which give us different things at different times. This includes Principles 29 Pneumatics and Hydraulics which is used in the imaginary solution below for an umbrella.

40 Principles Solve All Contradictions

Both physical and technical contradictions can be solved with the 40 Inventive Principles but the two routes to locating the relevant principles are very different.

There is the very technical looking Contradiction Matrix for solving technical contradictions and there are the powerful Separation Principles for understanding and solving the different types of Physical Contradictions; each Separation Principle offers a selection of the 40 Principles which help solve or guide us to solution concepts to overcome particular physical contradictions.

The Contradiction Matrix is only for solving Technical Contradictions and shows which of the 40 Principles solve a particular technical contradiction but offers no solutions for Physical Contradictions – shown as the blank green diagonal boxes.

Improve this one without making this one worse

39 Technical Parameters		Weight of moving object 1	Weight of Stationary Object 2	Length of moving object 3	Length of Stationary object 4	Area of moving object 5	Area of stationary object 6	Volume of moving object 7	Volume of stationary object 8	Speed 9	Force (intensity) 10	Stress or pressure 11	Shape 12	Stability of the object's composition 13	Strength 14	Duration of action of a moving object 15	Duration of action of a stationary object 16	Temperature 17	Illumination Intensity 18	Use of energy by a moving object 19	Use of energy by a stationary object 20	Power 21	Loss of energy 22	Loss of substance 23	Loss of information 24	Loss of time 25	Quantity of substance 26	Reliability 27
1	Weight of moving object		-	15 8 29 34	-	29 17 38 34	-	29 2 40 28	-	2 8 15 38	8 10 18 37	10 36 37 40	10 14 35 40	1 35 19 39	28 27 18 40	5 34 31 35	-	6 29 4 38	19 1 32	35 12 34 31	-	12 36 18 31	6 2 34 19	5 35 3 31	10 24 35	10 35 20 28	3 26 18 31	3 11 1 27
2	Weight of stationary object	-		-	10 1 29 35	-	35 30 13 2	-	5 35 14 2	-	8 10 19 35	13 29 10 18	13 10 29 14	26 39 1 40	28 2 10 27	-	2 27 19 6	28 19 32 22	19 32 35	-	18 19 28 1	15 19 18	18 19 28 15	5 8 13 30	10 15 35	10 20 35 26	19 6 18 26	10 28 8 3
3	Length of moving object	8 15 29 34	-		-	15 17 4	-	7 17 4 35	-	13 4 8	17 10 4	1 8 35	1 8 10 29	1 8 15 34	8 35 29 34	19	-	10 15 19	32	8 35 24	-	1 35	7 2 35 39	4 29 23 10	1 24	15 2 29	29 35	10 14 29 40
4	Length of stationary object		35 28 40 29	-		-	17 7 10 40	-	35 8 2 14	-	28 10	1 14 35	13 14 15 7	39 37 35	15 14 28 26	-	1 40 35	3 35 39 18	3 25	-		12 8	6 28	10 28 24 35	24 26	30 29 14		15 29 28
5	Area of moving object	2 17 29 4	-	14 15 18 4	-			7 14 17 4		29 30 4 34	19 30 35 2	10 15 36 28	5 34 29 4	11 2 13 39	3 15 40 14	6 3	-	2 15 16	15 32 19 13	19 32	-	19 10 32 18	15 17 30 26	10 35 2 39	30 26	26 4	29 30 6 13	29 9
6	Area of stationary object	-	30 2 14 18	-	26 7 9 39	-		-		-	1 18 35 36	10 15 36 37		2 38	40	-	2 10 19 30	35 39 38		-		17 32	17 7 30	10 14 18 39	30 16	10 35 4 18	2 18 40 4	32 35 40 4
7	Volume of moving object	2 26 29 40	-	1 7 4 35	-	1 7 4 17	-		-	29 4 38 34	15 35 36 37	6 35 36 37	1 15 29 4	28 10 1 39	9 14 15 7	6 35 4	-	34 39 10 18	2 13 10	35	-	35 6 13 18	7 15 13 16	36 39 34 10	2 22	2 6 34 10	29 30 7	14 1 40 11
8	Volume of stationary object	-	35 10 19 14	19 14	35 8 2 14	-		-		-	2 18 37	24 35	7 2 35	34 28 35 40	9 14 17 15	-	35 34 38	35 6 4		-		30 6		10 39 35 34		35 16 32 18	35 3	2 35 16
9	Speed	8 28 13 38	-	13 14 8	-	29 30 34	-	7 29 34	-		13 28 15 19	6 18 38 40	35 15 18 34	28 33 1 18	8 3 26 14	3 19 35 5	-	28 30 36 2	10 13 19	8 15 35 38	-	19 35 38 2	14 20 19 35	10 13 28 38	13 26		10 19 29 38	11 35 27 28
10	Force (Intensity)	8 1 37 18	18 13 1 28	17 19 9 36	28 10	19 10 15	1 18 36 37	15 9 12 37	2 36 18 37	13 28 15 12		18 21 11	10 35 40 34	35 10 21	35 10 14 27	19 2		35 10 21	-	19 17 10	1 16 36 37	19 35 18 37	14 15	8 35 40 5		10 37 36	14 29 18 36	3 35 13 21
11	Stress or pressure	10 36 37 40	13 29 10 18	35 10 36	35 1 14 16	10 15 36 28	10 15 36 37	6 35 10	35 24	6 35 36	36 35 21		35 4 15 10	35 33 2 40	9 18 3 40	19 3 27		35 39 19 2	-	14 24 10 37		10 35 14	2 36 25	10 35 3 37		37 36 4	10 14 36	10 13 19 35
12	Shape	8 10 29 40	15 10 26 3	29 34 5 4	13 14 10 7	5 34 4 10		14 4 15 22	7 2 35	35 15 34 18	35 10 37 40	34 15 10 14		33 1 18 4	30 14 10 40	14 26 9 25		22 14 19 32	13 15 32	2 6 34 14		4 6 2	14	35 29 3 5		14 10 34 17	36 22	10 40 16
13	Stability of the object's composition	21 35 2 39	26 39 1 40	13 15 1 28	37	2 11 13	39	28 10 19 39	34 28 35 40	33 15 28 18	10 35 21 16	2 35 40	22 1 18 4		17 9 15	13 27 10 35	39 3 35 23	35 1 32	32 3 27 15	13 19	27 4 29 18	32 35 27 31	14 2 39 6	2 14 30 40		35 27	15 32 35	
14	Strength	1 8 40 15	40 26 27 1	1 15 8 35	15 14 28 26	3 34 40 29	9 40 28	10 15 14 7	9 14 17 15	8 13 26 14	10 18 3 14	10 3 18 40	10 30 35 40	13 17 35		27 3 26		30 10 40	35 19	19 35 10	35	10 26 35 28	35	35 28 31 40		29 3 28 10	29 10 27	11 3

Technical Contradictions
I make something better (light table)
BUT
Something else gets worse (too weak)

Physical Contradictions
OPPOSITE SOLUTIONS
Benefits / Features
(LONG/SHORT)

TABLE MUST NOT BE HEAVY...
© Oxford Creativity 2007

Contradiction Matrix

Solve Technical Contradictions with the Contradiction Matrix

When analysing patents, Altshuller's team created the 39 × 39 contradiction matrix (below and there is a bigger version in the appendix) and initially mapped all the solutions to over 1200 contradictions – this matrix has been analysed and updated but today remains almost identical to the one Altshuller's team first built. We use this matrix to see which of the 40 Principles are relevant to our problem. So when we improve one thing and find something else gets worse then we have uncovered a contradiction and we can use the matrix to guide us to which of the relevant solutions in the 40 Principles will help us solve our technical contradiction.

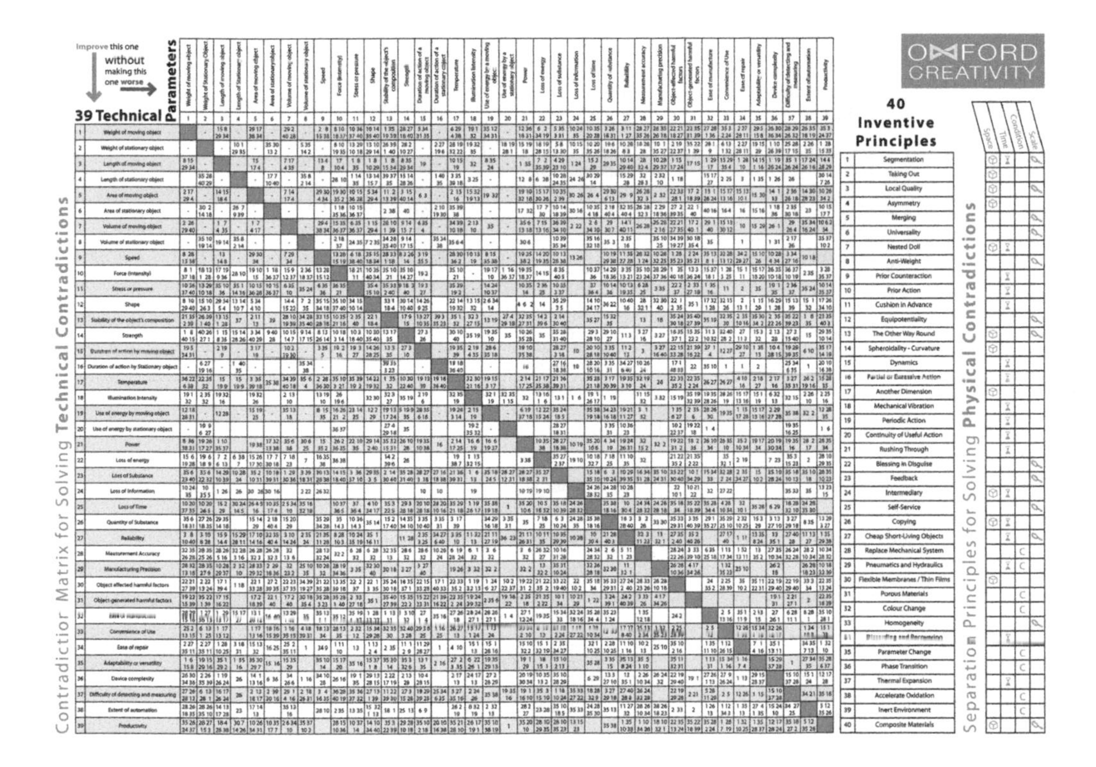

Contradiction Definition from the 39 Technical Parameters

(See full list with detailed description and definitions.)

Each matrix axis is made up of 39 Technical Parameters which were uncovered by studying patents to identify these 39 most widely used and important characteristics of technical systems.

They range from simple to complex features and functions, and between them can describe any engineering contradiction.

> **The Contradiction Matrix**
>
> The matrix is probably the best known TRIZ tool. It is a 39 × 39 asymmetric matrix and was built by analysing thousands of Russian patents – but it has remained almost unchanged after analysis of all other global patents.

The Matrix maps patent solutions to contradictions to guide us to which of the 40 Inventive Principles solve particular technical contradictions. The matrix answers the question:

How can I improve X without making Y worse?

and guides us to up to four principles for each contradiction.

The 39 Technical Parameters

Shown at vertical and horizontal axes of the matrix.

1	Weight of Moving Object	14	Strength	27	Reliability
2	Weight of Stationary Object	15	Duration of Action by Moving Object	28	Measurement Accuracy
3	Length of Moving Object	16	Duration of Action by Stationary Object	29	Manufacturing Precision
4	Length of Stationary Object	17	Temperature	30	Object-Affected Harmful Factors
5	Area of Moving Object	18	Illumination Intensity	31	Object-Generated Harmful Factors
6	Area of Stationary Object	19	Use of Energy by Moving Object	32	Ease of Manufacture
7	Volume of Moving Object	20	Use of Energy by Stationary Object	33	Convenience of Use
8	Volume of Stationary Object	21	Power	34	Ease of Repair
9	Speed	22	Loss of Energy	35	Adaptability or versatility
10	Force	23	Loss of Substance	36	Device Complexity
11	Stress or Pressure	24	Loss of Information	37	Difficulty of Detecting and Measuring
12	Shape	25	Loss of Time	38	Extent of Automation
13	Stability of Object's Composition	26	Quantity of Substance	39	Productivity

The 39 Technical Parameters describe features/functions of engineering systems

See Appendix for more detailed descriptions of the 39 Technical Parameters.

Understanding the 39 Technical Parameters

The Contradiction Matrix is for situations when we have two apparently dependent or linked features and a solution improvement to the system means one needs changing or improving. When we *improve* or change a parameter (such as strength) then another linked parameter such as weight may get *worse*. The Worsening Parameter (=extra weight) is the output we get as a consequence of changing or improving the other parameter (more strength).

We select on the vertical axis of the Contradiction Matrix the parameter we want to change or improve and identify the other Technical Parameter on the horizontal axis, which is linked and consequently gets worse. We can express this contradiction as a simple graph, which typically takes this form:

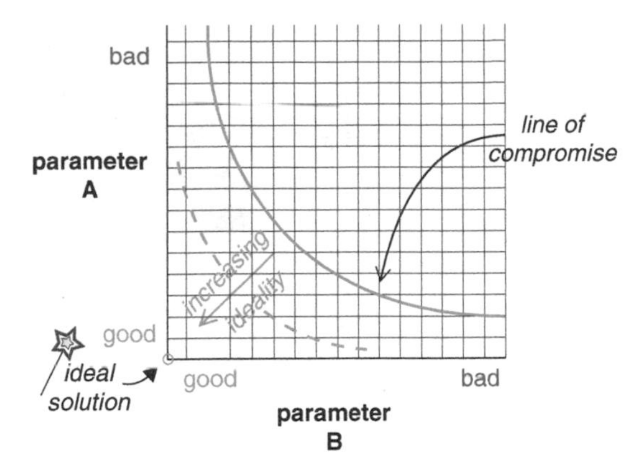

As we try to improve one parameter (strength) the other (weight) becomes worse, and vice-versa. It would appear that we are constrained to remain on the blue curve, which defines the set of possible compromise solutions available to us. What we would like to be able to do is to move off the blue compromise curve to a more ideal solution, i.e. in the direction shown by the green arrow. To achieve this we need to find a way of breaking the link between the two parameters by solving the contradiction.

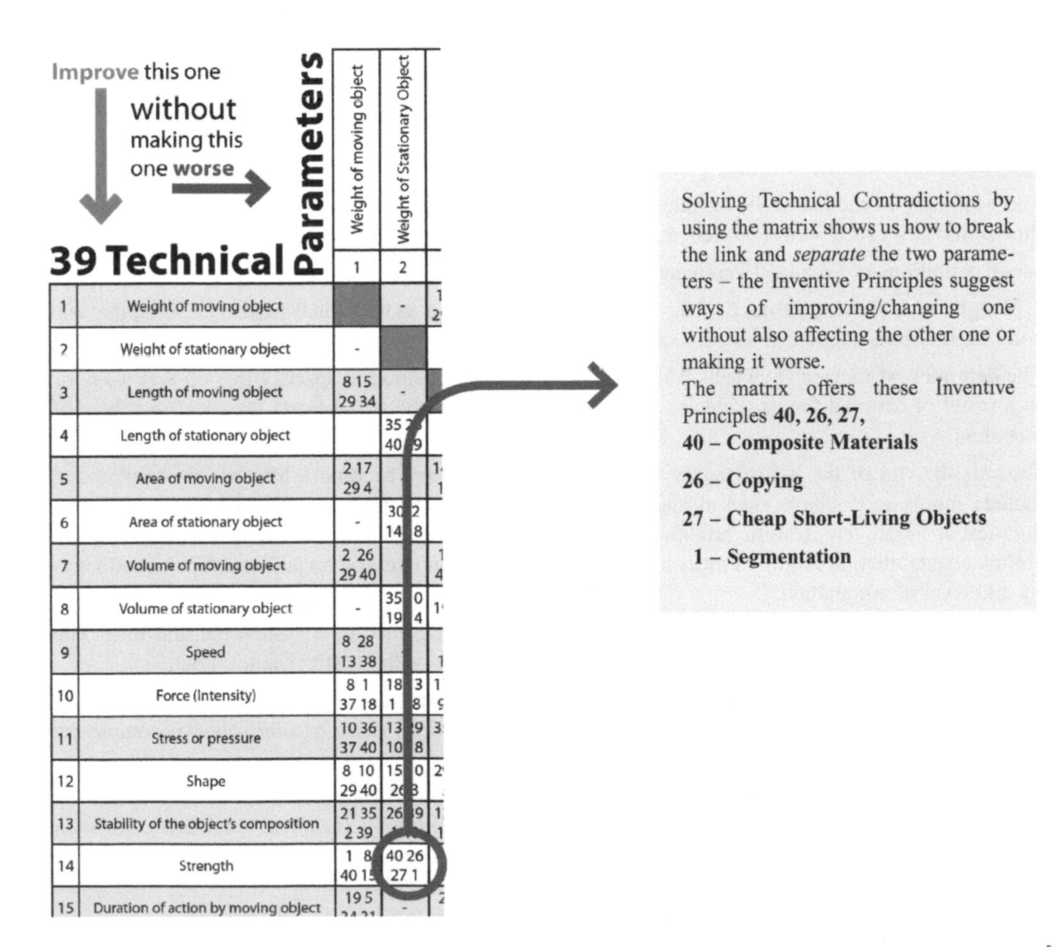

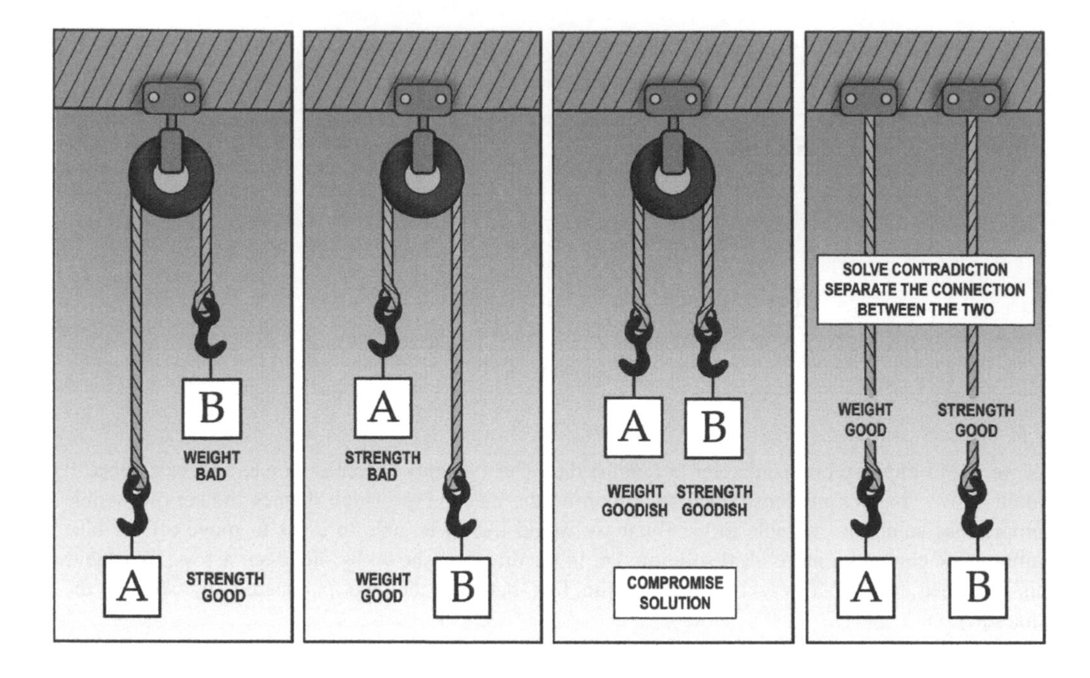

The beginning of the list of 39 Technical Parameters describes simple inputs (vertical axis) or outputs (horizontal axis) such as weight, length, area, volumes and shape for example.

Some definitions of Technical Parameters:

1. Weight of a Moving Object = The mass of the object, in a gravitational field. The force that the body exerts on its support or suspension.

The definition of moving = Objects which can easily change position in space, either on their own, or as a result of external forces. Vehicles and objects designed to be portable are the basic members of this class.

Towards the end of the list of the 39 Technical Parameters they become a little more complex and include inputs and outputs such as speed, forces, loads, stresses, pressures, use of energy, stability, duration of action, temperature, brightness, use of energy, power, reliability, convenience of use, repairability, adaptability, accuracy of measurement, accuracy of manufacture, manufacturability, productivity and level of automation.

39. Productivity = The number of functions or operations performed by a system per unit time. The time for a unit function or operation. The output per unit time, or the cost per unit output.

The list also includes clearly identifiable harms (outputs we don't usually want) such as device complexity, waste of energy, waste of substance, loss of information, waste of time, control complexity and other harmful outputs.

When using the matrix regularly, familiarity with the 39 Technical Parameters builds over time. It is worth looking at and learning (if possible) the meanings of all the 39 Technical Parameters, so when problem solving with the matrix, time is not wasted looking up meanings. This makes us more effective as valuable fast-thinking time is often needed when problem solving – if we don't have to hesitate to check unfamiliar terms our brains keep moving and working in powerful creative mode.

Using the Matrix

With TRIZ we uncover conflicts, and then solve contradictions with the 40 Principles: to solve a technical contradiction we use the matrix to separate the links between two parameters. Of all the TRIZ tools, beginners find the power of the Contradiction Matrix the easiest to understand and the hardest to actually use. This is because once we have uncovered our contradiction we cannot see how to match the things we've got to improve (and the things that consequently get worse) with the 39 Technical Parameters.

The 39 Technical Parameters are derived from the patent database and each example had to be reduced to its simplest, most general term. In the list of 39 Technical Parameters at the back of this book we describe what each one means and give some synonyms to help in this difficult and unfamiliar task. As in all difficult tasks we have to practice until we succeed and keep practising until it becomes a little easier (this step is always challenging even when we are experienced TRIZ practitioners). One way of making it simpler is not to try and find the perfect exact contradiction but try several that approximate to our contradiction. Try several similar contradictions and see which principles are suggested – and when we get some principles occurring more than once, then these principles are a good place to start and probably offer powerful solutions to our problem.

Sometimes special care is needed and we cannot assume always to understand what the Technical Parameters mean just from their titles. For example, 27. Reliability is defined as: 'A system's ability to perform its intended functions in predictable ways and conditions.' This means predictability and not necessarily a certain robustness which is sometimes assumed in the term reliable.

For some problems this mapping of real-world descriptions of a problem onto the 39 Technical Parameters used by the matrix can be a little tricky: the Technical Parameters are very general, and a certain amount of thought and careful consideration may be necessary to map the specific 'real world' problem parameters to the matrix general Technical Parameters.

Understanding which benefit is the most important is a powerful starting point for finding the really relevant solutions. In the dancing lady cartoon below we assume that strength is the really important benefit she wants and the loss of lightness is a problem. Essentially she needs to make the table *stronger*, but she doesn't want it to subsequently get *heavier*.

The Contradiction Matrix can be used for many different kinds of problems. To successfully use the matrix we have to match the solution to our problem to the 39 Technical Parameters and we then have to ask the right question: when I apply this solution (such as improving strength) what then gets worse – which Technical Parameter is adversely affected? So when we design a table to be stronger, our question might be: how can we get the strength we want without increasing weight?

The matrix contains all the answers of how to solve the Technical Contradiction – how to get more strength without increasing weight. Using the Contradiction Matrix to identify which TRIZ Principles show us how to get strength without increasing weight suggests the following solution triggers:

40	**Composite Materials**
26	**Copying**
27	**Cheap Short-Living Object**
1	**Segmentation**

To understand and apply the suggested solutions from the triggers we are given we also have to ask: what are all the benefits (functions) we want from the system? And what is the priority of those benefits?

40 Composite Materials suggests using a strong but light material and together with the trigger of Segmentation might suggest how to save weight by only putting strength where we need it.

27 Cheap Short-Living Object suggests something like a cardboard box – which assumes that durability is not a requirement. We need to understand all the requirements of our system (and the priority of benefits) to judge whether this is an appropriate solution or not.

26 Copying suggests that we replace a strong and heavy material with one which is strong but lighter. We might use a material that looks like wood but is actually a stronger copy material. This principle suggests that we *copy* the heavy material with one which gives us the strength we need but has a more appropriate weight. Another example of using this principle *copying* would be in testing something like a drill which is to be used in remote areas of inaccessible and very hard rock. The tests could be on a copy of rock – artificial rock which simulates natural rock in terms of strength in that it loads the drill with high torque, but could be made of a much lighter composite material which is more easily available than remote rock and easier to handle. Or even copying could mean that we use a video of our lady dancing instead of putting the strain on a small table.

1 Segmentation represents most of the modern design of tables today: segmented tops for different sizes; segmented table legs for ease of manufacture and assembly; segmented joining of different materials for extra strength.

TRIZ links solutions from the 40 Principles (answers/solution triggers) to the right questions when we use the Prism of TRIZ below:

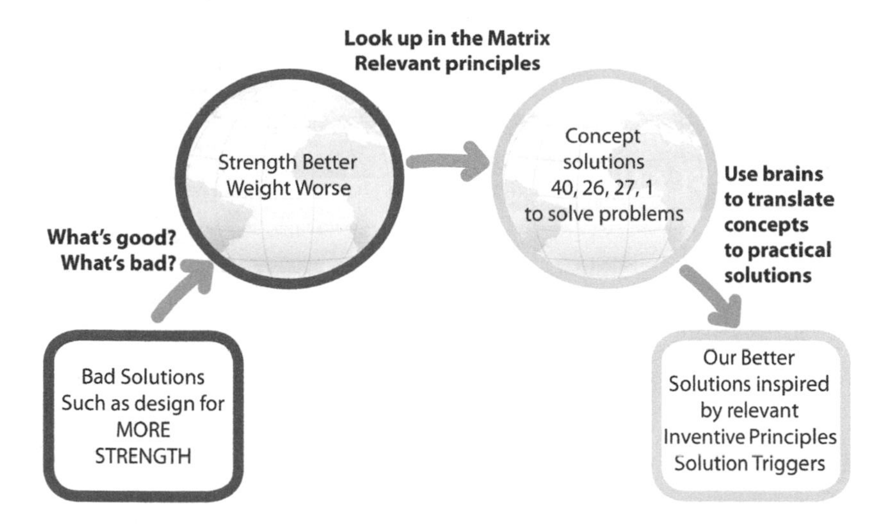

Logical Steps for Problem Solving Using the Contradiction Matrix – Start with Bad Solutions

The logic of solving contradictions with the matrix is a process which requires us to think of a solution – we are changing one parameter because we are trying to improve the system and our 'bad' solution involves changing something. The solution is bad because something then gets worse; the solution is good because something has got better. The contradiction has been created by a 'bad solution' – something has been changed which has good and bad consequences.

We can define or uncover contradictions by taking all our bad solutions and analysing what is good and what is bad about them.

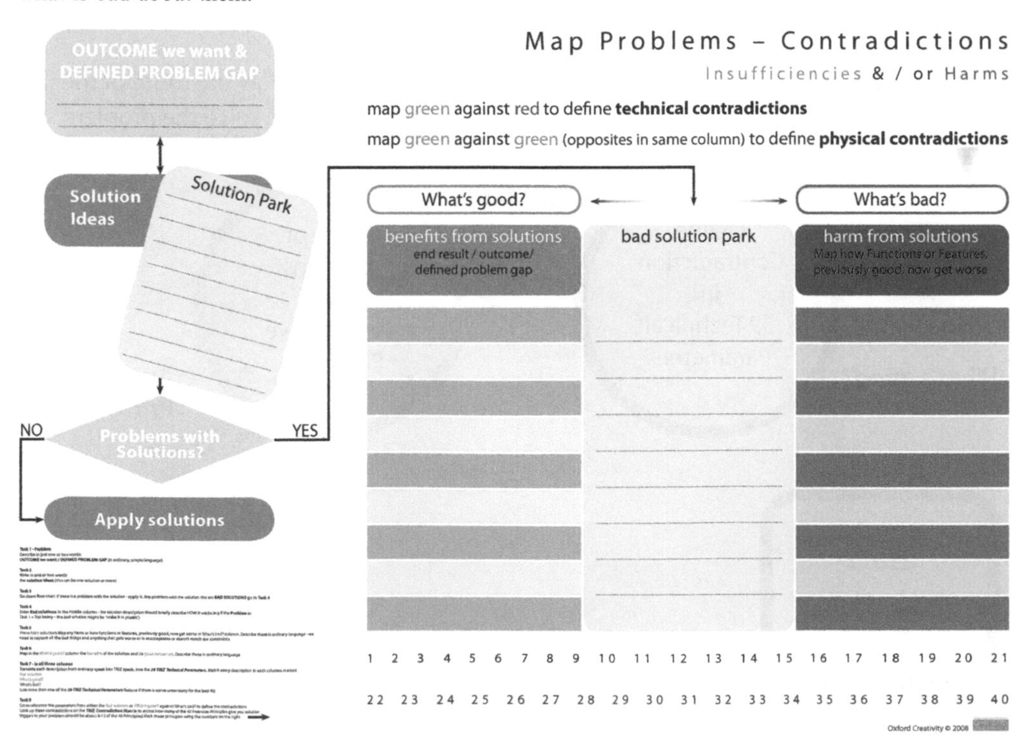

Solving Problems

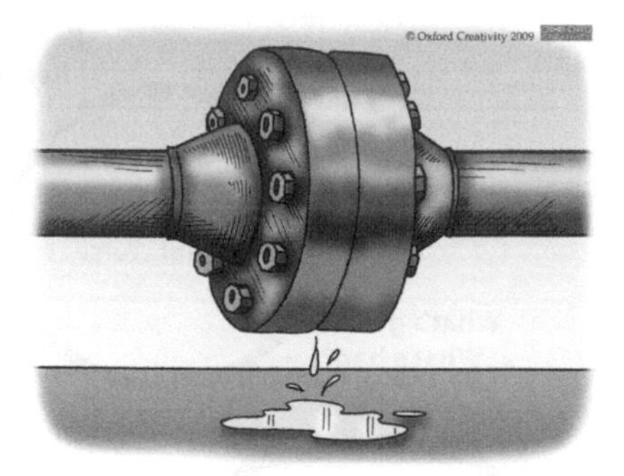

Defined Problem = LEAKING FLANGE

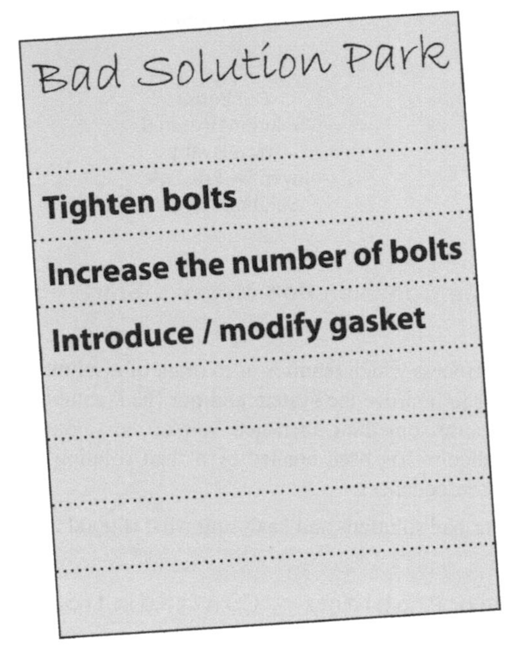

We can usually think of solutions BUT …
Whatever we do to solve the problem makes something else get worse

Tighten bolts
makes it harder to undo etc.

Increase number of bolts
adds weight, complexity etc.

Introduce / modify gasket
adds complexity

How can we match these to the 39 Parameters?

We need to follow the route of the Prism of TRIZ to solve the problem

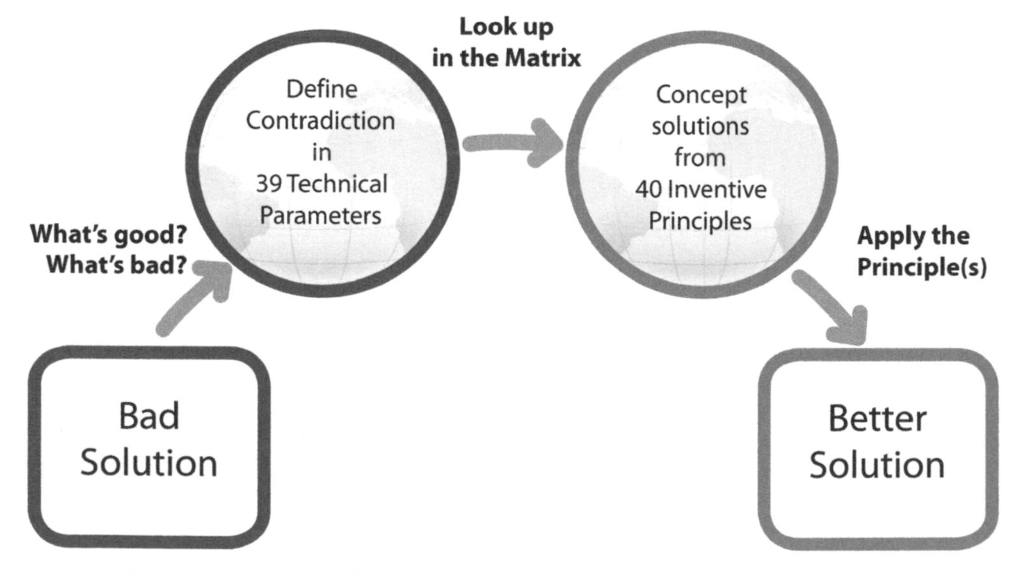

To Uncover and then Solve Contradictions - we start with SOLUTIONS

Immediate solutions come to engineers when faced with problems 'top-of-the-head' solutions for the leaking flange problem often include:

Increase the number of bolts

Tighten Bolts

Add gasket

These solutions are the starting points.

WHAT'S GOOD? What's Better? **Benefits we want** End result / outcome define problem gap	**BAD SOLUTION PARK**	**WHAT'S BAD?** **Solutions** create harms? Map how Functions or Features, previously good, now get worse map in 39 Parameters
NO Leakage	Increase number of bolts	Higher No. of bolts
NO Deterioration	Tighten Bolts	Bolt Torque too high?
Greater Stability	Add Gasket	Ease of Operation is worse
Greater Reliability	Increase number of bolts	Ease of Operation is worse

We now have to find contradictions to solve what do we match against leakage on the matrix? Which of the 39 Technical Parameters describe our problem? What do we want? We want to stop the leak – or at least reduce leakage but 'leakage' does not appear on the 39 Technical Parameters. Looking to match the term 'leakage' onto the terms offered on the 39 Technical Parameters we have several choices including Loss of Substance, Stability and Reliability. In this case rather than agonizing as to which of the three to use, it is normally worth trying all three and trying three contradictions on the matrix.

WHAT'S GOOD? Benefits we want End result / outcome define problem gap	**BAD SOLUTION PARK**	**WHAT'S BAD?** Solutions create harms? Map how Functions or Features, previously good, now get worse map in 39 Parameters	Map GREEN against RED Look up in MATRIX write down which of the 40 Inventive Principles are suggested
STOP Leakage **23. Loss of substance** **13. Stability of object**	Increase number of bolts **2. Weight of stationary object**	No. of bolts **2. Weight of stationary object** **36. Device complexity**	
STOP Deterioration **27. Reliability**	Tighten Bolts **10. Force** **33. Convenience of use**	Bolt Torque **33.Convenience of use** **34. Ease of repair**	
GET Stability **13. Stability of object**	Add gasket **13. Stability of object's composition**	Ease of Operation **33. Convenience of use**	
GET Reliability **27. Reliability**	Increase number of bolts **2. Weight of stationary object**	Ease of Operation **33. Convenience of use**	

How Can We Match Our Solutions to the 39 Technical Parameters?

When looking at the solutions to the flange problem we can map the better/ worse aspects of the problem onto the 39 Technical Parameters as in the table. This gives us good starting points for a number of contradictions.

What gets better? In the table we have translated what we want (reduce leakage and deterioration, more stability and reliability) into the 39 Technical Parameters as Loss of substance, Stability of object's composition, Reliability and Force.

What gets worse? We have translated our solutions which make the system more complicated, heavier and harder to use as Weight, Device complexity, Convenience of use and Ease of repair.

This then gives us a number of contradictions to solve.

The logic of the sequence of solving contradictions by starting with solutions is shown in the flowchart below. This can be used for either Technical or Physical Contradictions.

We uncover contradictions by looking at solutions:

Seeing what is *good* about the solution asking: How does it make the problem better?

Seeing what is *bad* about the solution asking: How does it make the problem worse?

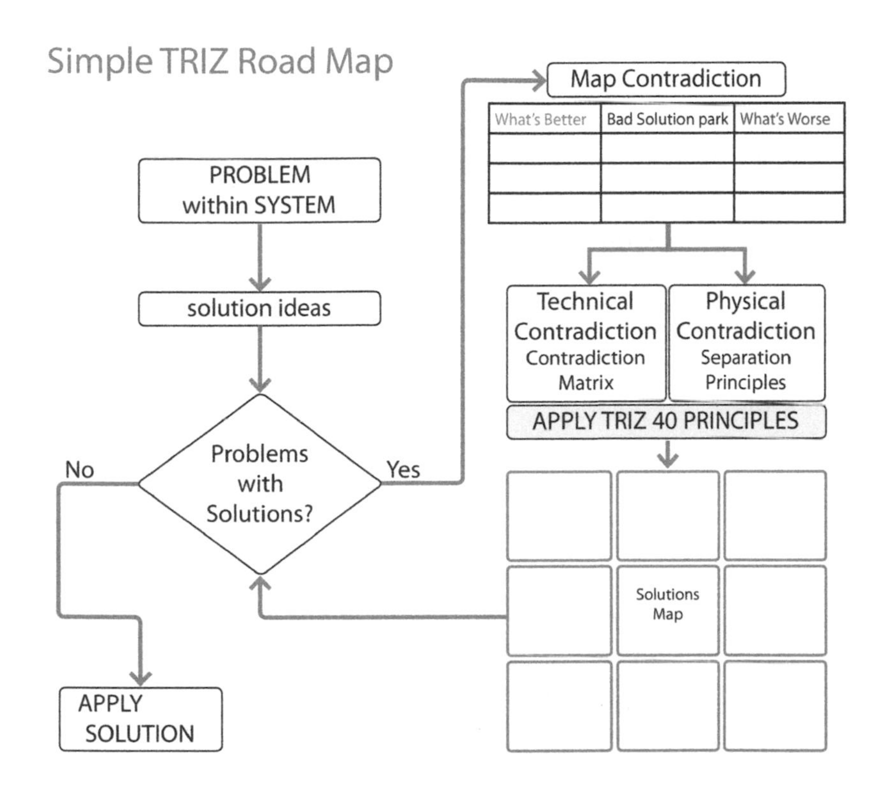

Solving Physical Contradictions

How to get opposite requirements – quick simple solutions to difficult problems.

Understanding Physical Contradictions gives us great clarity of thought which is immensely powerful because they help us ask the right questions, such as: 'under what circumstances (including where and when) do we need these contradictory requirements?' The way we solve these opposite requirements is by understanding how we can *separate* what we want, in order to get opposite benefits at different times or in different places or under certain conditions. The guide to achieving these opposite benefits/ solutions is the TRIZ Separation Principles to show which of the 40 Principles to use.

Separation Principles – separate opposite requirements

Separate in Time	One solution at one time, the opposite solution at another
Separate in Space	One solution in one place, the opposite solution at another
Separate on Condition	Opposite solutions in the same place and at the same time One solution for one element - the opposite for another
Separate by System	Separate by Scale (to Sub-system or Super-System) Switch to Inverse System Switch to Another System

Separate in Time

We want opposite benefits at different times.

We want a plastic bag to be *large* and *small* (large when full of groceries but small before and after use).

We want something *there* when we want to use it but *not there* when not in use.

Suggested Inventive Principles to Separate in Time

1 Segmentation
7 Nested Doll
9 Prior Counteraction
10 Prior Action
11 Cushion in Advance
15 Dynamics
16 Partial or Excessive Action
18 Mechanical Vibration
19 Periodic Action
21 Rushing Through
24 Intermediary
26 Copying
27 Cheap Short Living Objects
29 Pneumatics and Hydraulics
34 Discarding and Recovering
37 Thermal Expansion

Board Pointers need to be long to point and short enough to fit in a briefcase. These opposite benefits are achieved by applying **Principle No 7 – Nested Doll**.

We are surrounded by systems which give us opposite benefits at different times such umbrellas (**Principle No 15 – Dynamics**) and aircraft wings (long for takeoff, and then pivot back to short for high-speed flight).

Separating in time – seeing that we want opposite needs at different times – helps us solve many simple but ingenious problems. In sandblasting, we have a problem that dirty sand accumulates and must be cleared after the cleaning is completed. How do we get the abrasive quality when we want it and get it to disappear by itself after it has been useful? One effective solution is to use dry ice chips as the abrasive – *there* when you need them then *not there* afterwards.

There are lots of examples of using ice when you want a system – there and later not there (ice pigeons for shooting at are there when being shot at but then not there afterwards as they disappear by melting). Or if we want to lower something heavy to the ground slowly, we can place it on an ice stand: if ambient temperature is above freezing then the stand just melts away.

Physical Contradictions are Everywhere in the Real World

Opposite benefits or solutions – in the following example we can separate in time as we want a pile that is sharp and not-sharp.

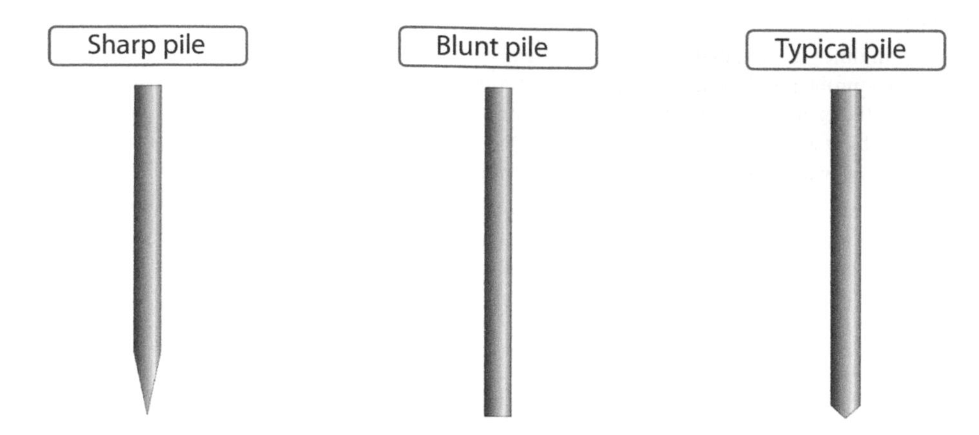

The feature sharp makes it easier to drive it into the ground; the opposite feature blunt is good for load bearing. Therefore we want our pile sharp and blunt at different times.

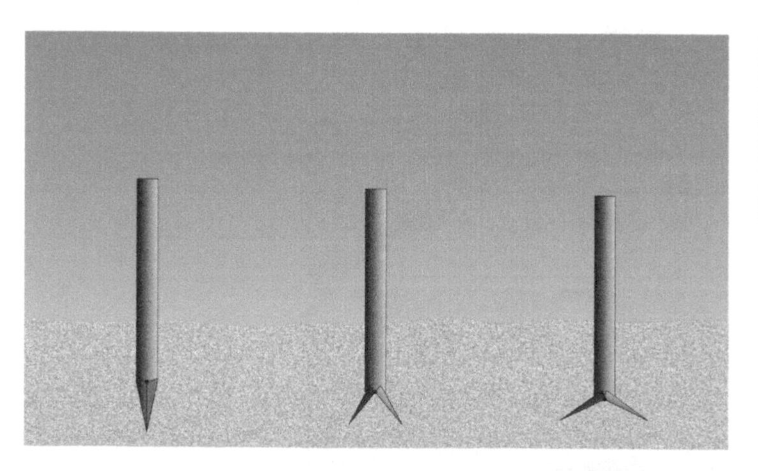

A simple solution for separate in time. The pile changes from sharp to blunt – pile is easy to drive in and can carry great load.

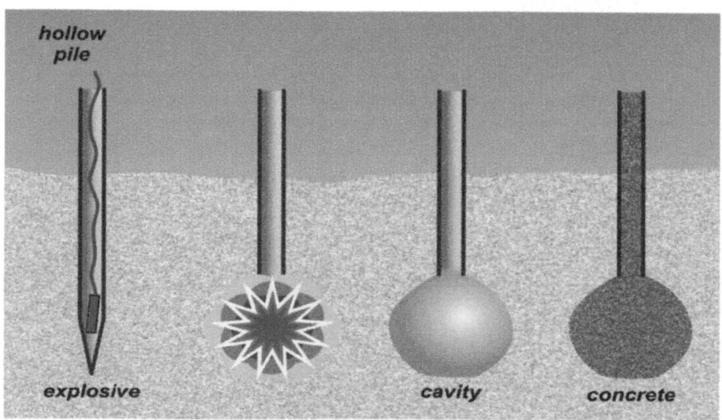

Another solution to this pile problem using separate in time – an explosive charge is lowered into a hollowed out pile, set into the ground. The resulting explosion forms a cavity into which concrete is poured. Thus the concrete pile is easy to drive into the ground and very good at bearing a load.

Physical contradictions help us understand how to solve problems by using both time and space to see clever solutions. The Separation Principles are very simple guides to solving problems. In some ways Physical Contradictions help us visualize the problem by separating out the conflicts between different times and different places. Simple mapping of a system in time or space steps (and visualizing its use at different times and places) gives us great clarity of understanding and helps us see obvious solutions.

Simply by seeing how to separate in time can help with all engineering problems.

At a recent in-house TRIZ course a simple fire-alarm control panel was described as having to be large and small. For fifteen years it had been made large. Why and when does it have to be large was asked? It had to be large when installed to allow the detector wires to be connected easily on site by someone unfamiliar with the device – but it would be better to be small when in use.

They had always designed it for installation rather than use. They realized they had a Physical Contradiction: they wanted the device to be large for installation but small when in use – so they could separate in time.

When they identified the relevant Inventive Principles this took them straight to a very simple (but to them innovative) solution of using hinged layers so it could be made large and therefore easy to install – but then be folded down to become small. They said the simple solution had saved them at least £300,000 and had been eluding them for many months.

Different team members provide different skills at different times and places

Separate in Space

Present in one place; absent in another.

We need to ask is there somewhere in space where the benefit is not needed?

We want a cup for hot drinks to be **hot** and cold:

cold where we hold it;

hot where it contains tea or coffee.

For separate in space we need to ask is there somewhere in space where hotness is not needed? Or where coldness is not needed?

What solutions do the relevant principles suggest?

Suggested Inventive Principles for Separate in Space

1 Segmentation
2 Taking Out
3 Local Quality
4 Asymmetry
7 Nested Doll
13 The Other Way Round
14 Spheroidality/Curvature
17 Another Dimension
24 Intermediary
26 Copying
30 Flexible Membranes and Thin Films
40 Composite Materials

1 Segmentation: Relevant layers?

40 Composite Materials: The right material for the cup?

4 Asymmetry: A handle?

7 Nested Doll: For thin plastic cups – another holder?

Similar questions can be asked when using a plate for hot food – does the rim ever need to be hot? It is a classic separate in space example – as we only need heat in the middle of the plate (**No 3 Local Quality**).

Another separate in space challenge is solved when submarines separate sonar detectors from their own noise by pulling sonar detectors at the end of several thousand feet of cable.

A very simple everyday example – a knife illustrates successful use of separate in space – sharp in one place and blunt in another.

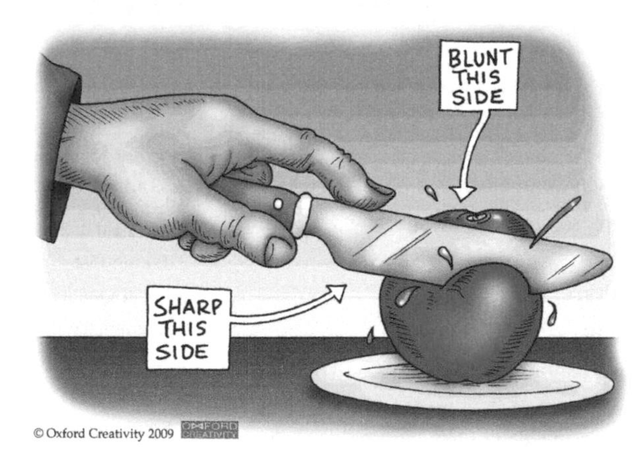

© Oxford Creativity 2009

Separate in space can also work for management problems:

1. Have a calmer team working in a different place from the problem (Houston/Apollo).
2. Experienced charity crisis teams at home – advising those on the ground at disaster situations.
3. Sales-people on road or at client don't need all the detailed information with them (just access).

All these simple management examples of separate in space look obvious but numbers 2 and 3 above were both devised by TRIZ teams – working on problems that seemed insurmountable at the beginning of the session but easily solvable at the end – an example of simple but clever solutions being found systematically by TRIZ.

Separate in time and space is used for bridges when we want them there for land traffic but not there when there is water traffic. Sometimes the permanent solutions are very simple but expensive such as two bridges at different heights (separate in space) but the separate in time bridges which change and move are rightly famous and appealing.

Separate on Condition

When we can't separate in time or space then we need to see how we can separate on condition when we need to separate opposite solutions in the same place and at the same time. The opposite solutions are achieved according to some condition of the component or some feature. The solution works on some elements but not others – one solution for one element; the opposite for another. For example a kitchen sieve is there for (collects) spaghetti but is not there for water (allows it to flow through).

The 'hit and miss' fence is there for privacy but not there for the wind (**31 Porous Materials**).

Another example, ideal windows are see-through for the people inside (give us a good view outside) *but* not see-through for people outside – they can't see in.

Separate on condition with music for older people which repels young people

Suggested Inventive Principles Separate on Condition

28 Replace Mechanical System

29 Pneumatics and Hydraulics

31 Porous Materials

32 Colour Changes

35 Parameter Change

36 Phase Transition

38 Accelerated Oxidation

39 Inert Atmosphere

A bogus 'beware of the bull' sign alarms strangers passing a remote house, but is understood and ignored by those living in the area who are familiar with the property.

Polar bears are black and white at the same time and in the same place. A polar bear is black for warmth and the same colour as its environment (white in snow) for camouflage.

Present/absent – water is soft if entered at a low speed. However, if one jumps into the same water from a height of 10 meters, the water feels considerably harder. Thus, the speed of the body's interaction with the water is the condition to be considered when applying this principle. Frothing up the water at the surface makes diving safer. A teabag also solves many contradictions and demonstrates separate on condition in that it has to allow water to flow in and out for the tea leaves to flavour the water but contain the tealeaves – keeping them separate from the water. Also it must contain the leaves when the teabag is thrown away.

One contradiction a teabag design has not yet solved is how to allow water in and out when in the cup, but keep the water completely contained when out of the cup to be thrown in the bin.

Separate by System

(also known as 'Alternative Ways')

If none of the above are appropriate to our problem then we need to find alternative ways of separating by system.

This includes:

- Separate in Scale
- Transition to Inverse System
- Transition to Alternative System – try a different system.

Separate in Scale

a) Transition to the Super-system – to achieve a particular benefit at the super-system even when we have the opposite benefit at the system level.

In areas at risk of earthquake, buildings have different natural oscillating frequencies. By connecting them with cables (**5 Merging**) they become one system and damp the

Suggested Inventive Principles for Separate in Scale

a) Super-system

5 Merging
6 Universality
12 Equipotentiality
22 Blessing in Disguise
33 Homogeneity
40 Composite Materials

b) Sub-system

1 Segmentation
3 Local Quality
24 Intermediary
27 Cheap Short Living Objects

vibrations between them. (In a complex dynamic system an approximately tuned damper may reduce the vibration intensity of a component not directly connected to the damper or even remotely located.)

b) Transition to the Sub-system – one value at the system level and the opposite value at the sub-system/component level

Bicycle chain is rigid at the subsystem but flexible at the system level.

Putting fires out – transition to subsystem – use a spray of water droplets rather than a large amount of water. What we want is water evaporation to take heat out of the fire. Water droplets give us a large surface area exposed to the heat, whereas water in bulk has lower surface area and can cause extensive water damage.

Transition to Inverse System

Try the opposite or anti-system. Try turning something 'the other way round' to solve the Physical Contradiction (**Inventive Principle No 13**).

Achieving all benefits by an inverse system applies to systems like wind tunnels or very small swimming pools with an artificial current for swimming. We stay still; the water moves. We can swim a long distance in a small pool.

Ha-ha – instead of a wall, have a ditch (there for the sheep; not there to spoil the view).

Suggested Inventive Principles

Inverse System
13 Other Way Around

Alternative System
6 Universality
8 Anti-Weight
22 Blessing in Disguise
27 Cheap Short Living Objects
25 Self-service
40 Composite Materials

Principle No 13 – The Other Way round

Exercising

I want to move (by running – good exercise).

I don't want to move (I'd rather be at home).

Solution: exercise machine.

Arranging a meeting

I want to meet with people in New York – I don't want to go to New York.

Solution: get the others to travel.

Transition to Alternative System – try a different system

Solve the Physical Contradiction by switching to a different system.

I'm a General. I want to visit troops at war to raise morale but I don't want to go to a war zone (too busy).

Solution: send a 'double' instead.

Physical Contradiction Examples

Fruit storage – we want air and we don't want air

When storing fruit, we want it to be fresh for as long as possible; carbon dioxide keeps it fresh but oxygen is harmful. Solutions for keeping fruit fresh include isolating it from oxygen and/or dousing it in an atmosphere of carbon dioxide. Air contains both oxygen and carbon dioxide so we want air and we don't want air.

We want to separate – we want the carbon dioxide near the fruit and the oxygen away from the fruit. Simply by stating the problem in these exact terms helps engineers come up with solutions.

Suggested solution Principles for Separate in Space

See what each principle suggests then seek further knowledge and resources to work out how to follow the suggested solution.

All these suggestions point to some kind of gas permeable film – some way of isolating the fruit from oxygen but not carbon dioxide – a search on the patent database reveals the existence of films through which oxygen can escape but carbon dioxide is retained.

Many food substances are packaged with carbon dioxide or another gas such as nitrogen used with crumpets.

This example has been in TRIZ literature for over 10 years. There are now patents applied for as late as 2006 – describing methods for optimal storing and transporting fruit, particularly bananas, and also to provide an appropriate atmosphere for ripening of the fruit. The fruit is placed within a suitably designed flexible bag, which allows the fruit to be open to the atmosphere, but is protected by a flexible cover, which is at least partially gas-permeable and separates the oxygen and carbon dioxide.

1 Segmentation – segment the air

2 Taking Out – take out the oxygen

3 Local Quality – have a layer near the fruit of just carbon dioxide

4 Asymmetry – encourage the air to be rich in carbon dioxide near the fruit and rich in oxygen away from it

7 Nested Doll – nest the gases

14 Spheroidality/Curvature – bubbles of carbon dioxide near the fruit?

17 Another Dimension – multi layers?

24 Intermediary – can we use an intermediary carrier of oxygen to remove it?

26 Copying – can we have some imitation fruit which protects it?

30 Flexible Membranes and Thin Films – isolate the fruit from the oxygen with a thin film

40 Composite Materials – find a material which is rich in carbon dioxide near the fruit

Finding Physical Contradictions

There is much description in the TRIZ literature about the challenge of locating physical contradictions. I find that the simplest approach is to begin by defining the Ideal Outcome to help map everything we want – all benefits – no matter how contradictory. Describing all the benefits we want within the ideal usually uncovers physical contradictions, as there are usually opposite benefits. To define how to get benefits we move into functions or features.

We may also begin our definition of a physical contradiction with a description of opposite features or functions. We may think we want something which is both small and big, or fast and slow, or sharp and blunt. Defining a physical contradiction can be by defining opposite benefits or features or functions.

Taking our problem description from features and functions back to benefits is generally useful as it gives us greater clarity of what we want and why. To check if it is a benefit – not a feature or function – ask if it contains a solution. Benefits do not describe how they are delivered, but simply what we want. Therefore when first seeking to define a physical contradiction it doesn't matter if we begin with benefits, features or functions.

For example we could describe physical contradictions in a toothbrush by opposite benefits, features or functions. Our ideal toothbrush has the opposite benefits of very effectively cleaning teeth but doesn't clean (hurt) gums. We could describe this as saying we want opposite features of a soft toothbrush and a hard toothbrush or that we want the opposite functions of removing material (removing plaque from teeth) and not removing material (from gums). All three of these give us the route into physical contradictions which can be solved with the Separation Principles and the 40 Principles.

This simple route to defining physical contradictions which can begin with an analysis of the problem, a definition of the ideal or a description of opposite features or functions is described in the physical contradiction map.

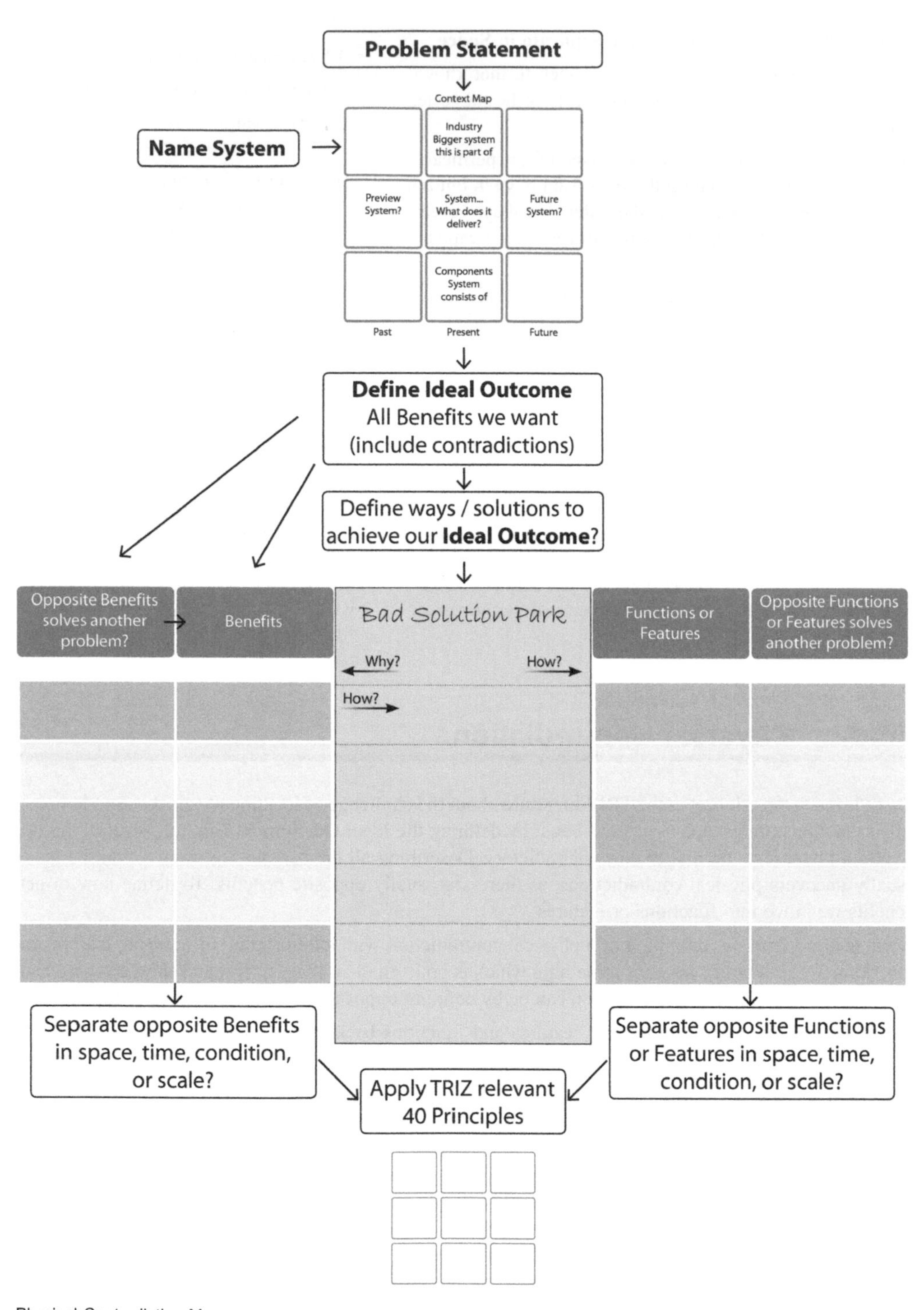

Physical Contradiction Map

Physical or Technical Contradiction

Use the Separation Principles or Matrix?

Design of a Board Pointer – we want it long and short. This is a Physical Contradiction (not covered by the contradiction matrix – we have a black/blank box if we map length against length) – we have to use the separation principles to solve a physical contradiction.

Do we want it long and short at the same time? No – then it is a Separate in Time problem.

How do we solve a Physical Contradiction of long and short (Separate in Time)?

We apply the principles which solve separating the requirements in: 1, 7, 9, 10, 11, 15, 16, 18, 19, 20, 21, 24, 26, 27, 29, 34, 37.

Principle No 7 – Nested Doll suggests a telescopic pointer

Using the Contradiction Matrix to Solve the Physical Contradiction of the Board Pointer

Could we devise a Technical Contradiction for the board pointer in order to use the matrix?

What do we want? Length. When? When pointing.

When don't we want it long? When transporting/ storing the pointer?

Is length really what we don't want? Or could we describe what we don't want as something not big? Perhaps what we don't want is a large volume or a large area rather than length.

It can then be re-defined as a Technical Contradiction, allowing us to use the 39 engineering Technical Parameters and the Contradiction Matrix.

This means defining our contradiction in terms that match the 39 Technical Parameters on the matrix.

So if we look at the Board Pointer and decide that we want it long then the Technical Parameter is very obviously 'length of stationary object'.

The Technical Parameter which gets worse could be area (or volume) of stationary object.

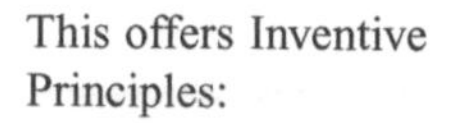

This offers Inventive Principles:

17 – Use another dimension

7 – Nested Doll

10 – Beforehand Action

40 – Composite Materials

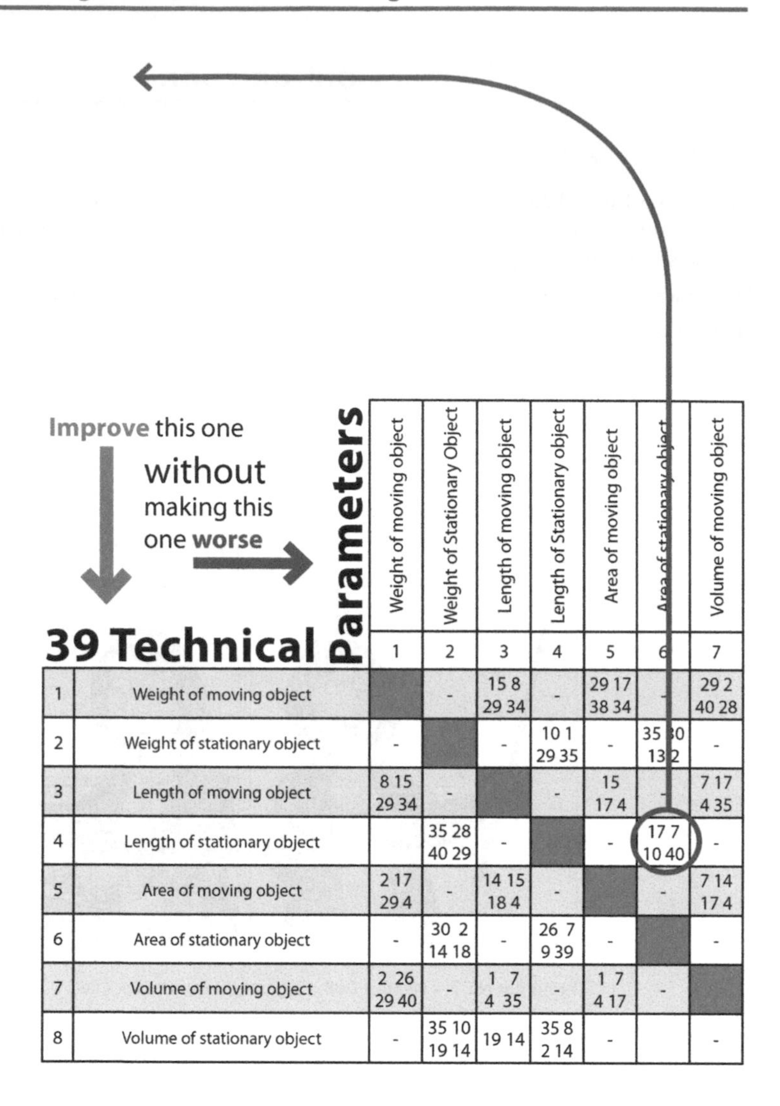

		1	2	3	4	5	6	7
		Weight of moving object	Weight of Stationary Object	Length of moving object	Length of Stationary object	Area of moving object	Area of stationary object	Volume of moving object
1	Weight of moving object		-	15 8 29 34	-	29 17 38 34	-	29 2 40 28
2	Weight of stationary object	-		-	10 1 29 35	-	35 30 13 2	-
3	Length of moving object	8 15 29 34	-		-	15 17 4	-	7 17 4 35
4	Length of stationary object		35 28 40 29	-		-	17 7 10 40	-
5	Area of moving object	2 17 29 4	-	14 15 18 4	-		-	7 14 17 4
6	Area of stationary object	-	30 2 14 18	-	26 7 9 39	-		-
7	Volume of moving object	2 26 29 40	-	1 7 4 35	-	1 7 4 17	-	
8	Volume of stationary object	-	35 10 19 14	19 14	35 8 2 14	-		-

Principle No 7 – Nested doll is suggested by both physical and technical contradiction methods, which offers a powerful prompt for solving this problem.

It is worth noting **Principle No 17 – Another dimension** is a solution idea for making something long and short such as shown by curly wire for telephones and electric razors.

Summary of Contradictions

Contradictions: whenever we invent, design or produce a system to give us outcomes we want then we also get outcomes we don't want.

There are two kinds of Contradictions – Physical Contradictions and Technical Contradictions.

These outcomes can be opposites (Physical Contradiction):

I want a big cup for cappuccino and a tiny cup for espresso;

or in conflict (I make something better and then something gets worse – Technical Contradiction):

I make my cup very thick and robust and the cup doesn't chip easily but it makes the coffee cold.

TRIZ Contradiction solving is the best known part of TRIZ and the easiest starting point for engineers. Most engineers recognize the power of the 40 Inventive Principles and the Contradiction Matrix and usually solve many problems with these tools. Once familiar with these tools they then often switch to using Physical Contradictions which often give more fundamental insights to the problem and faster solutions. All Technical Contradictions can be re-defined as a Physical Contradiction (and vice versa) which is normally a higher abstraction of the problem – a more general way of defining the contradiction.

All systems contain contradictions such as we would like a lightweight and extremely strong aircraft but the stronger we make it the heavier it becomes. The usual approach to solving this is to trade one characteristic off against the other – compromise. The TRIZ approach is to solve the contradiction – get strong and light weight. There are only 40 ways to solve a contradiction, based on all the patent analysis so far. These ways are known as the 40 Inventive Principles and they are a cornerstone of TRIZ.

Uncovering and solving contradictions requires careful thought but once we have uncovered contradictions TRIZ leads us to all the ways the world knows (40) to solve the contradictions. This leads us to good solutions where we forget compromise, and achieve all the things we want without getting any of the things we don't want. TRIZ Contradiction Solving helps all engineers both understand problems and find many good solutions because it helps them think clearly and gives simple, systematic methods for tackling any problem with conflicts.

Case Study: The Large and the Small of the Measurement of Acoustic Emissions in a Flying Aircraft Wing

From 7 years of Successful Problem Solving with TRIZ – Airbus and BAE Systems TRIZ their problems with Oxford Creativity

By Karen Gadd and Andrew Martin of Oxford Creativity

In 1999 in the Director's Box of the Reebok Stadium an historic three-day TRIZ course was organized to see if the TRIZ Tools and processes offered by Oxford Creativity would be useful and should be adopted. Those present were from both Airbus and BAE Systems; the organizers Bob Robinson and Pauline Marsh were so enthusiastic about TRIZ and they used it so effectively that they soon received a Chairman's Award for Innovation. This success and enthusiasm ensured that TRIZ was tested and tried in many other parts of both organizations.

TRIZ is now successfully applied in many places and in many areas in both organizations. TRIZ is an official and recognized route to innovation and problem solving as part of BAE Systems Engineering Life Cycle Management and anyone in BAE Systems can apply to do an Oxford Creativity TRIZ course through Xchanging. TRIZ has been used extensively for management and engineering problem solving.

Airbus have several TRIZ projects underway in the UK, France and Germany and are working with Oxford Creativity and Bath University to ensure TRIZ is more widely known and used by their engineers.

Below is the case study involving in flight test equipment.

Problem: The Measurement of Acoustic Emissions in a Flying Aircraft Wing

Problem Context

During flight testing of a commercial aircraft wing there is a requirement to measure acoustic emissions while the aircraft is in flight. The measurements are required to be very accurate; however, there are constraints imposed by the testing environment, which make the use of accurate sensors difficult. In particular, the test aircraft imposes a restriction on the weight and volume of the test equipment that can be used.

Before this problem was addressed using TRIZ, two possible approaches had been considered:

1. An electrical sensor was available. This was an excellent system that provided the required accuracy; however, it was too heavy to be used in flight.
2. A much lighter optical sensor was also available; however; this did not provide sufficient accuracy.

Discussion Of Problem

This is an excellent example of a non-trivial (and hence interesting) problem. There is a desirable output/characteristic that we need (in this case measurement accuracy) that unfortunately appears to be associated with something harmful, costly or unwanted (in this case, weight). In the language of TRIZ we call this a contradiction.

We can express this contradiction as a simple graph, which typically takes this form:

As we try to improve one parameter (accuracy) the other (weight) becomes worse, and vice-versa. It would appear that we are constrained to remain on the blue curve, which defines the set of possible compromise solutions available to us. What we would like to be able to do is to move off the blue compromise curve to a more ideal solution, i.e. in the direction shown by the green arrow. To achieve this we need to find a way of resolving the contradiction.

The identification and resolution of contradictions is a key element of the TRIZ problem-solving ethos. Inside all interesting problems we invariably find one or more such contradictions. If we can identify these contractions and resolve them we will have devised a high quality solution to our problem – not

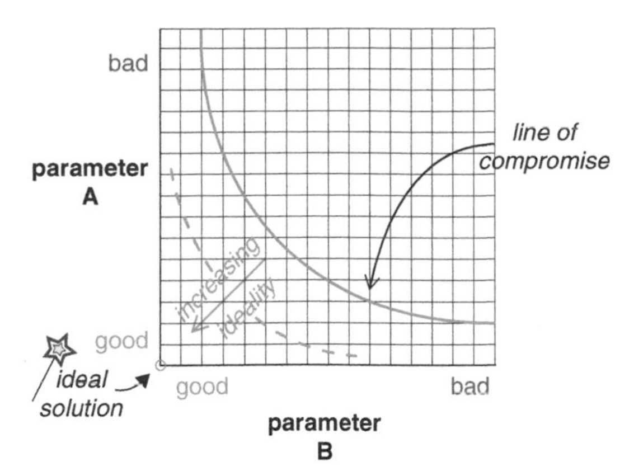

merely a better compromise, but an innovative solution that breaks free from the existing constraints and provides a step-change towards an ideal system.

In the case of this particular problem we are looking for:

A sensor solution that is accurate (like the existing electronic sensor, which is too heavy)

and

A sensor solution that is light (like the existing optical sensor, which is not accurate enough).

Our conceptual contradiction curve looks like this:

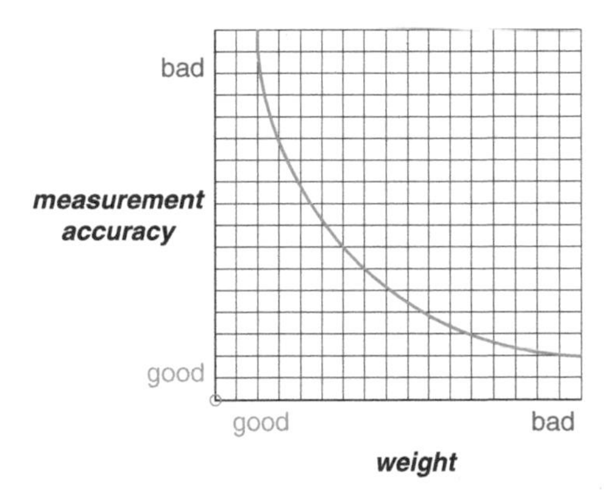

Ideally we want a sensor solution that has both good accuracy and good (which, in this case, means low) weight.

At first glance (if we stay on the blue curve) we appear to be asking for the impossible; however, it is important to suspend such judgements for the moment, to follow the systematic TRIZ method and allow the possibility that our particular contradiction(s) can be resolved – even if we have no idea of how it might be done. So we accept that it is possible that there is a solution that gives us the benefits we need without the associated costs or harms. With our minds open to the existence of such a solution, we can set out to find it.

Tackling the Problem

In this case, the Airbus/BAE Systems team decided to use one of the classic tools from the TRIZ toolkit for solving contradictions: the Contradiction Matrix.

As with much of TRIZ, this particular method provides a way of tapping in to the world's experience of problem solving – in this case the world's experience of resolving contradictions. The method is a three-step systematic process:

Step 1: Identify the contradictions in the problem, and classify them according to the nature of the contradictory system parameters.

Step 2: From the TRIZ toolkit, use a statistically derived look-up table (the Contradiction Matrix) to determine which generic Inventive Principles have been successfully used in the past to resolve contradictions of the same generic type as those in our specific problem.

Step 3: Take the generic Inventive Principles suggested in Step 2 and apply them to our specific problem. Through this final step it is possible to generate not just one possible solution, but usually a surprisingly large set of candidate conceptual solutions.

Step 1: Identifying the Contradictions

TRIZ deals with two types of contradictions: Technical Contradictions and Physical Contradictions. A Technical Contradiction is characterized by having different system parameters that constitute the contradiction, such as (in this case) measurement and accuracy. A Physical Contradiction, on the other hand, is characterized by having the contradiction derived from the same system parameter – for example an umbrella needs to be both large (when in use) and not large (when not in use).

In this case the BAE/Airbus team identified two Technical Contradictions.

Defining contradictions can be tricky and normally in TRIZ we recommend that you try several possibilities (don't agonize trying for the one perfectly defined contradiction):

We want a sensor that is *accurate* but we don't want it to be *heavy*.

For this contradiction we could ask why is it heavy. The answer may be to get enough power. This suggested the possibility of another type of contradiction in the problem:

We want a sensor that is accurate but we don't want it to require lots of power.

The Airbus/BAE team at Filton had identified both these Technical Contradictions:

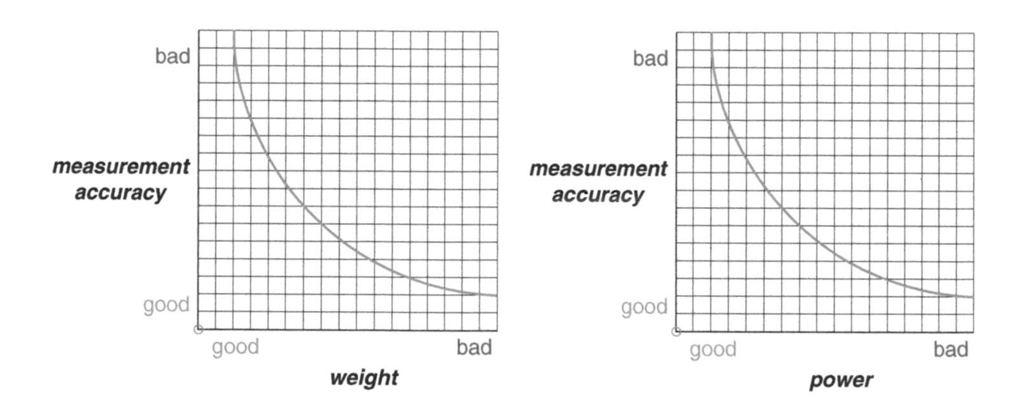

The Two Technical Contradictions expressed in terms of problem-domain parameters

Step 2: Using the Contradiction Matrix

Before the Contradiction Matrix can be used it is necessary to map the specific problem parameters (measurement accuracy, weight and power) onto the generic parameters used by the Contradiction Matrix.

For some problems this mapping of real-world problem-domain parameters onto the 39 parameters used by the matrix can be a little tricky – the matrix parameters are generic, and a certain amount of thought and careful consideration may be necessary to map the 'real world' problem parameters to the 'Matrix' generic parameters. In this particular case it was not very difficult to arrive at an appropriate mapping as:

Parameter 28 = Measurement accuracy matched well, as did

Parameter 1 = Weight of a moving object

and

Parameter 21 = Power

The full description of each of these parameters is:

1	Weight of moving object	The mass of the object in a gravitational field. The force that the body exerts on its support or suspension.
28	Measurement accuracy	The closeness of the measured value to the actual value of a property of a system. Reducing the error in a measurement increases the accuracy of the measurement.
21	Power	The time rate at which work is performed. The rate of use of energy.

Thus the two real-world contradictions, when mapped onto Matrix parameters become:

1. Measurement accuracy vs. Weight of moving object
2. Measurement accuracy vs. Power

Each of these contradictions are used with the Contradiction Matrix to obtain suggested Inventive Principles.

First Contradiction

Improving Parameter	*Worsening Parameter*
We want this to get better	but we don't want this to get worse
Measurement accuracy	Weight of moving object

These two parameters are used to cross-index the Contradiction Matrix to obtain the following four Inventive Principles that (statistically) have been found to be the most successful ways of obtaining better measurement accuracy without getting worse weight (for moving objects):

Inventive Principles

Measurement Accuracy vs. Weight of Moving Object

32 Colour Change

35 Parameter Change

26 Copying

28 Replace Mechanical System

Second Contradiction

Improving Parameter	*Worsening Parameter*
We want this:	but we don't want this:
Measurement accuracy	Power

(i.e. we don't want to have to provide lots of power)

In this case the Contradiction Matrix suggests the following three inventive principles:

Inventive Principles:

Measurement Accuracy vs. Power

3 Local Quality

6 Universality

32 Colour Change

Step 3: Apply the Suggested Inventive Principles

Having derived some suggested generic Inventive Principles, the team's next task was to apply them to their particular problem. This is part of a recurring theme in TRIZ: we take our specific problem, generalize it in order to access known generic solutions and then finally apply the general solution to our specific problem.

Principle 32 Colour Change

A – Change the colour of an object or its external environment

Use safe lights in a photographic darkroom

Use colour-changing thermal paint to measure temperature

Thermochromic plastic spoon

Temperature-sensitive dyes used on food product labels to indicate when desired serving temperature has been achieved

Electrochromic glass

Light-sensitive glasses

Camouflage

Dazzle camouflage used on World War I ships

Employ interference fringes on surface structures to change colour (as in butterfly wings, etc)

B – Change the transparency of an object or its external environment

Use photolithography to change transparent material to a solid mask for semiconductor processing

Light-sensitive glass

Smoke-screen

C – In order to improve observability of things that are difficult to see, use coloured additives or luminescent elements

Fluorescent additives used during UV spectroscopy

UV marker pens used to help identify stolen goods

Use opposing colours to increase visibility – e.g. butchers use green decoration to make the red in meat look redder

D – Change the emissivity properties of an object subject to radiant heating

Use of black and white coloured panels to assist thermal management on space vehicles

Use of parabolic reflectors in solar panels to increase energy capture

As **Inventive Principle 32 Colour Change** was suggested by the Contradiction Matrix twice, i.e. it was suggested by both of the two contradictions that were considered, the team gave particular attention to this as a solution trigger for their problem.

The next task for the team was therefore to understand the Colour Change Inventive Principle. This was important – the names of the 40 Inventive Principles are just convenient labels and it is important to appreciate the full definition of each principle before attempting to apply it. In the case of Colour Change, the definition (together with some examples) is:

At this point in the problem-solving process appropriate problem domain and technology domain knowledge is important.

The final solution that the team derived from this principle was to use of an electro-chromatic material to convert the signal from the electric sensor into a colour change that could be interrogated by an optical fibre. The team was able to quickly identify a suitable material that was already being used for adaptive camouflage applications.

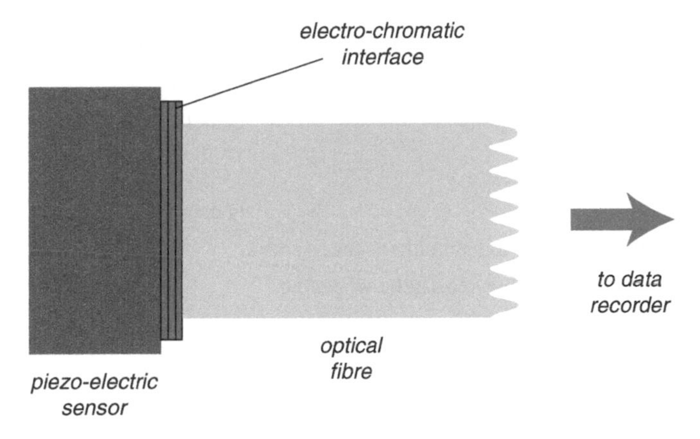

The final solution developed by the team

TRIZ pointed them quickly towards the right solution. Once we have identified the problems and uncovered the contradictions then we can use the TRIZ 40 Principles to help us locate all the relevant solutions. The unique power of TRIZ is that it includes such simple solution lists – all the freely available 40 ways/concepts the world knows to quickly solve problems containing contradictions.

Appendix 5.1
40 Principles: Theory of Inventive Problem Solving

1 Segmentation

A. Divide an object into independent parts:

- Different focal length lenses for a camera
- Multi-pin connectors
- Multiple pistons in an internal combustion engine
- Multi-engined aircraft
- Bullets/guns
- Compound eyes
- Somites
- Bullet proof glass – use many smaller sections together for a windscreen – so if bullet hits only a small section is damaged

B. Make an object sectional – easy to assemble or disassemble:

- Rapid-release fasteners for bicycle saddle/wheel/etc
- Quick disconnect joints in plumbing and hydraulic systems
- Single fastener V-band clamps on flange joints
- Loose-leaf paper in a ring binder
- Bailey bridge/bicycle chain

C. Increase the degree of fragmentation or segmentation:

- Multiple control surfaces on aerodynamic structures
- 16 and 24 valve versus 8 valve internal combustion engines

- Multi-zone combustion system
- Build up a component from layers (e.g. stereo-lithography, welds, etc)

2 Taking Out or Extraction

A. Extract the disturbing part or property from an object:

- Non-smoking areas in restaurants or in railway carriages
- Air conditioning in the room where you want it with the noise of the system outside the room
- (The contradiction here is noise vs coolness- the cooler it gets the noisier it gets- this solves the contradiction by putting the noise elsewhere)

B. Extract the only necessary part (or property) of an object

- Scarecrow
- Economy class on planes (travel but no frills)
- Sound of a barking dog (with no dog) as a burglar alarm
- Sealed windows with no openings (This involves understanding all the functionality and selecting only what you want- e.g. windows provide ventilation and light – with air conditioning you may not need windows which open)

3 Local Quality

A. Change of an object's structure from uniform to non-uniform

- Reduce drag on aerodynamic surfaces by adding riblets or 'shark-skin' protrusions
- Drink cans shaped to facilitate stable stacking
- Moulded hand grips on tools
- Material surface treatments/coatings – plating, erosion/corrosion protection, case hardening, non-stick, etc

B. Change an action or an external environment (or external influence) from uniform to non-uniform

- Introduce turbulent flow around an object to alter heat transfer properties
- Use a gradient instead of constant temperature, density, or pressure
- Take account of extremes of weather conditions when designing outdoor systems
- Strobe lighting

C. Make each part of an object function in conditions most suitable for its operation

- Freezer compartment in refrigerator
- Different zones in the combustion system of an engine
- Night-time adjustment on a rear- view mirror
- Lunch box with special compartments for hot and cold solid foods and for liquids

D. Make each part of an object fulfil a different and/or complementary useful function

- Swiss-Army knife
- Combined can and bottle opener
- Sharp and blunt end of a drawing pin
- Rubber on the end of a pencil
- Hammer with nail puller

4 Asymmetry

A. Change the shape or properties of an object from symmetrical to asymmetrical

- Asymmetrical funnel allows higher flow-rate than normal funnel
- Put a flat spot on a cylindrical shaft to attach a locking feature
- Oval and complex shaped O-rings
- Coated glass or paper
- Electric Plug
- Introduction of angled or scarfed geometry features on component edges
- Cutaway on a guitar improves access to high notes
- Spout of a jug
- Cam
- Ratchet
- Aerofoil – asymmetry generates lift.
- Eccentric drive
- Keys

B. Change the shape of an object to suit external asymmetries (e.g. ergonomic features)

- Human-shaped seating, etc
- Design for left and right handed users
- Finger and thumb grip features on objects
- Spectacles
- Car steering system compensates for camber in road
- Wing design compensated for asymmetric flow produced by propeller
- Turbomachinery design for boundary layer flows ('end-bend')

C. If an object is asymmetrical, increase its degree of asymmetry

- Use of variable control surfaces to alter lift properties of an aircraft wing
- Special connectors with complex shape/pin configurations to ensure correct assembly
- Introduction of several different measurement scales on a ruler

5 Merging/Consolidation

A. Bring closer together (or merge) identical or similar objects or operations in space

- Automatic rifle/machine gun
- Multi-colour ink cartridges
- Multi-blade razors
- Bi-focal lens spectacles
- Double/triple glazing
- Strips of staples
- Catamaran/trimara

B. Make objects or operations contiguous or parallel; bring them together in time

- Combine harvester
- Manufacture cells
- Grass collector on a lawn-mower
- Mixer taps
- Pipe-lined computer processors perform different stages in a calculation simultaneously
- Vector processors perform the same process on several sets of data in a single pass

Fourier analysis – integration of many sine curves

6 Universality

Make an object perform multiple functions; eliminate the need for other parts

- Child's car safety seat converts to a pushchair
- Home entertainment centre
- Swiss Army knife
- Grill in a microwave oven
- CD used as a storage medium for multiple data types
- Cleaning strip at beginning of a cassette tape cleans tape heads
- Cordless drill also acts as screwdriver, sander, polisher, etc

7 Nested Doll

A. Place one object inside another; place each object, in turn, inside the other

- Retractable aircraft under-carriage
- Voids in 3D structures
- Injected cavity-wall insulation
- Paint-brush attached to inside of lid of nail-varnish, etc.
- Lining inside a coat

B. Place multiple objects inside others

- Nested tables
- Telescope
- Measuring cups or spoons
- Stacking chairs
- Multi-layer erosion/corrosion coatings

C. Make one part pass (dynamically) through a cavity in the other

- Telescopic car aerial
- Retractable power-lead in vacuum cleaner
- Seat belt retraction mechanism
- Tape measure

8 Anti-weight

A. To compensate for the weight (downward tendency) of an object, merge it with other objects that provide lift

- Kayak with foam floats built into hull cannot sink
- Aerostatic aeroplane contains lighter-than-air pockets
- Hot air or helium balloon.
- Swim-bladder inside a fish
- Flymo cutting blade produces lift

B. To compensate for the weight (downward tendency) of an object, make it interact with the environment (e.g. use aerodynamic, hydrodynamic, buoyancy and or global lift forces)

- Vortex generators improve lift of aircraft wings
- Wing-in-ground effect aircraft
- Hydrofoils lift ship out of the water to reduce drag
- Make use of centrifugal forces in rotating systems (e.g. Watt governor)
- Maglev train uses magnetic repulsion to reduce friction

9 Prior Counteraction

A. When it is necessary to perform an action with both harmful and useful effects, this should be replaced with anti-actions to control harmful effects

- Make clay pigeons out of ice or dung – they just melt away
- Masking objects before harmful exposure: use a lead apron for X-rays
- use masking tape when painting difficult edges etc.
- Predict effects of signal distortion – compensate before transmitting
- Buffer a solution to prevent harm from extremes of Ph

B. Create beforehand stresses in an object that will oppose known undesirable working stresses later on

- Pre-stress rebar before pouring concrete
- Pre-stressed bolts
- Decompression chamber to prevent divers getting the bends

10 Prior Action

A. Perform the required change of an object in advance (either fully or partially)

- Pre-pasted wall paper
- Sterilize all instruments needed for a surgical procedure
- Self-adhesive stamps
- Holes cut before sheet-metal part formed
- Pre-impregnated carbon fibre reduces lay-up time and improves "wetting"
- Impregnated plasters
- Chilled cocktail glasses

B. Pre-arrange objects such that they can come into action from the most convenient place and without losing time for their delivery

- Manufacture flow-lines
- Pre-deposited blad
- Car jack, wheel brace, and spare tyre stored together
- Collect all the tools and materials for the job before

11 Cushion in Advance

Prepare emergency means beforehand to compensate for the relatively low reliability of an object ('belt and braces')

- Multi-channel control system
- Air-bag in a car /Spare wheel/Battery back-up/Back-up parachute
- Pressure relief valve
- Emergency lighting circuit
- Automatic save operations performed by computer programs
- Crash barriers on motorways
- 'Touch-down' bearing in magnetic bearing system

12 Equipotentiality

If an object has to be raised or lowered, redesign the object's environment so the need to raise or lower is eliminated or performed by the environment

- Canal locks
- Spring loaded parts delivery system in a factory
- Mechanic's pit in a garage means car does not have to be lifted
- Place a heavy object on ice, and let ice melt in order to lower it
- Angle-poise lamp; changes in gravitational potential stored in springs
- Descending cable cars balance the weight of ascending cars

13 The Other Way Around

A. Invert the action used to solve the problem, (e.g. instead of cooling an object, heat it)

- To loosen stuck parts, cool the inner part instead of heating the outer part
- Vacuum casting
- Rotary engines
- Test pressure vessel by varying pressure outside rather than inside
- Test seal on a liquid container by filling with pressurised air and immersing in liquid; trails of bubbles are easier to trace than slow liquid leaks
- Unstick frozen articles from each other by making them colder (rather than defrost)

B. Make movable parts (or the external environment) fixed, and fixed parts movable

- Hamster wheel
- Rowing jogging machines
- Rotate the part instead of the tool
- Wind tunnels
- Moving sidewalk with standing people
- Drive through restaurant or bank

C. Turn the object (or process) 'upside down'

- Tomato sauce bottle
- Clean bottles by inverting and injecting water from below
- Turn an assembly upside down to insert fasteners
- Garage pit

14 Spheroidality Curvature

A. Move from flat surfaces to spherical ones and from parts shaped as a cube (parallelepiped) to ball-shaped structures

- Use arches and domes for strength in architecture
- Introduce fillet radii between surfaces at different angles
- Introduce stress relieving holes at the ends of slots
- Change curvature on lens to alter light deflection properties

B. Use rollers, balls, spirals

- Spiral gear (Nautilus) produces continuous resistance for weight lifting
- Ball point and roller point pens for smooth ink distribution
- Use spherical casters instead of cylindrical wheels to move furniture
- Archimedes screw

C. Go from linear to rotary motion (or vice versa)

- Rotary actuators in hydraulic system
- Switch from reciprocating to rotary pump
- Push/pull versus rotary switches (e.g. lighting dimmer switch)
- Linear motors
- Linear versus rotating tracking arm on a record turntable ensures constant angle of stylus relative to groove
- Screw-thread versus nail

D. Use centrifugal forces

- Centrifugal casting for even wall thickness structures
- Spin components after painting to remove excess paint
- Remove water from clothes with a spin dryer rather than a mangle
- Separate chemicals with different density properties using a centrifuge
- Watt governor
- Vortex/cyclone separates different density objects

15 Dynamics

A. Change the object (or outside environment) for optimal performance at every stage of operation

- Foam or Shape changing mattress
- Gel fillings inside seat allow it to adapt to user
- Adjustable steering wheel (or seat, or back support, or mirror position…)
- Shape memory alloys/polymers
- Racing car suspension adjustable for different tracks and driving techniques
- Car handbrake adjustable to account for brake pad wear
- Telescopic curtain rail – "one size fits all"
- Easy to thread needle – eye opens for threading and closes again for use

B. Divide an object into parts capable of movement relative to each other

- Bifurcated bicycle saddle
- Articulated lorry
- Folding chair/mobile phone/laptop/etc
- Collapsible structures
- Brush seals

C. Change from immobile to mobile

- Bendy drinking straw
- Flexible joint
- Collapsible hose is flexible in use, and has additional flexibility of cross-section to make it easier to store

D. Increase the degree of free motion

- Use of different stiffness fibres in toothbrush – easily deflected at the edges to prevent gum damage, hard in the middle
- Flexible drive allows motion to be translated around bends
- Loose sand inside truck tyre gives it self-balancing properties at speed
- Add joints to robot arm to increase motion possibilities

16 Partial or Excessive Actions

If you can't achieve 100 percent of a desired effect – then go for more or less

- Over spray when painting, then remove excess
- Fill, then 'top off' when pouring a pint of Guinness
- Shrink wrapping process uses plastic deformation of wrapping to accommodate variations in vacuum pressure
- 'Roughing' and 'Finish' machining operations
- Over-fill holes with plaster and then rub back to smooth

17 Another Dimension

A. Move into an additional dimension – from one to two – from two to three

- Coiled telephone wire
- Curved bristles on a brush
- Pizza-box with ribbed (as opposed to flat) base
- Spiral staircase uses less floor area
- Introduction of down and up slopes between stations on railway reduces overall power requirements

B. Go single storey or layer to multi-storey or multi-layered

- Player with many CD's
- Stacked or multi-layered circuit boards
- multi-storey Car parks & office blocks-Stacking

C. Incline, tilt or re-orient the object, lay it on its side

- Cars on road transporter inclined to save space

D. Use 'another side' of a given area

- Press a groove onto both sides of a record
- Mount electronic components on both sides of a circuit board
- Print text around the rim of a coin
- Paper clip – works by pressing both sides of paper together

18 Mechanical Vibration

A. Cause an object to oscillate or vibrate

- Electric carving knife with vibrating blades
- Shake/stir paint to mix before applying
- Hammer drill
- Vibration exciter removes voids from poured concrete
- Vibrate during sieving operations to improve throughput
- Musical instrument
- Vibratory conveyor
- Mobile phones

B. Increase its frequency (even up to the ultrasonic)

- Dog-whistle (transmit sound outside human range)
- Ultrasonic cleaning
- Non-destructive crack detection using ultrasound

C. Use an object's resonant frequency

- Destroy gallstones or kidney stones using ultrasonic resonance
- Bottle cleaning by pulsing water jet at resonant frequency of bottles
- Tuning fork
- Increase action of a catalyst by vibrating it at its resonant frequency

D. Use piezoelectric vibrators instead of mechanical ones

- Quartz crystal oscillations drive high accuracy clocks
- Piezoelectric vibrators improve fluid atomisation from a spray nozzle
- Optical phase modulator

E. Use combined ultrasonic and electromagnetic field oscillations (Use external elements to create oscillation/vibration)

- Mixing alloys in an induction furnace
- Sono-chemistry
- Ultrasonic drying of films – combine ultrasonic with heat source

19 Periodic Action

A. Instead of continuous action, use periodic or pulsating actions

- Hitting something repeatedly with a hammer
- Pile drivers and hammer drills exert far more force for a given weight
- Replace continuous siren with a pulsed sound
- Pulsed bicycle lights make cyclist more noticeable to drivers
- Pulsed vacuum cleaner suction improves collection performance
- Pulsed water jet cutting
- ABS car braking systems

B. If an action is already periodic, change the periodic magnitude or frequency

- Improve a pulsed siren with changing amplitude and frequency
- Dots and dashes in Morse Code transmissions
- Use AM, FM, PWM to transmit information

C. Use pauses between actions to perform a different action

- Clean barrier filters by back-flushing them when not in use
- Inkjet printer cleans heads between passes
- Brush between suction pulses in vacuum cleaner
- Multiple conversations on the same telephone transmission line
- Use of energy storage means – e.g. batteries, fly-wheels, etc

20 Continuity of Useful Action

A. Carry on work without a break. All parts of an object operating constantly at full capacity

- Flywheel stores energy when a vehicle stops, so the motor can keep running at optimum power
- Constant output gas turbine in hybrid car, or APU in aircraft, runs at highest efficiency all the time it is switched on
- Constant speed/variable pitch propeller
- Self-tuning engine – constantly tunes itself to ensure maximum efficiency
- Heart paccmaker
- Improve composting process by continuously turning material
- Continuous glass or steel production

B. Eliminate all idle or intermittent motion

- Self-cleaning/self-emptying filter eliminates down-time
- Print during the return of a printer carriage – dot matrix printer, daisy wheel printers, inkjet printers
- Digital storage media allow 'instant' information access
- Kayaks use double-ended paddle to utilise recovery stroke
- Computer operating systems utilise idle periods to perform necessary housekeeping tasks

21 Rushing Through

Conduct a process, or certain stages of it (e.g. destructible, harmful or hazardous operations) at high speed

- Use a high speed dentist's drill to avoid heating tissue
- Cut plastic faster than heat can propagate in the material, to avoid deforming the shape
- Shatter toffee with a hammer blow
- Drop forge
- Flash photography
- Super-critical shaft – run through resonant modes quickly
- Bikini waxing (ouch!)

22 Blessing in Disguise

A. Harm into benefit – Use harmful factors (particularly, harmful effects of the environment or surroundings) to achieve a positive effect

- Use waste heat to generate electric power
- Recycle scrap material as raw materials for another – e.g. chipboard
- Vaccination
- Lower body temperature to slow metabolism during operations
- Composting
- Use centrifugal energy in rotating shaft to do something useful – e.g. seal, or modulate cooling air
- Use pressure differences to help rather than hinder seal performance

B. Eliminate the primary harmful action by adding it to another harmful action to resolve the problem

- Add a buffering material to a corrosive solution (e.g. an alkali to an acid, or vice versa)
- Use a helium-oxygen mix for diving, to eliminate both nitrogen narcosis and oxygen poisoning from air and other nitrox mixes
- Use gamma rays to detect positron emissions from explosives

C. Amplify a harmful factor to such a degree that it is no longer harmful

- Use a backfire to eliminate the fuel from a forest fire
- Use explosives to blow out an oil-well fire
- Laser-knife cauterises skin/blood vessels as it cuts

23 Feedback

A. Introduce feedback to improve a process or action

- Automatic volume control in audio circuits
- Signal from gyrocompass is used to control simple aircraft autopilots
- Engine management system based on exhaust gas levels is more efficient than carburettor
- Thermostat controls temperature accurately
- Statistical Process Control – measurements are used to decide when to modify a process
- Feedback turns inaccurate op-amp into useable accurate amplifier

B. If feedback is already used, change its magnitude or influence in accordance with operating conditions

- Change sensitivity of an autopilot when within 5 miles of an airport
- Change sensitivity of a thermostat when cooling vs. heating, since it uses energy less efficiently when cooling
- Use proportional, integral and/or differential control algorithm combinations

24 Intermediary/Mediator

A. Use an intermediary carrier article or intermediary process

- Play a guitar with a plectrum
- Use a chisel to control rock breaking/sculpting process
- Dwell period during a manufacture process operation

Eradicate toxic cane toads (who were introduced to Sydney, Australia in 1935 from Hawaii and have defied all attempts to catch them)

Trap the toads using ultraviolet disco lights. The lights attract insects which attract the toads)

B. Merge one object temporarily with another (which can be easily removed)

- Gloves to get hot dishes out of an oven
- Joining papers with a paper clip
- Introduction of catalysts into chemical reaction
- Abrasive particles enhance water jet cutting
- Bouquet garni in cooking

25 Self-Service

A. An object must service itself by performing auxiliary helpful functions

- Halogen lamps regenerate the filament – evaporated material is re-deposited
- Self-aligning/self-adjusting seal
- Self-cleaning oven/glass/material
- Self winding watch + Watch face self heating
- Odour eater
- Anti bacterial Tupperware
- Distributor
- Lizard's tail
- Software patches -Anti- virus
- Atomic clock
- Auto Headlamp levelling
- Self Loading rifle
- Tyres under golf course
- Dissolvable stitches
- Regenerative track steering
- Trees photosynthesis
- Parthenogenesis
- Self pollinating plants
- Thru colour panels on Saturn cars
- Penguins feet (stay frozen)
- Rock boots –wear away in a way that maintains grip
- Customer data
- Wikipedia
- Self righting boats
- Bone
- FLAP valve
- Dynomo on bike
- Regenerative braking
- Self closing door
- Condensing boiler

B. Use waste resources, energy, or substances

- Use heat from a process to generate electricity: Co-generation
- Use animal waste as fertilizer
- Use food and lawn waste to create compost
- Use pressure difference to reinforce seal action

26 Copying

A. Replace unavailable, expensive, fragile object with inexpensive copies

- Imitation jewellery
- Astroturf
- Crash test dummy
- Transfers instead of tattoos

B. Replace an object, or process with optical copies

- Do surveying from space photographs instead of on the ground
- Measure an object by scaling measurements from a photograph
- Virtual reality/Virtual mock-ups electronic pre-assembly modelling

C. If visible optical copies are used, move to infrared or ultraviolet copies

- Make images in infrared to detect heat sources, such as diseases in crops, or intruders in a security system
- Use UV as a non-destructive crack detection method
- UV light used to attract flying insects into trap

27 Cheap Short-Living Objects

Replace an expensive object with a multiple of inexpensive objects, compromising certain qualities (such as service life, for instance)

- Cheaper tools – screwdrivers
- Disposable serviettes, nappies paper cups/plates/forks/cameras/ torches /etc
- Throw-away cigarette lighters
- Matches versus lighters
- Sacrificial coatings/components

28 Replace Mechanical System

A. Replace a mechanical system with a sensory one

- Reversing sensors on cars vs. bumbers
- Natural Gas odorisation
- Replace a physical barrier with an acoustic one (audible to animals)
- Add a bad smell to natural gas to alert users to leaks
- Finger-print/retina/etc scan instead of a key
- Voice activated telephone dialling
- Movement sensors replace light switches
- Fire alarms: triggered by smoke/ heat
- Retinal recognition for security
- Voice recognition software instead of typing
- Traffic lights instead of traffic policeman
- Lights and bells rather than secure barriers at rail crossings
- Baby alarms
- Security Systems

B. Use electric, magnetic and electromagnetic fields to interact with the object

- To mix 2 powders, electrostatically charge one positive and the other negative
- Electrostatic precipitators separate particles from airflow
- Improve efficiency of paint-spraying by oppositely charging paint droplets and object to be painted
- Pick up tools
- Magnetic bearings
- Field activated switches
- Boot lock replaced by electrical coil
- TV remote
- Maglev over wheels/rails

C. Replace stationary fields with moving, unstructured fields with structured

- Early communications used omni-directional broadcasting. We now use antennas with very detailed structure of the pattern of radiation
- Magnetic Resonance Imaging (MRI) scanner

D. Use fields in conjunction with field-activated (e.g. ferromagnetic) particles

- Heat a substance containing ferromagnetic material by using varying magnetic field. When the temperature exceeds the Curie point, the material becomes paramagnetic, and no longer absorbs heat
- Ferro-magnetic catalysts
- Magneto-rheological effect – uses ferromagnetic particles and variable magnetic field to alter the viscosity of a fluid
- Ferro-fluids – e.g. Magnatec oil – stay attached to surfaces requiring lubrication
- Ultrasonic sealing (medical/plastic bags)

29 Pneumatics and Hydraulics

Use gas and liquid parts of an object instead of solid parts (e.g. inflatable, filled with liquids, air cushion, hydrostatic, hydro-reactive)

- Transition from mechanical to hydraulic or pneumatic drive
- Inflatable furniture/mattress/etc.
- Gel filled saddle adapts to user
- Hollow section O-rings
- Hovercraft
- Gas bearings
- Acoustic panels incorporating Helmholtz resonators

30 Flexible Shells & Thin Films

A. Use flexible shells and thin films instead of three-dimensional structures

- Use inflatable (thin film) structures
- Taut-liner trucks
- Tarpaulin car cover instead of garage
- Store energy in stretchable bags – accumulators in a hydraulic system

B. Isolate the object from its external environment using flexible membranes

- Bubble-wrap
- Bandages/plasters
- Tea bag
- Heat curtain instead of solid door
- Segmented plastic vertical strips used in refrigeration and hospitals

31 Porous Materials

A. Make an object porous or add porous elements (inserts, coatings, etc.)

- Drill holes in a structure to reduce the weight
- Cavity wall insulation
- Transpiration film cooled structures
- Foam metals
- Use sponge-like structures as fluid absorption media

THE ADVANTAGE OF THIS PANEL IS THAT IT KEEPS YOUR PRIVACY BUT LETS THE WIND THROUGH...
© Oxford Creativity 2008

B. If an object is already porous, use the pores to introduce a useful substance or function

- Use a porous metal mesh to wick excess solder away from a joint
- Store hydrogen in the pores of a palladium sponge. (Fuel "tank" for the hydrogen car – much safer than storing hydrogen gas)
- Desiccant in polystyrene packing materials
- Medicated swabs/dressings

32 Colour Changes

A. Change the colour of an object or its external environment

- Colour-changing paint or suncream
- Temperature-sensitive dyes used on food product labels to indicate when desired serving temperature has been achieved
- Electrochromic glass
- Light-sensitive glasses
- Camouflage
- Employ interference fringes on surface structures to change colour (as in butterfly wings, etc)

B. Change the transparency of an object or its external environment

- Photolithography used to change transparent material to a solid mask for semiconductor processing
- Light-sensitive glass

C. In order to improve observability of things that are difficult to see, use coloured additives or luminescent elements

- Fluorescent additives used during UV spectroscopy
- UV marker pens used to help identify stolen goods
- Use opposing colours to increase visibility – e.g. butchers use green decoration to make the red in meat look redder

D. Change the emissivity properties of an object subject to radiant heating

- Use of black and white coloured panels to assist thermal management on space vehicles
- Use of parabolic reflectors in solar panels to increase energy capture
- Paint object with high emissivity paint in order to be able to measure its temperature with a calibrated thermal imager

33 Homogeneity

Objects interacting with the main object should be of same material (or material with identical properties)

- Gin ice cubes – Make ice-cubes out of the same fluid as the drink they are intended to cool
- Food containers made of edible substances – ice cream cone
- Container made of the same material as its contents, to reduce chemical reactions
- Friction welding requires no intermediary material between the two surfaces to be joined
- Temporary plant pots made out of compostable material
- Human blood transfusions/transplants, use of bio-compatible materials
- Join wooden components using (wood) dowel joints

34 Discarding and Recovering

A. Objects (or part of them) disappear after completing their useful function or becoming useless (discard them by dissolving, evaporating, etc) or modify them directly during the process

- Dissolving capsule for vitamins or medicine
- Bio-degradable containers, bags, etc.
- Casting processes – lost-wax, sand, etc.
- Female spiders eat the male after mating
- During firing of a rocket, foam protection is used on some elements; this evaporates in space when shock-absorbance is no longer required

B. Restore consumable/used up parts of an object during operation

- Self-sharpening blades – knives/lawn-mowers etc.
- Strimmer dispenses more wire automatically after a breakage
- Self-tuning automobile engines
- Propelling pencil
- Automatic rifle

35 Parameter Changes

A. Change the physical state (e.g. to a gas, liquid, or solid)
  - Transport oxygen or nitrogen or petroleum gas as a liquid, instead of a gas, to reduce volume

B. Change the concentration or density
  - Liquid soap
  - Abradable linings used for gas-turbine engine seals

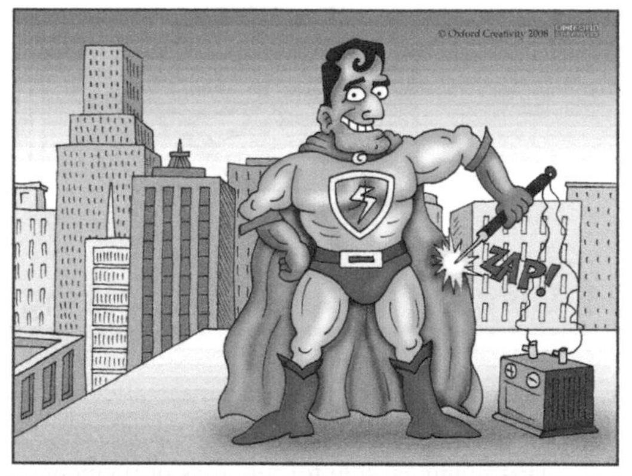

C. Change the degree of flexibility
  - Vulcanize rubber to change its flexibility and durability
  - Compliant brush seals rather than labyrinth or fixed geometry seals

D. Change the temperature or volume
  - Raise the temperature above the Curie point to change a ferromagnetic substance to a paramagnetic substance
  - Cooking/baking/etc.

E. Change the pressure
  - Pressure cooker cooks more quickly and without losing flavours
  - Electron beam welding in a vacuum
  - Vacuum packing of perishable goods

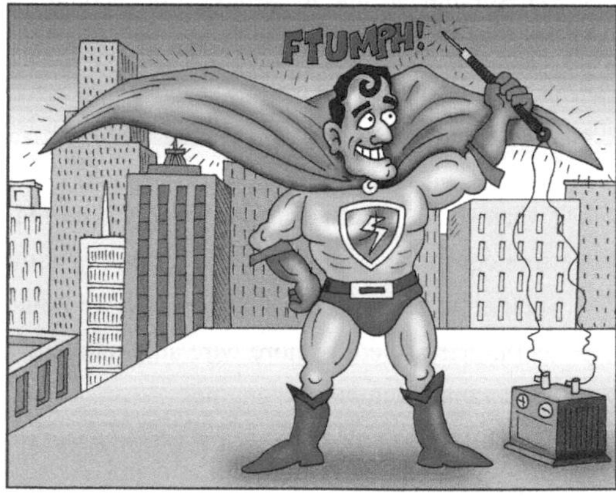

F. Change other parameters
  - Shape memory alloys/polymers
  - Use Curie point to alter magnetic properties
  - Thixotropic paints/gels/etc
  - Use high conductivity materials – e.g. carbon fibre

36 Phase Transitions

Use phenomena occurring during phase transitions. (e.g. volume changes, loss of absorption of heat etc.)

- Latent heat effects in melting/boiling
- Soak rocks in water, then freezing causes water to expand – thus opening fissures in rock, making it easier to break
- Heat pumps use the heat of vaporization and heat of condensation of a closed thermodynamic cycle to do useful work
- Volume expansion during water-to-steam transition
- Superconductivity

37 Thermal Expansion

A. Use thermal expansion (or contraction) of materials

- Fit a tight joint together by cooling the inner part to contract, heating the outer part to expand, putting the joint together, and returning to equilibrium
- Metal tie-bars used to straighten buckling walls on old buildings
- Thermal switch/cut-out
- Shape memory alloys/polymers
- Shrink-wrapping

B. Use multiple materials with different coefficients of thermal expansion

- Bi-metallic strips used for thermostats, etc
- Two-way shape memory alloys
- Passive blade tip clearance control in gas turbine engines
- Combine materials with positive and negative thermal expansion coefficients to obtain alloys with zero (or specifically tailored) expansion properties – e.g. cerro-tru alloy used in the mounting and location of fragile turbine blade components during manufacture operations

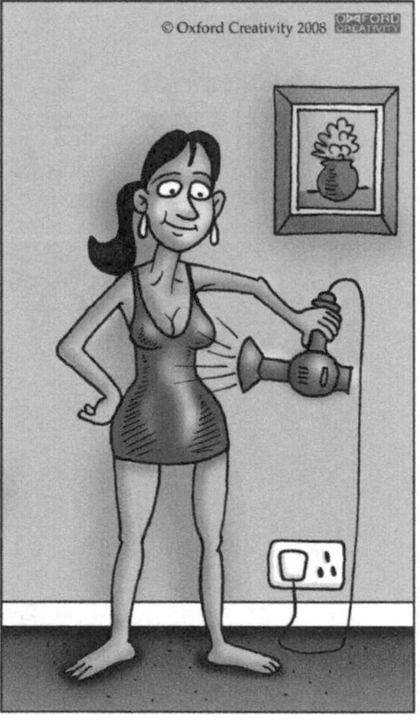

38 Accelerated Oxidation

A. Replace common air with oxygen-enriched air

- Scuba diving with Nitrox or other non-air mixtures for extended endurance
- Place asthmatic patients in oxygen tent
- Nitrous oxide injection to provide power boost in high performance engines

B. Replace enriched air with pure oxygen

- Cut at a higher temperature using an oxy-acetylene torch
- Treat wounds in a high pressure oxygen environment to kill anaerobic bacteria and aid healing
- Control oxidation reactions more effectively by reacting in pure oxygen

C. Expose air or oxygen to ionising radiation

- Irradiation of food to extend shelf life
- Use ionised air to destroy bacteria and sterilise food

D. Use ionised oxygen

- Speed up chemical reactions by ionising the gas before use
- Separate oxygen from a mixed gas by ionising the oxygen

E. Replace ozonised (or ionised) oxygen with ozone

- Oxidisation of metals in bleaching solutions to reduce cost relative to hydrogen peroxide
- Use ozone to destroy micro-organisms and toxins in corn
- Ozone dissolved in water removes organic contaminants from ship hulls

39 Inert Atmosphere

A. Replace a normal environment with an inert one

- Prevent degradation of a hot metal filament by using an argon atmosphere
- MIG/TIG welding
- Electron beam welding conducted in a vacuum
- Vacuum packaging
- Foam to separate a fire from oxygen in air

B. Add neutral parts, or inert additives to an object

- Naval aviation fuel contains additives to alter flash point
- Add fire retardant elements to titanium to reduce possibility of titanium fire
- Add foam to absorb sound vibrations – e.g. hi-fi speakers

40 Composite Materials

Change from uniform to layered/composite (multiple) structures

- Aircraft structures where low weight and high strength are required
- Composites in golf club shaft
- Concrete aggregate
- Glass-reinforced plastic
- Fibre-reinforced ceramics
- Hard/soft/hard multi-layer coatings to improve erosion properties

Part Three
Fast Thinking with the TRIZ Ideal Outcome

The Ideal Solves the Problem

Simple Steps to Fast Resourceful Systematic Problem Solving

The Ideal describes the perfect state, a perfect result; imagining the Ideal Outcome is a powerful tool for helping us understand the real problem and everything we want and are seeking when problem solving. The Ideal in TRIZ also points us in the right direction for finding all good solutions, breaking any psychological inertia (for ourselves and everyone else) which may fixate us on one poor solution and blocks mental access to the myriad of other possible solutions.

Seeking descriptions of the Ideal is a stimulating first step (or one of the early stages) for problem solving. Attempting to define the Ideal stimulates ideas, helps everyone think clearly, move in the right direction and focus on the desired end point and outcomes. Whatever problem we are tackling if we begin by imagining the Ideal version of the thing we want (by simply affixing 'Ideal' to its name) then we get quick understanding of the best possible outcomes. This applies to a whole range of problem challenges such as Ideal solution, Ideal client, Ideal customer, Ideal machine, Ideal process, Ideal attitude, Ideal price or Ideal outcome (or whatever is relevant to the situation we are working on) – asking everyone to focus on an extreme and perfect end point helps both problem understanding and problem solving.

The Ideal as a Concept Has Four Major Roles in TRIZ Problem Solving

1. **Finding solutions** – using the TRIZ prompt 'the Ideal solves the problem by itself'.
2. **Locating useful Resources** – the Ideal helps us seek, identify, recognize and mobilize free and available resources. (see Chapter Seven)
3. **Breaking Psychological inertia** and subjugating constraints.
4. **Understanding and accurately defining what we *really* want (the Ideality Audit – Chapter 8).**

TRIZ for Engineers: Enabling Inventive Problem Solving, First Edition. Karen Gadd.

The Ideal is a very simple but effective aid to clear thought and has many uses and descriptions in TRIZ technical problem solving: the terms Ideal Outcome, Ideal Final Result, Ideal Solution, Ideal System, Ideal Machine, Ideal Process, Ideal User Manual, Ideal Resource, Ideal Design, Ideal Substance and Ideal X Factor etc. all appear within the TRIZ literature. These terms often overlap but are not interchangeable as they cover the different levels in problem solving, from ultimate goals, to benefits, to functions, components and resources etc. As such they are not all the same, but by defining an ideal they are all simple pointers to help us define what we really want at each level and often suggest to us ways of how to get it. The Ideal Outcome gives us a quick understanding of everything we want – benefits, without solutions. Moving to the definition of the Ideal Solution then delivers answers and ideas of how we obtain all those benefits.

The Ideal Outcome is the ultimate Ideality with all and only benefits delivered (with all benefits sufficient and only present when required) and in addition needing no inputs (costs) and also without any outputs we don't want or problems (harms) of any kind. Moving towards an Ideal Outcome with only benefits is a theoretical end point but very important in TRIZ problem understanding and solving.

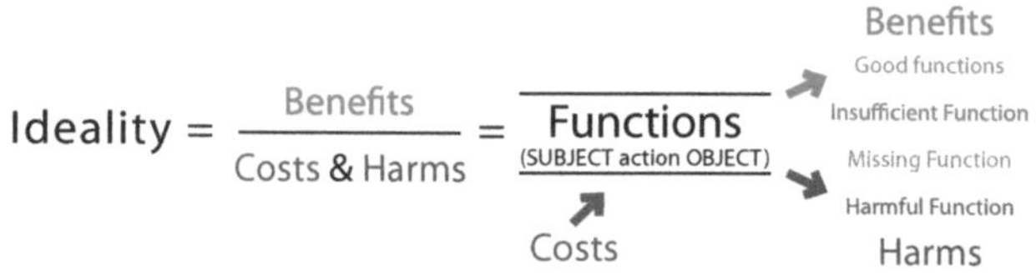

$$\text{Ideality} = \frac{\text{Benefits}}{\text{Costs \& Harms}} = \frac{\text{Functions}}{\text{(SUBJECT action OBJECT)}}$$

Benefits
Good functions
Insufficient Function
Missing Function
Harmful Function
Harms
Costs

Benefits are delivered by functions and the Ideal has only good functions, and all the functions we want. Functions are delivered by resources so by defining the Ideal we are focussing on the only functions we want and can therefore narrow down our resource hunt fairly quickly and effectively. Defining the Ideal is a very fast form of problem solving – it helps us recognize essential needs and helps us locate the functions and the resources to deliver them.

By forcing ourselves to think of the extreme point of view of the 'Ideal', we also break out of our psychological inertia, as we refuse to compromise or accept 'pragmatic' solutions. This is particularly important in problems that we are very familiar with: when we can no longer see what we don't see. Using the Ideal helps give us a new perspective on the problem and come at it with fresh eyes.

System We Want – the Acceptable Ideality

Much of normal, everyday problem solving involves getting the right functions (and getting the functions right) to deliver acceptable Ideality. We want to improve the 'system we've got' and transform it to the 'system we want' which delivers more benefits (but usually not all our needs), with acceptable costs (inputs) which we are prepared to invest into the system, and inevitably some harms (problems) which we are prepared to tolerate in this definition of acceptable as opposed to perfect.

Problem solving with TRIZ helps teams aspire to reach and if possible get beyond this acceptable Ideality – for better or higher Ideality – which means aiming to achieve all the benefits we want *and* fewer costs *and* fewer harms. Taken to extremes the ultimate Ideality is the *Ideal* with *only* benefits with no costs and harms. This is the direction of all TRIZ problem-solving – towards this Ideal – a theoretical destination where the ultimate goal is achieved. This powerful and simple concept shows us the right direction our problem solving should take towards this final place. In this chapter we explore the clever TRIZ short-cut problem-solving routes achieved by defining the Ideal and then using this as our starting point to search for solutions as close to it as possible from available resources.

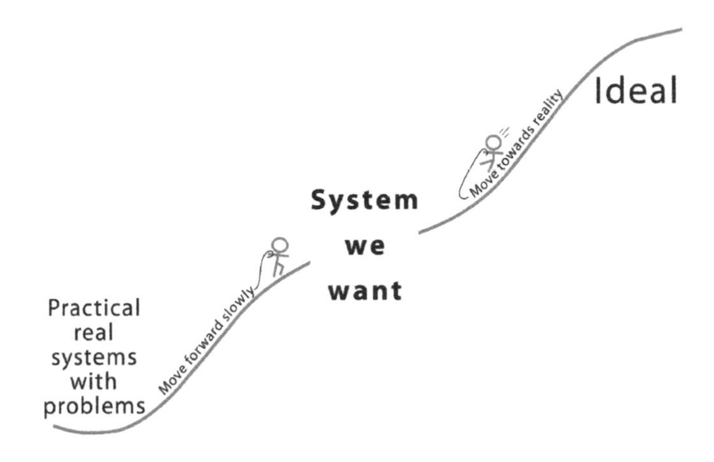

To many new TRIZ users the important central place given to the concept of the Ideal, can be puzzling. It is such a simple tool which takes a few moments to understand and yet it is so central to TRIZ problem solving and problem understanding. The Ideal is a great thinking tool, so fundamental to TRIZ, and yet so easy and powerful. This is because in the complicated and detailed routes to problem solutions the Ideal is like a lodestar or compass, and keeps us pointing in the right direction – never losing sight of the final destination. Altshuller who described it as the Ideal Imaginary Solution or Ideal Final Result (IFR) said, 'The IFR can be likened to a rope, which a mountain climber clutches to ascend a steep incline. The rope is not being pulled upwards, but it provides support and prevents sliding backwards. You only have to let go of the rope and you inevitably fall ...' *CAAES* p.97.

Ideal – Solves the Problem Itself

In TRIZ this is a very simple and powerful way to locate solutions by stating, 'The Ideal Solution solves the problem by itself' and look for ways of achieving it. This is particularly useful when there is no obvious system or answer and stretches our thinking.

Starting with the imaginary Ideal Outcome and Solution, which first defines and suggests ways to deliver all the benefits we want, may sound wacky and off the wall to hard-nosed, practical engineers but it is the most effective and simple tool successfully used in TRIZ problem solving. Starting an engineering problem-solving session by asking everyone involved to define their Ideal Outcome (all the benefits they want both big and small) and then brainstorming (either with or without TRIZ solution tools) for systems to deliver these, inevitably delivers fast, powerful results and high-octane thinking from everyone involved. The TRIZ trends tell us that all systems move towards self systems as they become more perfect and evolve – and if we look at our most necessary and simple historical inventions they often are some kind of ingenious Ideal self system.

Traffic Control Systems – Ideal Self systems

In the latter part of the nineteenth century technology advances were bringing big changes and as systems evolved and cars replaced horses, then road systems (super-system) had to adapt to harmonize with the new sub-systems (engines need fuel, tyres need tarmac etc.) and the environment had to adapt to the new cars, bicycles, fire engines etc. Traffic management was needed and everyone was looking for traffic control – preferably for automated self systems to cover all times and places.

As cars, traffic and road systems developed it became obvious that accidents, particularly at crossroads, might occur and needed good solutions. One solution employed by dare-devil young male drivers (particularly pilots apparently) employed the strange logic that as unmarked cross roads were a danger, it was logical to spend as little time as possible passing through them, and they would race through them at top speed hoping that no other vehicles were also there. (This is applying both TRIZ Principles 21 Rushing Through and 13 Other Way Round – go faster not slower.) This is clearly not an Ideal Solution.

A practical early solution was to put a policeman in the middle of the busier crossroads to direct the traffic, but this was not popular with either drivers or policemen and was an occasional but not a constant solution. Following the TRIZ Trend on Self Systems (less human involvement) the next step was to replace the policeman with a machine. This was invented by Garrett Morgan and patented in 1923, and like most first attempts to replace the human with a machine needlessly replicated human actions, as it had arms and signs saying stop and go. This had been preceded by two notable inventions still in use today. The first borrowed a system which already existed (for analogy ask, does this already exist – is it a life and death problem elsewhere?) and the analogous system was the railway system which already had traffic lights and had already developed the red, amber and green system. This was first adapted to be used on London roads in 1868 with gas lamps. The second is the roundabout which first appeared in 1903 in the UK in Letchworth. Roundabouts are closer to self systems, and an interesting development was the eventual emergence of very big roundabouts and very small mini-roundabouts. It is said that, once accepted and understood, the mini-roundabout has very high Ideality and is close to an Ideal Solution – nearly all the benefits of other traffic control systems, very low costs of installation and maintenance, and vey few harms and problems – the traffic controls itself at a road interchange.

Ideal Solutions are the basis for many self systems – such as self-repairing systems (surfaces, plastics and other materials, bolted joints, paint on cars, tyres, fuel tanks, etc.) and self-closing drawers, self-adjusting wings on aircraft, self-emptying containers, and a myriad of others. This is very simple and powerful route to moving from the small improvements of a faulty system to defining near 'perfect' solutions which stimulates clever engineering thoughts and ideas – and then moving back from the Ideal to practical reality. In problem-solving sessions it works almost every time and creates many clever directions for solutions to real and messy which problems.

Simple steps include:

- Imagine an Ideal Outcome. (If we had a magic wand what benefits would our total solution deliver?)
- Brainstorm Ideal Solutions which deliver all these benefits.
- Step back from these 'Ideal wacky solutions' towards real ones. This process can be helped by applying the simple TRIZ creativity triggers which help us use the Ideal to find solutions. These include looking for existing life and death analogies (Is there another industry or area where solutions exist, particularly if they are life saving and deal with similar but critical situations?) and try Principle 13 – the Other Way Round to find good self systems/ solutions.

Successful Self Systems often demonstrate the power of the Other Way Round

Define the Ideal – and Then Find the Resources to Create It

Most clever solutions use available resources in an innovative way. The extreme destination, the *Ideal Outcome* has only benefits and uses no inputs; but if we take just one step back towards reality the next best option is that no new or extra inputs are needed for something close to an *Ideal Solution* which is created by using available and free resources. This means that the solution has cost nothing as no new resources are added but it has solved the problem.

Quick Ideal Problem Solving

1. **Define Ideal Outcome** (quick understanding of everything we want – benefits – no solutions). Imagine achieving an Ideal by waving a magic wand which gives us everything we want.
2. **Seek ideas for Ideal Solutions to solve the problem.** Brainstorm Ideal Solutions which deliver all identified benefits – include all imagined solutions even if they are beyond current technologies (later we can step back from these 'Ideal wacky solutions' towards real ones).
3. **Identify the main functions of the Ideal Solution.**
4. **Seek available resources to provide the functions and create the Ideal Solution.**

The trick with available resources is to look hard at the system and all its components, their features (small, big, tall, dense, clean, cool, right shape, smell, transparency, absorbing qualities etc.) for any resources which are already there, needed and might deliver the Ideal. Also when looking at input resources look hard at the *free* resources of the environment, the harmful resources etc. to see if any of them can be mobilized to deliver what we want. An example here might be Cat's Eyes, which use available resources from the car's headlights to provide lighting in the road.

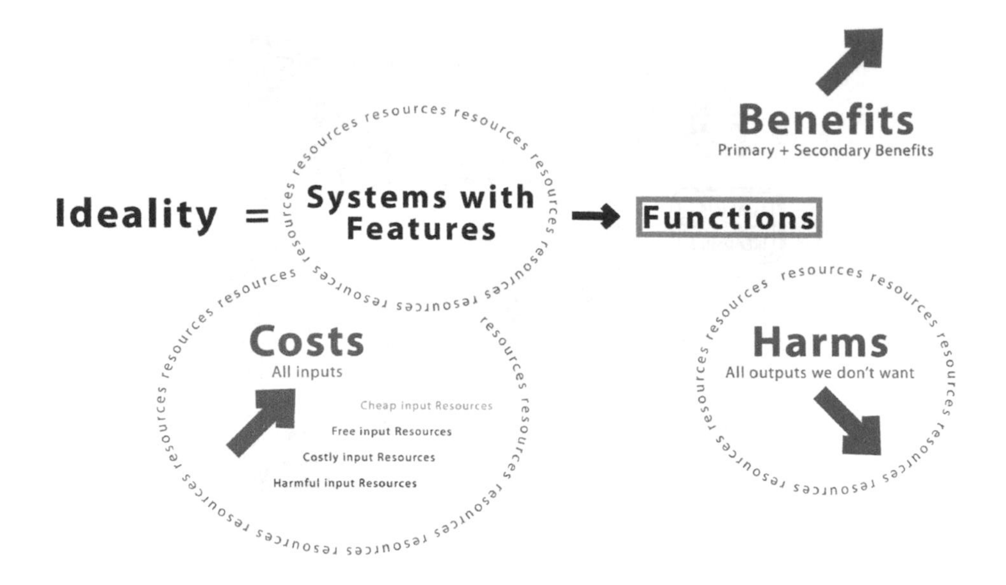

Ideal and Resources

The simple concept of the Ideal Resource or Substance – or Ideal X Factor – is helpful when we don't know what it is, but we know what we want it to do. We can define all its functions and say that they are to be delivered by a mysterious X Factor. These simple ideas of Ideal Outcome, Ideal Solution and Ideal Resource – X Factor – are theoretical solutions to help us visualize the outputs we want, and find ways of delivering them preferably with no or minimum inputs (costs) or harmful outputs. In Altshuller's problem-solving process they are the final solution and we locate resources by defining the X-Factor – which solves the problem.

Role of the Ideal Solution in the Problem-Solving Process

Once we have understood all our needs the Ideal helps us define a solution – the Ideal solves the problem by itself – so the first solution to look for in TRIZ is one which offers ways to solve using available resources.

Ideal Solution Requires No Inputs

A close to an Ideal Solution comes from existing resources. The Ideal Solution is a quick definition of how functions deliver all the benefits of our Ideal Outcome and delivers all the right functions (which deliver benefits) but has no costs and no harms. Defining the Ideal Solution, helps give us some indication of everything our system should do and help us define the direction of our work. (My Ideal kitchen chopping board is always clean, doesn't mark, doesn't slip, is good for cutting small things, collects juices of meats and foodstuffs, doesn't get contaminated with food smells such as fish or garlic, is safe, easy to find, hygienic.) How can we find our perfect/Ideal Solution? Often we cannot visualize this Ideal Solution so we have to think about a real, existing and imperfect system and then take that to the extreme of Ideal to see how to improve it. We can also do this for any parts of the system, any components and interacting systems – and ask what is the Ideal?

Ideal Resource or Ideal Substance – Ideal X-Factor

An Ideal Function is made out of existing resources. In TRIZ this is more simply described as a mystery X-Factor – which gives us what we want/solves the problem but requires no extra inputs as it is made of existing resources – it solves the problem with no costs, no harms

'The tendency toward the ideal by no means signifies moving away from the reality of the solution. In many cases the ideal solution can be fully implemented ... The clear thrust to the ideal is necessary not only in formulating the IFR but at literally all stages of solution of the problem', *CAAES* p. 84.

Ideal – Using Free Resources to Attract Customers

Finding great solutions from within available resources applies to any kind of problem including marketing problems. The Ideal marketing solutions for a small bread shop might be that customers are aware, reminded and desire their products and seek them out. A free and available resource to achieve this for a bakery is the smell of freshly baked bread which draws customers into the shop. In TRIZ terms this delicious smell was an unused output (a harm in TRIZ) and here was converted into a benefit.

Use what you've got – to get what you want – attracting customers

Genius, Resources and Ideal Thinking

When stories are told of great inventors they nearly always contain clever mobilization of resources often from imagining an Ideal System. There are stories of some personal devices they have made for their own convenience which shows how they use Ideal thinking to solve problems, big and small, and how they cleverly use the resources they've got to solve problems. Apparently James Dyson, one of our greatest living inventors, has a fence round his tennis court that is not the usual rectangle shape with annoying corners for tennis balls to congregate in, but has hexagonal tennis court fence which bounce all balls back to the centre of the court where they can be easily retrieved. This is using shape as a resource – one of the features to solve the problem. Henry Ford complained to his friend Edison that his garden gate needed maintenance as it was hard to open – too stiff. Edison informed Ford that when opening the stiff gate he had just pumped a gallon of water out of the well. Each of Edison's visitors raised a gallon of water to Edison's roof holding tank when they pushed through what they thought was just an old and rusty gate (*rzollinger.wordpress.com/page/2)*.

TRIZ Helps Us Think Like Great Inventors Who Cleverly Use Resources

The purpose of the TRIZ Creativity Tools like the Ideal is to help anyone think like a great inventor and come up with great solutions using Quick Ideal Problem Solving. When we want to achieve clever solutions and be as inventive as Dyson or Edison we would just apply the above four steps, starting with the Ideal Outcome (quick understanding of requirements including self-returning tennis balls, and water moved from the well to the attic tank), think about how to do this and then seek resources from everything in and around the system. This works for anything from pumping water to getting everyone's attention at boring meetings with disengaged participants.

©Oxford Creativity 2007
GOODMORNING!

Clever use of simple resources can be very effective

Ideal Solution/Machine/User Manual to Uncover All Required Functions

Once we have defined our Ideal Solution we also have to think about how it is used (if this is relevant). How we use our system (how it interacts with humans) requires some kind of user manual: we can take the definition of these interactions to perfection by defining our Ideal User Manual. This is not overuse of Ideal – defining our Ideal just helps us jettison constraints and 'I can't do that' thinking. Adding Ideal in front of the Outcome, System, User Manual, functions and solutions is a simple but powerful thinking tool to help understand them, uncover contradictions and often see how to improve them. Defining our Ideal User Manual helps us develop more simple, self systems by understanding exactly what steps need to be taken, allowing us to get rid of needless or complex steps, and questioning whether these steps need to be taken by the user.

The Ideal Method or Ideal User Manual defines the perfect way of how we could use the system with no costs, no set-up, no learning curve, no energy or time, no harms, no maintenance, with a self-regulating process etc.

Ideal Solution and Ideal User Manual for Public Toilets

Recent research looked at how to redesign male urinals for public lavatories to reduce the considerable cleaning costs. One element is how they are used, and the Ideal User Manual would suggest accuracy of use as significant in cleaning costs. What X-Factor could be help? One solution is to use the men themselves as a resource– as shown below. The X-Factor involves motivating the men.

Effective Resources can be found in the strangest places

Systems – Get the Right System and Get the System Right

Identifying the Ideal Solution is the part of the TRIZ problem-solving process which begins the powerful process to use the various TRIZ tools to identify and solve problems by getting the right system right. The Ideal Outcome helps ensure that we understand our needs and the Ideal Solution helps us identify the best system; however if real constraints prevent us from having the best system, then we have to use TRIZ to make the best of the system we already have and eliminate as many problems as possible.

System
Which delivers
What we want

Choosing which system to fulfil our requirements needs an understanding of the range and kinds of systems we are free to choose and what we can get with the resources available.

TRIZ helps us improve the systems we have, and when appropriate move to much better ones. If we are defining the ideal for an existing system or solution then we begin with the Ideal Outcome to outline everything we want – even if these benefits seem impossible or contradictory. We then can look for solutions and systems to deliver what we want. Our ideal for transport may be described as effective, no cost, comfortable, safe, enjoyable, exciting, fast etc. before we have chosen a system. The systems available may only cover some of these benefits the Ideal has identified. For example using a car may cover many of the benefits but be slow in traffic, whereas a motorbike will be fast in traffic and probably more exciting but is less safe than a car. Our Ideal Outcome would give us all the benefits not just some, but once we have specified our Ideal System we have made a choice and limited our options to that system. A car offers different things to a motorbike and an Ideal Motorbike covers different benefits to an Ideal Car, but each system would be defined as a perfect manifestation of their type. The Ideal System is Ideal in every way, provides all functions which deliver we want no matter how difficult to obtain but it covers these in the context of the chosen system. For short sightedness one can have spectacles, laser surgery or contact lenses. The Ideal System in each case is very different from each other – but takes us to perfection within our chosen route.

Ideal Outcome to Help Us Appropriately Ignore/Subjugate Constraints

Defining the Ideal Outcome ensures that we understand essential requirements, and that constraints have not prematurely pointed us towards some particular, limiting solutions and systems. We are striving to understand how to solve the right, real problem, not just deal with some symptoms, or limit our thinking by working within constraints which may or may not be pertinent. Although sometimes the inputs such as time and money (and constraints – space, safety etc.) may later prevent us from adopting the best engineering options and systems, it is important that we separate them at this stage. Unless we are perfectly clear about the main things we want then we cannot problem solve effectively and locate all the good solution options. The best engineering solutions (the Ideal) are well worth capturing and recording at the front end of projects along with the reasons they are being rejected. Defining the Ideal with TRIZ helps deliver this clear definition – an innovation audit which is useful for future projects. Sometimes after constraints have changed and managers and customers, who have themselves rejected our best options and solutions, ask us later why we didn't think of them – the TRIZ innovation audit records all the possible good solutions – and shows everyone what has happened and why.

Innovation must be allowed to happen – don't look for new solutions if you are scared to try them

Too Much Innovation?

The unquenchable corporate thirst for innovation is sometimes tempered with a nervousness of trying anything new. Recently TRIZ was extensively used on a military project as the UK government's agency requested more innovation – many new, innovative workable concept solutions were proposed but it was decided that a 'pragmatic approach' rather than an innovative one was needed and the next generation system was fairy indistinguishable from the last one – as this was felt to be well understood and safer. Better new potential systems had been identified and recorded using the Ideal for the next change of government directive. Recording the detailed definition of the Ideal is an important part of auditing the whole innovation process for revisiting later. This ensures that the selected results are not the only recorded output and that all good ideas, solutions and possible new systems are available for future innovation sessions.

Ideal Outcome to Solve Problems

When we begin problem solving we often follow a route as shown in the chart below. After setting the context of the problem in 9-Boxes and capturing the problem statement, we define our Ideal Outcome. When this is undertaken by a problem-solving team it always produces a lively session and while defining the Ideal Outcome the team will be coming up with solutions and filling up the 'bad solution' park. This often suggests that we want contradictory/opposite benefits and opposite functions to deliver them. For example we might want a very short, high car for certain circumstances and a much longer and flatter car in other circumstances. This is just a physical contradiction and all the ways of solving it are contained within the Separation Principles and the 40 Principles. The chart below shows the simple logical steps which gives us a simple route for problem solving taking in the Ideal Outcome as an important early step in the process.

Future systems will use TRIZ Solutions to give us everything we want in one system

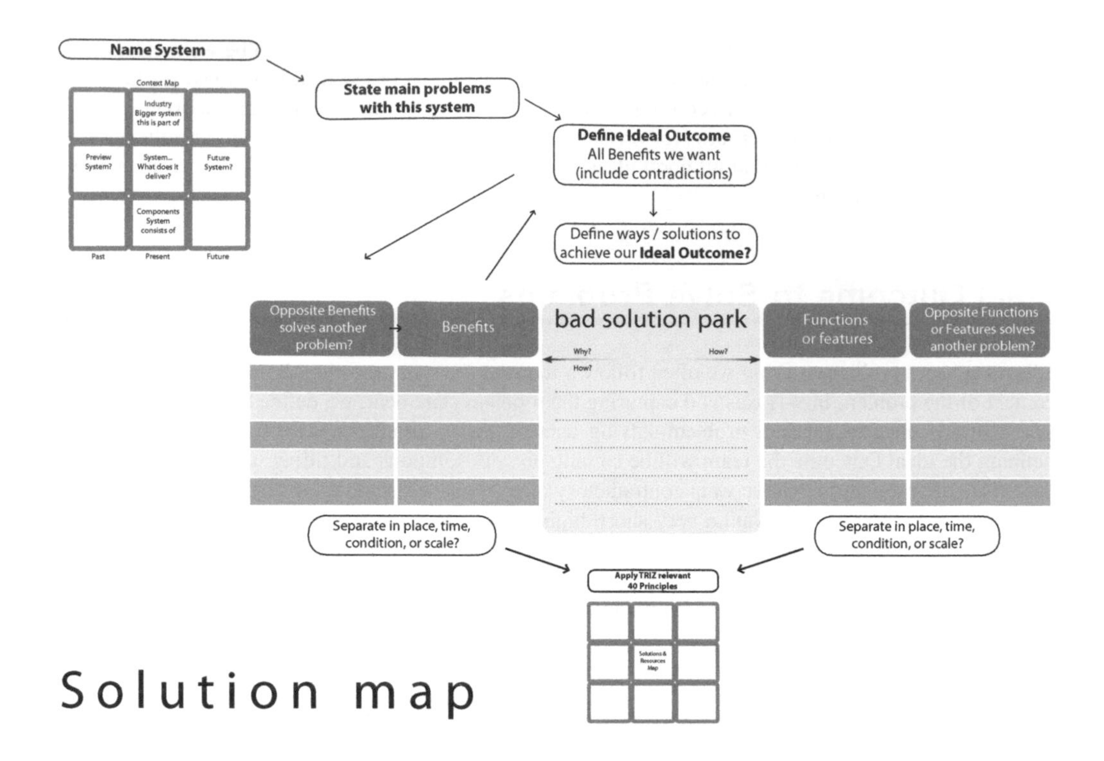

Ideal and Constraints, Reality and Problem Solutions

'Art lives by constraints and dies by freedom,' Leonardo da Vinci.

This could equally well read Engineering Invention and Problem Solving lives by constraints and dies by freedom. We often need the constraint of limited time, an urgent need or a crisis to focus our efforts and resources – from passing exams to finishing projects – constraints and deadlines are useful to get to solutions. However sometimes their very existence can often prevent us from not even trying to reach good solutions, or even understand what we want. The TRIZ approach of defining the Ideal Solution is a very simple way of overcoming the negative 'can't do it' approach which can haunt much engineering problem solving. Constraints are important and not to be ignored but their all-pervading negative effect sometimes needs to be temporarily removed to allow some clear thinking about all possible solutions.

It is often helpful to list all the constraints at an early stage simply to get them out of our heads – we may later discover they may not all be real constraints but we need to jot them down early in the process. Also some 'bad' inputs may be bad at some time and good at others – just because something has a negative effect initially doesn't mean that at some other time such as later or earlier it offers something good (and vice-versa). The famous classic bell curvc on stress illustrates how stress can be good for getting us going for exams, finishing difficult tasks etc. but eventually becomes bad if there is too much of it. The Ideal resource thinking helps shake off the 'good' and 'bad' tags by looking at everything there is at different times, different places, within different scales, and thinking about when and where it is good and how to use it. The classic something was good (worked well) initially doesn't mean that to keep always using more and more of the 'good' thing will keep adding benefits. Recent overdoing diets and excessive exercise have shown their harmful effects; in companies something that has worked well may then be endlessly and inappropriately applied, as corporate anorexia demonstrates.

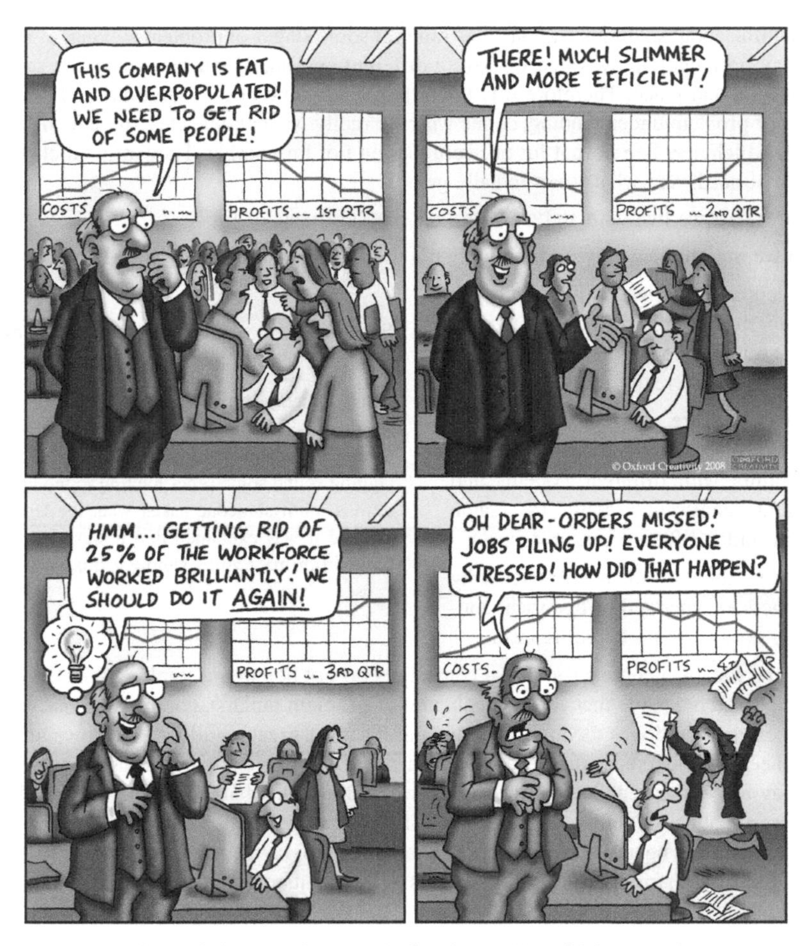

'Low hanging fruit' solutions may be easy the first time and harmful the second

Constraints = Restrictions on How We Deliver (Not What We Want/Don't Want)

Constraints are very real and present in every problem-solving session – both in official lists and others in participants' heads and memories. Constraints are restrictions on which system we can have and how we deliver benefits. Constraints are the reality of what we can have – not what we want. Real constraints must be understood and respected – imagined constraints need to be recognized, and removed. Also different solutions have different constraints,

Different system has different constraints

In 1960 my father was a young engineer on a UK trade delegation to Russia. Amongst the goods displayed was a UK refrigerator – a small rectangular white box with a door on the front, about one metre high. Russian delegates were fascinated by its smallness and shape, as Russian fridges then had curved tops and were much larger. When asked the point of the flat top it was explained that it was designed to fit neatly under a worktop or the top could be used to store other kitchen items – at this the Russians all laughed and said you couldn't ever put anything on top of a fridge because it would fall off.

so we need to think about solutions without constraints at some stages of problem solving, as the familiar constraints may no longer apply, when we have a different system, or when we have the same system in a different time and place. Therefore it is important to separate needs, solutions and constraints at the beginning of problem understanding, define what we really want and be clear about constraints (real not imagined) which may be relevant for one solution but not for another. They might dictate any gaps between our real needs and what we eventually get.

Constraints are what we live with and work around, they are facts of life.

We try when we are problem solving to initially put them aside (fairly briefly) for three reasons:

- to understand what we really want (*what we want* may be very different from *what we get* but that doesn't make any difference to what we really want);
- to check we are solving the problems with the right system (there are different constraints with different systems and different constraints with the same system under different circumstances);
- to ensure that we are not restricted by imagined, exaggerated or historical constraints which are no longer relevant.

Same system with different constraints in different places and at different times

At a problem-solving session in Scotland the problems related to exporting some machinery to a factory in Singapore. On the list was how to stop the lagging round some pipes melting in the hotter conditions, until someone asked what the lagging was for and it was explained that it was protecting the pipes from freezing. A simple 9-Box on constraints is useful here.

Starting with the Ideal Outcome ensures that constraints do not box in our thinking –for that reason we have to be able to step outside them at the beginning of problem understanding – because overcoming constraints is often what drives us on to think up and discover great solutions. When we are choosing a system we need to understand the benefits we want and their priority, including what price we are prepared to pay to get the benefits we want, Once we understand what we want, we must then assess whether we can adapt and mobilize the resources to achieve it or whether we must choose a less costly system with perhaps less benefits (e.g. less durable) and/or more harms to get the system we want (with the minimum specified harms) – there will be costs – the inputs required and ideal resource thinking helps us minimize them.

The Ideal Helps Test Our Real Constraints

Defining the Ideal helps us make sure that our constraints are real ones, for example when one big limitation is the budget. What would happen to constraints on cost if there were a crisis? Would the budget constraints go out of the window? Would money, resources, time etc. suddenly be made available? If so, are the front-end budget constraints real, and could a case be made for more up-front investment to avoid probable problems later?

Constraints do not decide *what we want* only *what we may deliver*, so it is valid to understand requirements before allowing any thinking about constraints. This is a difficult mental exercise for most of us, as we may be suppressing the acknowledgement of valid requirements, because subconsciously we may have made a judgement that they are unobtainable. We have a mental note of the constraints and may be reducing the listed requirements to what we believe is deliverable and acceptable. Mixing up requirements and constraints is a common pitfall caused by psychological inertia and the simple, mental trick of the Ideal Outcome together with the magic wand is aimed at breaking it and helping us clearly understand what we really, really want.

Inputs and constraints may dictate which solutions we finally adopt; constraints limit how we deliver, but they should not influence our clear thinking about what we want.

For example on a nuclear clean-up site our requirements may be that we want a suitable nuclear-rated building in twelve months time in which we can safely process contaminated nuclear materials. One constraint may be that it takes 5 years to design, authorize and build such a nuclear-rated building, with shielding to the outside environment. An initial solution (far from ideal) may then be forced on us for the first 5 years of processing to locally shield all production units, within the building, to protect personnel against nuclear contamination. We have started problem solving assuming we cannot get what we want because of a constraint which may not be real. Authorization and planning may be speeded up if this project is important enough. But if we start off with the belief that we can only solve this problem with local shielding to contamination, then we have not even bothered to understand and fight for the best option (our Ideal Outcome) but headed off to implement an imperfect, expensive, unwieldy and impractical solution. We have let constraints defeat us at an early stage of solution selection.

So what are constraints?

Constraints influence how we do things

They do NOT influence what we want/don't want.

Think about

Size

Speed

Delivery

Safety

Amounts and timing of money

Life of system

(which system we choose)

e.g. if choosing where to live between a house or a mobile home? We might want a big house which lasts several hundred years but constraints may force another choice

Start with Only Requirements – Initially Forget Both Systems and Constraints

We can list our constraints at any point in the problem-solving process, but we should not normally consider their influence until we are choosing the system we will adopt. Ideal Outcome is not about systems just needs, and we should always separate *requirements* from the *system*. Don't think about systems while defining the needs, as it is important to start with just requirements, and to forget the systems at this stage of problem solving. The famous example (urban myth) of the American Space Pen illustrates this. Was the requirement:

I want to a pen to write in space?

or

I want to write in space?

The question mythically asked was, 'How can we create a pen that will write in space?' The wrong question because they mixed up requirements with a particular system – a pen – and the Space Pen was developed at great expense. The myth continues that the Russians asked the right question as they started *only* with requirements and no systems, 'We want something to write in space' – and then chose allegedly chose a pencil. When we have no system (only requirements) then we are more likely to be able to identify and select the right existing system at the appropriate time or if it doesn't exist we need to invent a suitable system to meet the needs.

When companies are negotiating with their customers each party will have a system they visualize as the solution. Starting with requirements rather than systems can save time and phenomenal amounts of money, particularly in large-scale, long-term projects such as defence procurement.

Happy customers get what they want – not what we have

Ideal, Constraints – and the Appropriate Levels of Problem Solving

Having worked out what we want with the Ideal Outcome, the Ideal Solution can help us identify a level of problem solving unimpeded by constraints. If our problem was to establish the ideal way of getting to work, what we want could be dictated by a number of requirements. It could be comfort and privacy and we may wish to minimize the stress and amount of travelling time to work in the morning. We might have a whole range of needs and once we think about each one we can conjure up Ideal Solutions to each – from avoiding rush-hour traffic gridlock with solutions such as flexi-time or living near our work, or travel by bus which has the use of uncongested bus lanes or use a motorbike or bicycle to enable us to weave through stationary traffic.

Each solution uses different systems which deliver different primary benefits and which one we choose may depend on our authority and ability to implement them. If the company belongs to us and we own extensive assets, we may be able to move the company nearer to us. If we have no authority and few assets we may not be able to access or initiate a convenient solution such as flexible working hours to avoid busy times or a have chauffeur-driven car. The Ideal Outcome can help us define what we want (not waste too much time in travelling to work?) and then the Ideal Solution helps us choose the appropriate level where we ourselves are able to implement a solution – where we have a level of our authority to act on a problem. The flowchart below shows this sequence. The Ideal first shows us what we really want. It then helps define a practical level of problem solving influenced by constraints. We use the chart below to define the appropriate level at which our constraints and resources allow us to find clever but practical solutions to fulfil our needs.

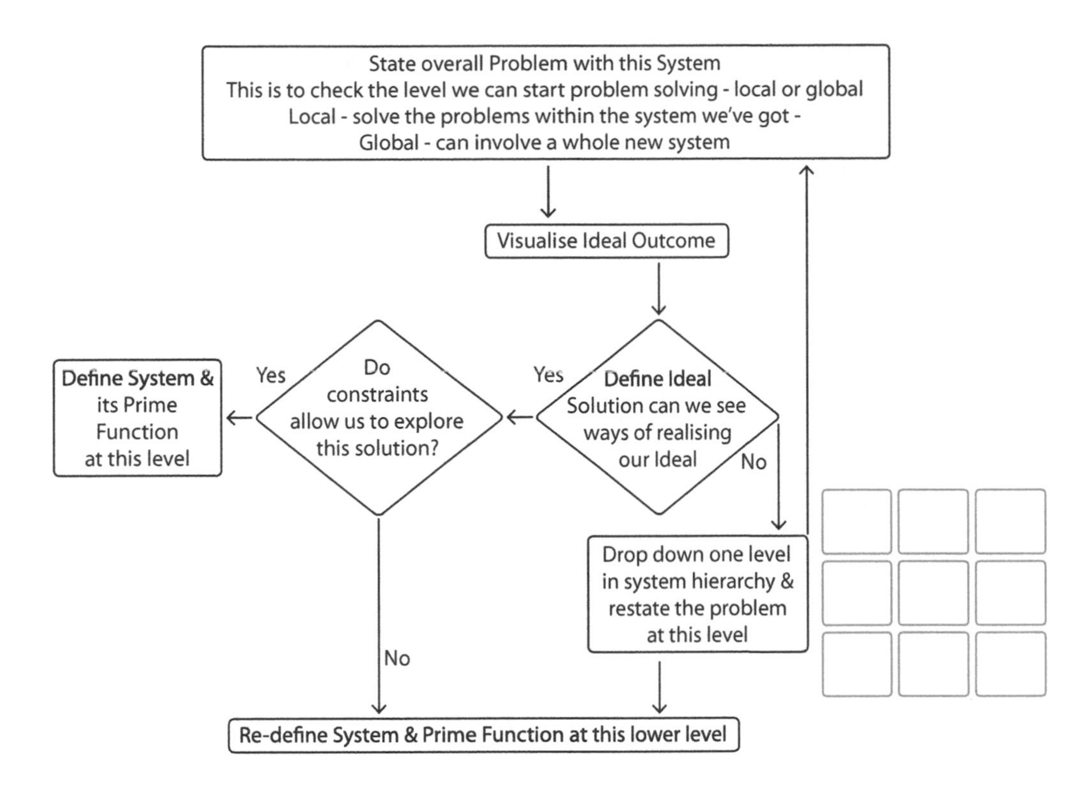

We can try the sequence of the flowchart above on a simple problem such as a bumpy ride to work on a bike and work out which Ideal Solution (at which system level) is within my grasp, my authority and constraints. My Ideal Outcome may tell me that I want utter comfort while cycling to work and I know that there are many solutions at the many different system levels. It is good to understand all the options, as well the likely areas of action, be aware and map constraints such as authority, opportunity, budgets etc. rather than ignore them. Remember constraints may not always be permanent or even real – they may be imaginary or falsely assumed as in 'think out of the box'.

Where and When Do I Solve the Problem of Having a Bumpy Ride to Work on My Bike?

- Rebuild new smooth bicycle tracks
  - Mend the holes in the roads.
  - Authority/resources to do this?
  - No? Then go down to system level.
- Buy bike with good suspension
  - Budget for this?
  - No? Go down to sub-system level.
- Buy gel saddle/ softer tyres

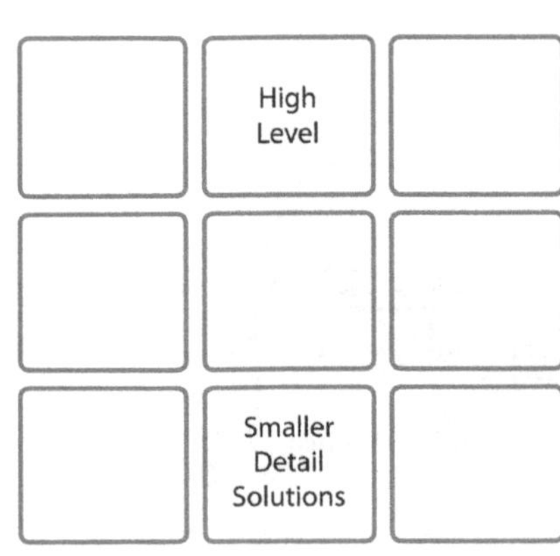

- Or I could use time by doing something before or after the bumpy ride
  - Make bum less sensitive (before).
  - Have a soft chair /big cushion to sit on (after).
- Or find alternative systems
  - Take a bus – Work at home – Corner-shop Offices – Beam me to work Scottie.

The definition of all these elements of the Ideal Outcome is useful when starting to analyse and solve problems. It is always worth trying to define all the elements of the Ideal Outcome for a system that matters. For example, one challenge facing us in the industrial world is the carbon footprint issue and the necessity of travel – particularly just for meetings and management control. When needing to travel to a meeting the definition of the Ideal Outcome may not be perfect travel for everyone – this ‘travelling’ is a solution to help solve the real problem. When defining the ultimate goal of the meeting we may be looking at communication, consensus, discussion and decision making and there may be solutions which involve no travel at all. We get high-level clear thinking by defining the Ideal Outcome and the way to check we are dealing with the ultimate goal is to check it doesn’t contain any solutions. It then helps us understand and be clear about understanding and defining and solving the real problem.

Once we have decided that travel is inevitable and necessary for a situation then the problems of human transportation – moving ourselves over small or great distances – defining the Ideal gives us a clear benefit statement. The Ideal/ultimate goal may be defined in many ways such as moving efficiently and quickly from one place to another, and there may be many systems to help us achieve this. Meeting the needs for transport are systems such as boats, animals, and machines – including bicycles, cars, trains, aeroplanes, spaceships or even theoretically by teleportation. All these relevant systems have their associated super-systems of pathways, roads, railway lines, canals, Starship Enterprise etc. and are all part of a transport system. Our goal, our need is to move from one place to another and we have a choice of systems. Once we select a system we are often dealing with the problem-filled details or at thr problem-development stage of our work. Whatever stage we are working on we should never lose sight of the Ideal and frequently remind ourselves of the highest level of needs that system is fulfilling and the real and fundamental purpose of our system (Prime Output). This is delivered by progressive systems as technology develops (as predicted and mapped by the TRIZ Trends of Evolution in Chapter 9). Ultimately the Ideal Solution delivers everything we want without existing – the example below tracks the technological progress of detergent for washing clothes. The TRIZ Trends predict that solid, monolithic systems will become more segmented until their functions are provided by some kind of field effect as shown below. The TRIZ Trends map and predict systems increasing Ideality – to the ultimate point of the Ideal Solution which gives all functions without existing so predicts that the Ideal detergent is no detergent for such a situation.

TRIZ Trends suggests practical ways of getting ideal solutions

Conclusion: Ideal Outcome Prompts Us to Understand Requirements and Simultaneously Find Solutions

The Ideal Outcome helps us identify our needs correctly, and this enables us to define and understand the system we want, so that we can we begin the process of problem solving with TRIZ and locating available resources to give us what we want for the minimum or no new inputs. The Ideal Outcome helps us define all and everyone's requirements. At the beginning of any problem-solving session the Ideal helps us deliver an efficient summary of what everyone wants. This efficiency is appreciated by technical teams because gathering requirements is a notoriously hard step for engineers, as they want to get on with problem solving and don't like lots of questions about understanding the problem. They prefer to cut straight to the answers and often, even beforc have properly understood the question, they offer solutions. Using a Bad Solution Park enables essential early stages of both problem understanding and problem solving to happen simultaneously when necessary.

Defining the Ideal Outcome works well for engineers, as it offers a very quick, approximate but reliable way of dealing with the question 'what do we want?'. It is fast and accurate, and complimentary to the other requirements-gathering toolkits which are often used prior to a TRIZ problem-solving session. More importantly it stimulates debate and controversy, as it fires their imaginations and prompts some clever thinking, which creates effective, quick and enjoyable sessions. In live TRIZ problem sessions this step always has everyone alert, engaged, argumentative and thinking clearly about requirements and also simultaneously producing solutions. It always delivers impressive results including some with a slightly wacky and theoretical bent.

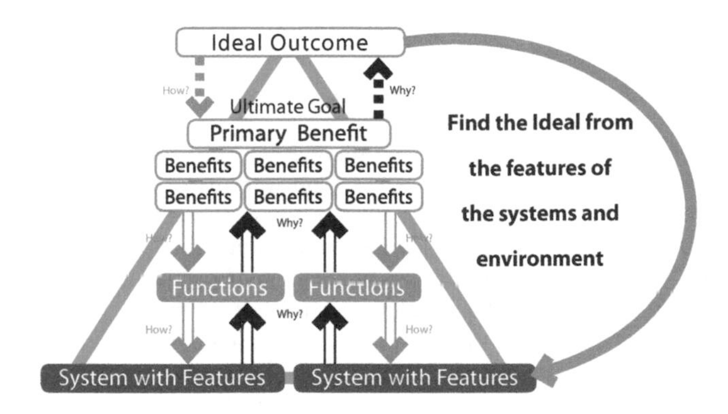

Problem-solving sessions are usually deemed successful when we have all the appropriate good solutions, which deliver what we want (the outputs we want) for the least cost (inputs) and harms (outputs we don't want). Finding a good range of solutions often covers both the short-term and long-term requirements. Knowing all possible solutions enables us to select the most appropriate solution options for the conditions we face. We may have to select a necessary quick fix for the short term, but be able to use a much better solution in the longer term. Knowing the best options over time is important even if we can't immediately use them.

Once we have understood all our needs with the Ideal Outcome – then we use 'the Ideal solves the problem by itself' as a prompt to look for ways to find clever solutions using available resources. The relationship between Ideal and resources is important, the Ideal Outcome shows us our benefits and we get the benefits from the Ideal solution/functions which are provided by the right resources. The Ideal Solution provides only good functions (no harmful, no insufficient, no missing functions) and works perfectly and easily by using available resources.

7 Resources: The Fuel of Innovation

Using Resources – How to Become a Resourceful Engineer

Resources are the fuel of innovation in all problem solving, but particularly in TRIZ which directs us to actively seek to use resources cleverly, by making them work hard and realize their full potential. TRIZ helps us deliver innovation, finding new ways and new systems, by simply and consciously mobilizing resources to minimize inputs for the maximum output. Innovation is variously defined as the creation of new goods and services, or the new application of existing knowledge or simply a new way of doing something. In practice TRIZ, which fosters innovation, makes us resourceful and able to deliver new or improved systems by the smart use of existing resources. Resourceful can be defined as able to use the means at one's disposal to meet situations effectively. Being able to recognize and mobilize appropriate resources is fundamental and essential to TRIZ. Once we automatically begin to search for and cleverly apply the right resources, this enables creativity, problem solving, invention and uncovering new ways, new uses, new development of established systems, and transforming familiar, proven technologies, for new applications, to make them more useful, less expensive and with fewer problems or harms.

Clever use of available resources has always delivered brilliant innovation for products and processes in all aspects of business and engineering. Think of any innovative device, old or new, any great

TRIZ for Engineers: Enabling Inventive Problem Solving, First Edition. Karen Gadd.

invention or new successful product and at the heart of its success has been the harnessing of some resource already present in the problem area, for example:

- the compass (which using the ever-present earth's magnetic field was previously cited as one of the most useful inventions for mankind)
- the sundial (clever and simple, and uses sunlight), or for more contemporary problem solving
- a new neckband which tells you if you are about to get sunburnt (uses the amount of sunlight you have been exposed to).

All these inventions match some set of needs to a system – and important elements of the system providers come from available resources. TRIZ problem solving shortcuts and formalizes this relationship and replaces the clever brainwave, moment of inspiration, with smart systematic searching for nearby resources, to match needs which have been cleverly and logically mapped. One first rule for finding the right resources in TRIZ is always search for solutions to problems close to the problem area.

Resources close to the problem area often offer great solutions

One of the powerful aspects of TRIZ is how to find and pick up the right resource to use at the right time. This is not left to chance or luck but like every other aspect of TRIZ problem solving we can systematically find the right resource, from everything available, but without long-winded or tedious searches and endless lists. TRIZ offers us ways of systematically heading for the right answer and the right resources, by ensuring that we first ask the right question by understanding and defining our needs and the functions which deliver them with accuracy and precision.

Use the Resources We've Got

This is almost a TRIZ Mantra – good awareness and clever use of available resources is essential to TRIZ problem solving. This is especially true of our engineering talent which is one of the best resources of all. Although using all resources intelligently is almost self evident in all problem solving, TRIZ processes particularly demand a systematic approach to being aware of everything that is easily available. 'Resources we've got' (substances, energy fields, attitudes, colours, spaces) can be defined as anything:

- available to us, in or around our system
- already there
- we can use for free, before bringing in other inputs
- very close to the problem area
- anything in the environment (sunlight, gravity, ambient temperature, pressure etc.)
- right down in the detail of the components (down to molecular level)
- problems, causes of problems or harmful things.

Transforming Harms

Recycling, composting, vaccinations, homeopathy, competitiveness, university and school examinations all take harmful elements and transform them into good: looking hard at the harmful elements to create good has been the source of much inventive success. This approach is fundamental to TRIZ. In every harm something useful is lurking: mobilizing the good or useful within anything harmful means that instead of quickly removing, dismissing or avoiding harms we should examine, analyse and then use them. Looking for usefulness in obviously annoying, painful nasty things takes a conscious effort. But even in crass mistakes, as in the cartoon below, some would see the harmful fears of uncertainty about job prospects or security (much in evidence in today's world) as a useful way of applying commercial realities to overpaid bankers, or more sadly and more likely for squeezing more work out of oppressed staff in a call centre – applied deliberately or by accident.

"Mr Frimley, sir, can I have a word about the motivational artwork..."

Even Harmful resources can sometimes deliver results

Minimize Inputs

The TRIZ systematic approach to minimizing expenditure of costly resources is to avoid needlessly investing and squandering anything by first using what we've already got; this approach chimes with today's mood in many parts of the world. Clever resource use is being sought everywhere now to make best use of our available energy sources with high-profile projects to efficiently harness wind and wave power and sunlight. TRIZ approaches are fundamentally helpful to this.

Locating and Defining Resources

To logically locate resources we have to see how broad the term is, referring not just to inputs but to every part of the system and everything around it, and interacting with it as a potential resource. But although the term resource is very general, and applies to many parts of the system, it can be accurately defined. Resources are anything which help provide the functions which give us benefits. Obviously this includes any inputs into the system, but also includes the way the system is used, time and process steps, the context, the environmental aspects and the surroundings of the system. Resources therefore include any aspect or feature of the system which helps to provide the necessary functions. For example if we wanted a saucepan which somehow informed us when the liquid inside was boiling, this could be provided in many ways. One could be an input of adding a small metal disc which rattled when the liquid boiled, so that we could hear when the water boiled. Another resource would be something which we could see, but obtained not by adding anything but from some feature of the saucepan itself to deliver this. One solution would be to have a see-through saucepan: the feature of transparent material such as glass, would provide the function of see-through, and that feature is the resource which has solved the problem in the picture below.

Features of systems are useful resources

Resource Hunts – Focussed by Functions Which Give Us 'What We Want'

In TRIZ the hunt for resources is not an unfocussed trawl of everything that's available but follows, and is focussed by, an understanding of the necessary functions which deliver our benefits. Satisfying needs with solutions which absorb available and least-cost resources is a conscious part of TRIZ Thinking. Even when we are given a ready-made solution it is helpful to check it against both our benefits and our available resources to ensure that it is an appropriate solution for both short-term and long-term expenditure and benefit delivery.

Knowing 'what we want' is an essential step in resource hunts and gives us a better starting point than opting for a ready-made solution with insufficient understanding of the functions we need and the real costs of delivering them. In both business and ordinary life, available systems are thrust upon us as starting-point solutions, whether they are other people's solutions, legacy systems, available products or innovation temptations. Checking the resources they absorb (money and all other inputs) should be examined to check the match of the benefits delivered for the costs incurred. Resource hunts are undertaken to ensure solutions which cost the least and deliver the most, i.e. use the minimum available resources and yet fulfil exact needs. This means checking that we don't opt for solutions because they are there, and accept systems which give us partial needs or things we don't want or more than we want. In TRIZ we are seeking a perfect fit of our exact benefits for the minimum inputs and we undertake formal TRIZ resource hunts to match these together.

Today in a global market we are bombarded with messages offering us ready-made products and goods, some of which give us all the things we want, but others excessively or only partially fulfil our needs (and some offer answers to unfulfilled needs we had no inkling we wanted). We are faced with a multiplicity of ready-made off-the-shelf products for everything we do, we use, we wear, the food we eat, domestic appliances, entertainment etc.

Resources and Make or Buy Decisions

In industry 'off the shelf' is global, visible and accessible for many technical systems, and can be an intelligent resource but only if it gives us the benefits we want for the right costs. The opportunities for off-the-shelf buy-in of sophisticated products and components, all with fast and accessible information on the Internet, make it harder to make, rather than buy, solutions and products. In some defence industries the switch in balance, from making everything on site (such as in past years in the Barrow Shipyards) to a high buy-in of components and systems, is a reflection of the changing global markets. Unless we begin with the correct perspective of 'what we want', followed by an intelligent resource hunt then we may be tempted on occasions to inappropriately opt for ready-manufactured products which are a poor match for our needs and bad value. The drive for efficiency and cost savings, (and a relaxation of rules and practices about only buying sensitive defence components within your own country) has dangers in that it encourages buying ready-made solutions which may be partially understood and can produce compromise solutions. Choosing off-the-shelf components needs great clarity of understanding (and a good accurate definition of their ideality) to ensure that we don't choose something which fulfils some needs, but not all, and may offer extra superfluous functionality at extra cost and complexity. Otherwise instead of locating the exact and appropriate answer to technical challenges, we may bolt together a number of ready-made systems, reflecting the downsides of our new world of vertical, industrial integration, which may result in a woeful deskilling of key industries. This is unfortunate especially when we have the funding, the will and the expertise in projects such as in new-generation submarines or military aircraft.

Products give us some of 'what we want'

Consumer goods are an interesting place to study the fulfilment of partial wants. Recently when buying a chopping board for the kitchen I was offered a fantastic selection of boards, each offering one or two of the six functions I wanted, and some offering functions I hadn't known I had wanted. With TRIZ I had identified I wanted a chopping board which was tough enough to hit repeatedly with a sharp knife, stable and wouldn't slip, hygienic (doesn't absorb and retain nasty foodstuffs such as raw chicken blood), suitable for all foodstuffs (does not retain flavours from garlic or fish), with a useful ridge to capture juices when carving cooked food, and also washable and not too heavy for me to carry to the sink and then fit in the sink to be washed. I had in mind a fairly heavy wooden board with a ridge round the outside, big enough to be useful and small enough to be washed in the sink.

In the shops the available products suggested I needed five different chopping boards each of which offered one or two of the functions I had identified plus one folding chopping board which had a function I hadn't thought of (efficiently pushing all crumbs and small scraps together to easily throw away efficiently without mess into a bin). One very light board had an ingenious under-surface to give it high grip, another set of thin, different coloured boards were marked for different foodstuffs such as fish or onions, one folded up to be put away and one was hinged for extra functionality.

Needs – the Beginning of Any Process – Engineering or Otherwise

In this world of a confusing array of off-the-shelf choices, TRIZ resource approaches help by starting us at the correct end of the process to help us efficiently and rapidly work out what we really, really want (and hopefully the minimum we want, a 'less is more' approach) and then seek how to achieve it with the best solutions with the minimum inputs. There has been a curious problem in this area without TRIZ because finding and developing solutions is seen as the fun part for engineers and one they are always eager to proceed to. However this most important, complicated and uncertain task of finding the right solutions is often given insufficient time in design and invention processes, with pressure to quickly be offering solutions (see Dave Knott's Rolls-Royce case study in Part One). TRIZ has proved a valuable addition for large companies with mandated processes for every step of design, selection and manufacture. When TRIZ is added to the mandatory steps, it helps engineers step back from the necessary detail and see the wood for the trees. It also helps them add TRIZ resource hunts to use the mandated requirements capture outputs to locate the full range of good, appropriate and available solutions. Resource hunts, focussed by requirements capture, are an integral part of this.

The simple relationship between requirements, solutions and resources is shown below

REQUIREMENTS / NEEDS

How?

SOLUTIONS / SYSTEMS

How?

AVAILABLE RESOURCES

Requirements, Solutions and Resources

In all problems the important first step is to get *requirements* accurately defined before choosing any solutions, but this does not mean that needs capture should be divorced from solution suggestions. (In our TRIZ sessions needs capture is always simultaneous and awash with good solution ideas which are properly parked in the Bad Solution Park for later assessment). To define what we want while thinking of solutions, and then look through all solution possibilities (including all ideas in the solution park) is a better process than timetabling requirements capture (and allowing no simultaneous solution ideas) and then timetabling just problem-solving ways of meeting the requirements. Worst of all is to just start looking for solutions with no conscious link or pre-work on requirements. Defining all needs provides a focus, otherwise confusion reigns, because when engineers are inventing/designing a new system they have an exciting opportunity, but a bewilderingly wide choice of options if unmatched to the requirements/needs they should fulfil.

TRIZ Helps Engineers Balance Ingenuity and Time to Encourage Innovation in Design

TRIZ location of Resources via the Ideal helps pull together and give a clear overview of the steps between what we want (all benefits) and the best ways of how to get it – the right directions for designing, inventing and/or choosing the right solutions to fulfil all benefits. When all stakeholder requirements have been fully captured from other toolkits such as QFD, Value Engineering or Six Sigma then the new system design process is much assisted by applying TRIZ at this stage. The TRIZ 'quick and dirty' requirements capture through the Ideal Outcome is also helpful here as a quick check before any assumptions are made about solutions and new systems.

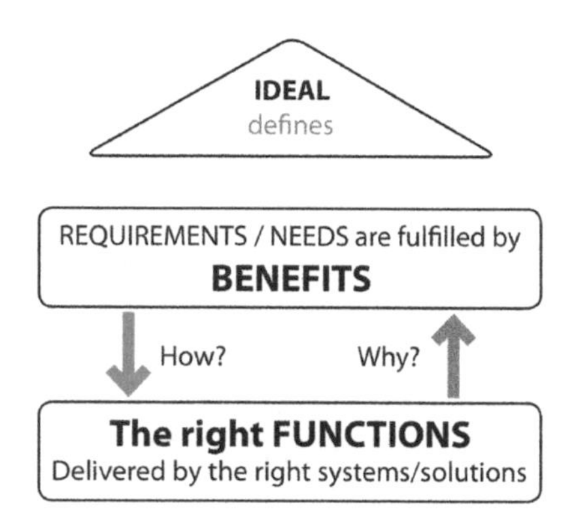

Functions = Solutions to Give Us What We Want to Deliver

Functions give us the solutions (which give us both benefits and harms). These come from resources as either inputs into the system or something within the system and its components. Other resources include the environment, time, process steps combinations of systems and the fields that join them.

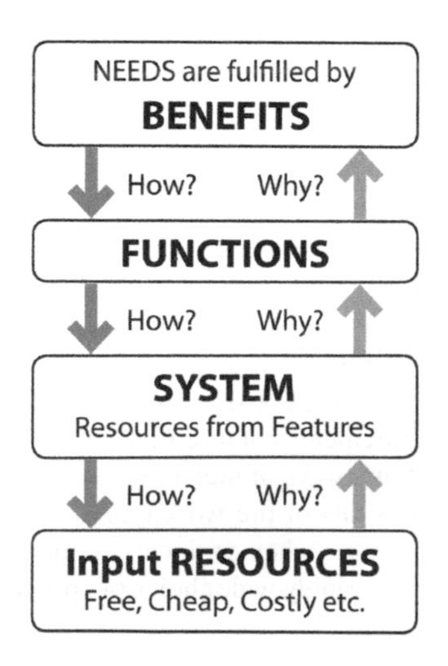

Therefore resources are not just inputs but also system features and system interactions in time and space. Understanding how broad the term 'resources' can be is important for all kinds of problems. When we talk about mobilizing resources we are looking at everything within and around our system which can be used. Ideally we start with everything that is already there, before looking to bring in anything new to solve the problem. If we know what we want (all the benefits) then we can look to see what functions could provide the benefits and what resources could provide those functions.

For example if the problem we want to solve is to avoid sunburn on our skin, the prime benefit we want is undamaged skin. There are many ways of solving it and many functions which could provide this benefit. We could protect our skin by covering it with a sun-block, or sunshades, or clothes, or sunscreen lotions. Or we could change skin in some way so that it is not sensitive to the sun by becoming very tanned, or taking some pre-tanning medication. Or we could use time as a resource by being exposed to the sun for the correct time period which prevents harm by having something inform us of impending sunburn. There are many resources to solve this problem of sunburn and these reside both within us and our skin, in our environment, in new inputs of sunscreens etc. Solving a problem is normally by cleverly combining available resources already in the system, in its environment with any input resources only as necessary. The aim is to minimize the input resources by cleverly using the existing resources. New clever solutions include: a sun-cream which uses UV light to activate it (it only works when the sun is shining on us); devices such as neckbands which tell us when we have had our maximum dose of sunlight on our skin; and clothing which is cool, light but prevents UV light from harming our skin (it solves the contradiction of not having to be thick and opaque to be protective).

Life and Death Solutions are Good Resources

TRIZ encourages us to use existing resources, especially other people's solutions and to look for existing solutions amongst those who need the solutions most – for those to whom the answer is life or death. Sunscreen solutions from Australia are an example of good solutions being developed as the ozone damage and an outdoor life create higher risks of skin cancer. In World War Two the US Air Force had a problem with fuel consumption that limited the range of their flights in the South Pacific. The man who had the answer was Charles Lindberg who in 1927 had been the first to fly solo across the Atlantic. Colonel Lindberg had wanted to rejoin the Air Force after Pearl Harbour but had been rejected partly because of his celebrity status, and worked in industry instead.

By 1944 Lindbergh had become a consultant with the United Aircraft Company helping them with field testing of their F4U Corsair fighter. In the spring of 1944 Lindbergh was in the South Pacific teaching Corsair pilots how to dramatically decrease their plane's fuel consumption and increase the range of their missions. (www.eyewitnesstohistory.com/lindbergh2.htm)

Lindberg's expertise got him back in combat. His focus on imparting his particular knowledge and skills on how to effectively use resources to double or even triple the flight distance on one tank of fuel was the key to his success in peacetime and war as an ace pilot.

Attitudes, Mood, Fears and Enthusiasm are Significant Resources Too

Clever resource use is important not just for good technical solutions but for ordinary life, and for business and management problem solving including dealing with people, when their enthusiasms, fears, motivators and de-motivators, likes and dislikes are resources which can be used to get results. This is the basis for many culture change management tools, which stimulate emotions, competiveness, uncertainty etc. to create a willingness or desire to see certain management cultural issues or solutions implemented.

Clever TRIZ resource use means getting what we want, for as little input as possible, by using what is already there. In the picture below people living in a remote and picturesque house want to discourage walkers from coming too close – a big fence or wall is expensive and far from Ideal. (The TRIZ Ideal System gives you want you want but does not exist.) Using resources within the walkers themselves, (fear of bulls) is an almost Ideal System. (The Ideal Bull does not exist!)

Forget fences – Arousing fears can be very cheap and effective way to keep out trespassers

TRIZ Problem Solving Using Resources

These can be summarized in the diagram below which helps us step through all available resources to match the defined needs. Resources are either inputs or are already there within or around the system and its use.

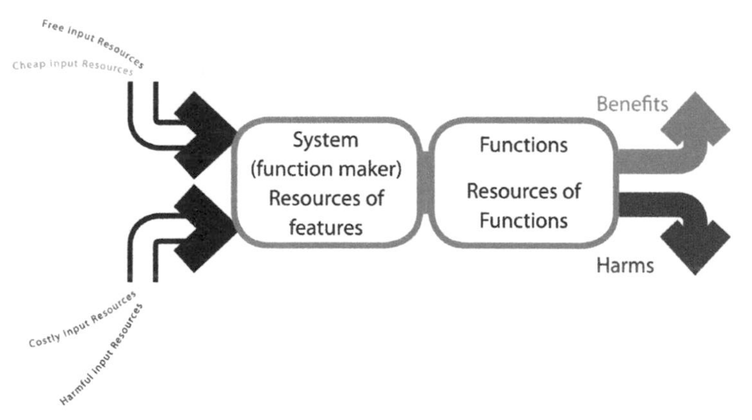

A detailed definition of Useful Resources is anything that can be used to give us what we want or solves our problem, that already exists in or around our system, and may not be currently used to its full potential. We need to systematically search for anything which may be usefully applied including all the things that are being partially used, or not used at all, or have never been used before.

Resources are:

All possible inputs

- materials (good and bad, cheap and expensive or even free)
- fields, actions, movement, interactions between components

- time and space
- engineering and scientific knowledge and experience (relevant information and data)
- harmful inputs.

Anything in the problem area

- systems, substances and components ... anything which can be used
- environmental and natural resources acting on and around the components – ambient temperature, air, sunlight, gravity, pressure, warmth, noise, smells, pressure, light, cold, energy in the system and environment
- spaces, voids and absences (important and easily missed!)
- any action, reaction or happening at the time of the problem ... mundane, or even invisible, because always been there
- harms and the causes of the problem – anything that appears at the same time and place as the problem.

  Harms we don't see as resources as we are usually trying to get round them. Using and transforming the bad things is one of the most powerful resource solutions. Look for all the *bad, annoying, nasty things.*

All features of our system and of every component (big and small, obvious and concealed)

- shape, behaviour, colour, transparency, symmetry, surface finish, smoothness /roughness, internal properties, magnetism, material properties, ease of use, cheapness, robustness etc.

All system interactions (components within systems and with other systems)

- any features of the fields/actions which create the interactions; fields acting on components – this is broader than the normal definition of fields as it includes pressure, information, temperature, gravity, weather conditions, smells, noise, energy carriers and flows – so heat, sound, smell, fear, enthusiasm, irritation are all fields and important resources.

Scale – context and detail – zoom out to the context and zoom in to the detail

- the environment – look at everything within and without the system
- details down to molecular level

Time

- preparation to perform some task
- changes occurring during process steps
- any time left to perform a task,
- before, during, after
- past, present, future
- how something may vary over time (e.g. phase changes)
- parallel operations (multi-tasking)
- any effect that time has on a system, e.g. some chocolates leave the factory in their packets in an unfamiliar state to how they are enjoyed by the consumer (the transport, storage and distribution time is a resource that is used to allow certain sweets to become crisp (having left the factory soft) and some fillings of chocolates to become runny – having left the factory in a viscous, sticky state.

Old Timers know about effective resource use

Resource Hunt

Undertaking a resource hunt is to ensure that we use everything appropriate and available – that we match the functions we want to resources. We ask how can we combine, transform, eliminate, change anything or everything in the system to get everything we want? We must remember to always include the causes of the problem – the harms, the harmful inputs and try and transform harmful outputs to something useful. For example blizzard survival sleeping bags claim the harmful outputs of their rustling noise keeps climber awake when they are looking out for rescue. Resources thus include everything in the system itself, problem areas, its details and components and anything in the big picture, and in the surroundings, in the environment including pressure, cold/heat, knowledge, gravity, weather conditions, smells or noise, anything that appears at the same time as the problem, anything that is there is counted as a resource and may enable solutions. This also includes the reserves which are often invisible until you look for them. We need to mobilize the best resources by systematically looking for things that are:

- in or around the system
- somewhere close to where they are needed and available when they are needed
- cheaply and easily (for very little cost) available and the causes of the problem.

In nature sometimes competition for food could come from your siblings: available resources can be made available by transforming your immediate family into a food source, which occasionally happens with tadpoles.

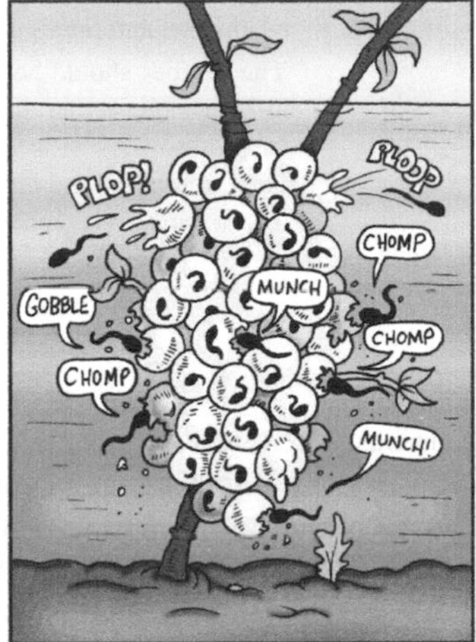

Many small mouths create an effective eating machine – weaker siblings are then a good food source

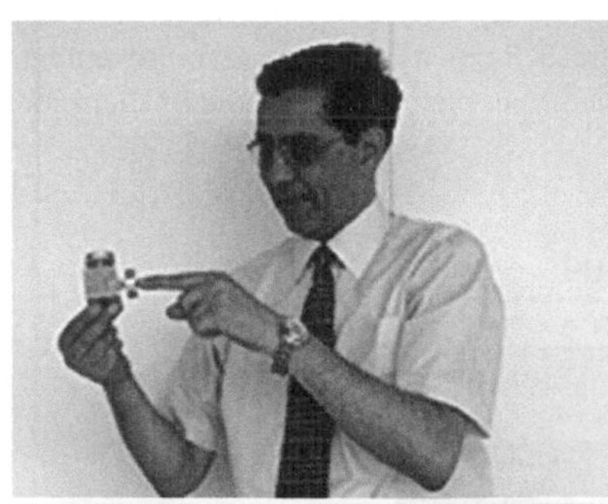

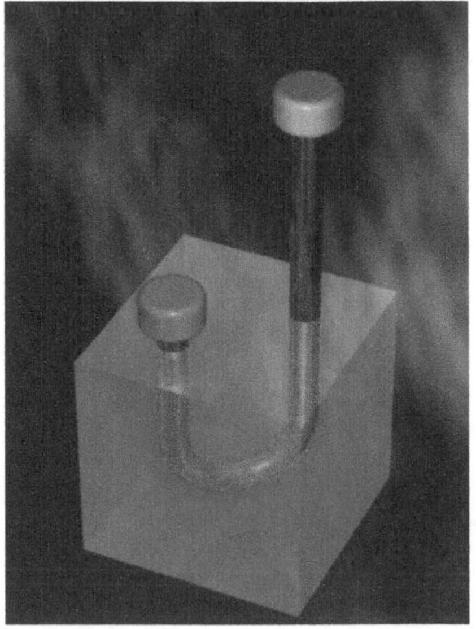

A superb invention example of using Sub-system resources

A brilliant invention for the perfect shock absorber uses the simple resource of the material properties when metals are forced to deform. By forcing a solid material to deform as it passes round a bend in a closed channel, it is possible to construct compact, low-cost and infinitely re-usable devices that absorb large or small amounts of mechanical energy.

This is using the resources in the sub-system level by searching all the features and functions of all components and their materials – zooming down to the detail level (molecular if necessary). A Press release from Bath University in 07 September 2005 stated:

'A simple but clever idea by a Bath engineer could revolutionize the way that safety devices across the world are constructed. Dr Fayek Osman's new concept could mean that devices such as train buffers, safety barriers and aircraft undercarriages will be much more efficient and cheaper. Dr Osman, of the Department of Mechanical Engineering in the University of Bath, UK, has developed a concept for devices that can absorb enormous impact and yet still remain intact so that they can be used again.

'Normally the impact of a crash on a safety device such as a train buffer will deform it so that it cannot be used again and must be replaced. Even ordinary stress on devices that absorb less dramatic impacts, such as aeroplane undercarriages, can wear them out quickly. Dr Osman's idea is that safety devices should be as simple as a piece of metal in a channel with a bend in it. During a crash, impact forces the

metal down the channel, and the energy of the crash is absorbed by the metal as it travels around the bend towards the end of the channel. The channel can then be turned around so that the next impact strikes the metal at the end of the channel and forces it back to its original starting point.

'The devices should be cheap to make and would require no replacement after an impact. Their uses could include shock absorbers in artillery pieces, car bumper mounts, joints in bridges and cranes and shock absorbers in buildings in earthquake zones.'

Using Super-system Resources

Resources include everything all around us which we could use but have to consciously look for or remember when solving problems. Super-system resources can be the easiest to forget because they include everything surrounding us in our environment, such as the air we breathe, gravity, knowledge, sunlight, temperature, pressure, substances or energy fields of any kind. These are all freely available to us.

We normally start by trying to find resources that are fairly close to the problem area – the system we are working on. An important part of TRIZ is not to overlook a resource because it's normally regarded as harmful. We must also remember things that are so obvious it is easy to forget them. Many brilliant solutions come from using existing free but very mundane resources. Resources are key to improvement in most systems and many resources are often left unused because people can't see some of them, or they can't see their relevance or they see them as harmful.

Even the simplest environmental resources – hills – can be very effective solutions

Resources and Hazards

Resources solve problems right at the roots of problems – the chains of causes of problems are important in problem solving and asking WHY? will generally take us to the fundamental cause of a problem. Many people think of problem-solving tools as finding-root-cause toolkits – and although such toolkits are important elements in problem solving they are only one in many. Locating with accuracy what is causing a problem is important and once we know the cause the solutions can sometimes be obvious – but will always need resources. For example at Amsterdam's Schiphol Airport, airplane collisions with birds grew from 3 incidents per 10,000 flights in 2003 to 5 in 2004. A KLM Boeing 737 skidded off a runway when a bird hit its engine as it took off. 'The nuisance is a regularly re-occurring subject of discussion with Schiphol. They could damage a plane's engines and place a flight at risk,' KLM spokesman Hugo Baas said. The Airport tried putting out dog patrols, alarm pistols and using tapes with shouting noises to get rid of the birds, and then someone asked why the birds were there? It was investigated and found that the birds were feeding on the mice in the grass near the runways. Mice attract birds such as buzzards and herons – once this link was established the problem has become how to get rid of the mice. The next question was what resources are available in Amsterdam? One resource is knowledge of mice and it is known that mice hate the smell of tulips. Dutch tulip growers were asked to plant hundreds of thousands of flowers at Amsterdam's Schiphol Airport to deter mice. The result was a lovely display of tulips and their smell drove away the mice, and with no mice there should be fewer birds.

Mice hate tulips – dangerous birds leave airport because there are no mice to eat

Resources When in Peril

In perilous situations when there are limited resources, especially of time, smart thinking about resources, use of everything useful and readily available can be a matter of life and death. Using the causes of the peril as a resource is a very clever transformation of harm into good. In the picture below the hole in the ship is hard to deal with because of the force of the water rushing in which could be defined as an input harm. Using the sail to temporarily patch a ship on the outside uses the harmful force of the water pressure to hold it in place. Transforming, using the harm for good, is an old sailor's trick.

The film Apollo 13 showed a famous example of a closed system, with restricted resources, a limited time window for survival, and showed how the crew were rescued and delivered by clever application resources, a smart combining of everything relevant available, especially experience and brain power. Using harms as a resource is almost second nature by those accustomed to perilous situations.

In Honduras in 1998 after Hurricane Mitch many people died from exposure to the elements and lack of water and shelter. In some areas people were helped by carefully considered TRIZ expert advice on resources, given from Rolls-Royce TRIZ experts in the UK by phone, advising how to quickly and effectively use the few available resources (especially both harmful and environmental) to deliver water and food to many in the immediate days after the storms. The resource used was the flooded river, large plastic bottles were filled with clean river water (or food or simple tools) at the head of the river. marked by small flags and floated down to reach people cut off by destroyed bridges and roads.

Four Simple Steps to Using Resources More Effectively

(= Quick Ideal Problem Solving as shown in Chapter 6)

1. What is the problem?
2. Define the Ideal – to list the essential requirements/needs to define benefits.
3. Uncover/define the functions which deliver those benefits.
4. Search for resources around the system to deliver those essential functions.

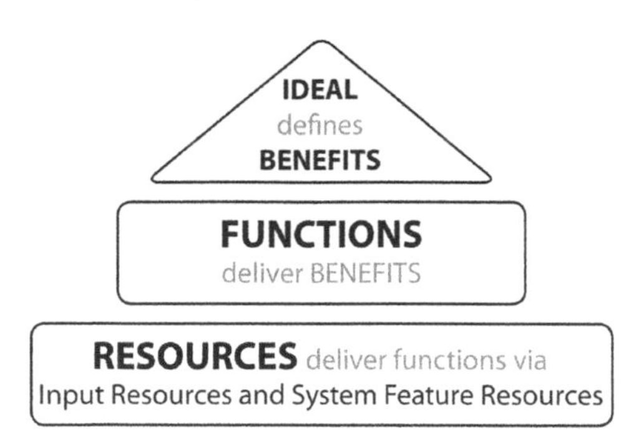

In essence TRIZ shows us how to move from the system we've got to the system we want by applying available resources in ways suggested understanding the essential functions we are seeking by the TRIZ solution triggers.

Where to find Resources? The answer is from everything in and around the system from available inputs, the components of the system, harms etc. These include 'system feature resources' such as each individual feature of components such as surface roughness or smoothness (roughness is a resource when we are seeking useful friction such as in Velcro; smoothness is a resource for systems such as a child's slide or a food conveyor chute). Other useful features can include colour (helping visibility in dangerous situations) – a white coat is safer than a black one at night. When seeking resources we look at everything in and around our system which could provide the functions we need to deliver benefits. For example many children's playground's are designed to be both fun and safe. New designs of trampolines in playgrounds have these at ground level over a hollow pit surrounded by soft sand. This design uses environmental resources of a low or zero height to ensure that if a child bounces off, they have little or no distance to fall and have a soft sand landing. The ground level design uses simple free environmental resources and the sand (although a costly input resource) ensures extra safety. Harmful inputs are also resources – such as pain which gives us feedback to prevent further injury.

RESOURCE HUNT to mobilise the resources we've got

SOMETHING (already there) WILL PROVIDE THE FUNCTIONS WE ARE SEEKING (including X-FACTOR)
We list all the resources in and around our system particularly those near the problem and needed only when the problem occurs. We try and use resources unchanged but otherwise in the order :-

- **UNCHANGED / RAW - as found**
- **COMBINED - with other available resources**
- **MODIFIED - changed to provide the needed function**

Assuming we have multiple resources - Resource A, Resource B and Resource C
and we also need a combination of functions - FUNCTIONS 1 and 2 - try each in turn
(Define exactly where and when the function is needed)

Ask the simple question of each resource in turn

Can Resource A provide Function 1?
Can Resource A provide Function 2?
Can the component (which Resource A needs to act on) provide the needed functions itself?

Can Resource B provide Function 1?
Can Resource B provide Function 2?
Can the component (which Resource B needs to act on) provide the needed functions itself?

Can Resource C provide Function 1?
Can Resource C provide Function 2?
Can the component (which Resource C needs to act on) provide the needed functions itself?

Do this with a team of experts/engineers with the relevant experience.
This simple exercise solves many problems elegantly, cheaply and effectively.

TRIZ Triggers Plus Resources for Practical Solutions

Minimizing inputs is achieved by smart use of available resources. Non-TRIZ problem-solving kits, which also champion resource use, exhort the use of long resources lists, extensive resource hunts etc. to ensure that the best resources for the job are found. TRIZ offers much cleverer, neater and quicker location of resources through the very clever, very focused TRIZ problem model which has a small number of TRIZ solution concepts to solve its problems. These TRIZ solution concepts (40 Principles. 8 Trends and 76 Standard Solutions) suggest solutions which help us look for the appropriate resources and which move us from the present system to a better system by applying these TRIZ concepts. This process may be repeated until we are happy with the improved system or ultimately until the Ideal Outcome is reached when Ideality has only the benefits we want with no costs and harms.

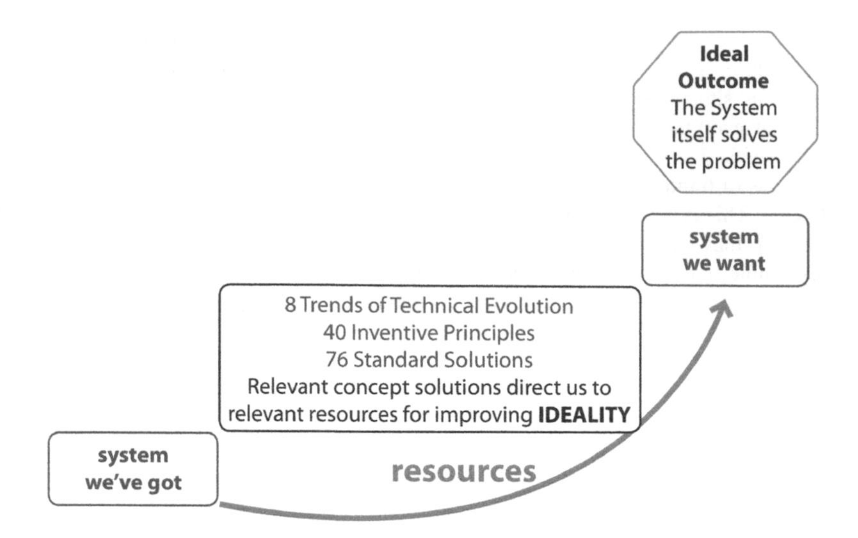

TRIZ is about the combination of resources and whatever the solution trigger suggests. One overlapping solution trigger which appears in all three lists is *segmentation* – when prompted by this we are being asked to look for resources which would solve our problem by using segmentation. Segmentation is the first of the 40 Principles for Solving Contradictions such as how can I have a door that is *always there* to separate two rooms or two environments but *never there* when I want to walk through it?

Segmentation has always been a solution for systems that have to be *there* and *not there* at different times such as scaffolding or staging – for problems of solid structures but one which can be moved from place to place. Available resources and solving contradictions with inventive principles including segmentation suggest design solutions which show us how to achieve contradictory benefits – such as easy to assemble and disassemble, light and easy to move from location to location but then strong and stable in situ. In essence TRIZ shows us how to define the desired move from the 'system we've got'

to the 'system we want' in one quick step by jumping straight to the TRIZ Ideal and looking for resources to deliver it, or in a few smaller steps by applying available resources in ways suggested by the TRIZ problem solving processes for solving contradictions, dealing with harms or boosting benefits. These lead us to solution triggers from the 40 Principles, 8 Trends or the 76 Standard Solutions which are implemented by mobilising resources.

Clever Solutions Use the Right and Available Resources

Imagine we have to move a large and heavy bell – our ideal would be that the bell is easily moved with no inconvenience to anyone and using no inputs. If we look at the TRIZ Contradiction matrix for how to improve *automation* without making *convenience of use* worse it suggests four solutions concepts from the 40 Inventive Principles, 1 – segmentation, 12 – equipotentiality, 34 – discarding and recovering, and 3 – local quality. It is the problem solver's task then to look for resources which use these concepts to solve their problem

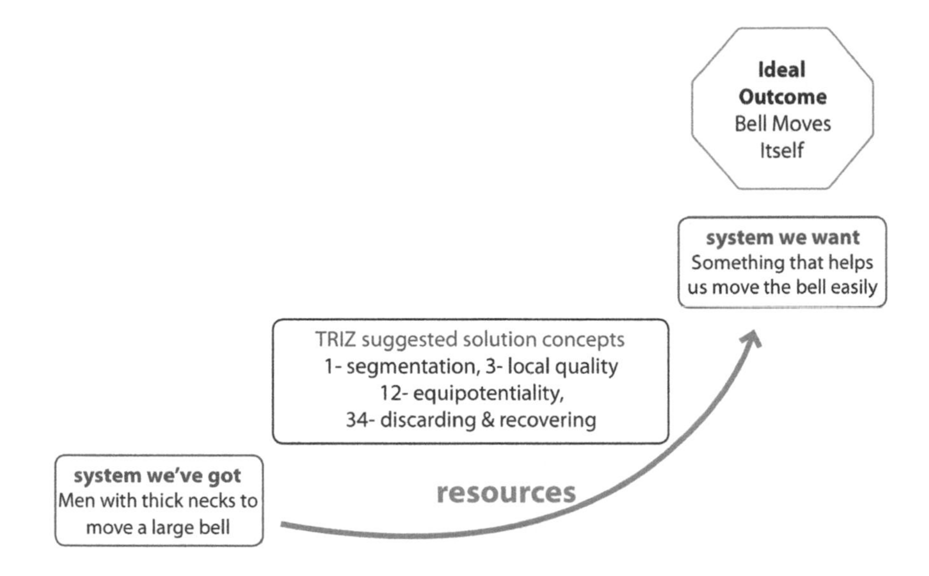

Think about available resources to solve the problem by applying each of the suggested four TRIZ Inventive Principles.

Resources (simple feature of shape) offer clever and simple solutions

Simple Steps to Resourceful Systematic Problem Solving

Like many ideas in TRIZ the simplest are the most powerful. Perhaps the grim beginnings of TRIZ in Siberian Gulags helped create a TRIZ culture of seeing the possibilities in everything around us as something which could be put to good use. The simple TRIZ way of looking afresh at resources with the Ideal Outcome focuses our thinking to help create shortcuts for simple problem-solving steps, taking us from the Ideal (for what we want) to resources (for solutions). There are three simple similar routines for solving problems by looking for resources and all use Ideal – functions – resources:

- Define the Ideal Outcome and all benefits, seek functions/features to provide benefits and then locate these from relevant resources.
- Define the Ideal System and identify its essential functions, deliver these by an Ideal self-system made from resources.
- Identify Ideal Solutions or Functions which are made from a mystery X-Factor, then try and create the X-Factor from available resources, which solves all problems.

The idea of the Ideal Outcome is that it gives us all the benefits we want. The Ideal System gives us all the functions which give us the benefits without any costly inputs – these in turn could be described as 'Ideal Functions' or 'Ideal Solutions' which need no inputs but somehow give us everything we want. How could we have functions with no costs or inputs? By using what is already there (existing resources) to get what we want. An Ideal Function is made out of existing resources and needs no extra inputs and therefore no costs. The simple TRIZ rule is to keep stating the Ideal solves the problem itself.

Quick Ideal Thinking

When we start defining our Ideal, two activities occur simultaneously – particularly when working in teams – we understand our needs and generate many solutions. Once we are into powerful solution mode then this can often get us thinking creatively and clearly. Such high-octane thinking moments should never be repressed – and all such ideas and solutions should be simply and immediately recorded in a Bad Solution Park. (All environments that are designed to be creative have places everywhere to record ideas – Leonardo da Vinci said always carry a small notebook and never lose any idea or solution that comes to you.)

Ideal Outcome, Delivered by Essential Functions – Look for Relevant Resources

This simplest of approaches follows simple steps Ideal – functions – resources

- define what we want (benefits) with the Ideal
- define essential functions to deliver them
- look for available resources.

Recently I have worked with a number of organizations on escape systems from vessels which have to be abandoned in high seas. The Ideal was defined as being delivered safely home. Before that stage some benefits were needed:

- escape
- survive
- be found alive and well by rescue and transported home.

The essential functions to deliver this after escaping were identified as

- float
- stay warm
- stay together
- be located by rescue
- sustenance – such as water and food and medication.

With such a simple list we were able to start problem solving by thinking about available resources. Looking for available resources would depend on the vessel we are escaping from – for submarines we are looking for something which can be stored within the tight constraints of space and can be removed through the small available openings by super-fit submariners. For a cruise ship there is room for lifeboats but passengers who may be too old or infirm to reach them or get into them. For an aeroplane landing on the sea there are other constraints of weight and access.

In any situation this simple approach of Ideal – functions – resources gives quick solutions.

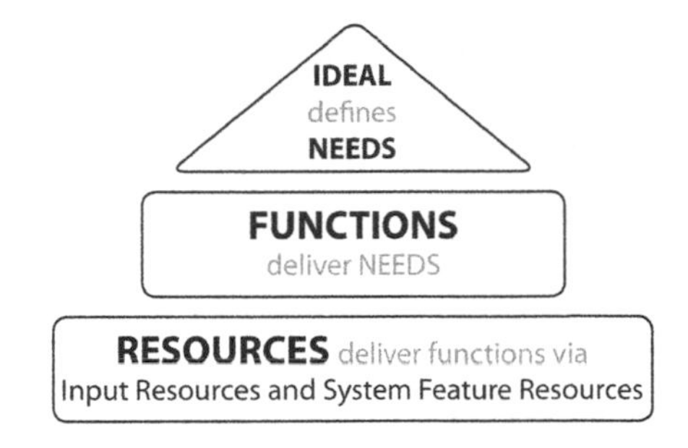

Starting with the Ideal is very powerful, very quick and very useful, using TRIZ by simply imagining an Ideal, to capture what we want, and inevitably thinking of fast solutions of providing the Ideal, and looking for those solutions in the resources, by searching the available resources to deliver the Ideal or as close to the Ideal as possible. Therefore defining the Ideal is a very straightforward TRIZ tool to point us in the right direction, to stimulate our thinking for understanding needs, and for imagining systems and solutions (good solutions reduce the gaps between what we want – *needs* – and what we have *current system*). Thinking about the Ideal Outcome is a very simple but systematic way of putting our brains in gear for both problem understanding of all our real, essential requirements and then helping to imagine some Ideal Solution to deliver the Ideal Outcome.

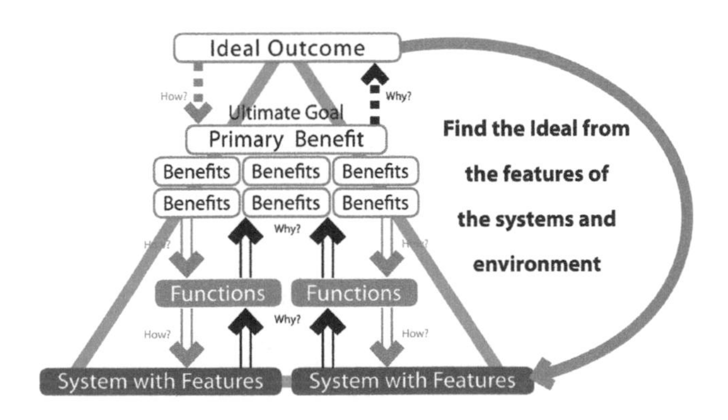

The Ideal sign is only there when there are floods

Ideal self systems – give us all our needs

Clever systems use available resources to provide self systems which give the functions we want, ideally giving us clever solutions only where and when needed, using available inputs to give beneficial outputs. They even mobilize harmful or problem inputs to be useful as in the snowman road sign.

One brilliant self system and a very practical example is the life-saving brilliance of Cat-eyes in roads. Light from cars guides the cars, with help from the Cat's-eyes which are always there, but only operational when needed. This using energy only when needed, and using the energy we've already got, is fundamental to our approaches to good carbon footprints.

The genius of Percy Shaw's invention of Cat's eyes in the 1930s was that they are also designed to be self-cleaning and use the occasional resource of cars running over them, to harmlessly push them down, to expel the useful and free resource of rainwater to wash them. They could be made more useful maybe by also indicating hazards by using more of their available resources simply changing colour to warn drivers when it is very cold and icy. New designs of cats-eyes use environmental resources of sunlight to power bright light-emitting diodes which illuminate the road ahead even without car headlights. Cat's eyes are a perfect system to demonstrate clever use of available resources as they are active when they need to be, and use all resources even the harmful ones of being run over by a car.

The Ideal Solves the Problem Itself – Ideal Self Systems

Seeking Smart Resources to Achieve Certain (Often Opposite) Benefits

The important TRIZ prompt for powerful solutions is to look for a self system – which itself delivers solutions using available resources, creating a self system which is simple yet effective. Self systems themselves deliver everything we want at different times, places and conditions. Self systems change and adapt for all the different process steps, e.g. *is there* when we want it but *not there* when we don't want it like the floods ahead road sign. Self systems solve physical contradictions (deliver opposite benefits) using resources to separate and still cleverly deliver the conflicting benefits.

Resources (material features) deliver different benefits in space

The constant, unchanging with time, system offers opposite things in different places – an Ideal smart food belongs with cartoon on next page plate is only hot in the middle – where it needs heat, but is cool on the edge where it touches human fingers).

YOU SHOULD TRY THESE 'SMART PLATES'. THEY'RE ONLY HOT IN THE MIDDLE!
© Oxford Creativity 2009

Even warmed plates don't ever **need** hot edges

Resources (voids/holes) deliver different benefits on condition

The hit and miss fence has holes in which allows it to respond differently to different elements – an Ideal fence lets the wind through, but not light, so gives privacy and does not get blown down.

THE ADVANTAGE OF THIS PANEL IS THAT IT KEEPS YOUR PRIVACY BUT LETS THE WIND THROUGH...
© Oxford Creativity 2008

The Hit & Miss Fence demonstrates how voids or holes are great TRIZ resources

Modern Solution to an age old problem – how to get into skin tight clothes

Resources in the dress (phase changes) deliver different benefits in time

The system changes – the Ideal dress is large and easy to put on and therefore doesn't need openings with zips or buttons, but becomes small and snug fitting when needed – it changes in time.

Ideal Self Systems – Ideal Resources Used to Design a Tomato Sauce Bottle

The traditional glass bottle for tomato ketchup remained popular for many decades after its first appearance in the nineteenth century. It used the feature/resource of see-through glass to market its lovely colour and to show how it contained pure tomato. However it always had the problem of being notoriously difficult to deliver the right amount of sauce in the right place. Traditionally after giving the bottle a good shake, a thump on the bottle's bottom shoots out dollops of sauce (a thixotropic fluid) – sometimes exactly where it is required on the food – sometimes on us or on the wall. The famous solutions to this problem were found by using simple available resources to improve the performance of the traditional tomato sauce bottle.

The **Bottle** or **Ketchup** itself moves/delivers dollops of sauce exactly when and where required.

In this system we have three main elements (or resources) of the person, the ketchup or sauce, and the bottle.

Using the TRIZ tool of the Ideal we would state the solution as follows:

The Ideal solves the problem itself = The bottle moves the sauce (we look for ways of doing this)

or

The Ideal solves the problem itself = The sauce moves itself (we look for ways of doing this)

In a resource hunt we try everything near and within the person, bottle and the sauce.

Some resource will move the sauce?

What is it? We look in resources to find it – but within the area we have specified.

The Ideal solves the problem itself = The bottle moves the sauce (we look for ways of doing this.)

The bottle will move the sauce. How?

What features of the bottle could we employ for this task?

The bottle has shape, glass, a top, thickness, rigidity etc. These bottle components and features will somehow move the sauce. How could any of the features of the bottle, or anything surrounding the bottle, move the sauce?

We ask the question about each feature in turn.

The solution *squeezy* was introduced about 100 years after the original glass bottle

Many clever resources are used even in simple things

The Ideal solves the problem itself = The sauce moves itself (we look for ways of doing this)

Person moves sauce or the sauce will move the sauce. How?

What features of the sauce could we employ for this task?

How could any of the features of the sauce, or anything in the surroundings, move the sauce?

The sauce has all the properties of a viscous liquid.

Something in the environment will move the sauce? How?

Sauce will move the sauce? How?

The sauce features will move the sauce? How?

The environment will move the sauce? *Gravity will move the sauce?*

The upside-down solution for bottles containing viscous liquid such as ketchup and hair conditioner were introduced at least ten years after the squeezy bottle appeared on the market.

Best Use of Resources – Overall TRIZ Philosophy

TRIZ teaches us to seek out and explore resources. Good engineers working at their best come up with good solutions cleverly using resources. Often they believe this to be instinctive as they are thinking fast and this re-enforces the widely held belief that 'A clever engineer can do for £10 what any fool can for a £100.' Engineers can do clever thinking, but not always to order, or to a timetable. The TRIZ approach to resources helps timetable smart engineer thinking, and helps us produce it to order when required. In this way TRIZ can help all engineers when their normal methods have only produced some mediocre solutions; what TRIZ then offers is the transformation of ordinary solutions into better ones, by reproducing the kind of clever ideas which come to us as a brainwave. The TRIZ process takes us through all the systematic steps which happen apparently instinctively, in, very fast, smart problem solving. Amongst these steps is the conscious, good use of resources and this is an integral part of this 'clever engineer' behaviour – the best solutions always come from available resources. Looking at very clever inventions it is the mobilization of what is already there which is often breathtakingly clever.

Simple sub-system resources to give us everything we want

New inputs are resources too but have to be paid for in some way, not just counted in money, but also in time, in transport, in complication, in future maintenance, in carbon footprint etc. Therefore a search of available resources should nearly always precede bringing in new resources. The TRIZ approach to problem solving is to be initially alert to all the resources that are fairly close to the problem area and not ignore a resource because it may be normally regarded as harmful or because it is so obvious that we don't see it – making it is easy to overlook. In science and engineering amongst the many brilliant solutions which come from using existing free resources are those which have accidentally mobilized the problem, harmful elements which have helped create anything from penicillin from dirty dishes, to melamine cups which are easier to clean when left unwashed for longer.

8 Ideal and the Ideality Audit

Ideal to understand Goals → Benefits → Outcomes → System → Functions → Solutions

Most things we seek in life begin with intelligent understanding of what we want, and where we would like to be, accompanied by problem solving to try and achieve these. Understanding both what we want and how to get it, takes great clarity of thought, organization of all the facts, an ability to recognize and obtain the relevant resources, and uninhibited brain power. TRIZ offers this simple tool – defining the Ideal – to help us to achieve all of these through a simple process of imagining some magical ideal way of getting everything we want, with no problems, no costs and no downsides of any kind. Picturing this theoretical Ideal helps us to clearly understand what we want, all the big things, our ultimate goals, as well as all our other smaller requirements – no matter how difficult or contradictory these benefits are in reality. Imagining the Ideal also often prompts us towards good outcomes to find many solutions. Hence the Ideal helps us to understand requirements and stimulates our brains to think up ingenious ideas; furthermore the Ideal helps us recognize and mobilise free and available resources to implement cost-effective, environmentally friendly, efficient solutions.

Defining the Ideal to find clever solutions and using good, available resources is in Chapter 6 – 'The Ideal Solves the Problem' which sets out the simple steps of starting with the definition of the Ideal to lead to resourceful systematic problem solving. In Chapter 7 how to find and apply resources to problem solving is explored. In this chapter we look at the power of the Ideal for capturing requirements effectively, if somewhat approximately, but then how to define practical requirements more carefully and accurately with an Ideality Audit. Undertaking an Ideality Audit is essential for problem understanding and to set the direction for problem solving and to log all the steps that have been covered. It is important to record the innovation process which helps us agree and define the benefits, costs and harms for the system we have, the system we want and the innovative directions we may wish to explore after defining the Ideal system.

An Ideality Audit helps us list the problem gaps we have to close in order to move from the current system to the system we want. Defining the Ideal both sets the direction of that journey and also defines all the benefits we would like to ultimately achieve – the top line of the Ideality Equation, the fundamental statement of TRIZ problem solving. The Ideality has a precise meaning in TRIZ as shown below:

TRIZ for Engineers: Enabling Inventive Problem Solving, First Edition. Karen Gadd.

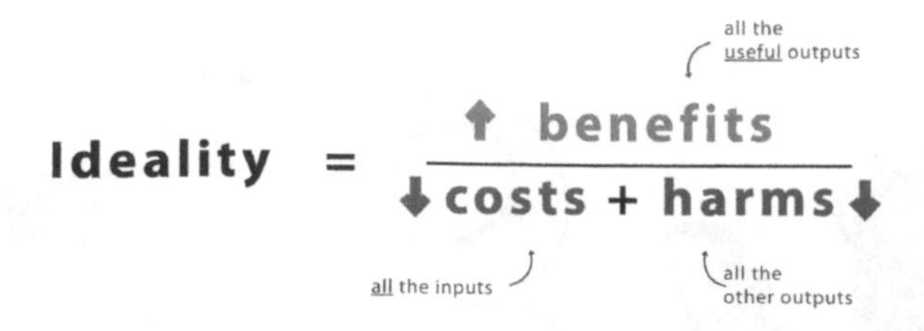

Ideality Audit

An Ideality Audit requires that we accurately record three separate elements – Ideality of the 'system we've got', the Ideality of the 'system we want' and the Ideal (all benefits no costs and harms). To solve problems we need to move from the Ideality of the 'system we've got' to the Ideality of the 'system we want'. By defining the Ideality of these two states, we can then locate all the gaps between where we are and where we want to be. But before we do this it is useful to map the right problem-solving direction by also defining the Ideal, which defines everything we want and the ultimate place we would like to arrive for the perfect solution. This process can stretch us to move further towards the Ideal than we first intended and look for more innovative solutions beyond solving our immediate problems.

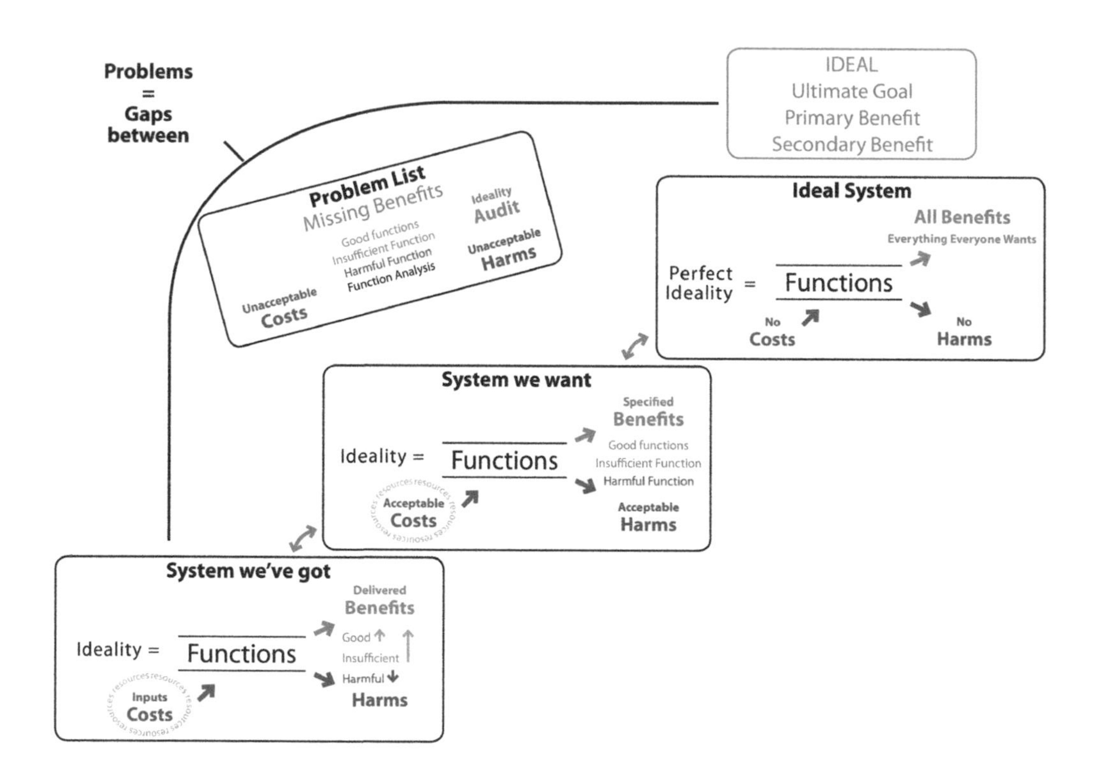

Ideal in TRIZ Comes in a Number of Names and Tools

- Ideal Outcome helps to understand requirements and is delivered by Ideal Solutions (no costs and harms).

- Ideal System delivers Ideal Solutions via delivery of all the right features and functions to fulfil all the requirements.
- Ideal Functions are made of available resources (or X-Factor) which deliver Ideal Systems.

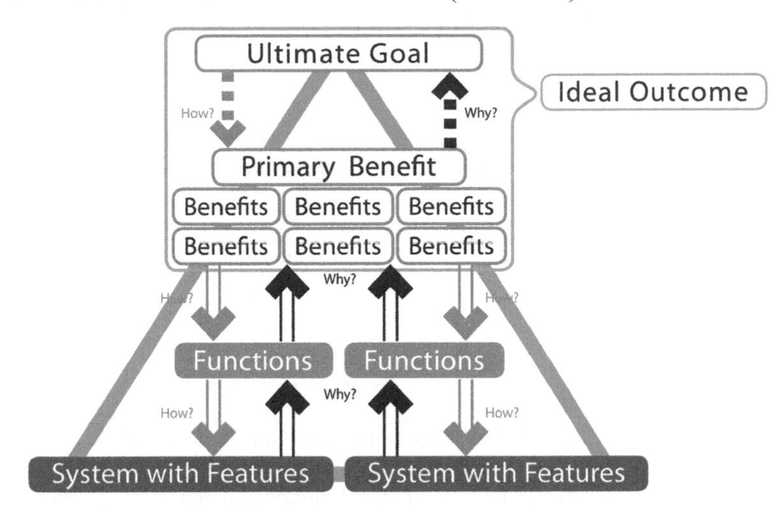

The Ideal Outcomes/Ideal Solutions can be distinguished from each other in that the Ideal Outcome is delivered by the Ideal Solution, which just magically, somehow takes us to a place where we have everything we want – this is the Ideal Outcome. Distinguishing between them may not be very important except that one defines a situation where we have arrived and have everything, and the other describes the change, the move to that place, the solution which changes and delivers everything we want. As they both overlap in giving us a list of everything we want they are often used interchangeably, and defining one helps define the other.

Ideal Outcome in the Bigger Picture

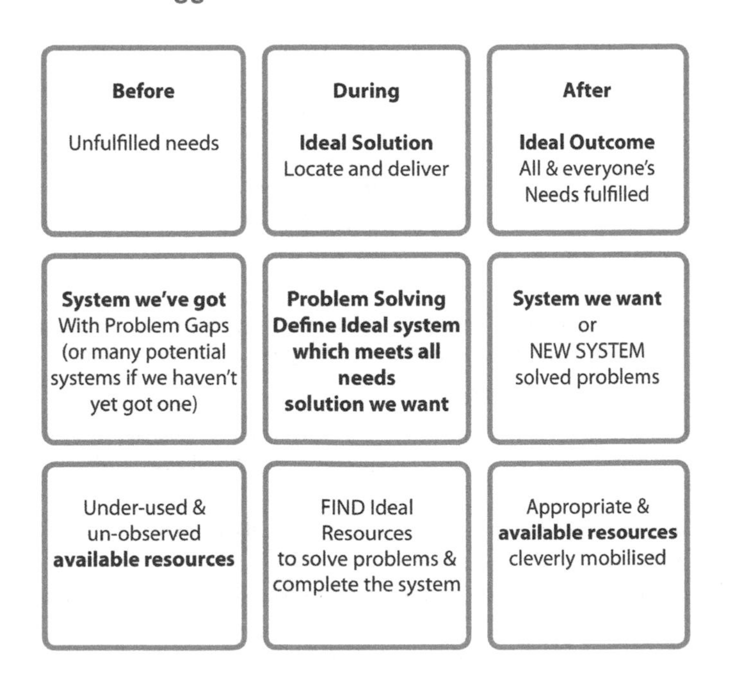

Just because the TRIZ Ideal processes are efficiently fast doesn't mean that we should skip defining all the levels of the benefits we want. Requirements have their own hierarchy from the ultimate goal, to primary benefits and secondary benefits which can be also divided into higher-level benefits (must haves) and lower level benefits (nice to haves). Accurate capture of these is important.

Ideality Audit Begins with the Ideal Outcome

The Ideal Outcome has a precise role in the Ideality Audit – it contains the ultimate goal, the reason the system or process exists, and all other benefits (secondary and primary benefits).

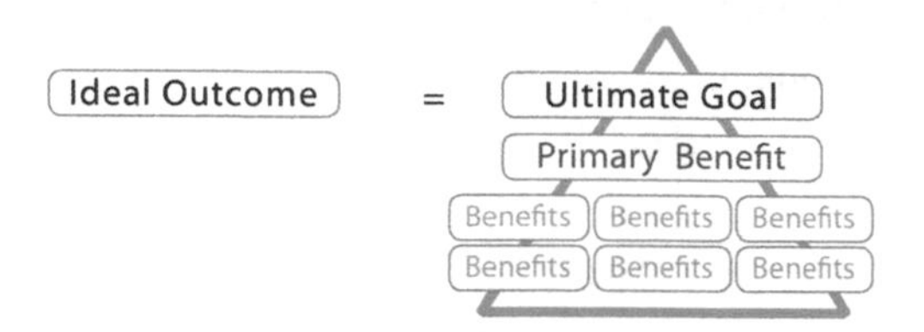

Defining the Ideal Outcome properly helps us to define the ultimate needs, all requirements of the ultimate *system we want,* and after this step TRIZ additionally offers greater system understanding by undertaking a function analysis to accurately, and in detail, understand and map any problems with the *system we've got*, especially any harms or insufficiencies of functions. (There are TRIZ Solution tools, especially the Standard Solutions, that guide us to solve the problems.) When we need to visualize all the benefits our Ideal Outcome should deliver, we use the *Benefit Capture Exercise* (see below) or alternatively one of the many other tried and tested routes used by a variety of requirements-capture toolkits. Combining these with the TRIZ logic gives us some simple and appropriate steps to begin problem solving by recording all the requirements.

Define Ideal Outcome (if you had a magic wand what would it give you):

1. Ultimate Goal (why the system exists)
2. Primary Benefit (the one, main essential benefit)
3. Other principal benefits as recorded in detail by Benefit Capture Exercise
4. What the system currently delivers
5. The gaps between current system and the Ideal Outcome

To initially understand the system's problem gaps – all the ways a system doesn't meet all our needs – we have to undertake this fairly simple mapping of what we want the system to do, compared to what it is actually delivering, and build the problem/Ideality gap list from the differences between them.

Think about Benefits we want from everyday systems – e.g. when building roads, who pays? Do they care about long-term benefits or if they last beyond our lifetime?

What benefits do we want from Roads? These might be:

- smooth/perfect surface, safe for driving in all conditions, no noise or pollution, no spoiling of landscape (put motorways in tunnels?), oil and debris removed without polluting the surroundings, safe environment when accidents occur;
- no barn owls/birds/deer/foxes/kangaroos in danger from roads, the right number and placing of roads to minimize congestion;
- the minimum number of roads spoiling towns and countryside, durable (minimum road works), minimum cost short term.

What contradictions are there between these requirements?

What constraints prevent us building the roads we should have?

Better outcomes usually require more up-front investment but better (cheaper) in the long run is not always the option adopted.

Benefit Capture Exercise

This is used by a team to map/record the complete Ideal Outcome of a system.

Defining the Ideal Outcome, capturing all needs/requirements/benefits both primary and secondary, must be undertaken without considering costs and harms at this stage.

Follow the simple five steps:

1. Everybody individually records everything they want from their system on stickies. They write down each benefit, function or feature they want (one per sticky). This helps record everything, everyone wants from a system. At this stage we ask people not to try and organize or sort them – just randomly record top-of-the-head ideas.
2. Everyone brings all the stickies together
3. All participants work together to sort the stickies into groups into categories – naming these gives an approximate principal benefits list. All the stickies are sorted under simple headings. Duplicates are discarded. Overlapping solutions (subtle differences) should be kept in the groupings.
4. Together everyone debates and identifies main/principal benefits by giving each *benefit group* a name.
5. By mutual consent the benefits are then numbered according to priority.

This exercise gives a simple list of the principal benefits we want; we must not forget to also define the very obvious main benefit – the ultimate goal and also the prime benefit of our system. The example below shows the results from a team gathering all the benefits they wanted from a car. They defined the ultimate goal as transport and prime benefit of a car as independent, enclosed, motorized locomotion.

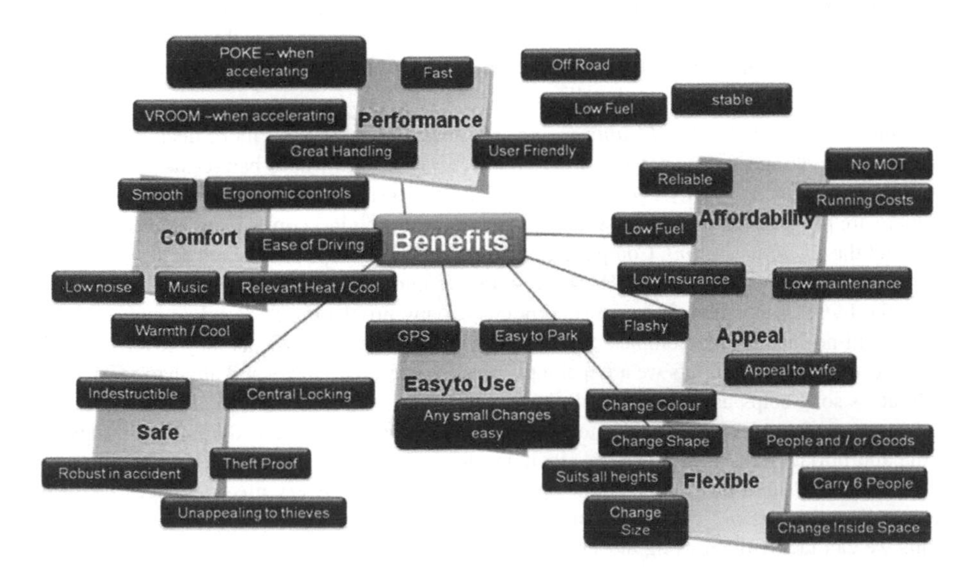

After capturing and giving some priority to the benefits the teams are then required to fill in the Ideal Outcome sheet (an A0 paper sheet. In problem solving sessions groups of 4–6 people each work on such sheets to summarise and define the Ideal Outcome for the problem in question.

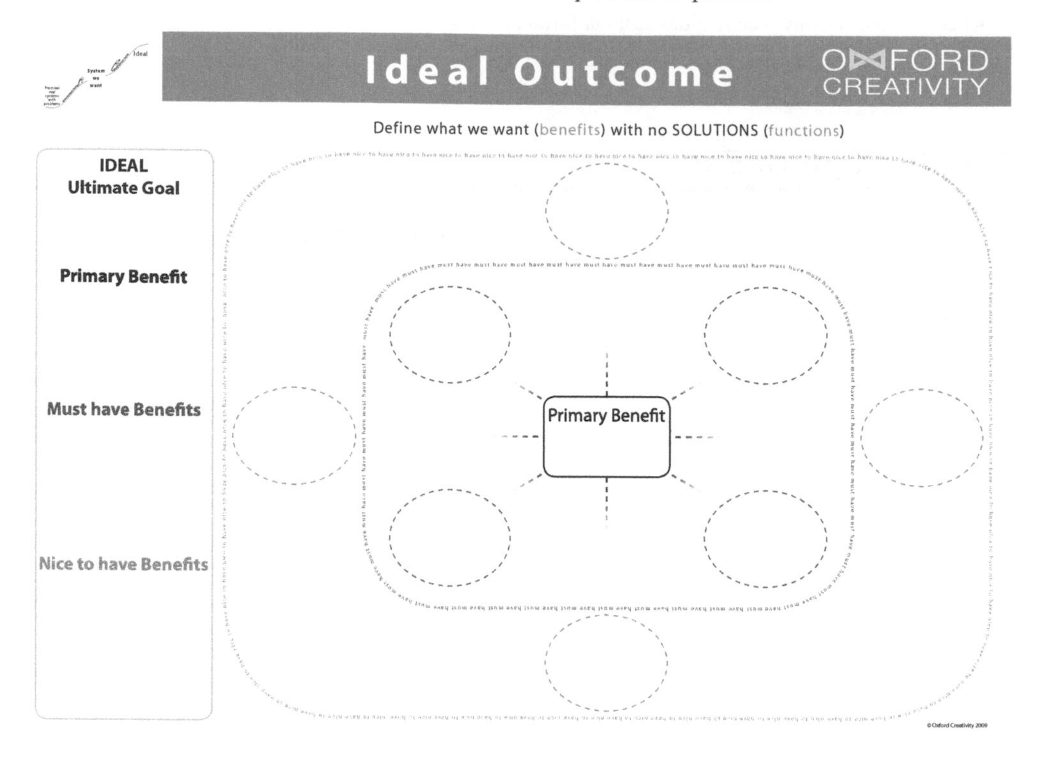

Undertaking an Ideality Audit

Successful, detailed and accurate capture of the Ideal Outcome gives us a clean start without irrelevant details, inappropriate solution directions and limitations. When we start problem solving we are usually hemmed in and concerned with constraints and problem boundaries, whereas this formal capture of the definition of the Ideal Outcome helps us think only about what we want, not just what constraints or the problem history are subconsciously telling us to choose.

We then undertake to define the *Ideality we have* it normally involves a more detailed study of the current system, looking in some detail at what's good (benefits) and what's bad (costs and harms). This can also include a function analysis map of the system (see Chapter 11) which helps us see at a glance where there are problems in the current system. The Ideality Audit is a comparison between the *Ideality we have* and the *Ideality we want*. Comparing these two helps us map, understand and prioritize all the challenges and problems of creating a better system. Defining the '*Ideality we want*', *the Ideality we have* and the Ideal Outcome are good starting points for any problem understanding and solving. This is very helpful no matter how complex or simple the problem or situation, as they cover both outputs we have and want, and the inputs we are making and are prepared to make, and any harms we have and will tolerate – so it helps us understand all the Ideality Gaps, the essential problems. If there are any unknown causes of problems which have to be located before solving then an additional search is needed using the appropriate root-cause tools (this includes 9-Boxes in TRIZ) before the problem solving can begin. Our Ideality Audit can help us draw up a Problem List – as every missing, excessive or insufficient benefit, unacceptable cost and unacceptable harm is a problem. Once we have identified the problems we can tackle them using the TRIZ problem-solving tools.

The definition of the *Ideality we want* is a step back to pragmatic reality as the summary of the *Ideality we want* succinctly includes all and only the benefits we want, the inputs are we prepared to make (costs) and the unfavourable or unnecessary outputs (harms) will we tolerate. By pinning down this *Ideality we want* we are beginning to understand the real problem in great detail.

IDEALITY AUDIT

First roughly define Ideal Outcome - Ultimate Goal & everything we want then define the following

IDEALITY of	SYSTEM we've got	SYSTEM we want	Ideal Outcome	Ideality Gap
Prime Benefit				
Other Benefits				
Prime Output / Function				
Acceptable costs / inputs				
Unacceptable costs / inputs				
Acceptable harms/problems				
Unacceptable harms/problems				

Include other relevant factors for problem understanding and solving including **Simple Drawings of System**

The definition of the Ideal Outcome is a good starting point in that it not only delivers everything we want, all and everyone's requirements, but as it has only good outputs (no problems) it also has no inputs (no costs). At the beginning of any problem-solving session the Ideal Outcome delivers a quick summary of what everyone wants but does not deal with any costs and harms.

No System Yet?

When we don't yet have a system and have to choose one or invent one, we can start with the benefits we want by defining the Ideal. Moving back towards reality we get the benefits we want from functions, and we have to understand and look for functions which are delivered by a system – so we are looking for the right system. Inventing is matching, finding a system to meet the unfulfilled needs.

For example if I want to travel to Paris fairly quickly from London (the primary benefit I want) I have look at functions and systems which will deliver this – I could fly, I could drive, go by train, use a ferry etc. One way of beginning to define exactly what I want is to define the other benefits and be more exact on what I want (be transported safely, economically, comfortably and quickly to Paris). I could then complete the definition of my Ideality (my Ideal System) by adding what inputs of time, money and effort I am prepared to put in, and make some assessment of the harms I will tolerate (crowded trains, seasickness, horrible treatment and delays at airports). Then I might be able to choose which system will suit me best.

Before we choose a system we need to simply define the benefits we want, and acceptable costs and harms: this defines our Ideality independent of any system. TRIZ problem understanding begins with an audit of the Ideality whether we have a system not.

Using the Ideal in Aerospace Problem-Solving Sessions

In one problem-solving session, when researching escape systems from aeroplanes in different crisis situations, we looked at how to escape from a partially submerged aeroplane. When asked to define the Ideal Outcome, someone shouted out 'Beam me up Scottie' and someone else shouted 'Beam me home Scottie'. These light-hearted solutions gave us some serious starting points as it suggested what sort of outcome we wanted – speed, just people moved, no damage etc. – as well as subsequent rescue from the sea to be taken home. It started the understanding and definition of all the requirements, without worrying about how to get it or without considering real constraints. Our Ideal Outcome was essentially that everyone was delivered quickly and safely out of the plane, and also then survived to be located by rescue to be delivered somewhere safe. This gave us a good high-level definition of the Prime Benefits of the solution we were seeking in a few minutes. Ideal Outcome is not therefore a normal requirements capture but a very simple and stimulating compementary process to all such exercises.

To define the Ideal Outcome, when facing a problem situation, we can also say 'Imagine you had a magic wand – don't tell us how you would solve the problem – just tell us the end result – What benefits would you want the magic wand to give you?'. The answer should rapidly and approximately cover all real and essential benefits. This simple trick assists the intelligent consideration we need to stimulate, to ensure that the answer is neither too precise, nor too vague and to guard against only being in solution mode. Asking a bunch of serious engineers to pick up and wave a pink magic wand is not as career limiting as it sounds: the principle it establishes is that there are temporarily no constraints; there is no 'we can't do that because …'. Ideal thinking via magic wands consistently delivers out-of–the-box thinking on requirements.

The nine dots think-out-of–the-box exercise was designed perfectly to demonstrate the danger of false constraints and to show people how they work within both real and imagined constraints. The nine dots exercise asks us to first draw three identical rows each of three dots, with the dots between each row spaced the same distance as the dots between each column to make a square of nine dots. There are then two stages:

1. Connect all of the nine dots with *five* straight lines without lifting the pencil from the paper. This is fairly easy and there is more than one way of achieving this.
2. Connect all of the nine dots with only *four* straight lines without lifting the pencil from the paper. Most people find this difficult as the solution involves us drawing much longer lines out of the square of nine dots than in stage 1. In stage 1 we could join up all the dots and stay 'in the box'. We therefore set ourselves false rules or constraints when asked to find another solution. Hence this teaches us to look 'outside the box' and be aware of only the real constraints.

The powerful thinking trick of the Ideal and the imaginary power of magic wands switches off limiting thinking within constraints (real or assumed) and delivers innovative and powerful thinking about requirements and solutions.

There are many splendid, thorough toolkits (such as QFD, Value Engineering, 8D, TQM and parts of Six Sigma) to help us capture all needs in systematic and detailed ways, and TRIZ problem solving is always enhanced by adding any such previous work from these toolkits to the problem-understanding stage. Even without other toolkits however the TRIZ Ideal analysis tools are very helpful for an efficient, accurate and fairly quick gathering of needs and there is an Ideal TRIZ tool for each level of accurate needs capture. It is valuable to include understanding of both the primary and secondary benefits in our needs capture otherwise a partial definition of requirements can lead to problems …

Systems may be ineffective unless they give us everything we want (purpose and other benefits)

In all cultures the way we learn through children's stories, folk tales and jokes are often frightening lessons about sloppily and partially defining the benefits we want and also about the dangers of mixing up requirements and solutions. In such tales the wise and the good are clear about what they want and define this clearly, whereas the foolish, the stupid and the impulsive confuse what they want with a half-baked solution and offer up ill-thought-out solutions rather than a clear and complete statement of requirements. The tales are based on the definition of their heart's desires covering only the primary benefit or being inadequately specified in time; they are mostly about not defining the secondary benefits properly.

Be careful what you wish for ...

Defining what you want is a tricky business and there are many myths and frightening gothic short stories about imprecisely and carelessly defining what you want and not accurately defining the secondary benefits (see '**The Monkey's Paw**' 'http://en.wikipedia.org/wiki/W._W._Jacobs') and also jokes about wishes being magically granted to surprised men by fairies or genies coming out of bottles and giving exactly them what they ask (no more, no less).

Thinking Up Solutions is More Fun Than Meeting Needs

Offering solutions rather than defining needs is a common trait of any management group particularly prevalent in creative (marketing people) and/or decisive (senior management). This is not surprising as it is their job to be seen to satisfy needs with their solutions. Working with TRIZ and marketing teams it is always hard to alert them to how each of them is in love with their own 'Bad Solution' and persuade them to define requirements by defining the Ideal Outcome. Defining the Ideal Marketing system with one company produced interesting results because they recognized that they are building on the company's reputation, previous investments, successful products etc. and that once a potentially successful product is recognized, developed and launched that marketing is still a cumulative and complex process.

The Relationship between marketing spend and good results may wrongly feel as random as winning an arcade game

One job of marketing is making us forget whatever our real requirements are, and get us to buy their solutions. Marketing have to convince us that their offered solutions are exactly what we want. A perfect marketing campaign ideally takes us straight to their proffered solution without us being consciously aware of our real needs and the many other solutions (competitive products) available to solve them. (Successful marketing would move us from a desire for something like a sweet snack to requesting a very particular confectionary bar.) This is often seen as pernicious in children when they are exposed and encouraged to embrace very particular solutions associated with cartoon characters or a film and their requests for presents can be for very particular solutions – as any parent desperately seeking a

Thunderbirds Tracey Island or a particular Barbie Doll before Christmas has experienced. Although anyone watching their children playing with empty present boxes knows that when playing they are happier with toys which are conceptual rather than some adult's vision of a detailed solution. In outdoor playgrounds simple equipment such as swings, roundabouts and slides are used the most, while some overdesigned and fairly useless equipment such as a detailed pirate ship is often ignored. Overdesigned children's playground installations are usually some adult's vision of what children want and their own solution – they rarely meet the simple needs of *children's* playgrounds.

Different Stakeholders Have Different Ideal Outcomes

TRIZ helps find solutions for different stakeholders with different needs

Frequent barriers to understanding problems can be the very many and opposing stakeholder requirements. When working with powerful customers such as government agencies and their suppliers in military projects or nuclear clean-up planning, these stakeholders have very different expectations, and summarizing them rapidly in a way that everyone understands is very important and helpful. One very powerful use of the Ideal is to capture each stakeholder's Ideal Outcome. These can be very contradictory but it is a way of summarizing differences and overlaps and understanding priorities.

Getting rid of one Stakeholders' harms may wreck some other Stakeholders' businesses (ask tobacco companies)

Ideality of All Stakeholders

Without TRIZ opposite needs of stakeholders can lead to incompatible and conflicting solutions

The Ideality and Ideal Outcome is often different for various stakeholders. Defining everyone's Ideal Outcome offers simple routes to capture the extreme reality of all benefits wanted by all stakeholders and can be the first and important stage in the Ideality Audit. Again there are many jokes about conflicting stakeholder requirements when each stakeholder has been magically offered their heart's desire and one final stakeholder wish conflicts and cancels out the others.

TRIZ Embraces Solution-Mode Thinking

No one who has worked on formal, step-by-step problem understanding can help becoming frustrated by the tedium of sticking rigidly to problem understanding and avoiding all solutions, or conversely by the rapid slide away from a complete understanding of the problem to prematurely choosing one solution to develop. One unique aspect of TRIZ is that while analysing the problem (and inevitably locating solutions) we can do both activities simultaneously. We can keep to and complete an accurate problem understanding (needs and system analysis) but also allow the capturing and parking of bad solutions as they occur to us – this is because unlike any other toolkits TRIZ has some problem-solving tools, concepts and processes which later can be applied to usefully interrogate and transform those bad solutions for good starting points to solve problems at the appropriate time.

Other problem-solving processes which mostly rely on solutions spontaneously occurring through brainstorming have this curious artificial dichotomy. In many non-TRIZ sessions first we must analyse and understand the problem in some detail (without solutions), and then we will have the fun problem-solving part, not when solutions occur to us, not when we are feeling most engaged and creative, but when someone puts us in a room and says 'problem solve now'. The TRIZ process does not do this and keeps us engaged and creative throughout and does not insist on us being good and doing our requirements capture, before we move onto the timetabled fun of problem solving.

Defining the Ultimate Goal and Prime Benefit

The TRIZ tool of Ideal Outcome has many aspects but primarily helps us to accurately define everything we want, and once we have defined all its aspects – the ultimate goal, the primary benefit, the high-level benefits and all other lesser benefits – then we can be fairly certain that we have an approximately correct gathering of needs. The differences between the ultimate goal and the primary benefit requires careful definition – which may seem theoretical and academic once a system has been chosen – but for absolute understanding of the main purpose and why we have a system, it is worth distinguishing between them as the following case study demonstrates.

Baby noise problem – In a TRIZ problem-solving session to reduce the noise made by a ventilator for premature babies, it was felt that parents were distressed by the unacceptable noise levels especially when they spent many hours sitting next to such machines. This problem was tackled and solved in a TRIZ Workshop, but at the outset we defined all the elements of the Ideal Outcome – taking the problem to its highest level. This produced some interesting new directions for research for new systems.

In defining the *Ideal Outcome* the benefit of not upsetting parents was added (delivered by the function of a *quiet* ventilator) and was just one of the many benefits identified as desirable for the new system and many other benefits were uncovered and recorded. The *Primary Benefit* was defined as 'safely and appropriately delivering air to the lungs through a tube placed into the baby's windpipe', while the *Ultimate Goal* was defined as 'the need to maintain adequate blood oxygen levels for the baby'. This was an interesting moment because it was then perceived by everyone present that a ventilator was just one system and in some ways quite a crude system to deliver appropriate blood oxygen to the baby. Solution suggestions then erupted in the room for getting oxygen straight into the blood without using the under-developed lungs including using the baby's skin, maintaining the umbilical cord (or an artificial replacement) and such ideas sent their research into new directions. All this was stimulated by properly moving from Primary Benefit to Ultimate Goal by asking the question – Why do we use a ventilator? And only when we understood that the ultimate goal was *how to get the right amount of oxygen into the baby's blood* and that by asking the right question, did they see the many possibilities for answers that may not involve a ventilator or even the baby's lungs.

Subtle Difference Between the Ultimate Goal and Primary Benefit

The differences between Ultimate Goal and Primary Benefit are sometimes hard to specify and may not be important. What are the Primary Benefits of a family holiday, compared to its Ultimate Goal? This varies from person to person but for many the primary benefits of a family holiday includes being enjoyably together while resting from work away from home – its ultimate goal may be to refresh or recharge individuals and their relationships to each other and there may be other ways apart from family holidays to achieve this. Generally Primary Benefits suggest an attachment to a chosen solution and system and some choices of benefits in systems to deliver it, whereas the ultimate goal is beyond the many solutions and systems which could deliver it. Ultimate Goals can be achieved in many acceptable and unacceptable ways – defining the primary benefits as well shows the preference, acceptable choices and essential benefits we require as well.

For example one Ultimate Goal of many people is to have a safe, enjoyable, daily journey which minimizes time for travelling to work – there are many solutions to this such as live near your work, travel when all the roads are clear, work at home etc.

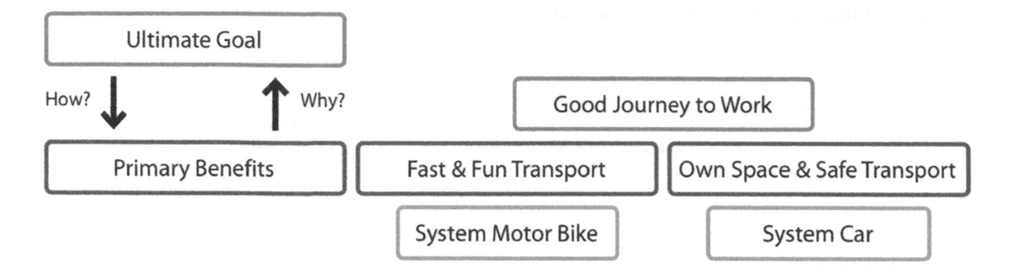

The Ideal Outcome helps us defines the Ultimate Goal even if there is no system. In these cases we can define the Ideal Outcome and many benefits. Our ideal for transport to work may be described as effective, no cost, comfortable, safe, enjoyable, exciting, fast etc. before we have chosen a system. The systems available may only cover some of these benefits the Ideal has identified – for example using a car may cover many of the benefits but be slow in traffic, whereas a motorbike will be fast in traffic and probably more exciting but is less safe than a car. Our theoretical Ideal would give us all the benefits not just some. After our Ideal Ultimate Goal we would next roughly define the main things we want – the Primary Benefits –and then think about all the other things we want – the Secondary Benefits. Defining the Primary Benefits takes us to preferences and towards practical choices, and may be fast and fun transport for some people, whereas for others it may be own space and safe transport.

Identifying Opposite Primary Benefits

Defining your Ultimate Goal is a useful step for reminding what we are trying to achieve but takes no account for preferences. Looking at opposite, different primary benefits can be a useful exercise as it helps us recognize more options for solutions. One ultimate goal I wanted shortly after getting married was 'Not to spend too much time on housework'. I discussed this with many friends and we recognized two extreme routes to achieving this – get other people to help me do it or do no housework at all (apparently after four years of no cleaning at all the house doesn't seem to get any dirtier – it reaches a stable state of filth). These opposite solutions (get housework done in a team/do no housework) made me realize that my Primary Benefits were clean house with limited personal time spent on housework. Benefits vary between different stakeholders and going the extra step to define the ultimate goal is generally very useful to break psychological inertia and understand clearly the main things we want.

Ultimate Goal Achieved by Opposite Systems from Opposite Primary Benefits

In our world of changing global powers, terrorism and warfare many people might cite safety as an ultimate goal for themselves and their families. If we think about safety from aggression or attack the Primary Benefits of different groups would vary from complete pacifism to high levels of protection through military hardware. In countries where guns are normal and legal for personal use the debates cover which routes offer the greatest safety for individuals

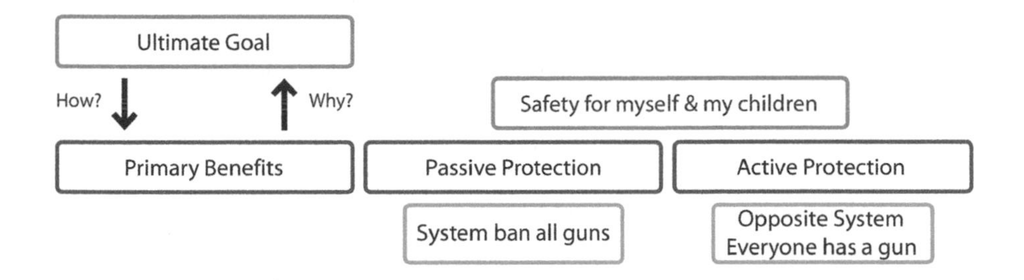

Identifying Real Goals – Owning a Submarine Fleet

Defining both the Ultimate Goal and Primary Benefit helps us understand why we have a system.

This is helpful in most situations, for example:

- What is our Ideal Outcome delivered by owning a submarine fleet? (Safety?)
- What is the Ultimate Goal delivered by owning a submarine fleet? (No one attacks us?)
- What is the Primary Benefit of a submarine? (Deterrent?)

These may not be the first questions we think about if involved in submarine design. When solving submarine problems we may be concerned with space, safety, footprint, capability etc. However understanding the high-level distinction between the ultimate goal of why a nation has a submarine fleet, and the primary benefit of a submarine may influence its design choices. When defining the Ideal Outcome for a submarine we would ask, 'Why have we got this system? What are all the benefits it gives us?' One general answer would be *safety*. This answer contains the ultimate goal, although in the UK it may be defined by different people in different ways, but one consensus answer for our Ultimate Goal has been to ensure that *no-one attacks us*. The way this is delivered, preventing attack is to deliver a *deterrent*, the prime benefit of the system – for a submarine. For some systems the Ultimate Goal feels too obvious to state, and for others it can be elusive but it is the highest purpose of our endeavours and should not be omitted when defining a problem.

The Primary Benefit of a submarine therefore may be to provide a deterrent and offers only one way/one solution of achieving its ultimate goal that 'no-one attacks us' – but Ultimate Goals can be delivered in many ways, including opposite solutions. Opposite ways of ensuring that no one attacks you could be making yourself too poor or unimportant or completely inoffensive to ensure that no one wants to attack you. This was the savage solution used in England in 1069 by William the Conqueror to crush rebellion and to stop the Scots invading the north of England. He laid waste the north east of England, burning all villages, destroying all flocks and crops and killing 150,000 people so that no one wanted to invade or attack that area as there was nothing left. Other solutions could have been to defend the north, attack the Scots, or create a terrifying deterrent to dissuade them from attacking; the 1069 solution to 'no one attacks' was leaving nothing to attack or invade was just one of the brutal, available solutions. For today's world our ideal outcome, safety, and our ultimate goal, 'no one attacks us', also have many worryingly brutal solutions. At the highest level we are concentrating on understanding what we want, to check that our chosen solution of *submarine* delivers this. Once we have chosen and commited to one solution to ensure 'no one attacks us' then we have limited our options and created constraints, and have to find ways of delivering the best system for the solution.

The ultimate goal has no constraints and no solutions or systems within its definition. So to check if you are at the highest level ask yourself – does this suggest any means or system of achieving it – if the answer is no then we have probably defined our ultimate goal – the term 'no one attacks us' contains no solutions or means of achieving it.

Ideal Outcome

When defining the Ideal Outcome for a system with a team of interested engineers it is often worth following these steps:

1. Ask everyone to simply note down on stickies every benefit, acceptable costs and acceptable harms they can think of (small and large – important and unimportant).
2. Group the benefits, and name each group (this gives us higher-level benefits).
3. Define the main benefit the system gives us (Primary Benefit).
4. Ask why? Why do we have this Primary Benefit – this gives us the Ultimate Goal.
5. Draw a simple diagram showing all elements of the Ideal Outcome as in 1-4.

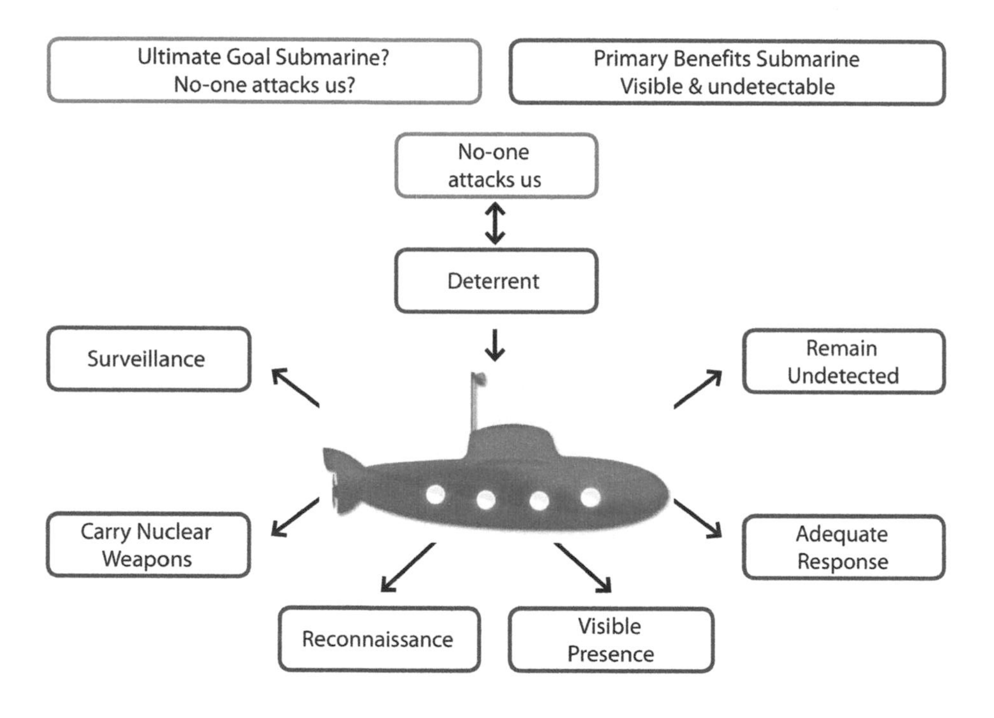

Ideal Outcome and Inventing

The simple TRIZ tool of the Ideal is very valuable for difficult challenges such as inventing. Invention is the providing of new systems to meet unfulfilled needs. The critical steps are a complete mapping of those needs and then a matching of the needs to the best possible new system (which may already exist in another area or industry). These new systems can represent clever and exciting inventions and can be anything from highest tech, state-of-the-art new systems at the forefront of science and technology (e.g. GPS) to another extreme of a number of old technologies in a new combination for the first time (clockwork radio). Ideal Outcome helps both visualize/imagine the new systems and define unfulfilled needs. Visualizing the Ideal helps engineers and scientists to undertake a 'quick and dirty' high-level needs capture (supplemented whenever necessary by a much more complete and thorough needs audit) and helps the matching of the identified benefits to unidentified functions. The system which doesn't yet exist can be defined as an Ideal System; this helps everyone understand the functions the system must provide, and provides some vision of the system they would like to create.

Invention, Ideal Outcome and Science Fiction for Gathering of New Solutions

Science fiction predicts future technologies: after at least 200 years of science fiction writing it is easy to map how good fantasies about the future have fairly clearly predicted our future needs. Writing science fiction is just an elaborate and detailed definition of the Ideal Outcome and shows the power of such thinking. Altshuller never made enough money from TRIZ to keep himself and his family so he had another career as a science fiction writer. This was under the name of Altov (Генрих Альтов) and he published books from 1958 until the end of his life in 1998 (some his books were co-written by his wife Valentina Zhuravlyova who also published her own science fiction stories). The Ideal Outcome is illustrated in their stories, for example in *Ballad of the Stars.* Industrial machines are constructed of very small, fine particles and their components made by using magnetic forces to keep them together. When the machines become obsolete, or need to be updated, the small particles can be re-programmed to form new machines (www.abebooks.com/search/isbn/0025017403). *This idea relates to both the Ideal and one of the main underlying assumptions of the TRIZ Trends that systems will become smaller until everything is made of tiny particles and is some kind of field effect.*

Prophecies come true just by being stated?

The foretelling of future events can simply make them happen, sometimes because ideas have been planted and people react to them. Some people claim that this is how science fiction shapes future technologies. One famous much quoted example is the influence of Star Trek. Everyone is waiting for technology to allow them to be 'beamed up' (teleportation) and the allegedly early mobile phone design was shaped by the famous Star Wars communicators which set our ideas of what they should look like.

Using the Ideal to Invent Systems

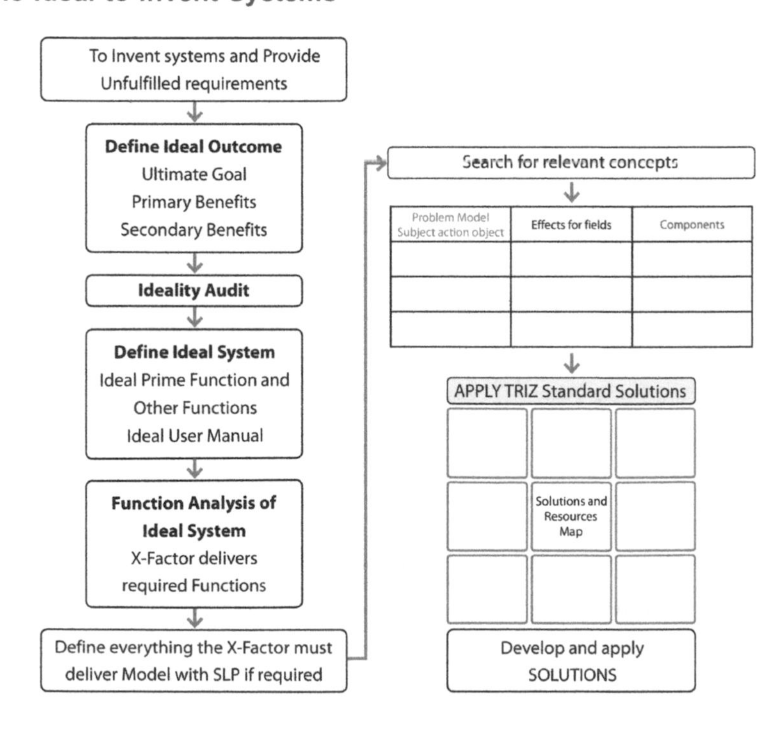

Using the Ideal to Understand What We Want and Then Achieve It – Windows for Houses and Offices

1. Problem Statement (includes constraints)
2. Map Context in 9-Boxes
3. Define Ideal Outcome (unfulfilled needs), Ideal User Manual
4. Ideality Audit
5. Define Ideal System (system to fulfil those needs)
6. Define Ideal Functions (delivered by X-Factor)
7. Seek *good* solutions (from Bad Solutions Park – filled while undertaking the previous six steps) delivered by available resources. Search for ways to apply resources by looking at Bad Solutions and previous solutions and analogous solutions, from the world's knowledge to find all the relevant ways of delivering the Ideal Functions.

1. Problem, Context and Initial Problem Statement + Constraints

I want to improve windows (remove their normal problems) and also ensure to that they help minimize energy use for houses and offices.

Typical problems – windows get dirty and need cleaning and maintenance, they can be fragile and get broken, they allow burglars in. Damaging heat, light and noise get in from outside, heat and noise escapes from the inside. They don't light the whole room, and they destroy privacy by allowing people to look in from outside.

Initial problem statement – I would like windows that give me everything we want: views, light where I want it, ventilation, complete privacy, no noise from outside, are completely safe from outside intruders, are unbreakable, provide a fire escape, retain and collect, store and deliver heat from outside when I want it, etc.

Constraints? Reasonable budget? Use windows that are currently available?

2. Map Context in 9-Boxes

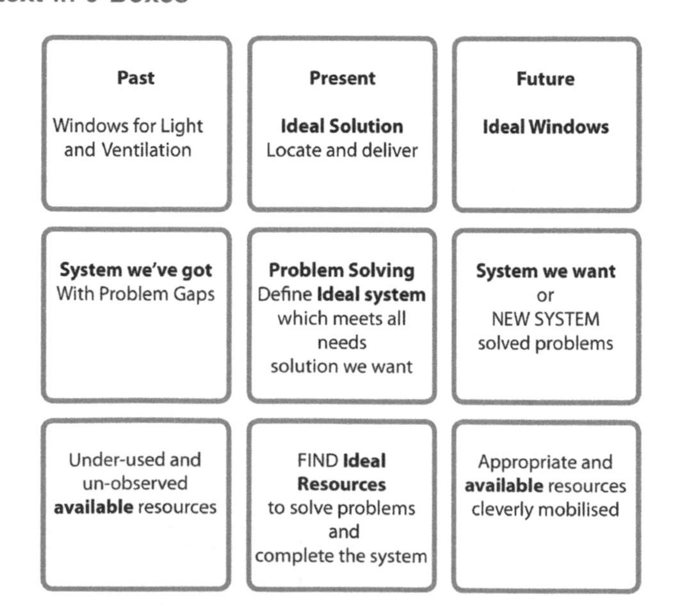

Past
Windows for Light and Ventilation
Present
Ideal Solution
Locate and deliver
Future
Ideal Windows
System we've got
With Problem Gaps
Problem Solving
Define Ideal system which meets all needs solution we want
System we want
or
NEW SYSTEM
solved problems
Under-used and un-observed available resources
FIND Ideal Resources to solve problems and complete the system
Appropriate and available resources cleverly mobilised

3. Ideal Outcome – What Do We Really, Really Want?

3.1 Ideal Outcome

Ultimate Goal = connection to outside (for illumination, views and ventilation when we want them) which enhances a room but creates no disadvantages of heat loss or gain, noise gain, loss of privacy, safety etc.

3.2 Primary Benefit (provided by Primary Function)

Connection to outside world for light, views etc.

3.3 Secondary Benefits

(provided by Secondary Functions) These include illumination,ventilation, good views, privacy, escape route, security, temperature control, sound-proof, safe for children, adds to the beauty/attractiveness of room and connected/open to outside

4. IDEALITY AUDIT

First roughly define **Ideal Outcome** (Ultimate Goal & everything we want) then define the following:

IDEALITY	SYSTEM we've got	Ideality of the SYSTEM we want
Prime Benefit	Connection to outside world (light, views, air)	Connects when and where required
Other Benefits	Illumination	Illuminates the whole room as wanted
	Brightness	Brightness but no dazzle
	Ventilation	Ventilation when and where we want it
	Good views	Enhances scope of views
	Privacy	No-one sees in
	Escape route	Instant access to outside when needed
	Secure	No access from outside
	Temperature control	No heat or cold penetrates unless required
	Sound-proof	No noise from outside
	Safe for children	Window material safe
	Adds to the beauty / attractiveness of room	Appearance delights user
	Connected / Open to outside	Instantly and only when required
	Limited heat control	Perfect heat control
	Lights up contents	Lights up with no fading or damage
Prime Output / Function	Sealed / open space for light, air, people to pass through as and when required	
Acceptable costs / inputs	Normal costs	Normal or lower costs
Unacceptable costs / inputs	Anything unexpected/ excessive +	High unexpected & upfront costs
Acceptable harms / problems	Cleaning, low maintenance	No cleaning or maintenance
Unacceptable harms / problems	Dangerous, noisy, too hot & bright	+ No fading or controlled heat changes

Include other relevant factors for problem understanding and solving include such as:-
Constraints (Imposed conditions, rules or regulations) Fits normal buildings / can be added to existing structures
Simple Drawings of System (if relevant)

5. Ideal System – What Does It Do?

5.1 Ideal Systems

Something in my wall which allows me to see out and enables perfect lighting and ventilation from outside to enter the room when I want it. Also acts as a fire escape and allows no risk or damage to any contents of the room. Controls any heat into and outside the room (collects the sun heat and distributes it inside or out when required).

System chosen is a Window? (Could also have been a camera? See-through bricks? Other systems???)

5.2 Ideal User Manual

It is easy to install with no maintenance, no cleaning, easy to open and close, fast and secure to lock/unlock.

6. Define Ideal Functions (Delivered by X-Factor)

6.1 Ideal Primary Function delivered by Primary System

Sealed/open transparent space for light which air can also pass through when required.

6.2 Ideal Secondary Functions of Windows Delivered by Secondary Systems

a. Stays clean
b. No excessive brightness in room
c. Allows light in to illuminate the whole room even at the back of the room
d. Keeps in heat in the winter and keeps out heat in the summer
e. Doesn't fade the furniture/contents
f. It can be open for fresh air-ventilation (original name comes from the Saxon 'wind hole')
g. Allows no dirt in or insects to penetrate inside
h. We can see out and but others can't see inside from outside
i. Is a simple fire-escape (easy and fast to open) but is secure against burglars (impossible for strangers to open/enter from outside?)
j. Stores daytime heat and releases it later/at night as required
k. Provides Sound proofing from outside noise
l. Secure and safe for children (they can't get out or break it)
m. Attractive and durable looking materials and design

7. Seek Good Solutions

(From Bad Solutions Park – filled while undertaking the previous six steps, delivered by available resources.)

Search for ways to apply resources by looking at bad solutions and previous solutions and analogous solutions, from the world's knowledge to find ways of delivering the Ideal Functions.

There are many solutions to improving windows and just a few are offered here to deliver some of the Ideal Functions being sought which are distinguished by clever use of resources.

7.1 Ideal Function = stay clean

Self-cleaning glass with a special coating uses resources of sunlight and rainwater The coating is photocatalytic and uses UV light to break down organic dirt, like bird droppings or tree-sap and loosen inorganic dirt. The coating is hydrophilic and helps water spread evenly to help rainwater wash away loosened dirt so that it dries quickly.

7.2 Ideal Function = no excessive brightness in room

This depends on where you live and what you want at different times of the day and in different seasons. The relevant resources include the size of the window and the resources of being south facing or north facing (or east or west). Resources such as blinds or curtains can help in this. One very clever design uses the resources of the angle of the light for global locations of the right latitude. This solution is an external shade which allows low winter sun into the windows, but blocks high summer sun.

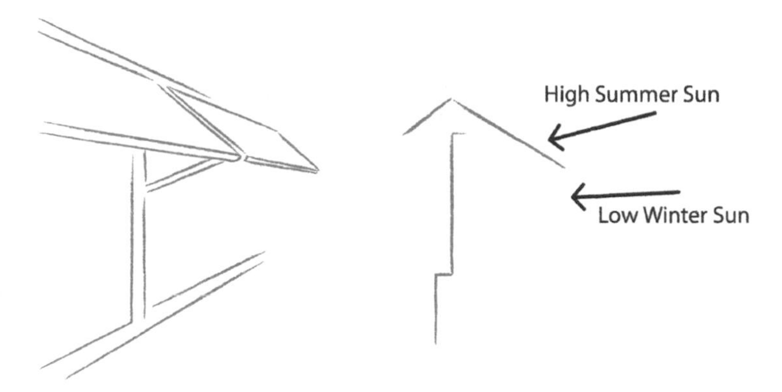

7.3 Ideal Function = Allow light in to illuminate the whole room even at the back of the room

At the end of the nineteenth century, daylight was a precious resource and needed to be used indoors effectively. Lux prisms came onto the market in 1898 and when added to windows directed light to the back and to the far corners of rooms The Luxfer Prism Company or the Radiating Light Company, was founded by James G. Pennycuick to distribute and popularize his 1882 invention (Patent No. 312,290 'an improvement in window-glass'.) This was simply the addition of horizontal prisms to the inside of square glass tiles. The prisms redirected sunlight from windows far into the room where light was needed, eliminating the daytime need for expensive artificial lighting and lighting wells. Luxfer prism glass was made famous by Frank Lloyd Wright, who designed some of his 'Iridian' prism tiles and used them in some of his buildings. Luxfer Prisms are now coming back onto the market and into general use to reduce the need for costly artificial light.

Clever resource solutions exist for the other functions we want.

Glass Companies have research projects which have developed glass to block and store outside solar heat and release it as and when required (to the inside on cold nights and to the outside on hot nights).

Glass can also solve contradictions of enabling us to see out but not allowing anyone to see in.

We can now filter the light into rooms so that it doesn't fade furnishings.

Double-glazing units are being redesigned to allow outwards opening from inside for escape and ventilation but no inwards access to make them burglar proof.

Conclusion

The Ideal is a simple and powerful concept to help us capture requirements. Once we know what we want, the Ideal also can help prompt our brains to come up with solution ideas and ways of realizing those ideas through resources.

The Ideal can help everyone capture their requirements – if there are different stakeholders, asking each to define their Ideal can highlight overlaps, contradictions and differences. For example if we were to ask the major stakeholders of UK Nuclear Clean-up to define their Ideal we might see at a glance their difficulties in contradictions and differences.

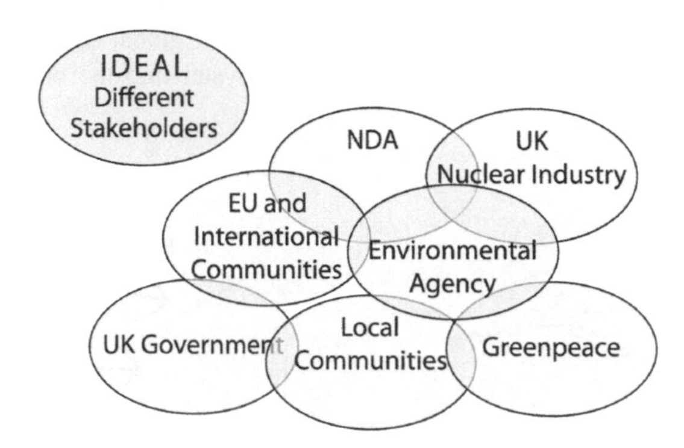

Such an exercise helps us capture all different stakeholders' requirements and highlights conflicts and overlaps. The Ideal is useful for all kinds of problems from dealing with problems with small components, to whole systems, to high-level strategic planning. Getting everyone to define their Ideal Solution helps discussions with everyone including supply chain/partners/customers and any other stakeholders.

When we have defined our Ideal Outcome, we can then check 'Does this definition contain a solution?' If the answer is yes, then we may have defined not only benefits but also a far from ideal solution which has limited our thinking to how we are going to deliver them. This may need us to think again, more precisely to remove any hidden solutions and only define the needs those solutions are fulfilling. Otherwise we can be trying to solve the wrong problem, because we have captured the wrong needs, or only captured part of the needs or more dangerously have put a solution into the question. This check for hidden answers ensures we only describe *what we want* and are not indulging in the creation of partial, sloppy solutions, which may have obscured the main needs, and instead given us apparently good but in reality bad solutions – such situations are the theme of many jokes.

When we have defined our Ideal Outcome and understood what we want – we can then look to see if a solution already exists. Useing the simple language of the Ideal is perfect for finding analogies and for any solutions and guides us to locate relevant resources to guide us to practical solutions.

The Ideal is a wonderful mental trick – a way of wiping away our old thoughts with irrelevant detail when we need to start with a clean slate to think without constraints about both requirements and solutions for problems. It can be used in a very approximate way to help understanding of broad needs and solutions, and also can be used in a very precise way to exactly capture requirements and define all the benefits we want.

Much engineering success depends on moving in the right direction towards the Ideal, getting systems right, selecting the right system, ensuring systems deliver what we want, or improving the one we've got, examining it for problems and/or checking that we have the right system working in the right way. Problem solving is closing the gaps between *system we have* and the *system we want* – to solve the right problems we have to understand and accurately map what we want and any gaps with an Ideality Audit.

Part Four
TRIZ, Invention and Next Generation Systems

System Development and Trends of Evolution

TRIZ Trends for Finding Future Systems

The TRIZ Eight Trends of Evolution are an essential element in the inventor's toolkit, and extremely useful to anyone who wants to develop new products, or find the next generation for any current systems or products. The TRIZ Trends offer powerful prompts for improving technical systems but also work well on process problems, general business situations, management and planning. They fulfil a very different role from most of the other TRIZ problem-solving tools which help us focus into particular problem areas and then find relevant solutions; by contrast the TRIZ Trends help us look more broadly at whole systems and their probable development, as if to lift our heads and look into the future. The TRIZ Trends can help us achieve competitive edge, more market share and find next-generation designs. As a standalone TRIZ tool they require no other or previous knowledge of the TRIZ problem-solving tools. They are simple to understand, powerful to use either for individuals or research teams and can be used in just a few hours to help deliver a vision of the many options to create new products, and evolve established ones.

TRIZ Trends show us the likely future of products and are the simple historical patterns followed by all technical systems as they develop. These patterns were uncovered by studying the evolution of systems from the patent databases undertaken by the Russian TRIZ community. The TRIZ Trends offer a broad and somewhat approximate forecasting toolkit, which can be applied to almost any product, and are fractal in that they help develop the whole system, as well as its individual components. The Trends help us predict the future of the whole and all parts of a technical system and show us how to intelligently follow and apply the knowledge from the past to improve current systems for future development.

For example one of the TRIZ eight trends is called 'Increasing Dynamism, Flexibility and Controllability' which shows us how systems change from their earliest stages of solid immobility to become more adaptable and flexible systems. The final stage is when complete freedom of movement is achieved by all the system elements becoming so small (down to molecular size) that the system eventually becomes a kind of field effect.

TRIZ for Engineers: Enabling Inventive Problem Solving, First Edition. Karen Gadd.

TRIZ TREND - Increasing Dynamism, Flexibility & Controllability

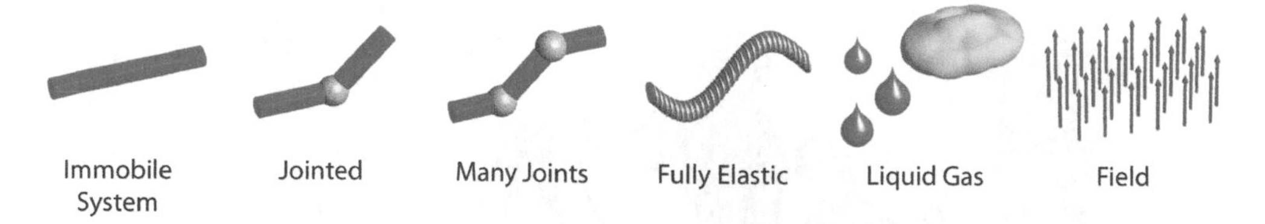

Board pointers and length-measuring systems (rulers) were first created as rigid wooden sticks, then they developed to become hinged, and then telescopic designs were used, and more flexible systems were developed leading to the field laser systems for pointers and measuring devices. Many other systems have become much more flexible but have not yet reached the final stage of field-effect solutions (as predicted by the TRIZ Trends). A view on the development of body armour is shown below and how it has become more dynamic and flexible throughout its long history. Although 'field' effect armour is imagined in the picture below and may not yet exist, the TRIZ Trends predict it will, and be developed and some time will appear on the market.

A knight in shining armour with TRIZ could have predicted that it would become, better, lighter and more flexible

Perfecting Products

The TRIZ Trends are simple, approximate guides to indicate the directions systems and products should follow to further meet human needs, through inventing, improving, perfecting and combining successful systems and technologies. They help us to predict how systems should evolve to anticipate, and more exactly meet, market demands. The Trends contain not just the approximate patterns for technical improvement towards perfection, but also show us the importance of achieving all the relevant benefits, increasing value, reducing harms and human inputs and finding ways for every need to be matched by improving systems. The Trends suggest ways of how to seek the right balance between benefits and harms and between inputs and outputs – how to improve successful products.

For example one very simple trend is to increase segmentation – suggesting a move from a solid structure to a segmented one – with increasing segmentation until we reach *nanotechnology* (predicted by TRIZ in the 1970s).

TRIZ TREND - Increasing Segmentation & Use of Fields

Segmentation has many sophisticated as well as low-tech, simple, practical applications for system development.

The TRIZ Trends of segmentation predicts the evolution of doors to the Ideal Door (there and not there)

With such simple prompts such as increasing flexibility or segmentation the TRIZ Trends help us to broadly understand the directions and approximate ways to deliver popular, economically balanced, technically intelligent products, and also show the changes of those balances with time.

TRIZ predicts that ideality increases and all systems become better, smaller and cheaper

Origin of the TRIZ Trends of Evolution

When Altshuller and his teams uncovered the fundamental truths about systems improvement – the TRIZ Trends – they gave us very simple descriptions of how to use these repeatable and recognizable patterns over time for technical development. When Altshuller identified the Trends he didn't just look at thousands of patents, but also studied successful technologies and products to see if he could identify and plot their life cycles. The categorization of the trends mapped all the simple patterns of evolution for technical systems which show us their probable future and development. Below are the eight trends as described and defined by Altshuller. They help us see the likely evolution and development of products, industries and processes and can provide the future directions for progress for successful systems.

In the bewildering unchartered world of new technologies the TRIZ Trends offer unique and straightforward repeatable patterns and they can provide a compass for the most likely successful future directions for products and processes to meet market needs. The Trends are therefore powerful for assessing and developing intellectual property (develop our own, and assess our competitors) and useful for patent strengthening and breaking. When we need innovation for new products and processes these TRIZ tools are simple and systematic to point us towards likely success for the future. But it is worth remembering that the trends are very general and define only the broad directions of evolution.

TRIZ Trends and Lines of Evolution

Trends = Patterns = Laws of Evolution

TRIZ literature is particularly confusing about the trends as so many different names are used in different books and articles. But generally

Patterns of Evolution = Trends = Laws of Evolution

(although sometimes the term Law may be seen as too bossy – implying it must be obeyed). These Trends may be subdivided into Lines of Evolution = Sub-Trends = Sub-Patterns. There are many different numbers of these patterns/laws when they appear in different TRIZ books and articles – 8, 10,

20, 30 – and there are close similarities in all the lists and confusing differences. The intelligent consensus seems to be that there are *8 Trends of Evolution* and the greater numbers described in various books are often just a more detailed (and sometimes confusing) mixing up of the original list of eight, by including some of their sub-patterns or the Lines of evolution as Trends. It probably doesn't matter much if we call it a line or a trend as long as all the important ways of mapping future patterns of technological development are included in the summary of Trends and their main lines of evolution. The term 'Lines of Evolution' refers to the sub-sets within the eight patterns and there are about 230–340 in total and the principal lines are discussed here. The Lines are a much more detailed version of each of the 8 Trends – hence the Lines identify specific stages of evolution contained within the various Trends/Patterns/Laws of Evolution.

Why Are They Useful?

The patterns of evolution are potentially very useful because they describe what has repeatedly happened in the past for successful technologies, and consequently guide us to what is likely to happen in the future. They can, in other words, be used to predict or at least suggest the next likely and beneficial technical development of a particular industry, for technical systems or its sub-systems. There are exceptions to all rules and strange human factors can disrupt the logical technical development especially when demand is not in line with technological evolution. Exceptions include when many portable radios got bigger (ghetto blasters) bucking the main trend that TRIZ shows us that everything gets smaller. Other systems such as Harley Davidson motorbikes etc. (noisy but appealing) and the historical and the heritage markets for clothing, furniture etc. which can defy technical logic with taste. Human factors can define markets as well as what is technologically possible – we may have historical cravings, or simply not be ready for a new development.

Frozen food was developed in the early twentieth century at a time when most people grew their own vegetables and there were no domestic deep freezers. The domestic demand for this great innovation of frozen food was low, for decades after its invention. The idea of preserving food by freezing was not new as it was tried by Francis Bacon in 1626 in his last experiment. Bacon showed that freezing food with snow slowed down the decomposition of chickens. Unfortunately while filling chickens with snow he caught a chill and soon after died. (Francis Bacon is sometimes called the Father of Science as he suggested that scientific theories should be based on observed facts rather than fitting facts to the theories we hold dear.) Altshuller and Bacon have much in common in their advocacy of using knowledge wisely and advancing knowledge sharing. Bacon suggested that science itself should be organized, and he promoted communication between centres of learning to share research and resources; as a result the Royal Society in Britain was formed. Bacon, scorned knowledge that did not lead to action, so he probably would have approved of TRIZ and also perhaps he would have been pleased when frozen chickens became a cheap and ready source of food on Europe in the second half of the twentieth century.

Evolution – Including Technical

Systems and technical products are analogous to living systems in that only the most successful and adaptive have within them all the codes to develop, survive, mature and thrive in hostile markets. Once exposed to the market and having survived launch and proved to be in demand, then the product itself tends to follow simple the 8 TRIZ Trends to meet further and changing needs. When companies apply these Trends they can anticipate and accelerate their product development to predict (and be the first to deliver) what the market is asking for next.

Evolution is about the survival of the fittest, and the TRIZ Trends of Evolution predict subsequent progress of the fittest and most robust products and systems which have survived birth (invention) and initial development to reach being launched onto their markets. Their development from this point is the territory of the TRIZ Trends to show the right directions for development to become robust systems in tough market conditions and technical environments, and how to adapt to changing needs and respond to competitive products. The Trends offer broad strategies for future planning – only products in protected environments such as those from near monopoly companies or government-funded contracts can produce less developed or weaker products which may survive without an adaptive strategy.

The TRIZ Trends show the progress of products which are not just technically successful, with a good Ideality balance of benefits to costs and harms, but whose market acceptance after launch has demonstrated that their outputs are useful, needed and recognized. Successful systems utilize, combine and develop technologies which anticipate and fulfil current human wants and needs.

Successful Products Meet Needs

Products meet needs – the market, the individual, user needs – and to do this they are invented, launched and then further developed. An important type of problem solving is matching products to previously unfulfilled needs (inventing) and then subsequently developing and improving the products so that they offer more benefits, cost less, have fewer downsides and usually appeal to more users and remain in demand (increase their Ideality).

The term *Product* or *System* is very broadly defined and can represent many things: whole technologies or the environments – the super-systems (such as a road), the whole system (such as a car) or just one of the components, a sub-system, such as a tyre.

The term *System* can also refer to a process or processes.

One essential test any product must pass is market acceptance. No matter how carefully researched, developed, designed, packaged and marketed, nothing is certain until it has been successfully launched, and shown to be useful, desired and needed. Every product has to prove that the functions it delivers relate to the benefits which the market currently wants. Once accepted by the market, and being used, and interacting with other users, systems are subject to feedback and criticism, which launches them on a journey of improvement, adaption and maturity that will continue until the need for the product disappears. Once a product is established then its progress and evolution follows set patterns of the TRIZ Trends for evolutionary improvement, and also in response to changing needs and developing technologies – predicting and anticipating that progress from launch until redundancy is the role of the TRIZ Trends.

Using the Trends for Practical Problem Solving

When companies use the TRIZ Trends to find new technologies, new products or new directions and developments of current products, their response is often similar – satisfaction that this toolkit has delivered rapid and useful results. There is also an appreciation that it is easy to apply to their product, systematic, simple and enjoyable and it has kept the whole team buzzing with useful ideas and directions to follow. One high powered Cambridge consultancy commented that they were able to involve everyone fully, and that the Trends kept them usefully at the task until it was complete. They commented that normally energy tails off and after a few good ideas from brainstorming they would fold the session

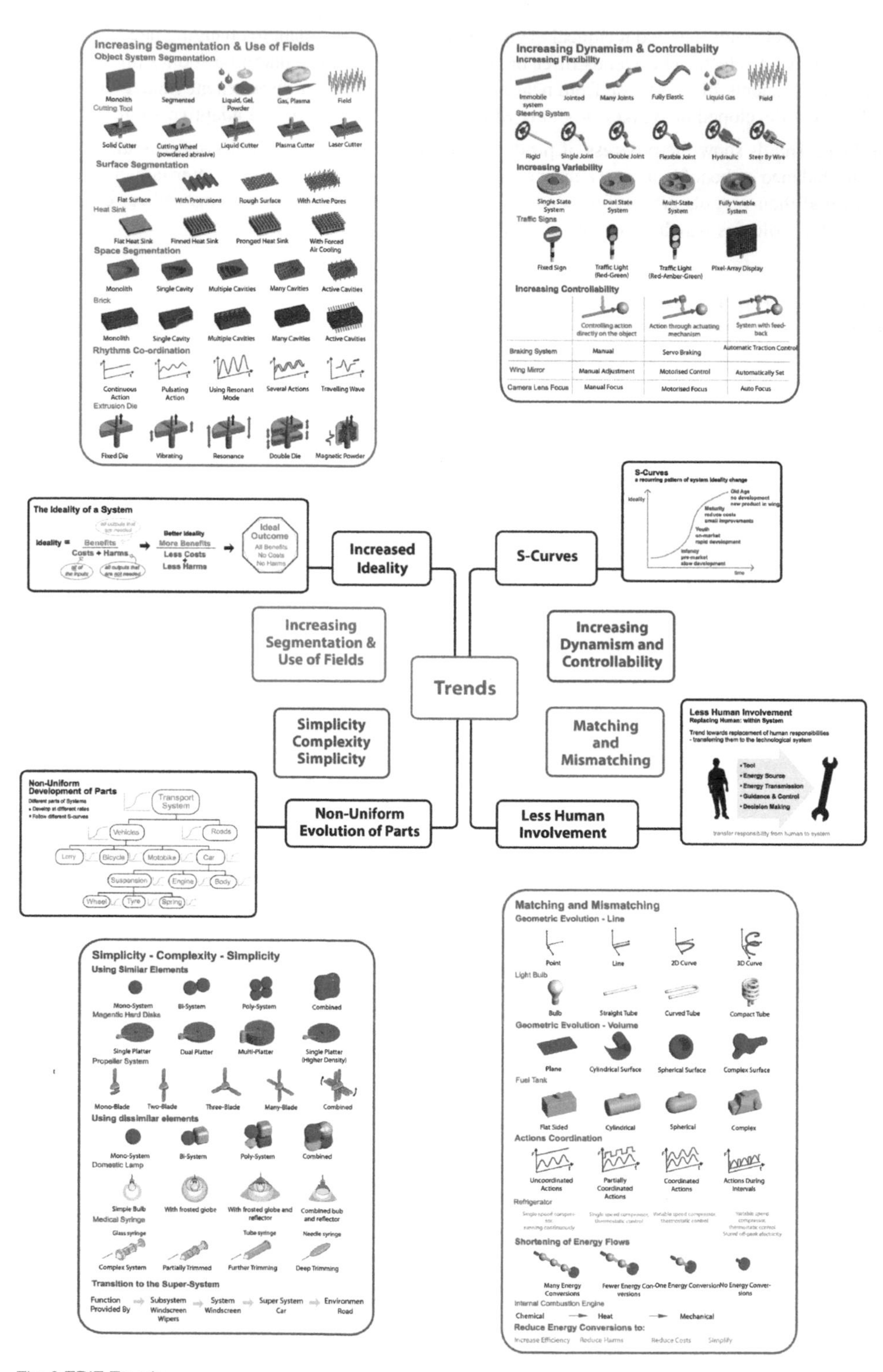

The 8 TRIZ Trends

and go with their first few good ideas. TRIZ Trends delivered a factor of five in multiplier numbers of ideas and the quality and usability of the solutions was much higher.

Each of the 8 Trends divide into Lines of Evolution – different versions with more detail as shown above.

The 8 Trends Map Natural Progression and Development

As systems develop over time they can be shown to:

1. **Increase Ideality** – become better and cheaper, achieve more benefits/functionality, while costs (inputs – everything we must pay for) and harms (outputs we don't want) decrease.

   $$\text{Ideality} = \frac{\uparrow \text{Benefits}}{\downarrow \text{Costs} + \downarrow \text{Harms}}$$

2. **Follow S-curves** – after being invented new systems improve slowly at first while being developed. After they come onto the market there are normally many improvements both in functionality and how they are made, accompanied by a reduction in cost. This is shown as a rapid increase in Ideality. Eventually this tails off until no further improvement is possible, and new systems are needed – with their own new S-curves.

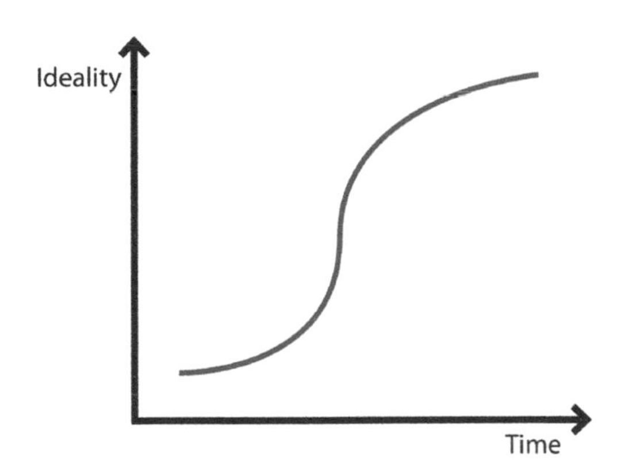

3. **Need less human involvement** – more automation and self-systems.
4. **Have non-uniform development of parts** – some parts of the system develop faster than others.
5. **Simplicity – complexity – simplicity** – start simple and then increase in complication and then simplify again (a cyclic/repeating pattern).
6. **Increasing dynamism, flexibility and controllability** – become more segmented, and as they become segmented they have more parts and therefore need more control.
7. **Increasing segmentation and use of fields** – use smaller and smaller parts until the parts are so small that together they have become a field effect (also called Transition to Micro-levels and Increased Use of Fields).
8. **Matching and mismatching of parts** – Matching function and functionality to all requirements not just the prime output to produce a system which finally delivers everything we want instead of just some of our requirements.

1. Ideality is the Most Fundamental of the 8 TRIZ Trends of Evolution

There are 4 lines of this Trend as shown below

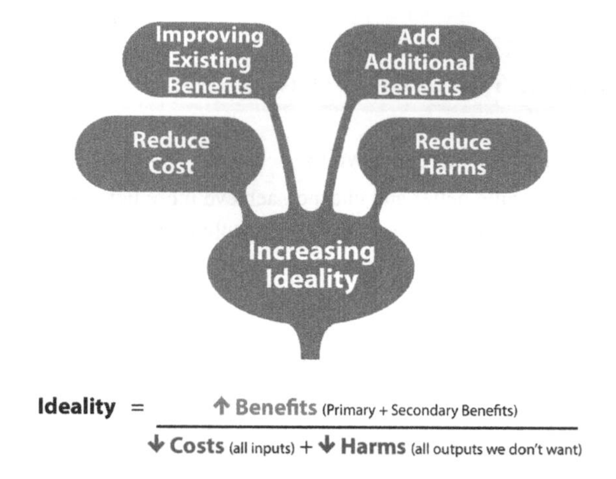

Ideality is a TRIZ term to measure the 'good' of a system. Although it is described as an equation and is the only equation in TRIZ – it probably isn't a real equation. The logic of increasing Ideality is wonderfully simple, and easy for anyone to grasp and use; it is achieved with more benefits and less costs and by removing any problems (any unwanted outputs – harms). Any system, product or process can be described by the Ideality equation. Its simplicity sometimes disguises its usefulness and rigour.

Why Ideality is an Important Starting Point

Ideality is a powerful way of thinking about and assessing any system, process or solution. Understanding the Ideality of a system helps us see how to improve an existing system, or invent a new one. The ultimate goal is an Ideal system which is one that gives us *everything we want* and gives us *nothing we don't want* and *costs nothing*. This Ideal is the goal to towards which we should always strive and points us in the right direction for problem solving.

Defining the Ideality accurately is a good start for any problem or system analysis and will always help everyone's understanding. The definition of a benefit is precise and important: a benefit is only a benefit if it meets needs. Hence benefits represent all the outputs that we need (no more, no less) and an output is only a benefit if we need it, otherwise it is a harm and part of the *unwanted outputs* from the system (even when neutral or not apparently harmful like the heat from a light bulb).

Costs are all inputs to the system – everything that the system requires (not just money but any inputs – materials, expertise, storage, transport, maintenance etc.). A system has inputs which enable it to deliver the outputs which can only be either benefits or harms – there is no neutral, middle category, of potentially useful outputs. It is only a benefit if it meets a need.

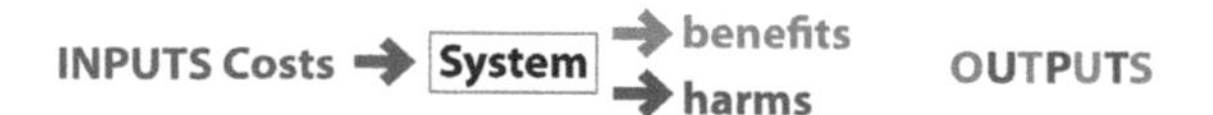

Products are invented to deliver benefits – which must balance with the costs and harms to ensure success

Reaching the Ideality Balance

Launching a product onto the market means reaching a point when the Ideality becomes positive and some Ideality Balance has been reached, and the benefits exceed costs and harms.

$$\textbf{Ideality} = \frac{\uparrow \textbf{Benefits} \text{ (Primary + Secondary Benefits)}}{\downarrow \textbf{Costs} \text{ (all inputs)} + \downarrow \textbf{Harms} \text{ (all outputs we don't want)}}$$

After invention systems will fly when Ideality becomes positive & benefits exceed costs + harms

The importance of Ideality cannot be stressed too much in TRIZ. It is a fundamental TRIZ measure of a system's success – its problems to be overcome, its costs to be reduced and all its benefits to be realised. Benefits is a very pertinent term in this context – in how they relate to the market needs of the product. Ideality Balance is only positive if benefits are higher than costs and harms. It should be noted that in TRIZ, *benefits* meet needs, any outputs which don't meet needs are *harms* – being offered empty or unused benefits doesn't wash with this TRIZ definition (benefits are only benefits if they fulfil a need). For example if a dishwasher has 12 programme options and I only want 4 – then the other 8 are *Harms* as they add cost and complexity and fulfil no needs.

Ideality only increases by better matching benefits to market needs, lowering costs or removing harms or problems. Ideality is the first TRIZ Trend – once the Ideality Balance has become positive and the investment has made it fly, then the system Ideality will increase following the TRIZ Trends.

When a system has been invented, developed and accepted by the market then this establishes that it has the right Ideality DNA to develop. This means that the system has the potential to fulfil all the needs it was created for, and its full development can be achieved. The product has within itself the key to how it can evolve to accurately meet changing future demands. Future directions of the system development within its market can be predicted by looking at just the system together with the 8 TRIZ Trends. This means that the system itself can predict its own future (the voice of the system or product) until its evolution is complete and the product is no longer needed or has been superseded by new technologies.

Towards the Ideal

TRIZ directs our problem understanding and solving towards a theoretical Ideal end point for Ideality (all benefits with no costs and harms). It helps us direct our efforts in the right direction towards this Ideal final point when we reach the perfect system which meets all the needs it was created for. Applying the TRIZ Trends activates the system's Ideality DNA to help the system move towards this perfection. When first accepted by the market a product is in its youth and may have many faults, but it has passed the critical test of successful market launch by demonstrating it is a system which meets needs – it has a positive Ideality Balance.

Ideality DNA – the Blueprint of Success

Ideality DNA encodes all the benefits the system should provide to meet needs (both primary and secondary) and its acceptable costs and harms (as it is representing a real system it will have to exist and some resources will be used for the system's existence), although the Ideal System is defined with no costs or harms only benefits. Once a system has achieved a positive Ideality Balance and is on the market, it succeeds because it fulfils some needs – with enough demand for those needs to make the product viable in the harsh commercial world. The Ideality DNA is some measure of all the needs the system could eventually meet.

Systems with inappropriate Ideality DNA may never succeed. The patent databases reveal many amazing inventions which, although ingenious, technically sound and would probably work well, were never successfully launched onto the market because the demands they met were so niche (only wanted/needed by a very few people) or they never actually met many needs (or in some cases any).

Sometimes apparently fantastic products are hard to find a match to market needs. Famous examples include stickies (Post-its®) which were erroneously thought to meet only very niche needs such as marking the right page in hymn books. Once the market is ready, and unmet requirements have been recognized, then the right product/technical system can be successfully launched and become accepted by the market and begin its journey of improvement. The creation, development, marketing of a sellable/useable product is great feat with many stages, and during those stages the right product's 'Ideality DNA' is developed and created. The Ideality DNA of a product holds within it the patterns of its future development and growth towards perfection of that system. The Ideality DNA can be read and activated by the TRIZ Trends.

The first launch of a product onto the market will be a far from perfect system – it will have lots of areas to improve, meet more requirements, reduce costs, minimize problems in order to move towards a perfect version of itself which is more exactly tuned to the needs it was created to fulfil. The TRIZ Trends plot the routes towards perfection. This means that the product (and its Ideality DNA) together with the TRIZ Trends can predict and guide us towards future success which in TRIZ is defined as improving Ideality. Improving Ideality is the most essential of the TRIZ Trends and represents the aim of all problem solving (including development of new products and next generation systems) and precisely means achieve more benefits *and* less costs *and* less harms.

Successful market launch reveals Ideality DNA – products then evolve following the TRIZ Trends

The improvement of Ideality is often the realization of benefits beyond the primary benefit, together with a reduction in costs and harms. For example the earliest motor cars brought onto the market in the early twentieth century amazed people by offering the primary benefit of assisted transport without animals or human power. Many other benefits were missing such as safety, comfort and reliability – all these would come as the car developed and improved its Ideality. The eventual realization of these benefits could have been foreseen and was in the car's Ideality DNA predicting that cars would eventually develop towards providing all the benefits demanded by its users, would reduce in costs and eliminate or substantially reduce all its problems. The benefits demanded will change and depend on time, place, markets and individual needs, and within the Ideality DNA is the blueprint of the perfect car. The first cars were not intended to be dangerous, uncomfortable and unpredictable – there were just missing benefits and unsolved problems to be dealt with. In today's marketplace, ecological, economic but high-performance cars which don't damage the environment are currently being sought, presenting a particular set of design and development challenges to the global car industry.

Anticipating the primary benefit which will make people buy your product is the genius of matching needs to systems. The first mobile phones offered the ability to make phone calls anywhere (theoretically, but the early ones had very limited range and they were very large and expensive) but they offered few of the secondary benefits available in most mobile phones today. Recognition of their Ideality DNA together with the TRIZ Trends perhaps could have helped in the helter-skelter rapid innovation seen in the mobile phone industry – especially as the type and level of demand was so unknown as it was all so new.

TRIZ Trends Point Us in the Right Directions but Market Exposure is also Essential

When a product reaches users they can try it, use it and respond in the time-honoured fashion to someone else's solution: they sum up what's good and what's bad about it, and work out how it fulfils their needs. Only by exposure to the customers or users will there be a realization, a true understanding of all user needs and how well and exactly the system matches them. Anything else is navel gazing, and could be very expensive navel gazing.

Ideality Steps from Invention to Perfection

- Invention (and all its stages) to reach an Ideality Balance
- Launch and market acceptance – revealing the Ideality DNA
- Market response to the system – feedback and improvement
- Following the Trends towards higher Ideality and perfect systems

The system will continue to meet more needs and the changing needs as it matures. Successful products deliver the right benefits with acceptable costs and harms (the right Ideality Balance). This technical balance gives us some measure of the 'goodness' of systems, and the practical success of technologies to meet needs. Systems require inputs (costs) to produce the functions we want, in addition all systems produce some outputs (harms) we don't want, and the balance between these improves as the benefits go up, while the costs and harms go down. Once invented, systems need investment (inputs of many kinds) to bring them to a point when they are practical enough to be launched onto the market. This is the point where they have a *positive* Ideality – and in a sense all the investment will now make them fly if the Ideality Balance is positive.

Failed products (including brilliant technical systems) may be good systems but they may have had insufficient backing to bring them to market, or have either misread needs or have a mismatch between benefits, cost and harms. Examples of products with a poor Ideality Balance (often too many harms) may be recumbent bicycles, airships and fuel cells.

Products that survive and flourish have successfully fulfilled the purpose of their creation and met some unfulfilled needs with the right balance of costs and harms – even if the market needs were not initially recognized. Some commercial successes are by an accidental meeting of needs and products. A well-known example includes the Sony Walkman which was an unexpected success, because Sony having financed their invention and development had not recognized, anticipated or predicted its particular appeal. Once an unpredicted need is uncovered and exploited, then the link between need and the function or feature which delivered it can be developed for subsequent versions using available technologies. The success of the Sony Walkman anticipated the iPod.

There is much investment in market research to uncover these mysterious user needs – and asking someone what they want is a bit like giving them a blank sheet of paper – which generally inspires blankness. Most people respond better by solving problems and finding solutions – and a solution is the matching of a system to needs. Whenever some solution idea, some invention, some new product is thought up it contains some cleverness, some technologies to meet needs – so it contains a *needs/user analysis* in its very creation, and this creates its Ideality DNA. Following the TRIZ Trends then gives the indications for technical system development and some forecasting of user needs. However once a product is launched onto the market it has the added advantage of market feedback to also help it develop in the right directions, and highlight any glitches or problems.

Many products are launched in the hope that they will catch some unfulfilled needs. There is a current fashion for innovations catalogues which offer us things we never knew we wanted (and in many cases we don't) – they all solve problems – often just one problem with a product. Upside-down tomato planters, curved shower-curtain rails for more space, high-tech specialist torches, see-through bread toasters etc. are all offered to see if they appeal and meet some needs.

2. The Power of S-Curves

Mapping system performance/Ideality over time gives us an S-curve.

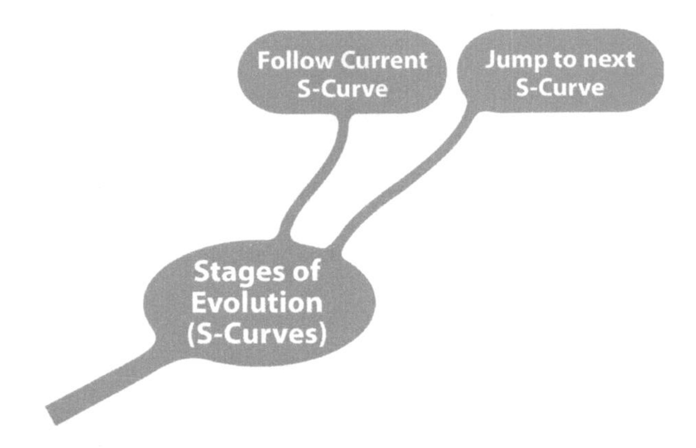

Systems Follow S-Curves

After being invented new systems improve slowly at first while being developed. After they have passed the major test of being successfully launched onto the market and are exposed to market forces, user needs and feedback, they change rapidly. There are normally many improvements both in functionality and in manufacturing methods; problems are sorted out and often this is also accompanied by a reduction in costs. This is shown as a rapid increase in Ideality. Eventually this tails off as there are no further improvements to be made, no further matching to user needs with this system, and any new needs, any step changes in benefits will only come from a new system with its own new S-curve. Hence systems improve over time at very different rates – slowly after invention, accelerating after the product goes on the market and then slowing down again when they are approaching a stage of perfection for this system and the applied technologies. This is normally in the shape of an **S** and known as an S-curve.

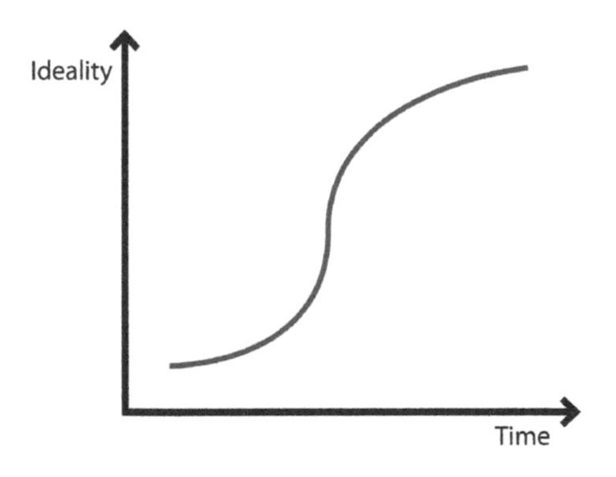

S-Curves show the development of a system from invention through development and launch onto market. They map the rapid system improvements as its usefulness and acceptance increases, and/or the price decreases and all glitches (harms) are dealt with – and show the eventual point where the system cannot improve any further and Ideality plateaus.

The importance of the knowledge understanding given to us in this simple S-curve should never be under-rated. The vertical axis is sometimes described in other ways –as in Value Engineering when harms are ignored as well as the important relationship between benefits and functions and the vertical axis describes only *value* (functions/costs). Benefits are provided by functions. The end point, the 'benefit' is what we are ultimately interested in. The way we get the benefit – the function – may change several times in the life of a system.

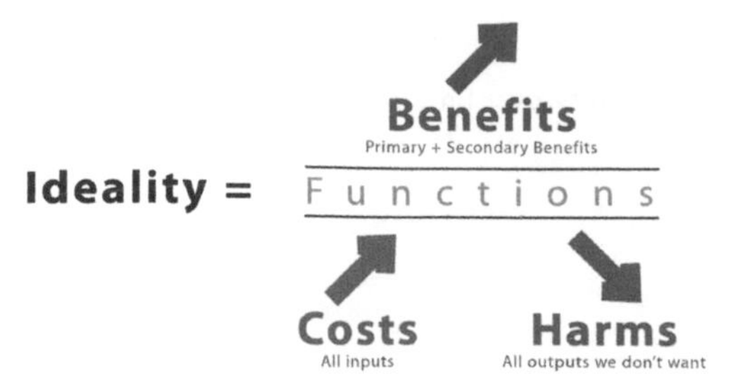

Ideality normally increases as all the system resources are used more effectively. Good resource use helps to eliminate problems and reduce costs, but there is a point where no further improvement is possible. The life of all technical systems, processes and industries can be represented by S-curves (as can most systems including biological).

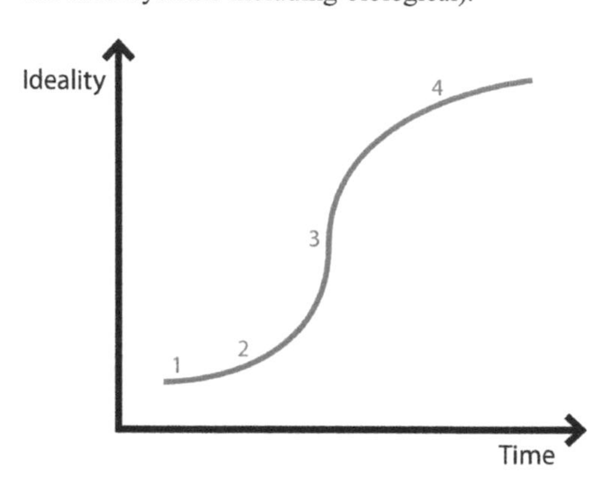

Each system component evolves independently and has its own S-curve, which represents the life of that technical system. The vertical axis is the Ideality, the horizontal is time.

To move up the S-Curve we must increase the Ideality which means increasing the benefits (the functions of the system) or decrease the costs or harms in the system.

S-Curves: 4 Stages

The S-curve shows how the main performance of a system changes over time. The evolutionary stages on an S-curve are:

1. Invent system and develop it to make it work properly and devise manufacturing processes.
2. Bring on to market (followed by rapid improvements due to market feedback).
3. Rapid increase in ideality – more benefits, less costs and less harms.
4. Keep a mature system on the market for as long as possible because this is nearly always a very profitable stage.

Who Buys or Uses Systems at the Bottom of the S-Curve?

Early enthusiasts, geeky technophobes, niche-market early adopters or those with very specialist needs will pay more than other people for new products which offer unique new primary benefits despite possible high costs and harms. The rest of the market will follow when the balance of benefits, price and technical problems has improved. This relationship of changing and various demands for technology is part of TRIZ S-curve analysis, an important element of the TRIZ Trends

Increases in Ideality Are Represented in Successive S-Curves for Successive Systems

For any system if we have achieved most of our benefits and minimized costs and harms, by using all the resources in the system, it is near the top of the S-curve. What then? The system seems incapable of improvement and we need a new system (the next S-curve) and need to make the S-jump! This is to either a new and better technology, an S-curve above our current S-curve (the red line) or an S-curve with lower Ideality (the blue line) but with the potential to improve and leave our current technology behind.

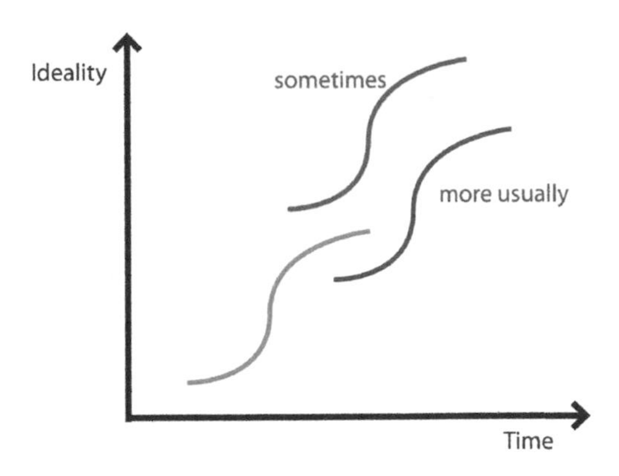

Making the S-Jump

The blue line shows the normal pattern of S-jump when the new system is inferior in terms of Ideality/ performance to the old system. However understanding of S-curves gives essential insight that new investment in the new technology must be made before the market for the old system disappears or competitors bring in the new technology first. The red line is rarer and much easier to make the right investment decisions about the new technology. The blue curve is harder because we are moving to what looks like a worse technology – and there may be many blue lines, many new and competing technologies to choose from. We may not always be able to choose the best technology because of market forces. We may have to choose a popular technology we know we can get onto the market most quickly.

Jumping to a new S-curve requires new attitudes, new demands and a search for new resources that may in turn be used for further development. These new resources may be complex or simple, old or new and be derived from other systems and markets. New resources – particularly of new materials – lead to new products, as with stainless steel razor blades. James Dyson caused vacuum cleaners to jump to another S-curve by applying an old and well-known technology (cyclone separation) to create bagless vacuum cleaners. Long-term development needs evolution of a whole technology or market forces – wartime brings many new technologies onto the market both in terms of big technologies such as weapon systems and domestic as with safety razors. Famous examples of dramatic new S-curves include electronic/digital watches replacing mechanical watches or CDs replacing vinyl records. Sometimes the new technology almost completely wipes out the old one as often happens with mobile phones or computers.

Not All Systems Improve Rapidly Over Time

Human tastes, market forces or technical inertia may halt a system on its S-curve (piano, fuel cell?). But generally improvements continue until the system reaches a highly evolved state when there can be no greater use of the system resources. After that, Ideality is only increased by making the S-jump to the next S-curve.

Where is our System on the S-curve?

The simple graphs below are very powerful aids to understanding where a system is on its S-curve. When we start using the Trends we need to think about everything round us and see where it may be on its S-curve. This helps general understanding about the potential to improve systems and helps us to find new and improved systems much more easily.

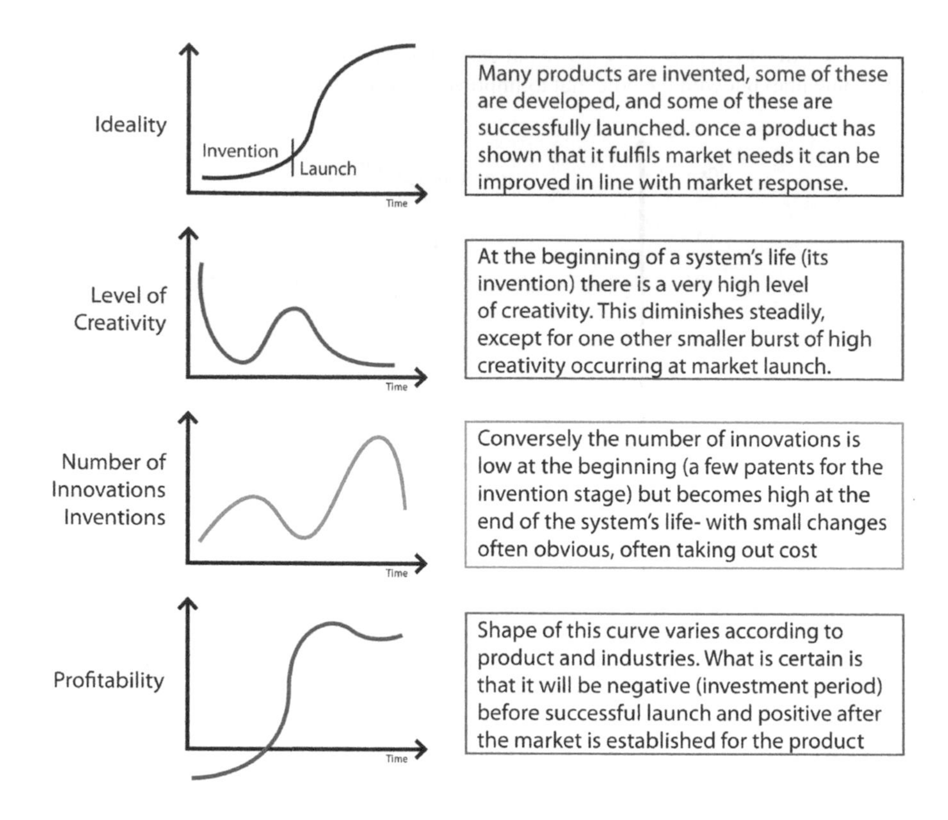

Ideality
Invention
Launch
Time
Many products are invented, some of these are developed, and some of these are successfully launched. once a product has shown that it fulfils market needs it can be improved in line with market response.
Level of Creativity
Time
At the beginning of a system's life (its invention) there is a very high level of creativity. This diminishes steadily, except for one other smaller burst of high creativity occurring at market launch.
Number of Innovations Inventions
Time
Conversely the number of innovations is low at the beginning (a few patents for the invention stage) but becomes high at the end of the system's life- with small changes often obvious, often taking out cost
Profitability
Time
Shape of this curve varies according to product and industries. What is certain is that it will be negative (investment period) before successful launch and positive after the market is established for the product

After invention in order to get a product onto the market we need development and this means that money and time need to be invested. This can only be recouped after the product is on the market and profitability normally occurs further up the S-curve. The longer the system stays on the market, without being replaced by the next generation, the more money it will earn. Once on the market hopefully money starts to flow back and at the top of an S-curve we normally see the most profits.

Launching Products at the Right Time

If the benefits of the system are tuned to user needs then the system will probably be successful but timing may be critical if your competitors are launching too. There are dangers of launching too early in order to grab market share, which may disillusion customers if the product has too many faults (unless you have a great unique selling point (USP) or near monopoly). There is also a danger in letting development engineers keep working to perfect a system and then launching a very good product too late, and letting your competitors establish market share while your product is still being improved. Market exposure often helps to sort out and indicate how to improve systems and find out what is really needed.

Military systems have particular problems as they would like to start using the products only when Ideality is high (top of the S-curve) and must miss the market exposure of the climb up the S-curve (full of useful feedback and reassurance that the problems have been eliminated or minimized). Some solutions are to find civil non-military applications for systems and let the market test them and hone them.

Which Side of the S-Curve Do You Want to Be?

For engineers and scientists the most challenging and sometimes most enjoyable work is after invention and at the early stages of a product's life – making it work.

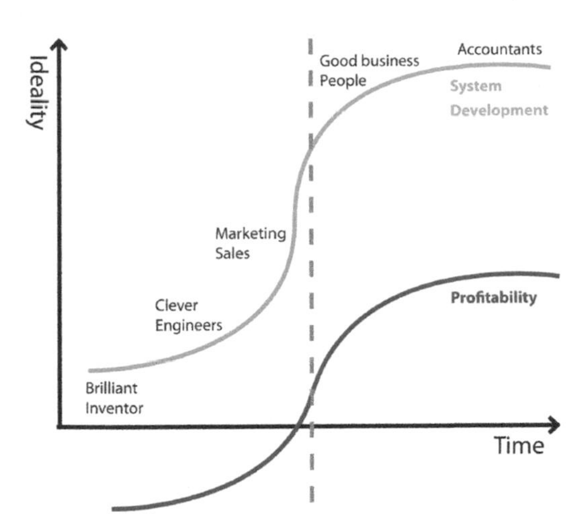

Where Do You Work on the S-Curve? (is all the fun on the left side of the dotted line?)

If you are a good engineer or scientist and now spend all your time in management, consider adding some projects near the bottom of their S-curve to your work portfolio.

3. Less Human Involvement

There are three lines of this Trend as shown below

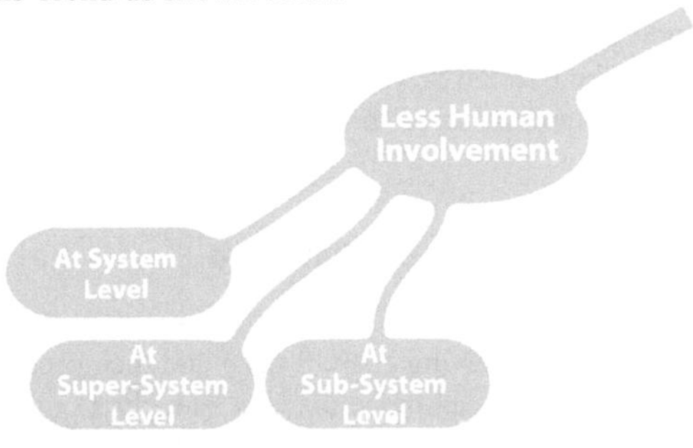

This Trend is about how all systems evolve to require less human intervention. It can be seen that developments in technology and science have given us machines and systems which free humans from many tasks. Everywhere in factories, offices, public places and at home, machines take over boring and repetitive tasks, and make our lives easier in many ways. Fifty years ago in Europe many women worked full-time in the house washing, ironing, cleaning and cooking (as well as bringing up children). Every week my Grandmothers devoted all day Monday to the hard labour of washing clothes, sheets etc. using mangles to remove surplus water and pegging out wet washing on clothes lines which they collected when dry and spent all the next day ironing. Today washing machines, tumble driers, powerful steamers and easy-care fabrics have made all these tasks much simpler and less labour intensive.

In the richer countries of the world industrial development has seen the human replaced by more and more sophisticated machines, and on a personal level we have the possibility of easier and more varied lives. For those with the means to afford it, we are freed from spending our time and energy on just surviving. We also have many sophisticated devices and luxuries which offer us good and varied education, cheap global transport, varied food and drink from almost anywhere in the world, easy communication, good healthcare etc. We can keep our houses clean and warm, our children well fed, well educated and mostly healthy and are offered a varied choice of careers, activities and entertainment. The systems and products we use are becoming … self systems

Self-Systems

Specifically the trend of less human involvement is about how the human action is replaced by technology. We all seek *self-systems* such as self-cleaning glass, self-repairing buildings or pipelines, self-balancing tyres, self-focusing lenses, self-tuning TVs etc. to keep human labour skill and error to a minimum. Our cars, houses, factories use self-systems to keep maintenance down and to keep us safe from mechanical failure and hazards.

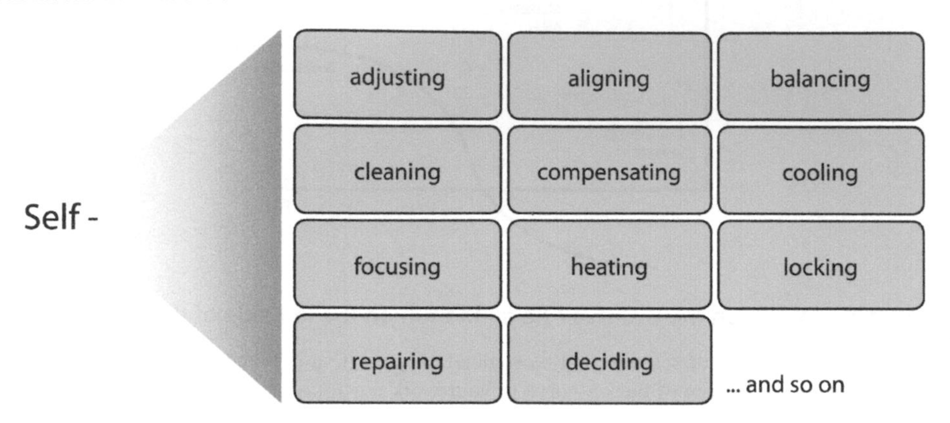

Self-systems are important in industries which deal with waste such as the nuclear or sewage industries. Self-cleaning systems are much pleasanter when dealing with the removal of human wastes – and there are hosts of patents for self-cleaning of sewage systems produced each year. Self-systems have existed since the beginning of technology and Roman sewers cleaned by rainwater are functioning in Rome and elsewhere. When building of the London Sewers began in the 1850s under the supervision of Joseph Bazalgette, egg-shaped tunnels were discovered to help self-cleaning and have functioned since then with minimum human intervention.

Self Seeking Insect Sprays Are Now Being Patented

Insect sprays are not considered to be a good thing for the environment and can become ineffective if insects become immune to them from receiving a small dose and then escape. It is a difficult problem to tackle because Insects are so small and can move far very quickly. Usually when we use a spray we have to use a wide spread of particles to ensure that enough of it reaches and covers the insect, this means that a large proportion (typically 90%+) of the insecticide misses the target and is therefore wasted, and worse, can be creating immune insects. Self-seeking sprays are being patented which are based on electrostatic charge differences and use a small amount of spray which seeks, finds and covers the insects – ensuring a fairly horrible death.

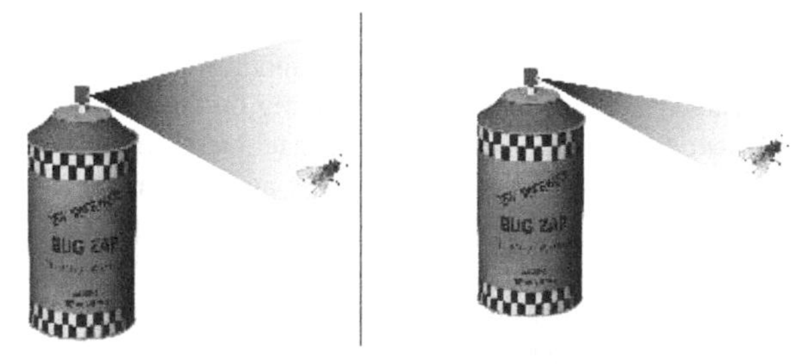

This trend of self-systems and less human involvement can appear to be more obvious and based on common sense but it applies to many systems and it is always worth looking at the evolution of the system in question and asking where will the human intervention and contact disappear next, and what sort of self-system will replace it.

A self-healing paint protector is now offered on some black cars. This is a special paint finish that self-repairs scratches on car surfaces, especially light scratches from accidental contact with twigs and branches from driving in narrow country roads in the UK. I have a black Lexus car with a very shiny and immaculate finish. It has been treated with a high-elastic resin which protects the inner layers of a car's painted surface. The outer protective layer is self-healing and the scratched surface repairs itself in a few days depending on the season and the seriousness of the scratches. The paint is also waterproof and has a higher resistance to scratches from automatic car washes, where apparently most scratches to cars are caused.

We live in an age where some craftsmen are trying to recapture lost human skills which were supplanted by machines. The trend of machines and technology replacing humans has been easily observable since the industrial revolution. Since the Luddites it has also been regarded as a mixed blessing particularly when machines replace humans who lose their jobs. The greater gains of the new technologies may not be shared evenly initially by all members of society.

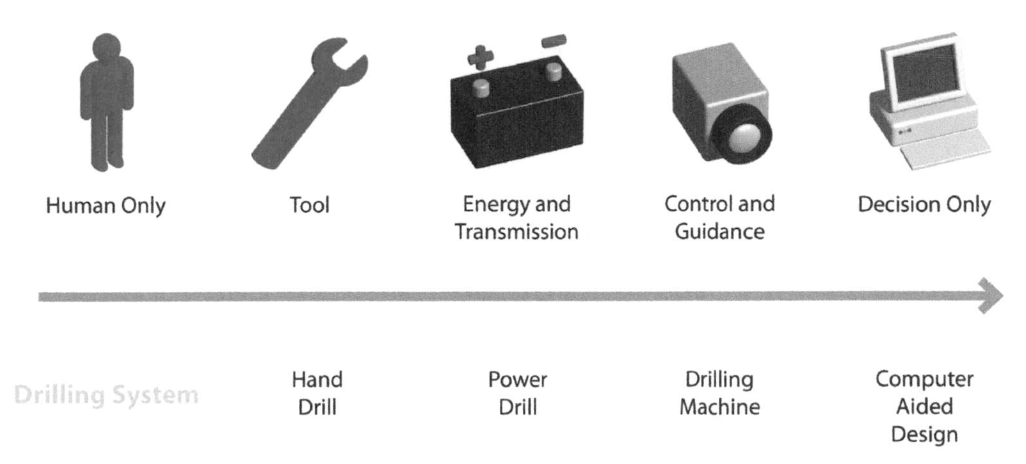

The trend of less human involvement is clear to trace in the history of machines: there is a clear line of development of the complete technical system with less human involvement from the human (with tool) providing the energy source, and the transmission and the guidance, control and decision making to sophisticated technical systems with little or no human involvement. The human is normally replaced in the order shown below:

1. tool (performs function)
2. an engine (energy source)
3. transmission (conveys energy)
4. guidance and control
5. decision making.

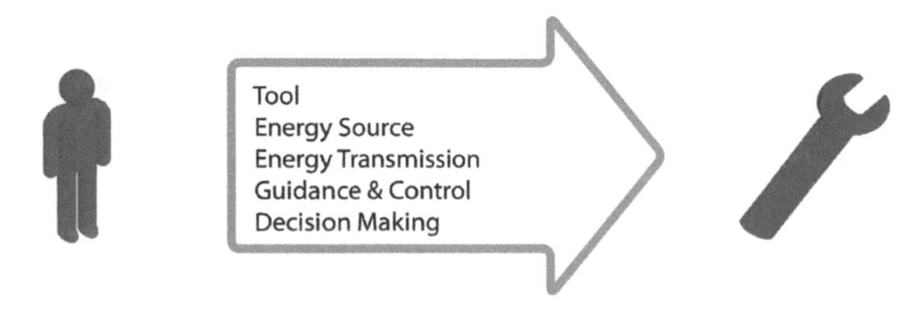

The human is replaced by the machine in a particular order from 1 to 5 as shown above. At first the human ceases to provide the function, but will still provide the other roles. Often the first systems maintain the human method of action (early windscreen wipers mimicked the human hand wiping the windscreen from the top but were driver operated by turning a handle). Next the machine provides the source of energy (next development was that windscreen wipers were powered by various means) and then transmission of the energy and finally control and decision making (windscreen wipers are completely automatic is some cars now). Many other systems are becoming completely automated or self-systems.

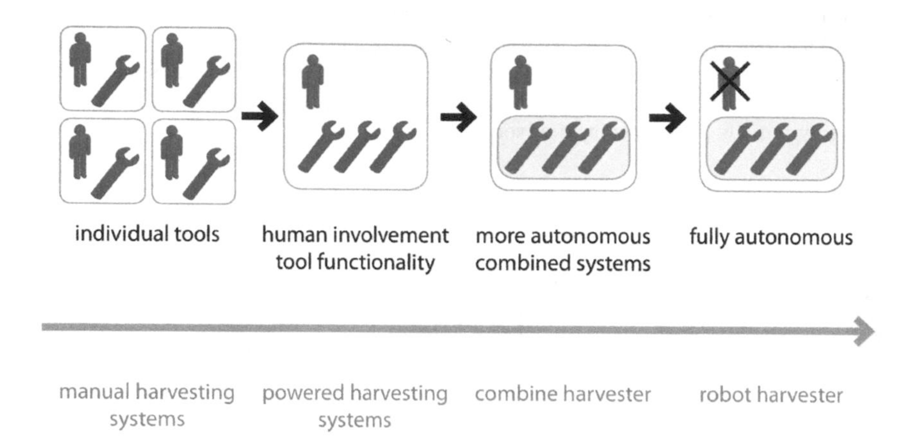

4. Non-Uniform Development of Parts

There are three lines of this Trend as shown below:

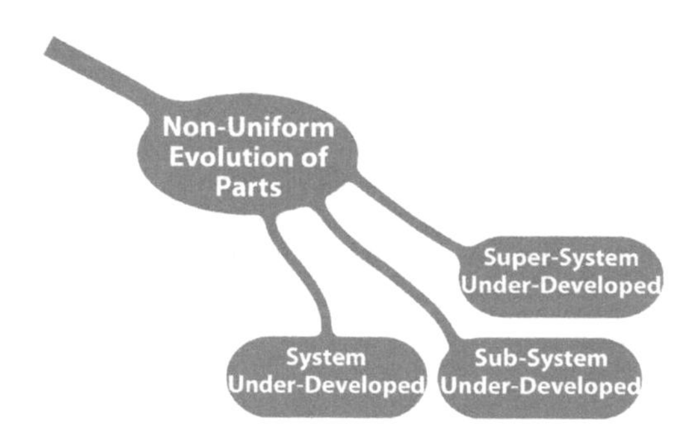

This Trend highlights that different parts of systems develop at different rates, follow their own, different S-curves and have reached various levels on their own S-curves, which may be very different to the S-curve of the whole system. It is also sometimes known as the 'Trend towards Increasing Coordination of Systems' (between components *and* between the system and the super-system *and* between the system and the sub-systems). It has three main Lines of Evolution which highlight where in scale we have underdevelopment:

- Super-system (such as African roads being unsuitable for European cars)
- System (e.g. personal computers and phones are needlessly complex and have a short life)
- Sub-systems (roads and cars are made as safe as possible, but drivers remain various and erratic and some systems such as the steering wheel or car airbags may kill short people).

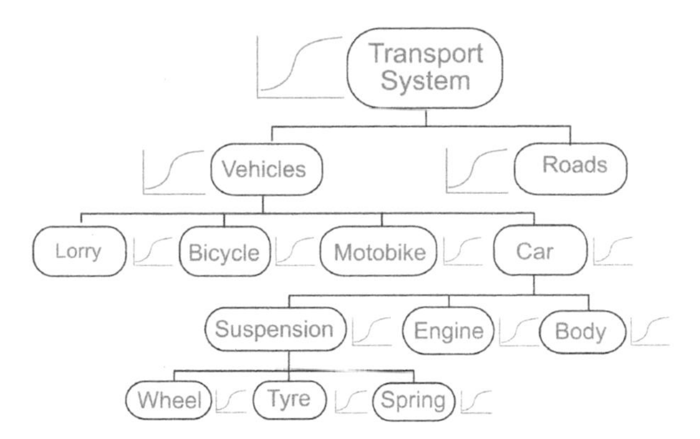

There is nothing in evolution or human effort that makes all systems develop evenly in the short or medium term. When developing and improving systems we tend to work on the parts we know well and understand, as we are comfortable and familiar with them and can see how to improve them. In contrast we neglect the difficult areas which inevitably become more of a problem. Sometimes chance or luck makes some parts develop before others. Psychological inertia and lack of strategy or a vision of the whole system and how it fits with its environment and its own components can keep a system from developing at its full potential. Whenever we look with a fresh unbiased eye at unbalanced systems we may see the problems – this is easier for outsiders, but those who work in the detail may find it hard to step back and see the obvious misfit in development. TRIZ literature talks of many examples including cargo ships which developed good speed but not sufficient braking systems to deal with the speeds they could achieve. Generally the more complicated the system the harder it is to see where the system is being held back by an uneven development of its parts. We may exacerbate this by tweaking (over-developing) the parts we know, like and avoiding those we don't.

Some inventions are long overdue – especially when they simply combine two old systems

Each system component evolves independently and has its own S-curve which represents the life of the sub-system. The vertical axis is the Ideality, the horizontal is time. To move up the S-curve we must increase the Ideality which means increasing the benefits (usually the functions the system is delivering) or decrease the cost or harm in the system. This means that components reach limits to their development at different times. The component that has hit its limits holds back the whole system.

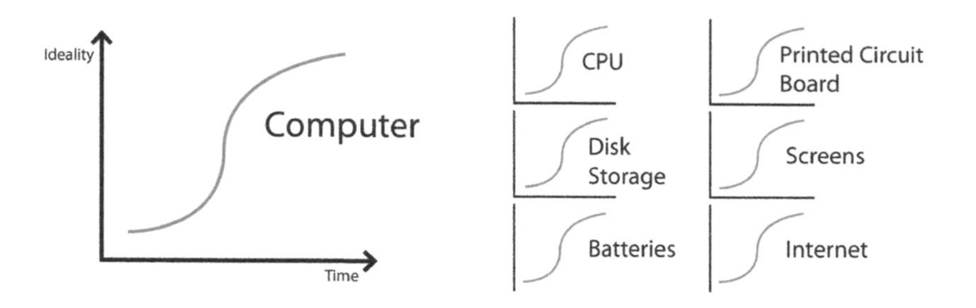

5. Simplicity – Complication – Simplicity

There are four lines of this Trend as shown below:

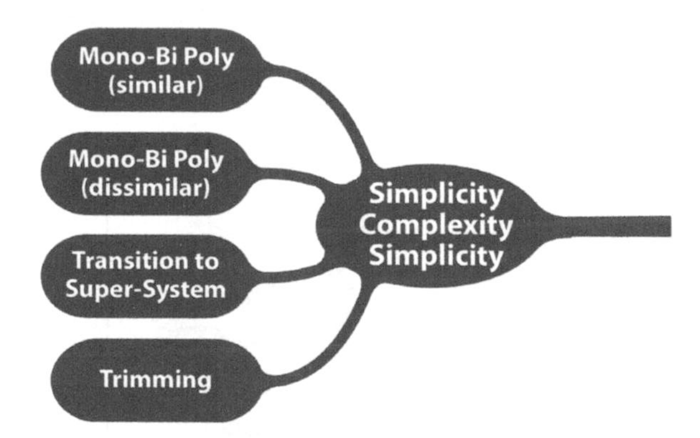

Many technological systems initially are fairly simple but then become more complicated (often we want more functions and this means more parts) and then later become simpler again as we learn how to reduce the parts but keep the functionality we want. Rarely is the complicating part of this cycle missed out, as humans seem to need to complicate before they can simplify. Therefore initially as quantity and quality of functions is increased, there are more parts; but then often, to reduce costs and harms, we try to simplify the system without losing the new, extra functions. The last part of this cycle is elegant design to give us the minimum number of parts with the maximum number of functions. In TRIZ this process of simplification is a very powerful route for solving specific problems and for improving systems. The Standard Solutions contain many prompts for simplification and in particular the very specific Trimming Rules suggest ways of removing components while keeping all functionality (see Chapter 11).

When we apply the Trends we are simply being offered directions to follow – we use other parts of the TRIZ Toolkit for detailed and specific suggestions for system simplification. Improvement is often fairly rapid when we first develop and evolve a product. There is an increase in complication as we add parts (and although we are also adding cost and possible new problems – benefits are rising rapidly) and ideality is higher. This is the steep part of the S-curve and when the rate of improvement starts to slow down we are reaching the top and flat part of the curve. Just because complication has initially delivered benefits it doesn't imply that adding more parts will always give us more benefits. We reach a point when we have to stop doing the same things and find new ways for increasing Ideality and this is where simplification normally cuts in.

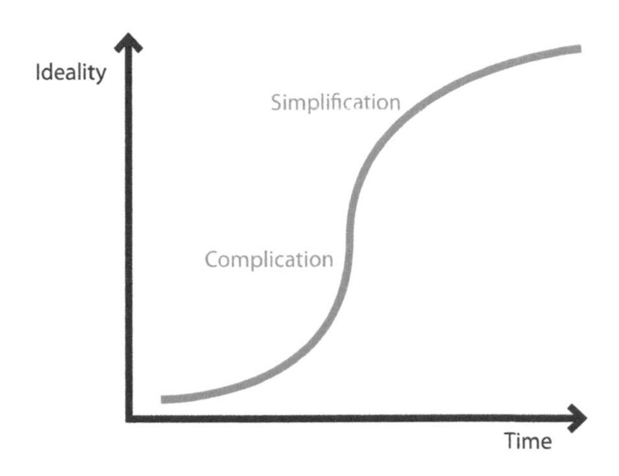

System Evolution – begin simple – become complicated and offer much more – become simple again without losing performance

Two Routes for Complication

Systems become more complicated as they develop and this Trend is thus also called 'Mono-Bi-Poly' which describes how systems complicate by going from mono systems to bi systems to poly systems.

Systems normally become complicated in two ways by adding:

- similar elements (as in a propellers or magnetic hard discs)
- dissimilar elements (as in a mobile phone, with camera, music and alarm clock etc.).

Similar Elements

Normally we complicate by adding more elements until there is no further advantage and Ideality starts decreasing instead of increasing. We can then recombine the elements and simplify again. When improving systems, it is a very obvious and easy step to double up what we have, and if that helps, add another element again until our systems reach a peak. The optimum number will be hard to predict and will probably change with technologies of materials and manufacturing methods. In the figure below the development initially by complication and later by simplification for magnetic hard discs is shown.

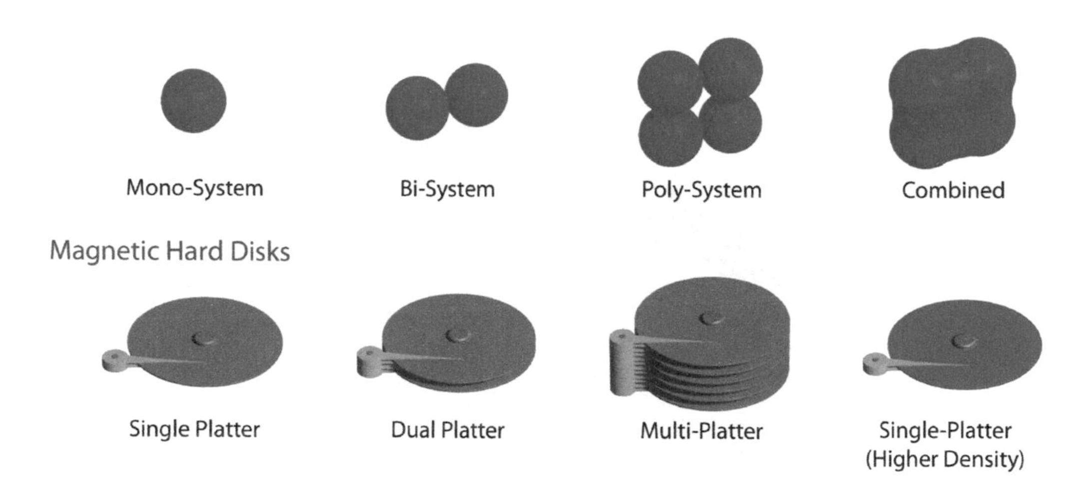

Dissimilar Elements

The complication of a system by adding dissimilar elements is obvious in many systems we use today. My first stereo system for listening to music was a small record player. I also had a separate radio. As the systems got more complicated I added an amplifier, a record deck, a tape recorder (reel-to-reel and cassette), two enormous speakers, a radio tuner and some headphones. These needed a lot of investment, specialist knowledge, special furniture and a great deal of space. But after complication comes simplification which occurs in four principal ways: the first two are the combining of the elements which were added in the complication stage, the third is by transferring the functions up a level to the super-system, and the fourth the most essential to TRIZ approaches, simplifying by reducing the system to an absolute minimum of parts while keeping all the functionality and benefits.

Combining very different elements can produce spectacular results

Four Routes for Simplification

We can achieve Simplification by:

- combining similar elements
- combining dissimilar elements
- transition to the super-system
- trimming out components but keeping all functionality.

Simplification by Combining Similar Elements

When we look at many everyday systems around us we can see this trend very clearly – the complication process by doubling up and then adding more similar elements until there is no further advantage. Simplification is now required and the system becomes something using all the elements combined back to a single element above.

Simplification by Combining Dissimilar Elements

This is shown in many modern systems such as in the development of mobile phones, which now combine many systems into one or the Ipad which offers functions from many systems in one product.

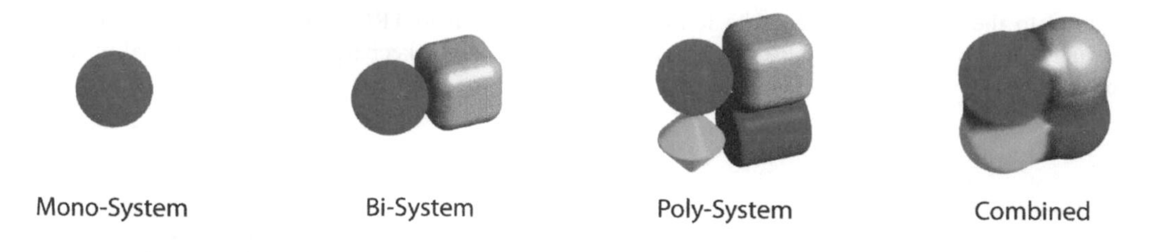

Simplification by Transition to the Super-System

This means that a system is simplified by the super-system taking on some of the system roles, examples include:

- Overhead lighting on roads and paths in towns eliminated the need for individuals to carry their own torches or lamps, well lit roads may eliminate the need for headlights on cars.
- Information available on the Internet has eliminated the need for individual information to be provided. This has proved very valuable to arts and music societies who used to print advertisements, posters, programme books etc. which they found expensive and hard to recoup their investment. Now they can post all the information on their websites and in 'what's on' websites and drastically reduce printing and distribution costs.

Trimming Out Components but Keeping All Functionality

When a system develops and moves up its S-curve, normally we add more components to deliver more functions to give us more benefits (*complication*), as we reach the top of the S-curve and enter the *simplification* stage, then we seek to keep all functionality to provide all benefits but take out as many parts as possible. The red line in the figure shows how part count increases as the system first develops and then peaks near the top curve of the S-curve and then is reduced as much as possible by applying TRIZ *trimming*.

The trimming rules of TRIZ are designed to reduce parts but keep all functionality (see Function Analysis Chapter 11 for full details) and are simple and powerful.

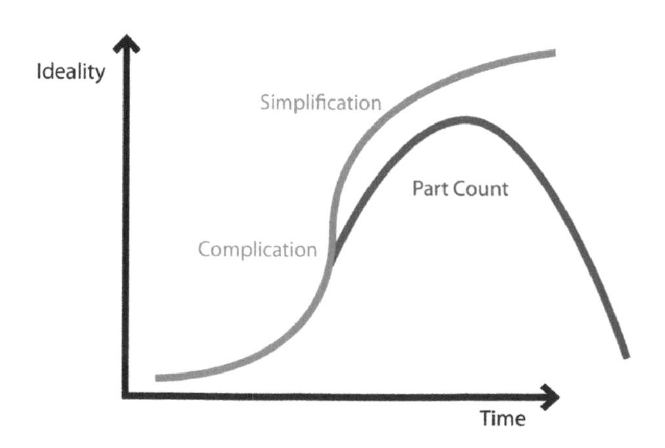

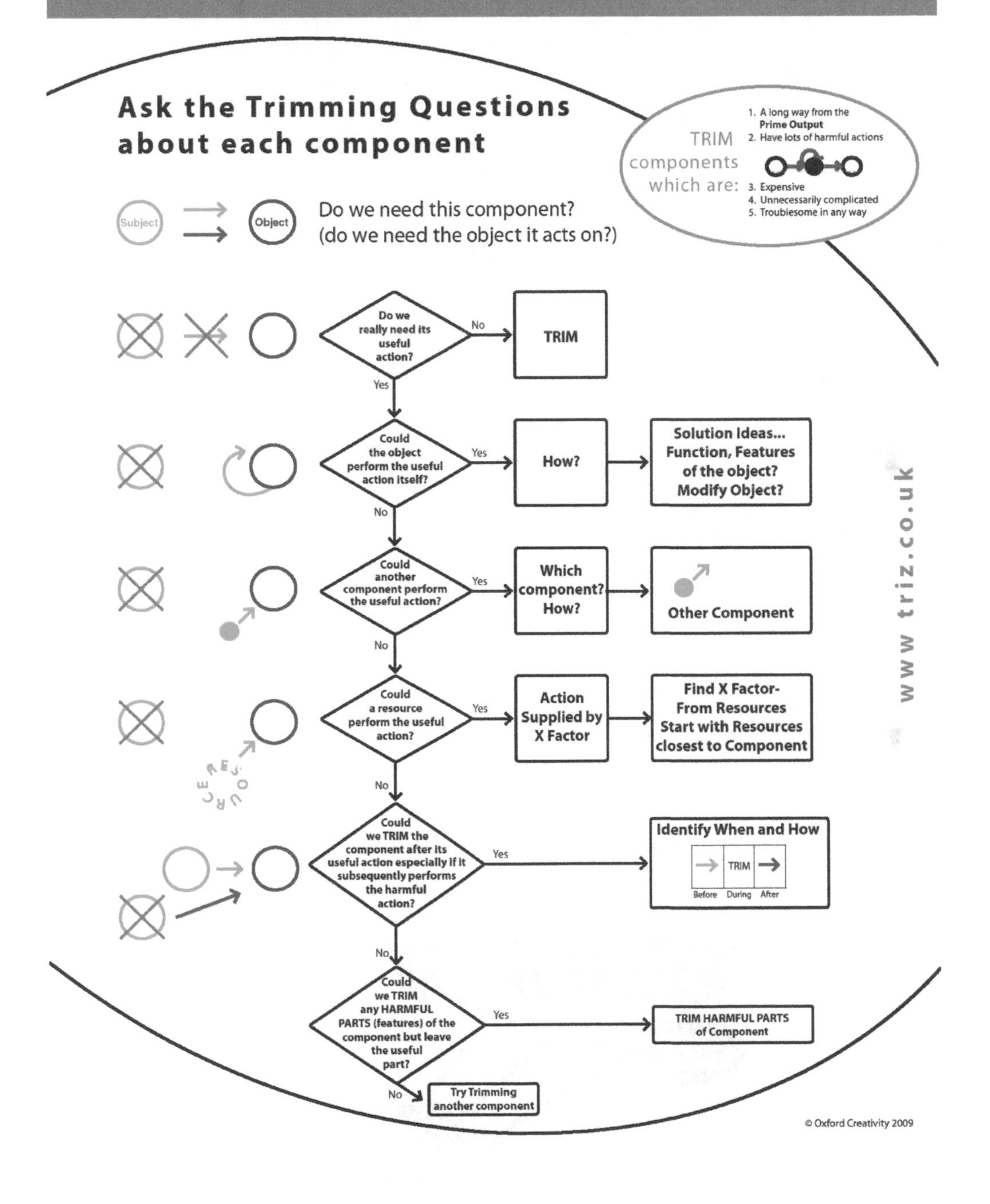
Trimming Rules
OXFORD CREATIVITY
Ask the Trimming Questions about each component
TRIM components which are:
1. A long way from the Prime Output
2. Have lots of harmful actions
3. Expensive
4. Unnecessarily complicated
5. Troublesome in any way
Subject
Object
Do we need this component? (do we need the object it acts on?)
Do we really need its useful action?
No
TRIM
Yes
Could the object perform the useful action itself?
Yes
How?
Solution Ideas... Function, Features of the object? Modify Object?
No
Could another component perform the useful action?
Yes
Which component? How?
Other Component
No
Could a resource perform the useful action?
Yes
Action Supplied by X Factor
Find X Factor- From Resources Start with Resources closest to Component
RESOURCE
No
Could we TRIM the component after its useful action especially if it subsequently performs the harmful action?
Yes
Identify When and How
TRIM
Before During After
No
Could we TRIM any HARMFUL PARTS (features) of the component but leave the useful part?
Yes
TRIM HARMFUL PARTS of Component
No
Try Trimming another component
www.triz.co.uk
© Oxford Creativity 2009

Trimming delivers everything we need with much simplified systems

Simplification or Trimming or Combining is a Very Important Design Step

TRIZ is obviously not the only toolkit for engineers which demonstrates the power of simplicity. In the many excellent toolkits which exist to help engineers and designers the process of making a system simpler but keeping all functionality is a challenge.

Can we keep all functionality and must some functions be kept separate or can they be combined? Such questions are asked and tackled by using toolkits such as LEAN or Axiomatic Design. (An axiom is a starting point and something that is not to be proved because it is a self-evident truth.) It rules include:

The Independence Axiom which = Good Design occurs when the Functional Requirements of the design are kept independent of each other;

Information Axiom = Good design occurs when the minimum 'information' content is achieved (good design = minimum complexity).

6. Increasing Dynamism, Flexibility and Controllability

There are three lines of this Trend as shown below:

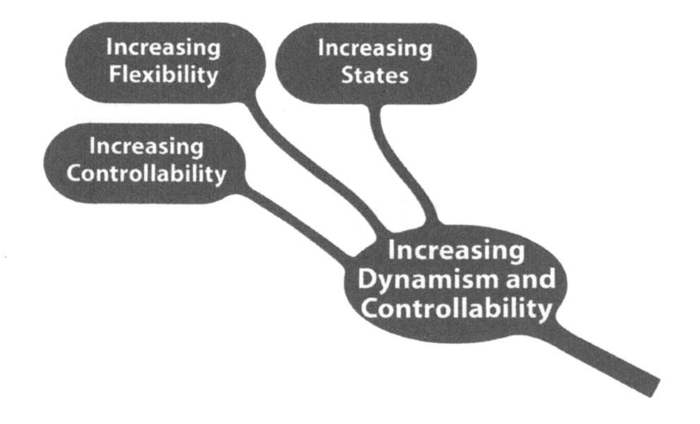

Increasing Dynamism, Flexibility and Controllability

Systems evolve to become more flexible with greater degrees of freedom. This means that parts of the system which were fixed become moveable. Increasing flexibility and controllability allows functions to be performed with greater flexibility and in a greater variety of ways. However when we make a system more dynamic and flexible we need more control. Hence the increase in controllability lines include:

- transition to more flexibility
- transition to more variability or more states
- increasing controllability.

Systems such as gears for bicycles and cars demonstrate how systems become more dynamic, more adaptive to their environment, and more adjustable at all levels, allowing more functions and offering a greater variety of options.

Increased Flexibility

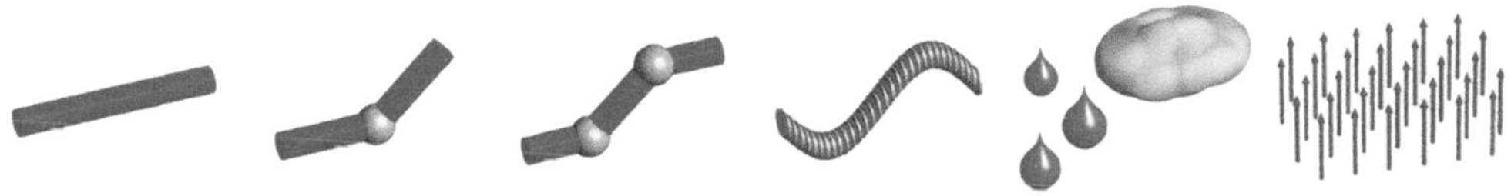

Immobile System | Jointed | Many Joints | Fully Elastic | Liquid Gas | Field

Steering System

Rigid | Single Joint | Double Joints | Flexible Joint | Hydraulic | Steer By Wire

Increasing flexibility at the mechanical level shows systems becoming dynamic by adding hinges or hinged mechanisms, using flexible materials etc. as shown in the figure – until some kind of field effect is reached. More flexibility however compromises stability and needs more control by more advanced, capable systems which can sometimes initially decrease Ideality by adding disproportionate cost and harms. The example above of steering by wire had been prevalent in the aerospace industry for a long time before it was considered by the automotive industry where for some time it was considered rather too expensive and risky for many applications.

Increasing variability and degrees of freedom can influence system performance and offers more flexibility between components, than in rigid, monolithic systems. When components are joined to give better flexibility then this is helpful for access in maintenance and repair and offers many advantages.

Another line of this trend is increasing variability of states as seen in the developments of many systems such as simple traffic systems as shown in figure below.

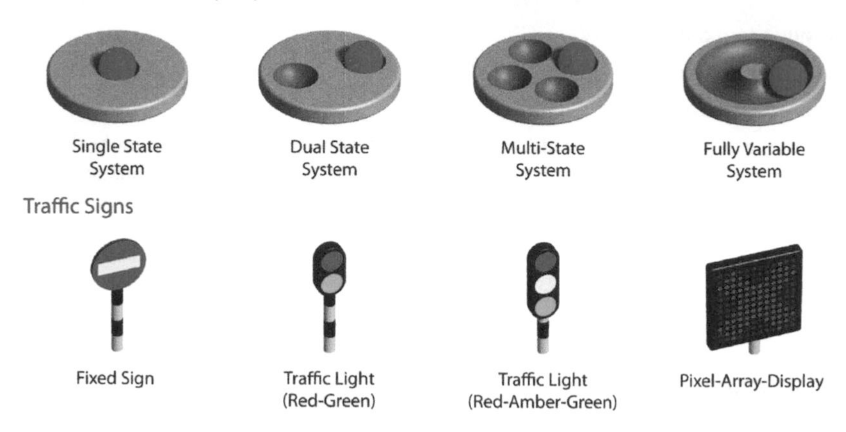

Increasing Controllability

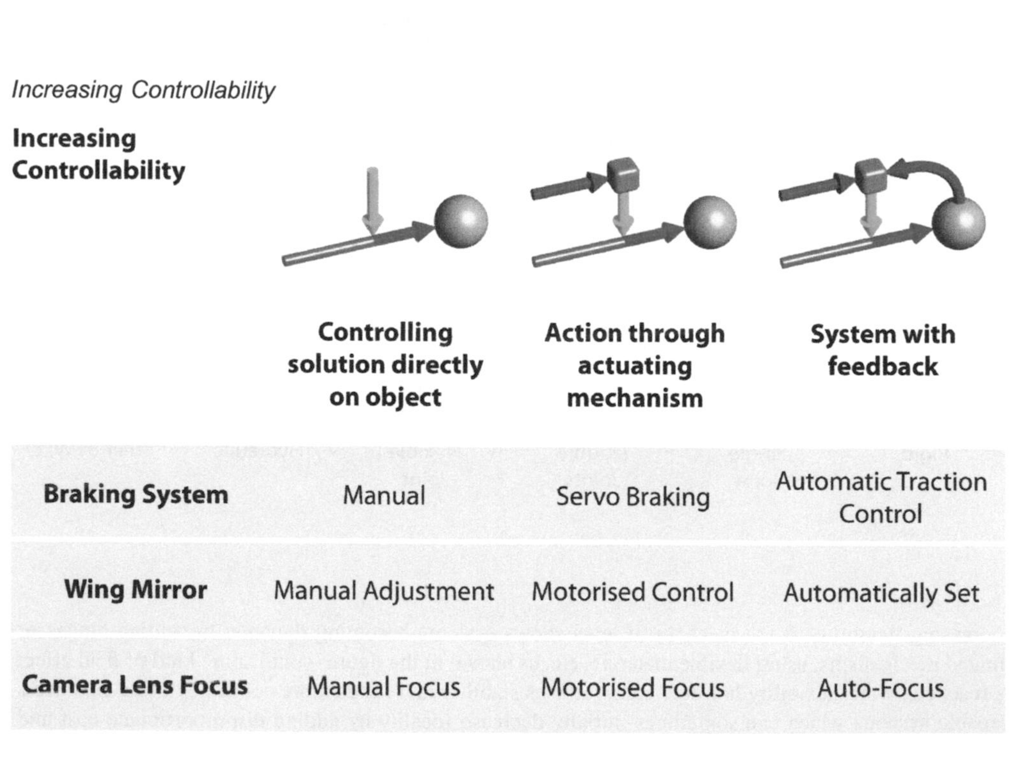

Braking System	Manual	Servo Braking	Automatic Traction Control
Wing Mirror	Manual Adjustment	Motorised Control	Automatically Set
Camera Lens Focus	Manual Focus	Motorised Focus	Auto-Focus

Adding control systems is an opportunity in a system's evolution to add guidance and control which give us options for more and varied functions and therefore many more solutions to problems.

7. Increasing Segmentation and Increased Use of Fields

There are four lines of this Trend as shown below:

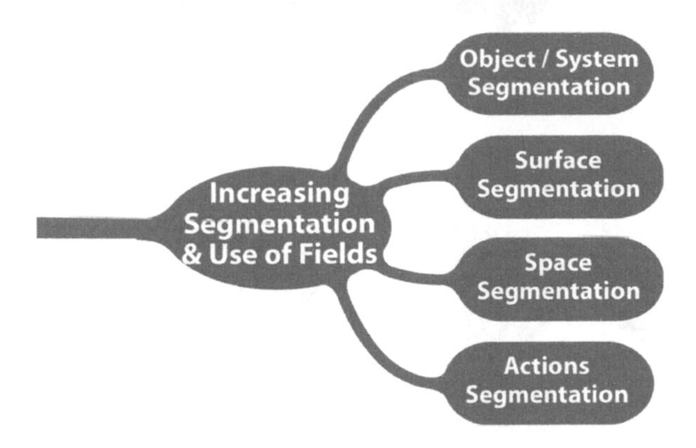

Segmentation is very fundamental to TRIZ approaches to solve technical problems. It is the first of the 40 Inventive Principles to solve contradictions and is used extensively to solve both physical and technical contradictions. Segmenting – making smaller, dividing into many parts – is a fairly obvious trend and usually very easy to understand and apply when looking at system development. Segmentation comes in various forms as shown below.

Object or System Segmentation with an example of the evolution of Cutting Tools

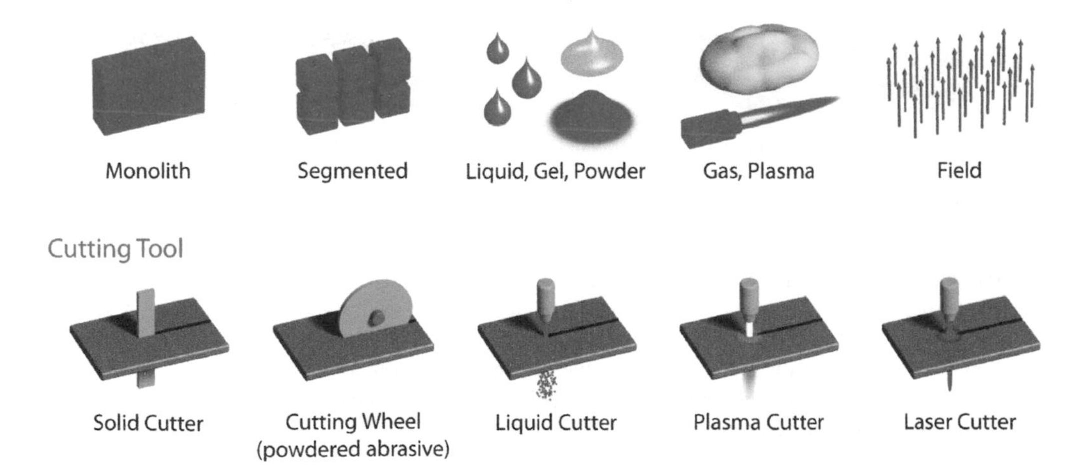

The advantages of moving from a simple monolith to something more sophisticated, by segmenting it in some way, is a trend we can see in many areas of life influencing much technological development. Segmentation of surfaces or 3-D materials we can witness in so many products around us – from the handles on toothbrushes and plastic razors (which show both) to cleaning materials and cloths, packaging goods, building materials and floor coverings.

Surface Segmentation

Flat Surface

With Protrusions

Rough Surface

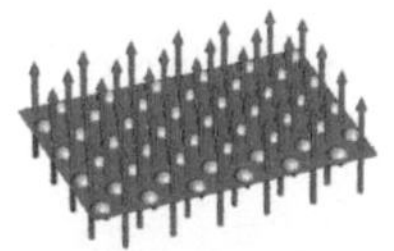
With Active Pores

Integrated Circuit Cooling

Flat Heat Sink

Finned Heat Sink

Pronged Heat Sink

With Forced Air Cooling

Space Segmentation

Monolith

Single Cavity

Multiple Cavities

Many Cavities

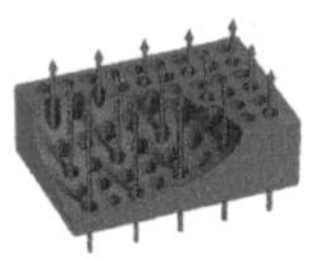
Active Cavities

Evolution of Materials used for many applications

Monolith

Single Cavity

Multiple Cavities

Many Cavities

Active Cavities

Segmenting Actions

This means moving from a continuous action or force to one which is divided up and delivered with different patterns and in different ways. When a system evolves there is segmentation of the actions as well as the parts. There are many examples of these and a simple example is in jet washers. A single jet delivers less effective washing power than a pulsed jet. Studying this trend showed several companies how to also use the natural frequencies of the objects being washed – beer bottles and coal trucks were two examples – pulsing the jet washes at the object's natural frequency, helped all the dirt fall off and meant that resonance could be used to help the process.

Rhythms Co-Ordination

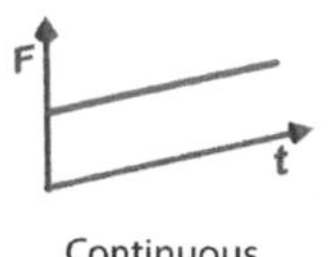

Continuous Action

Pulsating Action

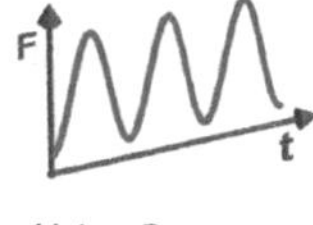

Using Resonant Mode

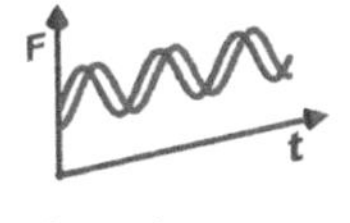

Several Actions

Travelling Wave

We can see further segmentation of actions such as moving from continuous operation to pulsed operation to sine wave to square wave to saw tooth operation to resonant operation to complex and travelling-wave operation. We can see this trend in electric motors in the move from continuous operation to square-wave pulse, to sine wave and now, in recognition that not all is linear (which would make a sine wave the ideal waveform), shaped waves are being used. New mathematical methods (wavelet maths, soliton waves) are playing an important role in helping progress along this trend.

8. Matching and Mismatching of Parts

There are four lines of this Trend as shown below:

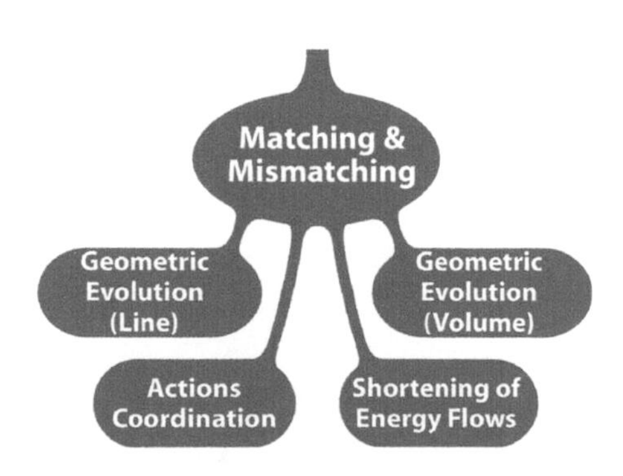

This Trend is concerned with evolving systems so that their parts, their features and their outputs improve or change to deliver all their required functions more effectively – even if this makes their manufacture more challenging. System elements improve or evolve to match (or mismatch) the best ways to deliver their functions. Achieving more benefits and reducing harms to make systems more effective may sometime increase inputs and costs. This may be because matching involves the development of the geometry of the system, evolution of the materials used, better manufacturing processes, improved timing of the system or good coordination of its various outputs, all of which may add cost and complexity, but still deliver significant advantages and therefore increase the ideality.

A system becomes matched to deliver all its benefits, not just its primary benefits. When a system is first invented usually it offers the primary benefits but not many other benefits. The system becomes matched to deliver all benefits as it evolves. For example the first mattresses were made of straw or feathers and provided the primary benefits of comfort and warmth. Modern mattresses are much better matched to all our needs and offer many other benefits. Depending on what we want these might include long lasting (20-year guarantee), utter comfort (no pressure on body joints), variable softness in different areas (matched to individuals sharing a bed) and many functions to provide all benefits, such as insect resistant, hygienic, easy to clean, waterproof (if required), needs no turning etc.

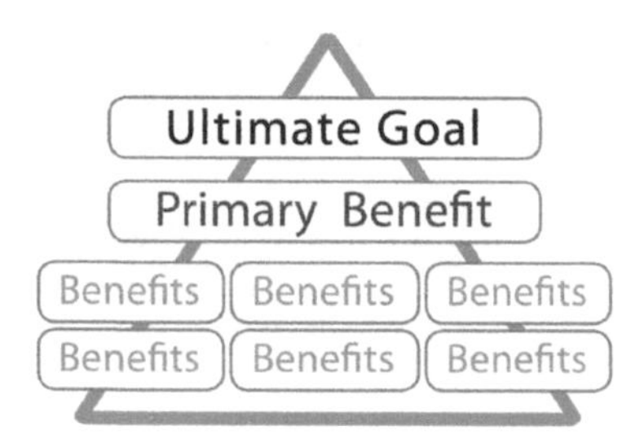

Mismatching

This is when a system is deliberately mismatched to improve its performance. Examples of this include using white noise to cancel out other unwelcome intruding noise (as in noise-cancelling headphones), any mismatching to avoid damaging resonance such as anti-resonant wave shapes used in switched reluctance drives to reduce noise levels and reduce bearing wear. One simple example might be when soldiers break step when crossing a bridge.

Geometric Evolution – Linear Evolution

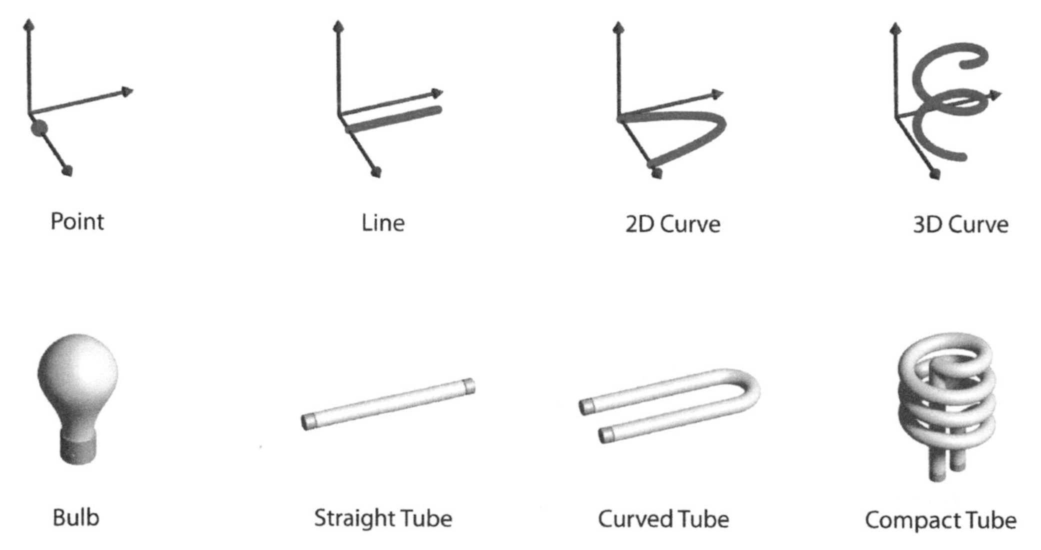

The system evolves from single point of simple light bulbs, to multiple point sources, to strip lights, to curved strip lights, to complex curved strip lights, to surface illumination etc. – matching the demand for light exactly to needs.

Matching elements can be seen in the development of both lighting and heating systems. Originally point sources such as fires, candles, lamps gave light and heat in a small local area. Evolution offered evenness of heating and illumination, giving the same levels everywhere in a room if required (although open fires, small lamps or candles for a certain, special atmosphere are still often used). Lighting and heating systems now use walls, ceilings and floors as delivery resources, especially as materials and technologies evolve.

Changing of Dimensions and Shape to Deliver Better Functions

This means moving through the following sequence – from points, to straight lines, to curved lines, to lines with multiple radii, to lines curved in multiple dimensions. Similarly surfaces move from flat to curved surfaces to multi-dimensional surfaces to complex-shape surfaces with different scales of details.

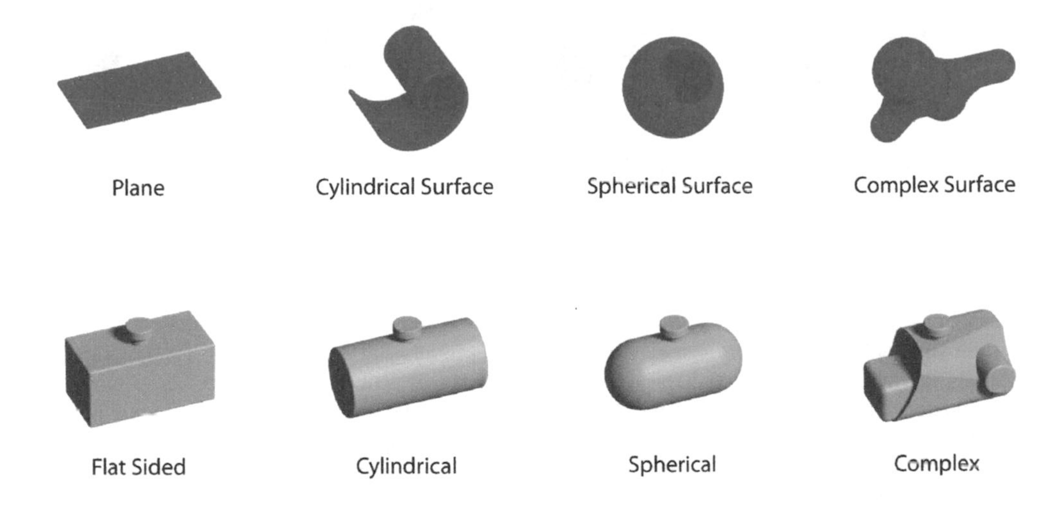

We have seen this trend in many systems and products as computer aided design (CAD) and computer aided manufacturing (CAM) have developed particularly since the 1980s. CAD has enabled many designers' visions to be realized, with detailed and complicated curved surfaces enabled to be accurately specified and created. Volumetric evolution can be seen in car body shapes and many other systems especially in recent years. The brilliance, innovation and power of modern CAD tools developed by mechanical engineers have been moving towards other disciplines. Architects and civil engineers are now moving to more sophisticated CAD tools to integrate their skills and effectively share information at all stages of complex construction projects, with geometric evolution, and particularly volumetric evolution, at the heart of their outputs. We can now see so many interesting and complicated shapes and curves appearing in new architectural buildings, with design becoming matched with cost-effective, sophisticated construction methods and processes.

Matching – Shortening of Energy Flows

The reduction of energy conversions to an absolute minimum is one factor in efficient utilization of energy available. The main advantages of are:

- increase efficiency
- reduce harms
- reduce costs
- simplify.

Energy is a commodity which is being truly valued as we face a global problem of appropriate use of scarce and limited resources. The quest for energy for industrial applications and for electricity to be delivered to homes and offices has led to an interesting history and evolution of energy use, distribution and energy conversions in many processes. The shortening of energy flows is an important trend both to map the efficient use and delivery of energy. For example in an internal combustion engine the main energy conversions are chemical → heat → mechanical and in a coal-fired power station chemical → heat → kinetic → mechanical → electrical.

Shortening of Energy Flows

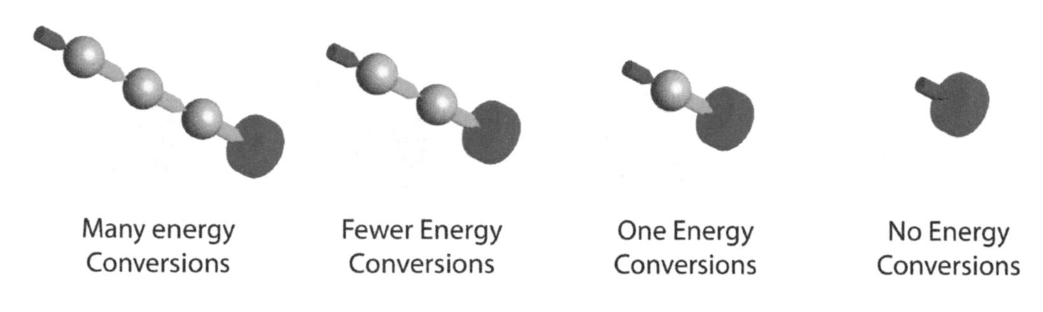

Matching of Actions

This matching of actions or functions could describe many systems and processes. For a simple heating system you could have uncoordinated actions of the heat inputs and the temperature of the room it is heating. Partially coordinated could be when a cut-off switch is activated above a certain temperature. Coordinated actions could be a heating system with a thermostat to measure the temperature outside the building, and another inside the rooms and a resulting heating system controlled to match the outside ambient temperature deliver the right level of heat at all times inside the rooms. Action during intervals could be a heating system which does everything the coordinated actions system provides, plus storing heat from the outside sources, such as sunlight and releasing it when it is appropriate.

This trend is saying that everything within a system or process will become better matched, better coordinated to give maximum performance, delivery of all benefits (not just the primary benefit) and higher achievement of Ideality.

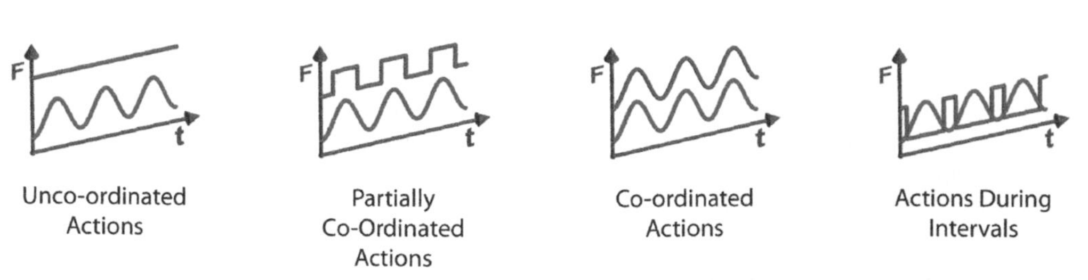

This trend also applies to teamwork, and increasing coordinated actions as teams learn to work well together.

Better matched and co-ordinated systems are important in technology and management

Ideality is Increased by Moving towards the Ideal Along Any or All of the TRIZ Trends

When we have established where our system is on for a particular trend, then there is probably an advantage in moving along the trend towards perfection – the Ideal. It is therefore likely that for systems to be able to improve they must consciously aim to follow the trends. This is likely to be true even if we cannot see what the advantage might be! The chart below is useful when consciously applying the TRIZ Trends to your products and will show you where to put effort to develop, evolve and increase the Ideality of your systems and their components.

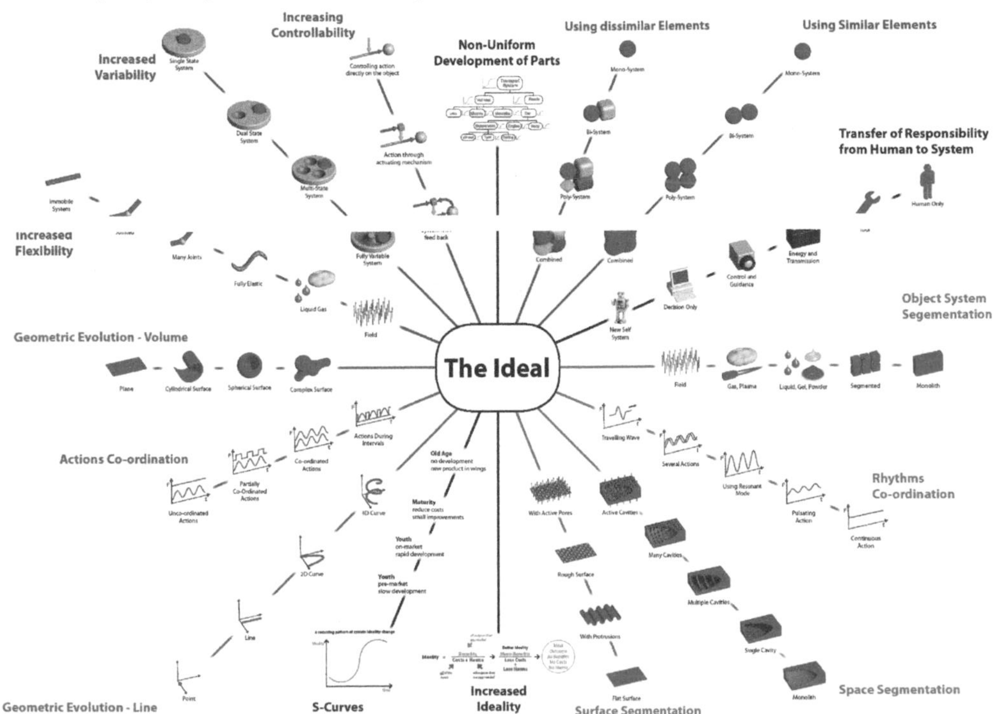

10 Inventing with TRIZ

How to encourage invention? This is a curious dilemma – and fundamental to the way we fund innovation, new ideas and invest in research and development of new products. There is a curious widely held view outside industry that invention is so random and so difficult to predict, that only chance and luck will help us. Training, supporting and encouraging individual inventors and investing in successful research is important to our future prosperity. The history of science and technology shows us their steady advances and how they have benefited every area of our life, and come from sustained searches for new areas. These searches need substantial investment of time and money from companies and individuals. Yet this essential, sustained long-term investment in our future has a popular image linked to risk and uncertainty with its outcomes viewed as prizes.

TRIZ simply offers a record of what was successful in the past – including invention – and uncovers the relevant simple, formulae for success, which can be applied to future invention. TRIZ shows us what we know about the patterns of past proven invention and helps us recognize and apply these patterns to future invention and new products.

Companies expose their research teams to TRIZ to support many of its principal activities including:

1. New products – TRIZ helps bring their work into focus, converging the many strands of useful research into actual products and overcoming problems. TRIZ can help teams identify future and next generation products, to match their newly developed technologies.
2. New areas for research – TRIZ work in R&D departments often involves those very tenuous beginnings of deciding where to diverge, where to look for promising directions. TRIZ as an innovation toolkit helps overcome psychological inertia when looking for promising areas for future work.
3. Creating patents, patent strengthening and challenging competitors' patents.
4. Problem solving – R&D by its very nature will be beset by problems – how to scale up, how to manufacture, how to improve yield, how to remove impurities, glitches or unexpected harms etc. TRIZ problems-solving processes and tools will often solve 'unsolvable' problems in R&D.

TRIZ for Engineers: Enabling Inventive Problem Solving, First Edition. Karen Gadd.

TRIZ Tools offer systematic routes to inventing and developing new products

TRIZ teams work well with R&D departments everywhere. One exciting session our TRIZ team all enjoyed was when a major company invited all its research teams from all over Europe to Linz for a three-day workshop to sample TRIZ and then to apply it to their most difficult and unsolvable problems. The session began in a very gloomy mood with the Scandinavian teams in particular declaring that it was all a waste of their valuable time but as the head of research from each country had attended, along with their brightest and best, there were great expectations for successful outcomes. None of them knew about TRIZ or had used the tools before. A very rapid training workshop was followed with six teams each working on one or two problems. These sorts of sessions are both difficult and challenging for the TRIZ experts who are facilitating, as they must communicate how to use TRIZ quickly, understand often complex, highly technical problems and guide the teams to the right TRIZ tools in order to help them find solutions. This requires a lot of concentration and skill, as the teams sometimes requires coaxing out of some common behaviours which can prevent the session being successful. These include:

- re-exploring the places where no solutions have been previously found;
- spending too much time in the detail of a 'comfort zone' of knowledge;
- reluctance to try a new approach (even when the old approach hasn't worked);
- refusal to use a different approach to another toolkit which an individual has already spent often considerable time and energy mastering (e.g. Six Sigma, RCA, requirements capture).

Ironically these problems can be magnified when the problem is particularly urgent or important. Problem-solving time is surprisingly rare and therefore valuable, and it is easy to fritter it away by going over old but safe ground. TRIZ helps us break out of this and in my experience TRIZ always works, when problem solving is actually allowed to happen.

In Linz the euphoria started to build in the last two hours of the third day. When I left at the end of the day with my colleague to catch our flights home, at least two teams didn't even look up to say goodbye as they were so deep in their solution discussions and the excitement of having solved these unsolvable problems. As TRIZ is a multiplier of an individual's problem-solving skills, it works particularly well with such highly focused teams of very-well-educated, clever people whose purpose is to solve problems in the exciting areas of research and development of solution concepts.

The corporate invention scene is very different from the world of the individual inventor. Inventors have the exciting reputations of being those who have succeeded against all odds to overcome tremendous difficulties – and are a great inspiration for young scientists and engineers. Traditionally inventors have usually been great engineers and good businessmen (mostly men for historical reasons – although many research departments today are often led and run by women) who can meet the triple challenge of identifying what society needs, creating systems to fulfill those needs, and succeeding in reaching and selling their solutions to society. Insight, technical brilliance, resilience, business skills and the ability to mobilize backing and resources have always been part of successful invention. Invention within large corporations is a different game and how most products reach us today. Present-day stories of famous individuals who have succeeded on their own as inventors (often against the large corporations) still makes for exciting reading.

There are of course many great but unknown inventors (present and past) who have produced great ideas for good and useful products but may have failed because the odds are so stacked against them. The famous inventors always have interesting tales of technological success, marketing triumphs preceded by many trials and setbacks. There are many common characteristics of great inventors no matter how far back in history we look: obsessive dedication, sustained hard work, ruthless ambition, an ability to see the big picture. Many interesting studies have been published including some from Altshuller who spent his last years working in this area; his findings are published in his books, which include *How to be a Genius* and *How to be a Heretic*, but are available only to those who read Russian.

No matter who we admire as our inventor heroes – Leonardo Da Vinci, John Logie Baird, James Watt, Frank Whittle, James Dyson – they all share certain behaviours in that they all worked tirelessly and steadily to overcome many obstacles to take their place in history. The years of daily hard work are less interesting to the storytellers because both invention and discovery of great truths will always be associated with great accidental and eureka moments. Despite the rigour and discipline of mathematics, science and technology, it is rather strange that we still often attribute many of science and engineering advances to trial and error, and one moment of lucky chance. Archimedes jumping out of his bath and running round shouting 'eureka' about his understanding of displacement has a lot to answer for. Too often we picture scientists achieving their breakthroughs simply from random flashes of genius. Fleming and penicillin, Newton and falling apples all strengthen our perception that scientists' breakthroughs owe more to a moment of luck, than hard work and great thinking.

TRIZ removes the randomness for Scientific Breakthroughs

The allure of the accidental discovery is belied by the history of science, mathematics and engineering which shows that we only progress when societies and individuals have the luxury of investing time and sustained effort into research. We get results when we are allowed time for thinking, and given time to search for and then develop new ideas, which can be transformed into the real products we need. Allowing time for this is hard for companies obsessed with measuring and checking everything, as the culture discourages any activity which is not audited.

However there is a curious contradiction in all creative activities, as this time and space 'to think' needs to be combined with a certain amount of pressure and deadlines. A famous example of the negative effect of total freedom from pressure is the Finnish composer, Sibelius, who was given a generous state pension and never composed again (and lived for a further 30 years). All creative endeavours – including invention, art and music – require hard graft, education, training, apprenticeships and experience in addition to thinking time. Some of our greatest composers such as Mozart, Beethoven, Haydn, Handel and Bach all were professional musicians and composers who worked hard at music on a daily basis for their whole lives, and had to meet deadlines, search for work, struggle to earn their living and deliver high-quality, new and great musical works. Anyone who has worked hard for an exam knows about the importance of an incentive to create motivation. Creativity needs the time, but it may also need the pressure of something like a deadline – this is only a contradiction, which like all contradictions can be solved with TRIZ.

An important aspect of giving time to be creative is being given the 'mental space', free from smaller pressures. To give another musical example, the composer Mahler was also a conductor, and his conducting job involved many tiresome administrative chores. He dealt with this problem by building a cabin near a lake, far away from the city, where he would retreat for the whole summer, and compose without being disturbed (even by his wife or children). Many engineers feel that their creative ability as engineers is being eroded because they have too many trivial tasks: 'corporate cost savings' requires them to be their own travel agent, secretary etc.; a sheer complexity of detailed systems imposed on them of anything from long meetings, to accounting for their time, uses up much of their energy and time. When working with teams to discover what is curtailing their ability to solve problems and be innovative, we use worksheets, illustrated by sets of pictures, such as below and ask them which one most represents their own situation. No teams have ever indicated that the first picture represents their current situation.

Inspired design and scientific success needs enough space, time and dedication

Given time to research, time to problem solve, time to think and understand is not to be given a holiday from normal work; time just for thinking needs to be part of the regular, pressurized, hard-working environment of any company. Dave Knott of Rolls-Royce UK presented a paper to a TRIZ conference about how the structured, systematic processes provides engineers with improved capability and motivation to innovate, but unless they are provided with the opportunity to use TRIZ, and an environment that nurtures innovation and creativity, there will be zero benefit.

Everything – from the first priest castes in ancient Egypt developing geometry to help mark regularly flooded Nile land for ownership, to the twentieth-century space race – shows how great achievements happen when we can afford to have professions with people with ability, education, time and space to develop new ideas and solve even 'unsolvable' problems. Freedom from just working to keep warm, safe, fed and watered is a luxury that has always come from wealth in societies (although we look to the future with a belief that technological advances will help us all achieve freedom from daily drudgery). Many societies with surplus and wealth have invested in progressing science, mathematics and engineering.

Places for professional learning and research such as universities have always provided this space for time and thought, and the payback on progress is much higher than chance or lucky breakthroughs alone could give us. Isaac Newton, one of our greatest scientists and mathematicians, did his greatest work while a Fellow at Cambridge University. But he said that the great plague of 1665–6 was 'the prime of my age for invention', when removed even from the demands of university life he moved to Lincolnshire and had two or three years of uninterrupted intense mental activity which produced his best work on *Philosophiae Naturalis Principia Mathematica* (Mathematical Principles of Natural Philosophy). Newton's Laws of Motion and his work in mechanics on momentum plus his many other achievements in abstract thought made him immediately acknowledged as one of Britain's greatest scientists and mathematicians (www.newton.ac.uk/newtlife.html). Yet it is not the unexpected freedom from lecturing at Cambridge for three years which is credited with helping him uncover the fundamental truths about gravity – but an apple.

'Unfortunately for science, Newton was feeling a bit peckish that day.'

Chance and luck (not sustained research and hardwork) cause scientific breakthroughs?

Chance, luck, playfulness, humour are of course all important in the discovery of new ideas and systems, because they break psychological inertia and join the creative side of the human brain with the systematic, analytical side. Accidents, luck and fun, which shift our perception and help us all see familiar things in a new way, are only a small but important part of the story. We often need that jolting step, that re-arrangement of very familiar pieces of a puzzle into a new pattern, that suddenly delivers what we want. When we have been working in a very sustained way and looking hard and long for the pattern and know what we want, letting go or achieving some wacky stimulus and accidental prompt helps us to finally recognize the significance or answer which was eluding us – but so does TRIZ.

TRIZ offers us useful Idea Creation tools and 9-Boxes to help create those moments of clarity and see the elusive connections but also suggests that when we follow the TRIZ process then accident and lucky big leaps to the answer won't be needed. TRIZ will take us close to the solution space where only small leaps will be needed. When we think of a lucky breakthrough moment we forget the many trials and dead ends, all the years of study and only focus on the great moments of great clarity of thought when either the right connections were made or the relevant answers or patterns were suddenly seen and understood. A sudden leap was made to the right answer – but that leap was preceded by sometimes years of working out in which direction to leap, years of trials and gathering information, until the solution direction becomes clear when the relevant knowledge is recognized and harnessed appropriately.

Seeing a problem in a new way is very important and fundamental to being creative and innovative; just stepping back and looking at it in a bigger scale, in context, or just lifting our heads from the detail all help us put together a complicated or simple answer. One simple TRIZ prompt is Inventive Principle 13 – Do it the other way round, and this powerful approach is recommended by the Rolls-Royce TRIZ group as a good starting place for any solution. In a case study on 9-Boxes on Australian highways, the simple TRIZ logic for both understanding and solving the problems is shown. In a project of great complexity and difficulty for building a complex highway interchange, clever solutions helped with its complete success – unexpectedly on time, and within budget.

Practical pragmatic problem solving as well as invention, research and uncovering of essential scientific truths all need the right knowledge, systematic approaches, hard sustained work and the confidence of travelling in the right direction. TRIZ supports all of these, as well as providing the jolts, the triggers to help us break out from a mental torpor (which can slow us and leave us bogged down in a useful place) and move us on and get to the answers, the solutions to our problems, which are often very close by. Accident, play and a chance glimpse of a new pattern all contribute to innovation and invention but TRIZ offers creativity tools to help us leap in the right direction when we need it. The 9-Boxes help us shift our perception in scale and time, and by making us focus right into the essential problem. TRIZ can show us the small number of right answers.

Very mundane Inventions

What do we call an invention? The first time a system is created which delivers something we want? Inventions can be after many years of trial and work such as an aeroplane or something very simple such as toilet paper, a pencil or a teabag.

For a long time both dustbins and wheels have been in use. They were not combined into a wheelie bin in Oxford in the late 1980's. After that time wives (rather than husbands) could easily put out the dustbins.

Putting wheels on something heavy like a dustbin (or a suitcase) which has to be moved regularly is not rocket science. This simple and (with hindsight) obvious innovation/invention saved many people from back problems and a weekly sense of annoyance and some family arguments!

Despite the pitfalls, and mistaken disproportionate importance given to serendipity, successful new inventions happen all the time and can be highly technical, exotic and complicated – or very mundane and simple. Their usefulness is measured by their market acceptance and the needs they meet. Matching needs is mostly deliberate but sometimes we invent systems by accident, or without understanding all the needs they can fulfil.

Clever Invention is often combining existing systems

How to Be a Great but Mundane Inventor with TRIZ

When there is no system available and we have to invent one, this is often done by thinking about some terrible, inadequate system to fulfill our needs, and then using the TRIZ tools to get a much better system – getting the right system and also getting the system right. We have to accurately define needs first but assuming this is accurate, TRIZ problem-solving tools take us from bad systems to better ones. Successful products achieve some clever better matching of unfulfilled needs by often using the resources we've got. This most TRIZ-like of approaches to problems can include improving old, tried and tested systems which are already out there and need a clever new approach. Many examples of really good invention have been when someone is determined to solve a problem and create a new product, and have done this successfully without TRIZ. But as we only hear of the successful outcomes there is a worry of how many failed products took up time, resources and inventiveness.

One question asked of TRIZ is can it help me systematically repeat success – can TRIZ map the steps of successful invention? The answer is yes and the steps are not that complicated at the highest level of the logic of how to invent. Whenever I see an invention I really like and think is clever, I map it in the TRIZ process – and I find that good invention always follows simple TRIZ logic and has a good use of resources as in the example below.

There was much need for the new invention of smart domestic fire alarms which plug into a light fitting and is called a 'Fire Angel'. The simple fire alarms are attached to existing light fittings and are high up on the ceiling where they need to be, and not very visible – more importantly instead of needing batteries which could fail, they use existing resources – and are powered by the electrical lighting circuit and have a rechargeable battery within them – so they work during a power cut too. This invention has nothing new except its outputs – it uses existing resources but is clever, useful, economic and reliable. It is better than previous fire alarms and made its inventors, Nigel Rutter and Sam Tate, successful entrepreneurs in their early 30s (they are both graduates of Coventry University which helped them at the critical early stages).

Rutter claims that after the stimulus of his design degree he decided to invent something and looked round one evening for inspiration and noticed the annoying battery operated, obtrusive and unattractive fire alarm attached to the ceiling with its wires hanging out. Together with Tate he decided to invent a new fire alarm for domestic use. Their Fire Angel is a system of high ideality – with lots of benefits and reduced costs and harms. In 1998 Rutter and Tate managed to win £55,000 of funding through a scheme run by Coventry University, which also helped them to file a patent and launch their product. (http://business.timesonline.co.uk/tol/business/entrepreneur/article4837139.ece)

I particularly enjoy this success story as I was Governor of Coventry University at the time and I was delighted to see the University offer practical help to our graduates, and help them create such a success by moving the two young graduates from being good designers to also being successful business people. Stories like this are encouraging because too often when not seen as a result of an inspired idea, invention is seen as a hard slog, with too many options, which need many trials, and whose success relies on discovering or locating the right combination of concepts and technologies. Intelligent application of TRIZ reduces those trials and can inspire this kind of activity.

Nanotechnology attracts many jokes

Accidental inventions are the stuff of legends and myths, and we like to think of great new products emerging by chance or luck. One famous good accidental invention, if somewhat ordinary, is the creation of crisps in the nineteenth century (so called in the UK – but known as chips in the US). Crisps were created by an irate chef in a restaurant in response to a customer who complained that his French fries were not thin enough. After thinner and thinner fries were rejected by the awkward customer, the chef sent out very fine, paper thin slithers of crisply fried potato, as a bad-tempered joke. Although these crisp, wafer thin slithers of potato could not be eaten as part of a meal with a knife and fork, the resultant crisps/chips were deemed delicious and became one of our most popular snacks. Fun as accidental invention may be, it can be misleading implying that a random flash in the pan will do. This invention came from dedicated hard work with the pan, and normally only those doing the relevant work and really understanding the significant technologies can recognize an accidental prompt for a great invention. The chef who accidentally created crisps/chips understood potatoes, and how to slice them and fry them, to create perfect French fries of the various sizes and thickness – the creation of this new snack was related to all his previous work and an extension of french fry skills, and needed his skill and experience.

TRIZ and Invention

Altshuller believed that he could help all inventors of systems – and within the TRIZ toolkit are many aids for the inventor:

- Problem understanding, the relevance of the history and future directions (9-Boxes).
- 8 Trends of Evolution show us all the ways to evolve systems towards perfection.
- The Ideal Outcome guides the direction our work should take and helps us clearly state essential needs.
- Uncovering and solving contradictions. Solve with the 40 Principles – helps us overcome conflicts when improving a system, and also shows us how to achieve opposite benefits.
- 76 Standard Solutions help for simplifying and improving systems and solving problems within systems, particularly when something is wrong with our system – when there are excesses, harms, insufficiencies or absences. Solving with the Standard Solutions leads to improved systems or to moves to the next system on the evolutionary map.
- Creativity Tools which deliver a mental jolt, a shifting of perception; there are 13 in all and they include Size-Time-Cost and Smart Little People
- Transforming top-of-the-head solutions to better solutions through mental clarity. TRIZ instils the discipline of shifting from a Solution Idea to a Solution Concept and then finding many more solution ideas and the known functions to deliver them.

To use these tools we need the TRIZ Problem Solving processes which help us identify the appropriate TRIZ tools to apply to solve various problems. If we need to improve a system, or choose the right system, then we have to ensure that we understand what would be best to deliver our needs. If constraints prevent us from having the best or a different system, then we need to make the best of the system we already have, and eliminate as many problems as possible.

All the above TRIZ tools are useful for creating and improving the systems to meet those essential needs plus the TRIZ creativity prompts for the critical moments of visualizing solutions (including Size-Time-Cost, Smart Little People and Ideal Solutions). Simple TRIZ disciplines and clarity of thought encourage absolute understanding of the difference between an original concept and solution idea. TRIZ can also help teams work well together by using everyone's initial 'bad solutions' (which contain many useful ideas) and combining them to make a better solution. This enables teams to see all the possible good solutions that can be systematically derived from an initial 'bad eureka' type idea by locating the concept behind it and deriving more ideas from those first concepts.

For example one famous TRIZ problem example is how to speed up the production of liqueur chocolates. The system which needs improving is the filling of chocolate bottles with thick liqueur. The liqueur flows slowly as it is viscous and sticky. One solution is to heat the liqueur to make it flow faster but this has a problem that the hot liqueur melts the chocolate bottles (a clear technical contradiction of making one thing better and something else gets worse).

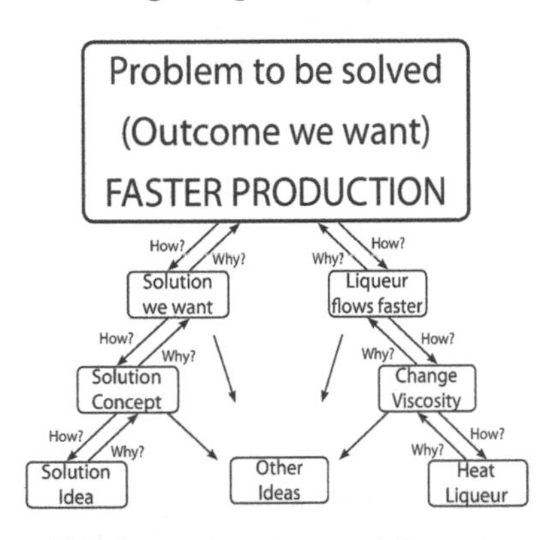

Despite inevitable discouragement many inventors' new concepts can find many usable solutions form creative teams

TRIZ helps us use concepts and translate them to many ideas to create new systems and combine existing systems to create new applications from technologies. This works bests with teams who will all have different ideas.

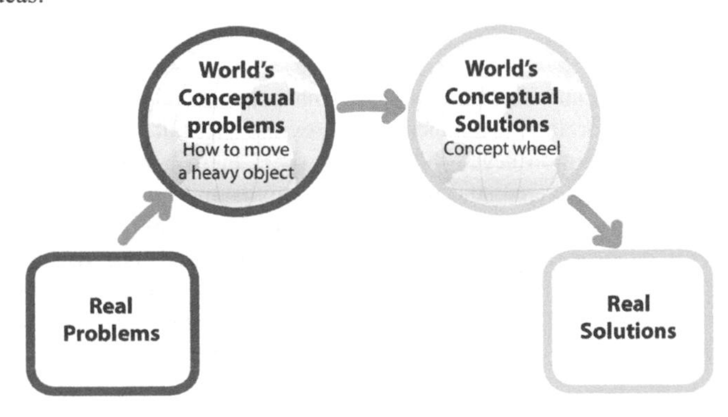

The wheel (a great invention) is a great concept: what to use it for, and how best to use it, needs understanding of what we want and how it can usefully combine with other systems. Each person in a team may have a different view from the miserable old man (who tries to squash the innovation). Their enthusiasm and useful and individual insights create many solution ideas from the one wheel concept of how it can be used. Inventing comes from both sides looking at *needs* and searching for systems to meet them, or looking at *systems* such as a wheel (particularly conceptual systems) and searching for *needs* they would meet. Both of these work well with teams.

Product DNA Predicts Future Systems

When a successful invention is brought to the market, it often contains within itself the potential to develop alongside market needs. The product has appeared at a time when it is needed, not by some amazing coincidence but because individuals and teams (including engineers and scientists) have recognized (either consciously or sub-consciously) and then met those needs with new technology. Both the needs and the system will develop together until the system disappears because it is no longer needed. This is why identical inventions occur simultaneously all over the world: technology, science and the ways they meet human needs are all developing together, and humans will be striving to meet needs and use new technologies and match each other. (Competition is a great motivator in science as in other areas.)

The inventing of a system is a fundamental recognition of all the needs the system will meet now, and in its future. There are many new ideas, inventions, products and their success depends on many factors including commercial recognition, sound backing and courage to be brought to technical usefulness and then to the market. Although we must mourn the many wonderful inventions which never see the light of day, it is important that we recognize that the ones that do succeed must offer appropriately priced benefits which match current, unfulfilled needs. If a system succeeds, then built into its DNA is the ability to ideally fulfill all the needs it was created for, and when it is developed by following the 8 TRIZ Trends of Evolution then the product itself will evolve as if anticipating future market demands.

This means that the system itself can tell designers and inventors about the right future directions and by applying the 8 Trends can point us to its related future products. This suggests that instead of just studying the market to look for promising new directions, researchers should also study the products themselves, as they carry their own successful future in their DNA. When invention occurs there has been at some unconscious level an understanding and anticipation of every need the system could fulfill. Inventors, when they create systems, have begun an evolutionary technical journey and although at the system's invention only some of the needs are met, later all the needs the system should fulfill can be realized by development and evolution. The first version of a product usually gives the prime benefits and some others – but potentially could fulfill all needs even if they are not recognized at the time.

The first cars delivered the primary benefits of assisted transport without animals such as horses. Early cars were dangerous, uncomfortable, incontinent in energy use etc. but they were not invented to be any of those things; their invention began the journey where these harms could be eliminated, the unfulfilled benefits could be delivered, and unacceptable inputs and costs brought under control by technical development. By developing cars according to the TRIZ laws of technical evolution we move closer to the perfect car, as more needs are met for less costs and harms. The idea that every system is first created to match/deliver a set of unrecognized, not yet understood needs is fundamental to the power of the trends. The system's DNA holds recognition of all the needs, and the TRIZ Trends are the directions systems should develop if they are to meet all those needs.

TRIZ helps us develop systems to give us everything we want

TRIZ directs our problem understanding and solving towards an Ideal end point, and helps us direct our efforts in the right direction, towards a final point when we reach the perfect system which meets all the needs it was created for. Applying the TRIZ Trends activates this system DNA to ensure that the system can move forwards to perfection – the end point is no system, the Ideal system is no system. This is the idea that ideally a system should be as small as possible, need as few inputs as possible; and no problems is a theoretical endpoint where the needs are still delivered but the system occupies no space, costs nothing and has no harms – an Ideal of delivering all needs for nothing.

Development of the Breathalyzer

Inventors solve problems: they satisfy needs by finding solutions. Those solutions normally involve combining an array of technologies and existing systems, but every good inventor has recognized a set of unfulfilled needs and searches for ways to meet them.

The development of the needs and the products were well matched for breathalyzers. All the needs were not recognized initially – in the 1920s and 1930s some argued that drinking enhanced their driving and asked what was wrong with drinking and driving; but despite some ambivalence, the law was eventually changed and clarified to make drink-driving an offence. This produced many arguments from drunk drivers who claimed that they weren't drunk, that any unsteadiness of gait, red eyes or an inability to stay awake was due to overwork not inebriation. Therefore a need was identified: to measure the level of alcohol in the system of a driver, in a non-invasive way. This became more pressing with the repeal of Prohibition in 1933/34, and alcohol became more readily available

The first breakthrough came from Professor Rolla Harger, who invented the Drunk-o-meter. This used the knowledge that alcohol in the deep lung breath is in proportion to the alcohol in the blood. Drivers suspected of intoxication had to blow into a balloon-like device which contained a chemical solution that changed colour according to the proportion of alcohol; the more alcohol present the greater the colour change. It was easy then to calculate the level of alcohol in the bloodstream. The Drunk-o-meter was somewhat cumbersome and needed recalibrating in between uses but it worked well because the dramatic colour change showed the amount of alcohol immediately to the policeman and the drunk driver. This often made the driver admit to alcohol and not invent a string of implausible other excuses (a common occurrence apparently).

Dr. Robert Borkenstein, a captain with the Indiana State Police, and a successful inventor, was responsible for the next generation device he named the Breathalyzer in 1953, which used chemical oxidation and photometry which was much more portable and easy to use. Roadside drink-driving analysis has since evolved into products that are very cheap, easy to use but very fast and accurate, so admissible as court evidence.

As the Drunk-o-meters and similar devices developed they matched all the emerging needs, such as instant proof of drunkenness which could also be accepted in court not just at the road side. Other needs were portability and a device that was foolproof and easy for a hard-pressed, non-scientist policeman to administer and record the results. It needed to clearly show instantly whether alcohol was above the legal limit (or acceptable limit as for some time there were no exact measures). All these needs were matched as the breathalyzer evolved – and its evolution followed the TRIZ Trends which could have led its first inventors from their simple devices to the current-day sophistication. The system is still evolving not just to inform of drunkenness, but preventing driving when drunk by breathalyzers being attached to cars and only allowing the car to be driven if the breath test was clear. All these needs were in the DNA of the first devices, even if the needs hadn't been initially recognized by the first users.

This suggests that product developers can look to the initial system to tell them about future markets and needs, and use the TRIZ trends to visualize the future product and the future customer needs it will fulfill. Listen to the product, as well as the customer, to design the right future products is the message of this part of TRIZ – the TRIZ Trends of Evolution are the ways systems get better and meet all needs.

TRIZ for Invention

TRIZ for invention is powerful: TRIZ directs us to the systematic routes to good solutions. It offers very certain arguments against the annoying culture that insists invention is a wildly random activity, and the province of mad inventors who arrive at their great discoveries by chance. TRIZ is sometimes resisted as inappropriate for invention, because it is a systematic toolkit. But TRIZ was created to take inventors straight to the relevant solution areas, cutting out many empty endless trials which are deemed necessary to eventually find the right answer. TRIZ is and has always been very useful for inventors and research teams, as it is systematic and auditable, and is not the realm of chance, but invention is probably the area where it is still only used occasionally and not daily as in more mundane problem solving.

Altshuller wanted to offer inventors an alternative to what he called the 'empty trials' and move surely and confidently to the areas where clever solutions can be located. TRIZ tools such as the Ideal Outcome both points us towards the place with the good answers, and helps break our psychological inertia, which has us looking in the wrong direction, or not looking far enough, or simply blocks our ability to see a logical answer or self-evident truths. Time and Scale and 9-Boxes offer us clarity of understanding and an ability to see the wood for the trees. The Trends of Evolution show us the directions to perfect systems.

Altshuller argued in *Creativity As An Exact Science* (pp. 273–5) that TRIZ offers these known formulae that help with both invention and the uncovering of scientific theories. Altshuller claims that the 'torments of creativity' are unnecessary as the rules for achieving great creativity exist, and are known, and could be applied to take away empty trials and fruitless searching. He cites the story of the Double Helix and great uncovering of the structure of DNA as an example of accidental connections delivering the truth and the final simple clarity. The story is always seen to support the view that scientific discoveries need more than hard work, intellectual rigour and persistence but also requires accidental connections to be made to achieve final, unexpected goals.

Crick and Watson's work in 1953 is sometimes seen as inspired, brash, intuitive and pioneering and it is often concluded that it was pure chance that they unified widely ranging work from many different areas; and that two ill-matched energetic scientists drew on the meticulous research of others to pull together all the knowledge to reveal the DNA. Chance and luck is primarily cited as the drawing together of the work on DNA by different people in different academic institutions. It is also commented that one of the greatest scientific papers of the twentieth century was only one page long. What is now known is that it was a race to unveil DNA structure, which would have been achieved by others anyway only later, and that Crick and Watson's great achievement was to get there first, by a combination of skills, knowledge and unorthodox approaches (model building) simultaneously trying to define the theory and join it to evidence delivered by the extraordinary work of Rosalind Franklin.

Rosalind Franklin of King's College London was a great biophysicist and X-ray crystallographer who produced the amazing X-ray diffraction images which revealed the structure of DNA. When she apparently didn't initially recognize their significance, this was attributed to the fact that this wasn't her area of expertise and her lack of interaction with others working on the same problem (as a female her interaction with other academics was curtailed: she wasn't even allowed in the KCL senior common room for coffee, where discussion and exchange of views is an important part of the process). Her competitive and somewhat unfriendly colleague, Maurice Wilkins, showed Franklin's work to Crick and Watson without her knowledge. Her high-resolution X-ray images of DNA fibres suggested a helical, corkscrew-like shape. Franklin, although subsequently well known for her work on DNA (and supportive, open and great friends with Crick and Watson until her early death in 1958), should have been cited as one of its key discoverers (and not just a female helpmeet) along with Crick, Watson and Wilkins, who received lifelong honour for their work including the Nobel prize. Both Crick and Watson said that Franklin's pictures were 'the data we actually used' to reveal DNA structure.

Watson's book *The Double Helix* was controversial and Crick tried to block its publication as it felt it showed them as scientists in a bad light, showing the personal controversies and unpleasantness that can plague academic competiveness. This great story of uncovering of such an important scientific truth had a number of moments of serendipity which scientific advances are always associated with:

- Franklin had produced the evidence in her pictures, but it took other people working separately on DNA, Crick and Watson, to actually recognize what they revealed.
- Crick and Watson only got together by chance. Crick accidentally overheard of Watson's work on a 'perfect cosmological principle' in a Cambridge pub – apparently the phrase intrigued him enough to consider the perfect biological principle.
- Teamwork came from different disciplines – Watson and Crick had very different but complementary scientific backgrounds in physics and X-ray crystallography (Crick) and viral and bacterial genetics (Watson).
- Serendipitous analogous thinking – Crick and Watson's famous stroke of inspiration in recognizing the double-stranded molecule, a double-helical structure, was apparently suggested to them while building the famous model from simple materials. From his knowledge on the symmetry of DNA crystals, Crick, an expert in crystal structure, saw that DNA's two chains run in opposite directions.

What this story tells us is that there a number of factors critical for scientific success:

- a vision of the truth (initially not necessarily completely connected to the facts);
- the importance of teams with different backgrounds and skills;
- an unbiased, fresh eye to look at the facts and evidence free from psychological inertia.
- an open sharing of progress and ideas within the scientific community may not produce a breakthrough for one person, but for many teams everywhere;
- a top-down (Crick and Watson) and bottom-up (Rosalind Franklin) approach of joining the facts and the theories.

This means that scientific and mathematical proofs can be founded by pooling global brainpower and that the Internet will help this. In his spellbinding book *Fermat's Last Theorem*, Simon Singh shows many pieces of the puzzle were assembled in different times and places from revolutionary France of the eighteenth century to twentieth century Cambridge – with work as far afield as Japan with many mathematicians working individually on this great conundrum. I enjoyed this book with my teenage children (in audio form in the car on many school runs) and this was the time I was first learning and using TRIZ and I was struck then at the importance of global teams and global communication. In the last ten years global communication has become a daily reality for many parts of the human race.

Somehow we have to overcome the contradiction of keeping the advantages that competition gives us and yet share the advantages and benefits of new discovery. Rivalry and competition may provide the extra spur of a deadline, and help create races and imperatives to be the first in discoveries, which may drive science and technology forward a little faster, but conversely having an open global community which shares every step forwards with others, may be the route to making advances faster and sharing those advantages with others. The debate on pharmaceutical companies' high prices for drugs which have been created after years of their investment and hard work in research to create the drugs will always be controversial. While companies fund and organize research they will need to benefit from it or they won't engage in it.

The Patents System itself tries to solve the contradiction of sharing knowledge and protecting it – and there are many debates on whether the patent system encourages a new global colonialization via intellectual property (IP). Some argue that the corporations who own the most important IP will be in effect owning too much of the world's resources in the future. In the controversial book *Protect or Plunder? Understanding Intellectual Property Rights* (Global Issues Series: Zed Books), Vandana Shiva argues that the word 'patent' although originally meaning open – to imply sharing of knowledge – has actually has always been associated with closed ownership and similar to the colonialization of first land and property, and patent colonialization is creating now the closed ownership of everything from human DNA to the DNA of next year's seeds. Her plea is not to let ownership by huge global companies replace the sharing of knowledge and create a grim future for the world.

Whatever the trends in patents ownership, it would be a tragedy if corporate dominance of IP led to the loss of our genius lone inventors and suppressed individual creativity in corporate cultures set by non-scientists. The opening of the world's patents via the Internet is a great boost for engineers and scientists everywhere. In TRIZ sessions one of my first recommendations to everyone is that if they are not using the world's patent databases on a regular basis then they are missing one of the wonders of the modern world. These amazing free resources (free to anyone with access to the Internet) offer all the published genius of so many great scientists and engineers, and are available to us all at the click of a few buttons. Less than a third of the knowledge from these sources is actually still protected and restricted and they are comprehensive sources of all the known solutions to problems – so a good starting place for any kind of problem solving. On I6th March 2006 I heard Vinton G. Cerf speaking on BBC Radio 4 and was thrilled by both his message and his over-the-top job title of 'Chief Internet Evangelist for Google'. His message was that all this easy access to knowledge was of less significance to many people, but to engineers and scientists (who unlike the others could understand most of it) it offers the chance to change the world for the better, and that they should wake up to the fact that the future was in their hands more than ever before. He said, 'The sharing of information by the Internet will now accelerate the rate at which new discoveries are made and new feats of engineering can happen.'

Interesting Gaps Between Inspirational Ideas and Scientific Proofs

Great scientific discoveries have a top-down, bottom-up approach – bottom-up building on observed data, and top-down from scientists' visions of the uncovering of the true theories. Newton's theory of cooling seems self-evident once it is stated; Einstein's theory of relativity was acknowledged after the bottom-up work on the Sun's eclipse gave scientific data to support the theory. Altshuller claims that the TRIZ helps both and that the top-down part could be helped by the simple TRIZ creativity tools – visioning the truth with the Ideal Outcome and encouraging and stimulating imagination.

Invention is not always a brilliant leap forward of science and technology but often a fairly pragmatic step of using proven products or systems in a new way to meet some human wants. Very few inventions come from ground-breaking scientific and technological advances, much more often they use proved and established technologies. Novel systems do emerge, of course, from specially adapted state-of-the-art technologies, but more often they comprise of a previously untried combination of some existing, well-known and trusted systems.

TRIZ makes problem solving directed, fast and effective

In early 2009 many of the long-established companies of the UK's great successful, eighteenth-century inventors have failed: Wedgewood, Royal Worcester, Viyella etc. All were created on technological brilliance, a recognition of what society wanted, and a brilliance and flair for marketing which reached people and delivered desired, high-quality products to many markets for over two hundred years. Common factors in their recent decline were cited as losing sight of people's needs (supplying a market for grand dinner parties that no longer existed) and attempts to reduce manufacturing costs by shifting production to China, where lower costs were accompanied by lower and inconsistent quality. As a result their customers who still bought their products, primarily for their high quality, melted away. Deliberately shifting a previously successful Ideality balance (established by successful launch into the market, and a long-term ability to flourish) is a dangerous business for established companies.

The Ideality Balance is fairly simple – it means ensuring that the right functions deliver actual (not imagined or assumed) needs, at the right price and with the minimum of harms and problems. An inventor seeks how to deliver the right functions with the right Ideality Balance.

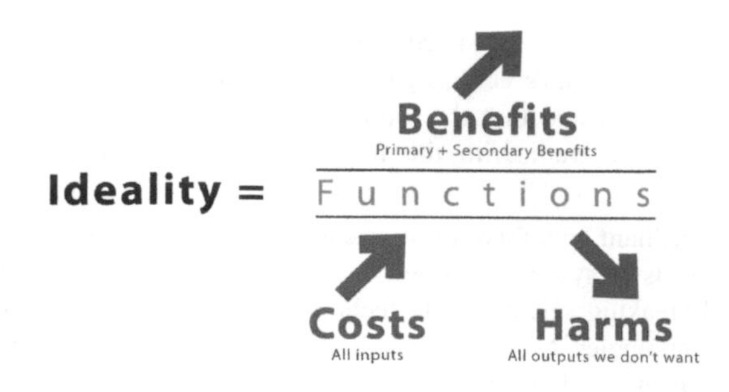

When we are certain of what we want – our needs – then we seek the best systems to achieve them; if the right system doesn't yet exist, then we need to invent it. A system delivers functions, and functions deliver benefits. Invention is the very first meeting of unfulfilled needs by a newly created system which deliver the right functions.

Successful invention also requires a very accurate assessment of the needs it is to meet. Some research and development follows mistaken interpretation about needs which may be very niche or exist no longer or have not yet been realized (frozen food was invented long before its demand). Forecasting the wrong needs can come from poor or inappropriate market research, or an incomplete understanding of the good market research, or from arrogantly assuming that everyone's needs are somehow intuitively understood. There have been many dominating managers with a misplaced confidence in their own ability for anticipating others' wants. Get the needs wrong, misunderstand what everyone wants and all kinds of products could be expensively developed and launched which very few people actually want. There are many wonderful and existing toolkits specifically for *needs capture* such as QFD and Value Engineering and other major general toolkits such as Six Sigma which have adopted these available and proven methods, to ensure that requirements can be understood and recorded. TRIZ complements all these toolkits well and itself offers the Ideal Outcome as a quick and useful way of roughly capturing needs. With an accurately defined set of needs we can search for products to satisfy them and if they don't yet exist then we can invent them.

TRIZ and All Routes to Invention – Creating Systems

When we have identified requirements but have no system then we are into the first of the *invent* modes and need to choose or find a system. People who recognize needs/requirements which don't yet have a system, and then can find the right one can become very rich by developing a new system, or adapting an old one to meet requirements in a new way. Sometimes invention happens because of new requirements – sometimes requirements stay the same but the system to deliver them becomes available through advances in technology, new materials or new manufacturing methods. The Internet provided possibilities for many new inventions and successful businesses like Facebook, eBay and Friends Reunited. In 1998 Friends Reunited did not exist. It was started in a back bedroom of a small semi-detached house by a couple who thought that they recognized an unfulfilled need – the need for connectivity. They created a system to meet those needs – the email-based Friends Reunited – and tapped into a huge demand to reconnect with formative friendships, enabling contact with old friends and the chance get back in touch with 'old school' society. Friends Reunited was the first commercially successful online social network in Britain – its founders sold it for £25 million in 2009.

TRIZ Helps with All the Major Routes to Invention

There are two fundamental routes to invention each with their own subsets as shown below. All these routes have seen many inventions both deliberate and some accidental.

1. Meet needs in new ways with new and old systems. We have some unfulfilled *needs* and these are met in new ways by:

   1.1 New systems – Seek new technologies to create new systems to solve problems and meet very specific needs (Whittle, Watt).

   Deliberate Invention by bringing together *new technologies* or *systems* in innovative ways to meet needs.

   1.2 Old systems – Combine old existing systems in a new innovative way to meet unfulfilled needs.

   Deliberate invention by bringing together *existing systems* in *innovative ways* to meet needs (as in the invention of the wheelie bin, the telescope, Bayliss and the clockwork radio, Dyson vacuum etc.).

2. Find new uses for systems, technologies, functions and features. We have some systems which we find *new ways of using* by meeting different needs from their original intention:

   2.1 New technologies – Apply new technologies in different industries from their original intention – look for new applications for new concepts to fulfil needs (e.g. GPS systems, microwave ovens, Internet).

   Deliberate invention by seeking *new uses* for *new systems or technologies* in innovative ways. Dyson product apply their expertise to 'moving air' to many product beyond the vacuum cleanre- with their new fans and Airblade for drying hands.

   2.2 Old technologies – Find a new use for established, old or redundant technologies, systems or functions (e.g. Wilkinson's sword).
   Deliberate invention by seeking *new uses* for *old systems or expert functions* in innovative ways.

The challenge is which of the four routes to adopt – one famous example of a modern invention is the Space Pen.

The urban myth is that the US showed the world great NASA technology in this invention and spent millions (1.1 above) while the Russians used a pencil (1.2 above). In fact NASA and Russian space technologies both used pencils initially which had many problems such as lead parts breaking off and floating around in zero gravity and the inflammable dangers of wood. An inventor called Paul Fisher invented the Space Pen independently (with no NASA funding) and created a pen that would write upside down, underwater, at any angle, at any temperature, on greasy paper or any greasy substance and in zero gravity. Both the USA and the Russian space programmes adopted the space pen for use.

1. Meet Needs in New Ways with New and Old Systems

Find or invent ways of meeting needs.

Pollution and every new flu scare send us looking for sophisticated safe solutions

1.1 Seek New Technologies to Create New Systems to Solve Problems or Meet Very Specific Needs

Bring together *new technologies* or *systems* in innovative ways to meet needs.

This most exciting realm of invention is right at the forefront of science, technology and invention. We have many problems facing us in the world: human poverty, starvation, sickness, global warming, energy supply etc., and much human ingenuity is directed towards these problems. Many pharmaceutical and food companies work to find positively healthy food and are seeking new ways to predict, treat and cure illness. Many energy companies are seeking ways to harness energy efficiently in a cost-effective way.

One recent story was the development of a 'super-microscope' using light as bright as a million-watt bulb which could help identify early signs of Parkinson's disease and identify changes in brain cells before the disease destroyed them.

Using a synchrotron – or Diamond Light Source (DLS) – at Harwell, Oxfordshire, the large doughnut-shaped particle accelerator (which is the size of five football pitches, and fires particles at just below the speed of light) focuses them into a beam less than a single cell in diameter. This allows researchers to observe iron levels in individual brain cells which are affected by Parkinson's, and for the first time provide treatment before irreversible damage occurs (BBC website 14 February 2009).

The use of DLS for medical breakthroughs was not in its initial publicity when it was cited for it abilities to analyse materials. One example was applying its nano-science beam-line to image magnetic materials and see how new magnetic materials work, which has many practical applications such as for making better computer hard drives.

Another example of an exciting new material with many applications form is Aerogel, an extremely low-density solid material with several remarkable properties, most notably its load-bearing qualities and its remarkable effectiveness as a thermal insulator. Also called *frozen smoke*, *solid smoke* or *blue smoke* because of the way the transparent material scatters light, it has some interesting features. One is that it is friable – so when touched softly there is no impact or mark; but by pushing harder, marks can be made and if pushed hard enough it will shatter.

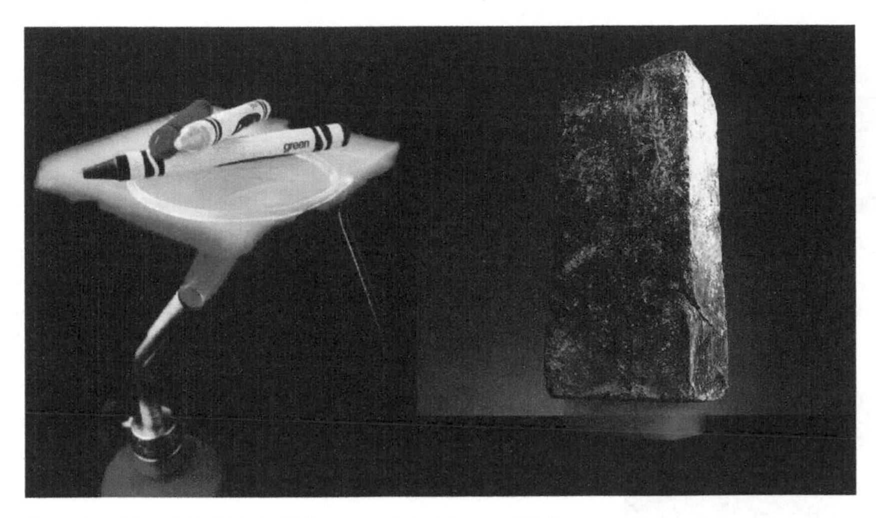

Reproduced from http://stardust.jpl.nasa.gov/photo/aerogel.html

The many applications and potential uses for Aerogel are just being explored and applied in materials for anything from sunroofs to space shuttles. But this is a remarkable material which will be part of many new inventions.

Challenge

Look at unfulfilled or badly fulfilled needs and define functions they require – look for new technologies which deliver these functions. Create a theoretical system from the new technology which meets all the needs.

e.g. Invent a perfect oven glove.

1.2 Combine Existing Systems in a New Way to Meet Unfulfilled Needs

Bring together *existing systems* in *innovative ways* to meet needs.

Great but mundane inventions – What drives their discovery and success? With a set of recognized unfulfilled needs how do we seek the best ways of delivering them? This often requires the innovative approach of seeing old systems in new ways – visioning how to use things in different or innovative ways than for what they were invented and designed. There are many famous examples of successful new products just recombining old technologies.

Ice cream cones – In 1900 ice cream was a great treat and served on the finest plates. In 1904 at the USA World's Fair held in St Louis, Missouri, the ice cream cone was invented by the lucky meeting of two food products.

It was very hot and one stall was selling ice cream which was very popular and they had enough ice cream but not enough dishes, which had to be collected and washed before being used again. At the neighbouring stall Ernest A. Hamwi was selling Zalabia – a kind of wafer-thin waffle from Persia which was less popular in the hot weather – and Hamwi suggested that he could help the ice cream stall and sell his Zalabia as well. By rolling the Zalabia into cone shapes they created delicious edible dishes which could hold the ice cream and be eaten alongside the ice cream removing any need to collect and wash the dishes. (www.ideafinder.com/history/inventions/icecreamcone.htm)

Dyson vacuum cleaner – James Dyson identified the need for vacuum cleaners with no messy expensive bags, and better suction. Like many great inventors Dyson showed that he combined the skills of a great businessman with those of a great engineer.

However the technologies used by Dyson from his Ball-barrow to the Dyson vacuum cleaner are hardly new. The principles of cyclone separation used in his vacuum cleaners was used in saw mills for at least 100 years before Dyson adapted it to get rid of the vacuum cleaning bags. One unintended good consequence of his vacuum cleaners is that you can see how much dirt and dust you have collected. This has made it popular with men.

Clockwork Radio – Trevor Bayliss the inventor recognized the need for portable devices like radios in areas where batteries are not easily available – like Africa. He adapted the old but reliable clockwork technology for radios and other things like torches. He despaired when designing his radio because he couldn't get it small enough – when discussing this problem in Africa with young potential customers they told him they didn't want small radios anyway – they wanted *big radios*. So he found that his designs met market needs.

Challenge

Look at unfulfilled or badly fulfilled needs and define functions they require – look for *old* technologies which deliver these functions. Create a theoretical system from the old technology which meets all the needs.

e.g. Invent the perfect solution for keeping a pair of socks together when being washed.

2. Find New Uses for New and Old Systems, Technologies, Functions and Features

2.1 Find New Uses for New Systems/Technologies/Functions

We have a *new* system etc. – let's look for some *different* needs it could meet.

TRIZ suggests many 'Other way round' solutions

Apply new technologies in different industries from their original intention – look for new applications for new concepts to fulfill needs. (GPS systems, microwave ovens, Velcro, the internet)

Challenge

Look at new technologies and examine the functions they deliver (e.g. Aerogel offers the functions heat insulation and the ability to hold great weights). Look for new applications for these functions.

2.2 Find New Uses for Old Systems/Technologies/Functions

We have an *old* system etc. – let's look for some *different* needs it could meet.

Find a new use for established, old or redundant technologies, systems or functions

Challenge

Look at old technologies and examine the functions the deliver (e.g. concrete offers the functions strength and elasticity). Look for new applications for these functions.

Systematic Routes to Invention

Ariadne's Thread – Don't Lose the Way!

The simple TRIZ tool of the Ideal is our Ariadne's Thread and will take us to the good solutions. Defining Ideal Outcomes enables research teams to do a quick future needs capture, and defining Ideal Systems helps understand the functions we want and a wish list of systems they would like to create. By applying the TRIZ toolkit in reverse we can move from benefits to available high-tech functions (new and old) which could provide them.

Capture Solutions – Top-of-the-Head Ideas

We can get a good idea of solving future problems initially by just jotting down very quickly all solution ideas which will help capture everything we want and why we think we have a problem in the most general way – top of the head – no great analysis. TRIZ problem solving, although systematic and thorough, often involves quick thinking and helps to keep us from diving into inappropriate details at the wrong time (and then often losing the thread). To keep the essentials in the forefront of our thinking we also roughly map our system in 9-Boxes capturing the appropriate amount of detail, so we can come back to it later knowing we won't overlook anything, but initially concentrating on the big issues.

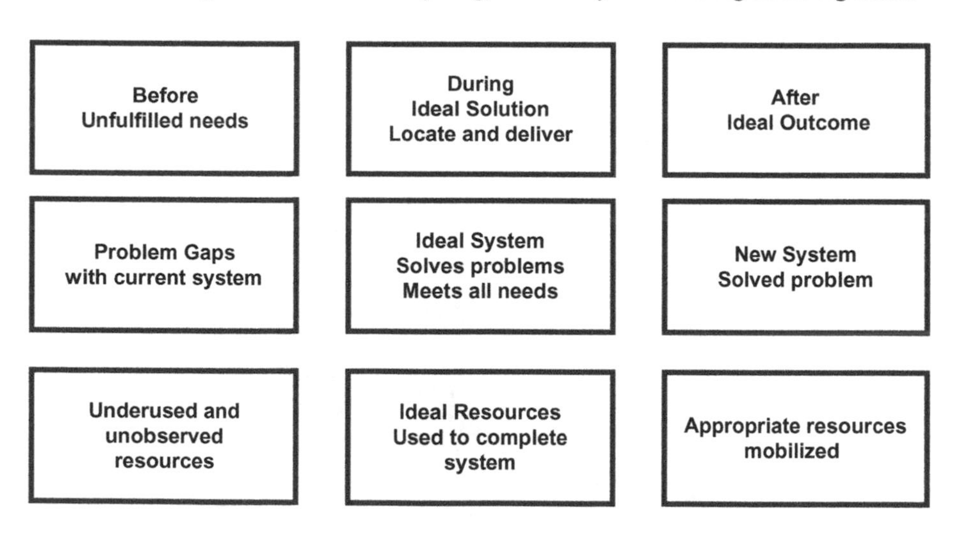

Flowcharts for Invention

New Systems – Seek Technologies to Create/Invent New Systems to Meet Very Specific Needs

Try inventing something mundane by following the TRIZ processes. Define all needs by defining the Ideal Outcome of a system which needs to be invented: Try inventing the perfect oven glove and the efficient sock pairer. Follow this simple route starting by defining the Ideal for each system.

The Ideal Outcome has a precise definition in TRIZ: it contains the ultimate goal, the reason the system or process exists, and all other benefits.

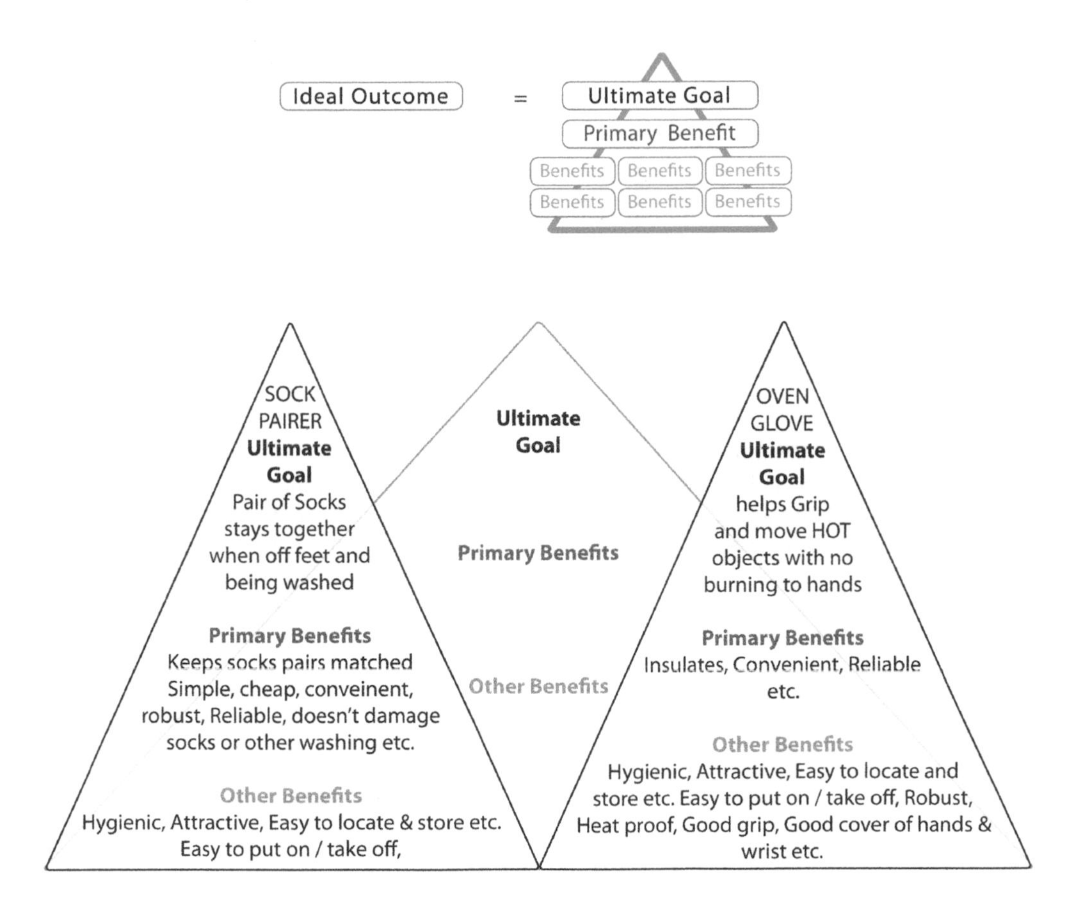

Ideal Outcome and Ideal System and X-Factor

Ideal Outcome – What do we really, really want, what is everything we want?

The Ideal Outcome of an oven glove includes all benefits.

BENEFITS

Hygienic, Attractive, Easy to locate and store etc. Easy to put on / take off, Robust, Heat proof, Good grip, Good cover of hands & wrist etc.

HOW?.....

FUNCTIONS

List functions which deliver the benefits

HOW?.....

Which Technologies deliver these **FUNCTIONS?**

Search Patent databases
Effects databases
etc.

Look for New or Old Systems, Technologies, Functions and Features which will deliver the function we have identified.

SYSTEMS, TECHNOLOGIES etc	WHY?.....	FUNCTIONS	WHY?.....	BENEFITS
	What functions do these systems use?	List functions of system Technology etc.	Who else needs these Functions? What benefits do they supply? Who is looking for these Benefits?	Search Patent databases and effects for examples of systems using the functions

Corporate Innovation and Invention is Poorly Rewarded

TRIZ has tools for inventing, perfecting and improving systems to develop next generation systems, and delivers relevant innovation to products and processes. Innovation and successful inventions are desired by all companies and seen as the lodestones for future riches but those who deliver both are often poorly rewarded for success and face problems for failure (the culture for forgiveness for mistakes is practised in truly innovative companies). The argument for bankers being amply rewarded is that their abilities and work contributed to profits. Some companies claim that their long-term, substantial profits derive from a small number of patents etc., yet there is no system to reward their in-house corporate creators, as bankers are. If successful invention was rewarded (on a scale like the bonuses in the banking system) it might have a phenomenal effect on the recruitment and staying power for building a bigger and more effective community of inventors.

Part Five
TRIZ for System Analysis and Improvement

11 Function Analysis for System Understanding

TRIZ Function Analysis delivers a Function Map of an engineering system (or any system) at one moment in time. This Function Map helps us understand the system, highlights any problems, contains a simple guide to identify the types of problems, and then enables us to precisely locate the TRIZ solutions to these problems. This is by a direct matching of the problems to the relevant, simple but powerful Oxford Standard Solutions, which give us all the relevant conceptual prompts for how to solve each of the revealed problems.

It is then the task of problem solvers to expand these TRIZ suggested answers (solution triggers) to find useful and detailed solutions to each of their problems. Working confidently with all the relevant conceptual solutions leads problem solvers simply and quickly to apply (add in) their relevant practical knowledge of the problem to deliver clever and practical solutions.

In this process TRIZ problem solvers often uncover very powerful new but existing solutions through analogous thinking; the relevant concepts direct them to all the possible answers to each problem. These are in a very general form and will contain all the solutions they already know as well as new solutions. This leads them to innovative ways of delivering required functions, or solving problem functions, with established solutions, already proven and tested but not previously known to them or used in their industry. These useful and established technologies are often outside their own area of expertise and offer their industry or company, routes to new innovative solutions.

TRIZ for Engineers: Enabling Inventive Problem Solving, First Edition. Karen Gadd.

Function Analysis and Maps for Problem Understanding

The aim of creating an engineering system is to ensure that the sum of the parts is greater than the individual components and that it is the right combination of parts, interacting in the best way, which will give us everything we want (see INCOSE definition of system). Function Analysis helps us understand and improve systems.

Even very simple products and processes can sometimes seem complex when we are trying to work out what is wrong and see how we can deliver all of what we want, with minimum cost and harm. When we need to understand exactly what is happening, then a simple Function Map helps us to see which components or elements contribute to the problems, how they all interact, and where to look for answers to solve their problems. Function analysis offers a simplified but detailed picture, a snapshot, of what is happening and is a very powerful aid for our understanding of all the problems, their interrelation, and their priority and helps us capture, log and communicate the essential problems to ourselves and others. The Function Analysis map enables us to draw up an accurate and complete problem list, so that the identified problems are categorized in a way that allows us to access the relevant Oxford Standard Solutions to solve them. It is not a flowchart and does not reveal flows or process steps, which would need function maps for each relevant point in time.

Definition of a System

A system is a construct or collection of different elements that together produce results not obtainable by the elements alone. The elements, or parts, can include people, hardware, software, facilities, policies and documents; that is, all things required to produce systems-level results. The results include system-level qualities, properties, characteristics, functions, behaviour and performance. The value added by the system as a whole, beyond that contributed independently by the parts, is primarily created by the relationship among the parts; that is, how they are interconnected (Rechtin, 2000). 2 Oct 2006 International Council on Systems Engineering (INCOSE) www.incose.org/practice/fellowsconsensus

Why Draw Function Maps?

Function maps take time and careful effort and have a simple but strict logic which needs to be understood, learned and adhered to in order to produce something useful. The systematic aspect of Function Analysis for analysing and mapping a system with problems is particularly valuable when we are stuck, faced with a difficult problem which we don't know how to tackle – either because of confusion with too much information or we simply cannot see the way forward. Function Analysis helps with complicated situations, when there are multiple problems and causes which are hard to pin down and define. By drawing a Function Map of the problem situation it will help us identify all the problems individually and rank them.

TRIZ Function Analysis can be very simple from just one functional relationship to something very complicated with many functions linked together. TRIZ Function Mapping helps us understand and highlight problem places for any system or situation – general or detailed, management or technical. It is always worth trying on simple everyday problems.

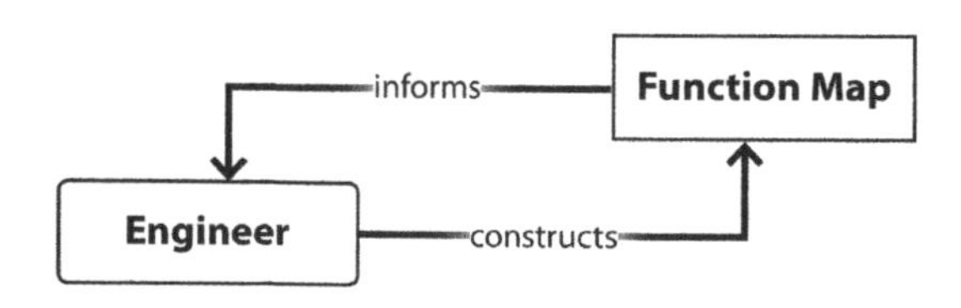

Why Use TRIZ Function Analysis?

Function Analysis helps us understand our choices and constraints in solving problems. Although building the Function Map involves careful thought and a rigorous approach it has many benefits. It helps us:

- understand the system with a simple visual Function Map;
- systematically highlight all the problem areas within a system;
- demonstrate how the system delivers functions and any problems (both small and great) are identified with the components and their interactions;
- reveal the complication of a system and how to identify which components could or should be eliminated or changed;
- sort out all the issues to see how to prioritize and solve problems;
- find all the possible solutions to all the problems – find the relevant TRIZ solution triggers to all the problems;
- communicate the real and important problems to others;
- provide an audit trail of what has been done and why to understand and then improve the system and as all information about the System is documented – it is available immediately and later.

What Can TRIZ Function Analysis Reveal at a Glance?

- all components and their interactions
- all functions – good, bad, insufficient, missing, excessive etc.
- a function snapshot of one moment in time of a system
- contradictions (when there is both a good and a bad action between two components)

It is worth remembering that even when we are drawing a Function Analysis map for a well-defined system with a well-recognized problem, additional and often obscured problems can be highlighted. These other problems only become apparent during the mapping process but when highlighted and then solved often offer additional benefits – sometimes more benefits than solving the original problem. Function Analysis is powerful for any kind of problem including management and business problems – any situation where we have a system and that system is not perfect then Function Analysis will help us highlight and understand any problems it may have. It is also useful just to understand interactions within a system before we start looking for problems. Below is a simple Function Map of a coffee cup, containing coffee on a table. It shows the simple useful functions provided by such a system with the green lines. The simple problems which may occur such as the ambient air cools the coffee, the coffee cup marks the table, the coffee cup cools the coffee, the coffee stains the cup are all highlighted in red.

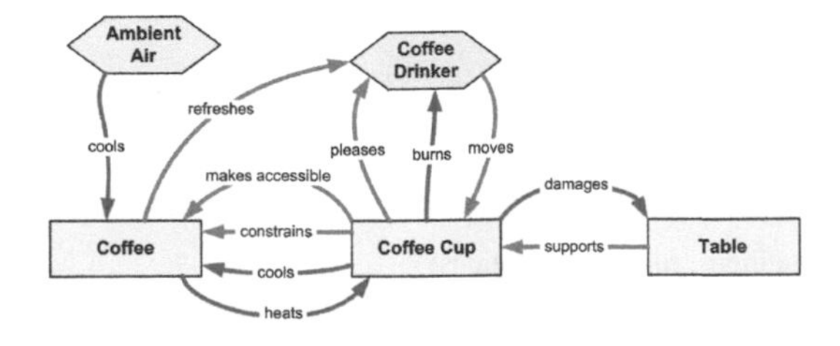

Basic Building Blocks for Problem Solving – Defining Ideality

It is difficult to assess whether a system has problems or not if we don't first understand exactly what we want from our system – its *Ideality* – what our system should be delivering, everything we want (benefits), what we want to put into it (acceptable costs) and everything we don't want (harms) within the terms of the system's constraints. *Functions* provide *benefits* and ill-defined or partially understood benefits means we cannot see whether we need all the functions, or what is wrong with the functions, see if they are harmful, insufficient, excessive or even missing.

We develop simple technical systems to give us what we want – to provide functions which deliver some benefits. A cup provides the function of holding a liquid so we can fulfil our need/benefit of drinking from it, and there are many other benefits we consider when choosing or designing a cup. The analysis of all benefits/needs/requirements is an important step in problem understanding and as the provision of these benefits comes from functions, a Function Analysis should be linked to a clear understanding of benefits.

Functions also require some inputs and nearly always produce some problems (harms/outputs we don't want). Improving a system means improving its Ideality and functions are revealed Ideality equation.

$$\text{Ideality} = \frac{\text{Benefits}}{\text{Costs \& Harms}} = \text{Functions}$$

Benefits
Good functions
Insufficient Function
Missing Function
Harmful Function
Costs
Harms

Improving systems involves moving from the 'system we've got' to the 'system we want' and for this we need to understand the Ideality of both of these. When we map the 'system we've got' with Function Analysis we show how well or badly the benefits are being delivered by the functions. We also highlight all the harms (problems) and can also map anything needlessly excessive. In TRIZ problem solving we are trying to minimize the functions and seek ways to get the most benefits for the least costs and harms.

IDEALITY AUDIT First roughly define Ideal Outcome – Ultimate Goal & everything we want Then define the following

IDEALITY	SYSTEM we've got	SYSTEM we want	Ideality Gap
Prime Benefit			
Other Benefits			
Prime Output /Function			
Acceptable costs/inputs			
Unacceptable costs/inputs			
Acceptable harms/problems			
Unacceptable harms/problems			

Include other relevant factors for problem understanding and solving include such as:-

Constraints (Imposed conditions, rules or regulations)

Simple Drawings of System

Don't Miss Out or Skip the Ideality Audit

Benefits are often assumed to be understood (dangerous assumption). Missing out the Ideality Audit makes it hard to define systems and problem directions accurately. Sometimes 'the system we've got' gives us the prime benefit but has so many other essential disadvantages that it is unacceptable.

For Problem Solving We Need Both the Ideality Audit and the Function Analysis

Problem solving is the closing of the gaps between 'system we've got' and the 'system we want'. We are establishing an Ideality Gap which can be fairly simple or very complicated; it can be at a high level or be detailed. In an Ideality Audit we can point to both the general and specific directions we would like to travel – where we are and the end points we want, the essential benefits we want, and how we want them delivered.

The Ideality Audit helps to see what we want and uncovers and roughly defines the benefit gaps, and gives broad directions of problem solving. An Ideality Audit is important before we start problem solving using Function Analysis as we first require some understanding of everything we want (and don't want).

After completing the Ideality Audit we can define the *problem gap* – which shows us what we want our problem solving to achieve – the closing of the problem gap. Function Analysis is a study of the system to look at its problems in detail and find ways of solving each of these problems by applying the TRIZ solutions tools.

IDEALITY AUDIT — HEAT EXCHANGER

Very roughly define and describe the Ideal Outcome – Ultimate Goal & everything we want

Ideal Outcome for Heat Exchanger = Instantaneous and Perfect 100% heat exchange with no losses, no problems of any kind etc.

IDEALITY	SYSTEM we've got	SYSTEM we want	Ideality Gap
Prime Benefit	Efficient Heat Exchange	More efficient Heat Exchange	% Efficiency
Other Benefits	Compact	Much smaller	Too big
	Reasonable cost	Lower Costs	Too expensive
	Robust & reliable		
	No vibration		
Prime Output / Function	Heating of one medium by another		
Acceptable costs / inputs	70% of the current costs		
Unacceptable costs / inputs	High maintenance		
Acceptable harms/problems	Some small heat losses		
Unacceptable harms/problems	Leakage, unreliability etc.		

Other Relevant factors for problem understanding and solving include such factors as:-

Constraints (Imposed conditions, rules or regulations). For the Heat Exchanger constraints could include that the media must be separated (must not mix) –and inlet and outlet conditions cannot change

Provide Simple Drawings of System whenever possible

Function Analysis of the Current System (System We've Got)

The steps for problem solving with Function Analysis of the current system (system we've got) are:

1. Create function list of all components and their interactions (prime system and its components as S-a-Os; see next section for explanation of S-a-O).
2. Draw Function Map
3. Draw other time-based Function Maps – system process steps (if relevant and necessary).
4. Record all system problems shown by S-a-Os (harms, insufficiencies, contradictions, or how to do something).
5. Prioritize problems and *trim*. After trimming refine problem list to show what is still wrong with the system.
6. Direct problem solving – apply relevant Oxford Standard Solutions harms/insufficiencies to the system (we normally try to deal with harms first as this often yields big improvements).

Step-by-Step Function Analysis – Drawing a TRIZ Function Map

Subject Action Object-SaO – Basic Building Bricks of Problem Solving

TRIZ Function Analysis maps entire systems in order to identify the problems with system functions. Function mapping slices the system into small simple units for one moment in time – each one a delivered function created by a *subject/action/object*. This is a simple three-word statement which describes a *subject* doing something to an *object*. Altshuller identified that the simplest (minimal) technical system can be described as having just two components with an *action/field* between them; when combined together they provide a *function*.

Functions – A function is delivered by a complete S-a-O – an action between two components or two or more components interacting with each other. Functions are the happenings in a system and why we have the system – functions are the ways we get the outcomes. For example a simple system of 'Hammer *hits/moves* nail' has an action of *hits* or *moves* – and that action results in a function (normally moved nail). The result – the function – can be exactly what we want (no problem) or it can be insufficient (weak), excessive or harmful. Problems with the functions can be solved by doing something to the hammer, and/or to the action – hits, and/or to the nail or even to the environment – such as the material the nail is being hammered into.

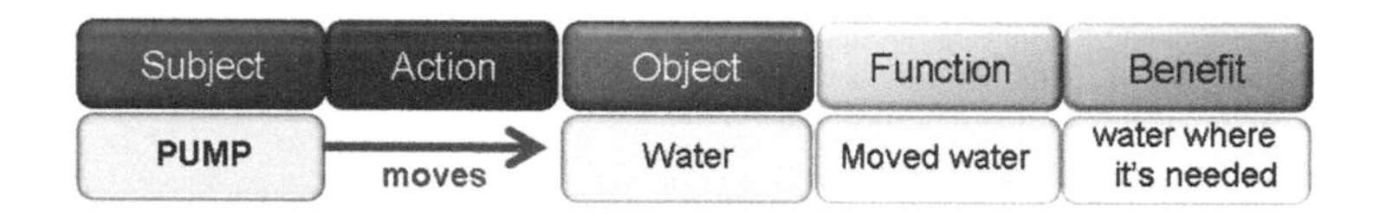

Some Definitions and Explanations for Function Analysis

Subject (tool) is the active tool or initiator of an action or influence within a system (S-a-O). The Subject is the function provider.

Object is the passive receiver of an action or influence within a system (S-a-O). It is changed in some way by this action or influence from the subject.

Action is provided by some kind of field between them. Hence the action is any influence within the S-a-O that causes the Object to change. Actions can only take place if a *Field* of some sort exists between the Subject and the Object

Subject action Object is a simple delivered function which allows us to describe the different types of problem as follows:

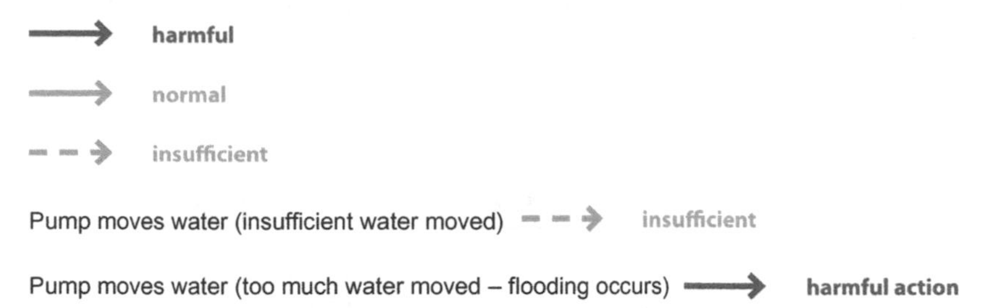

Field – Describes the nature of the effect, the energy causing the action. Technical examples of fields include mechanical, thermal, electrical and magnetic. Management examples include influence of any sort (e.g. reward, recognition, fear, love, inspiration, loyalty etc). Examples would include *hammer moves nail* (technical and systems) or *manager motivates salesman* (business and people). It is only when the functions of *moved nail* or *motivated salesman* are not achieved (or achieved in a way that causes a problem) that we use Function Analysis to examine what is wrong and the Oxford Standard Solutions to see how to put it right. These Standard Solutions provide triggers to suggest all of the ways of achieving these possibilities. The function is not the benefit/outcome/ultimate purpose we may be looking for; the function is what enables us to get that outcome.

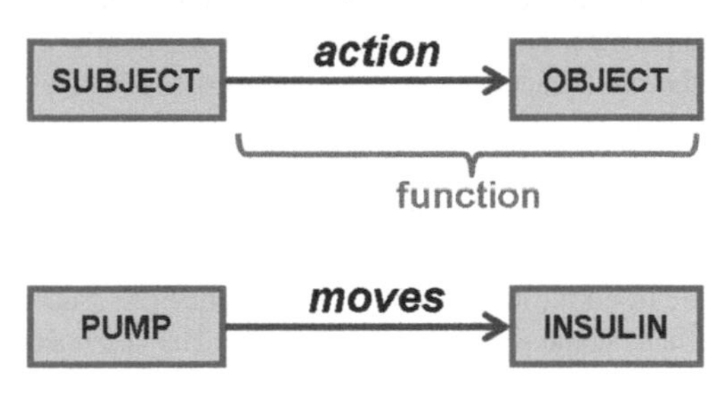

Function Map – All the *Subject action Objects* together

A TRIZ Function Map of all the components of a system shows what each component delivers to the system. It also shows exactly what each component is doing to other components and what it is having done to it by other components. Therefore TRIZ Function Analysis maps all the functions done by, and to, all the individual components and hence maps the outputs of the entire system. It identifies all the *functions* the components of the system delivers, by analysing the *actions* between the components and whether the outcome of the action is good and/or bad, sufficient, insufficient or excessive. The functional model is that one component is doing something to another component – something acts on something else. The active component is described as the subject, the passive component as the object and the doing is described as the action (Subject action Object or S-a-O).

Function Analysis for Understanding and Solving Simple Problems

A function is often something very simple, the result of one action between two components. For example how could we loosen an old screw in wood which appears to be stuck and not moving? The main components are the screwdriver, the screw and the wood, and there can be other components which are causing the problem such as rust or tar. A simple Function Analysis is shown below which shows that the screwdriver's action of moving the screw is useful (green line) but insufficient (dashed green line) and that the tar and rust are performing harmful actions (red lines) of stopping the screw from being removed.

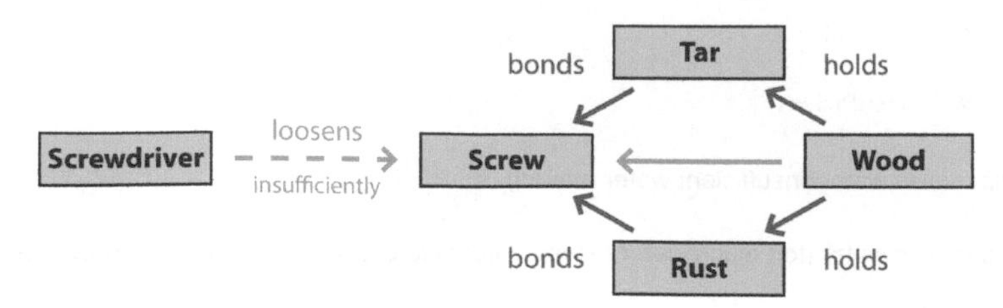

Solutions are probably needed to unstick/loosen the bonds between the screw and both the tar and rust. Looking at the Function Analysis gives some clues as to simple solutions, e.g. some forces/fields need to be applied to the screw to deal with both rust and tar (either through the screwdriver or from some other tool). A solution such as tapping the screwdriver could loosen the rust and adding heat to it would cause the tar to melt. Simple problems with simple solutions can be solved with common sense and experience; they can also always be solved with Function Analysis and the Oxford Standard Solutions. For this problem if we can't think of solutions then the Oxford Standard Solutions has a section for overcoming insufficient actions between the screwdriver and the screw (the first conceptual solution on the TRIZ list would suggest adding another field – including heat and vibration). TRIZ helps us solve problems (from the most complicated, difficult to the simplest) especially when common sense and experience have not yielded the right practical answers. Engineers will always try their own solutions first but TRIZ is there to help and supplement ordinary problem solving. Drawing a Function Analysis is always useful to help us understand the problem, and we can then use our knowledge and experience to solve it – and when we can't provide suitable answers ourselves we can then use all the solutions the world knows about in the Oxford Standard Solutions.

Function mapping – the picture of all the actions of all the components – shows both the functions we want and whether they are sufficient, and the ones we don't want. Hence Function Mapping is a simple visual guide to show us what is wrong. Any problems (contradictions, harms, insufficiencies and excesses) with those functions can be seen at a glance – a contradiction is shown when there is a red line and a green line between two components. When we are ready to draw a Function Map of a system we can make it as simple or as detailed as we like – it depends on the problem we are trying to solve. It is probably worth starting at the highest and simplest level and adding in more detailed maps as required. We probably don't need to capture everything unless it is relevant.

Functions, the results of the actions between two components, appear implicitly on the diagram – we simply show one component (subject) acting on (a field/force/action) another component (object) and this combination creates good or bad functions (harms); problem functions are denoted by the colour of the actions. When harms are created (when the action is harmful or features of the component are harmful or susceptible to harm) it is denoted in **red.** Good/useful actions are denoted in **green.**

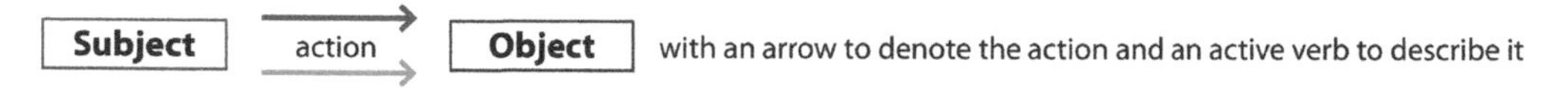

Good/useful features are also denoted in green; this is shown by marking the components as having good features with a curled green line. When there are problems with the components such as too expensive, or they contain some harmful features, these harmful or weak features are marked with red curly lines which goes back on itself onto the component. Insufficient features are marked with a dotted curly, green line. This additional information on the Function Map can tell us a great deal about the individual components, as we can mark features that are inherently good, insufficient or bad (or harmful) or expensive.

This can also show contradictions of good and bad features. One famous fashionable shoe manufacturer has claimed that by using finite element analysis to design very high-heeled shoes, the contradiction of comfort and height can be resolved – although this has made the manufacture quite complex and the shoes are very expensive.

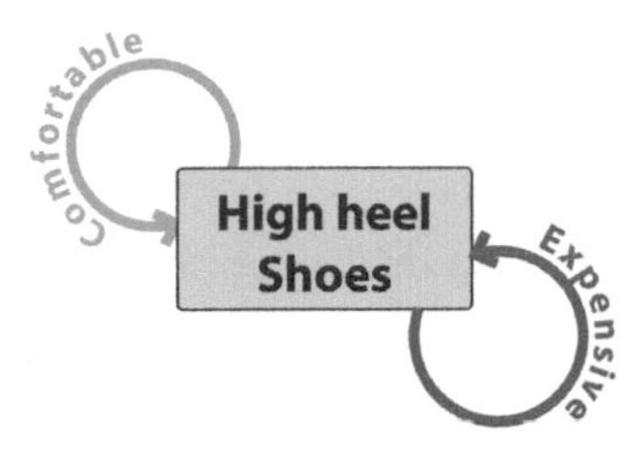

TRIZ of course offers many solutions to solving contradictions and within the Oxford Standard Solutions one suggestion for solving contradictions with high heels is to increase (exaggerate the contradiction) and have both opposite solutions.

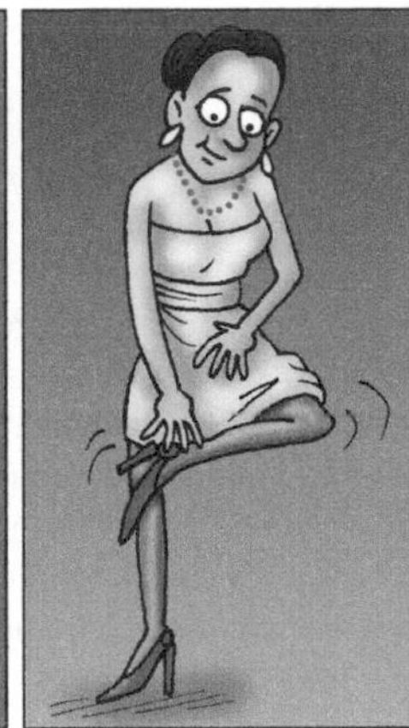

Even the contradictions of comfort and glamour are solved by TRIZ

Systems Develop to Deliver Benefits Better – Perfecting Functions to Deliver Those Benefits

All systems exist to provide what we want, our requirements, benefits, and these are delivered by functions which may develop, get better, as a system develops over time. For Function Analysis it is important to be able to understand how benefits are delivered by functions and not lose sight of the main benefits we want – delivering the prime function or prime output – which is important enough to be separately included in a Function Map. If we consider simple razors used by men for shaving their faces, the prime benefit of a razor is the safe removal of hair. The way it does this – the main function – is shaving or cutting hair. Function Analysis of a system involves studying all the components and their relationship with each other, taking each interaction and describing it as an S-a-O, drawing all these together as a Function Map, identifying all the problems in a problem list and then solving these problems one at a time using the Oxford Standard Solutions. Any system is made up of several S-a-Os. If we have a complex situation and problems are hard to define, then we can identify problems individually by drawing a Function Map of each part of the problem situation. This shows us the problem situation in detail to reveal problems singly, and helps to show their interactions and the results and outcomes. A minimal system is two components with an action between them. One component is active (the subject) and acts upon another passive component (the object). A razor is a simple system:

RAZOR cuts HAIR

There may be several functions delivered by this system, some may be what we want, some we don't want, or they may be what we want but inadequate or too much. Function Analysis may be very simple – from just one functional model (Subject action Object) to something very complicated with many functions linked together, or anything in between.

The table below shows a simple Function Map of the main functions delivered by a small, plastic razor.

For the razor the mapping of one or two of the components representing the essential system at a simple level shows us roughly how the functions provide all the benefits, and also shows us what is wrong, when and how the system provides those functions inadequately and when it has harmful functions – such as the razor cuts the skin. A razor exists to cut hair (prime function) – it may also have the harmful function of cutting the skin and another beneficial action of smoothing the skin. Problem understanding requires capturing all the functional relationships to enable the drawing of an accurate Function Map.

Subject	action	Object	Function	Benefit/Harm	Ultimate Goal
RAZOR	cuts	HAIR	Cut Hair (GOOD)	Removed Hair	Good shave
RAZOR	cuts	SKIN	Cut Skin (HARMFUL)	Damaged Skin	
RAZOR	smooths	SKIN	Smoothed skin (GOOD)	Soft feeling Skin	Smooth Skin

For the razor the mapping of one or two of the components representing the whole system shows us how the functions provide the benefits, and also to show us what is wrong, any harms, and when and how the system provides those functions insufficiently or inadequately (**smooths** – is an insufficient/too weak action and denoted by a dashed green line). Once we understand all the problems we can then choose how and whether to deal with the harmful, (including excessive), and the insufficient actions. The Oxford Standard Solutions have solutions lists for insufficient functions and harmful functions and guide us to all the solutions – giving us all simple, conceptual advice on how to deal with such problems.

Systems Develop in Response to Changing Needs

When razors were first developed this was done by a 'cut-throat' razor. Cut throat razors were crude and simple systems, they did deliver the function of shaving by cutting hair with a sharp blade, but they required skill and careful handling to avoid serious cutting of the skin, as well as the hair, and were not designed for self use – normally one person was shaved by another (by a barber or their valet).

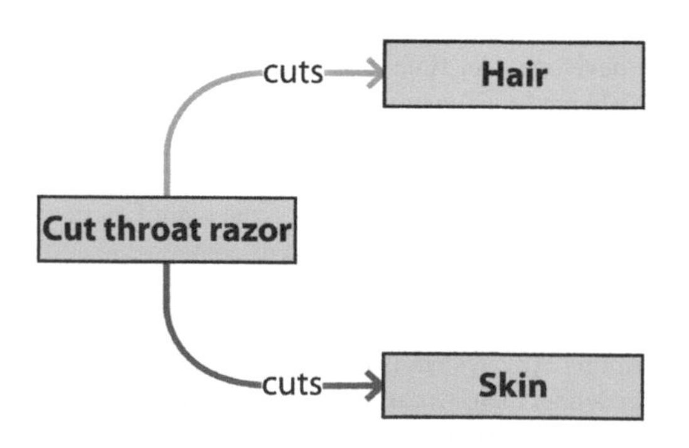

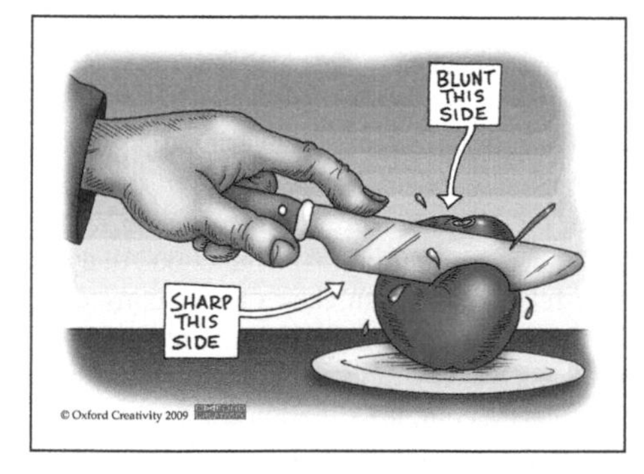

Even age-old simple systems have solved contradictions

The problem of holding a sharp blade had been solved long ago in the design of a knife by 'separating in space' the sharp side where the blade cuts from a blunt side but there are still problems of safety. Human ingenuity has been solving problems since we began using tools and creating systems – this ingenuity is captured in the Oxford Standard Solutions, and when systems need to change then a Function Analysis will reveal where and when we can apply the simple established rules for successful change.

In the early twentieth century until the outbreak of World War I, it was fashionable in western society for many men to have beards and the limited use of cut-throat razors had been the normal method for those who wanted to shave. The benefit *smooth/shaved face* was delivered (very carefully) by the function *cutting facial hair* with a cut-throat razor. As this took some care and skill it was considered a skilful process and a hazardous activity requiring experience, practice and a steady hand (belonging to another person). Safety razors were beginning to appear at the time of the World War I, but although they had been invented and available over a hundred years earlier had not been widely adopted.

In the 1790s a safer razor was invented by Perret to enable men to shave themselves; he also wrote *the Art of Learning to Shave Oneself.* This had a small impact. In the nineteenth century safer, self-use razors were made in Sheffield, England and in Germany and the first safety razor was patented in 1880. In 1901, the American inventor Gillette invented a safety razor with disposable blades. Only when this was commercially successful, and became widely available, were the majority of men able to safely shave themselves for the first time and not rely on a family member or barbers. In 1914 with the outbreak of war many bearded men joined the army. This was a problem especially as they were required to shave hair not just from their faces, but anywhere on their bodies prone to infestation with lice. Safety became very important. The main benefits of safe cutting of hair had to be realized quickly, in primitive conditions by men with no skill or previous experience with razors. The safety factor had to be added to the system itself and 3.5 million safety razors were issued to meet the minimum requirements which had now become – when using a razor for the first time – cut hair and don't cut anything else. The razor systems developed in response to changing needs, with a recognition of the necessary benefits and the location of all the functions to deliver them.

The prime function was still 'the safe removal of hair' and there were still cuts to the skin with the new safety razor, but they were relatively insignificant compared to the damage which could be inflicted by a cut-throat razor in inexperienced hands! Systems exist and develop to deliver what we want – but often they fall short of this until they become better developed. The Ideality Audit helps us uncover everything we want from a razor, and think about its development to meet needs, from the invention of the safety razor to the present day. Function Analysis highlights problem areas and where the work needs to be done to develop and improve product, and Oxford Standard Solutions suggest the answers.

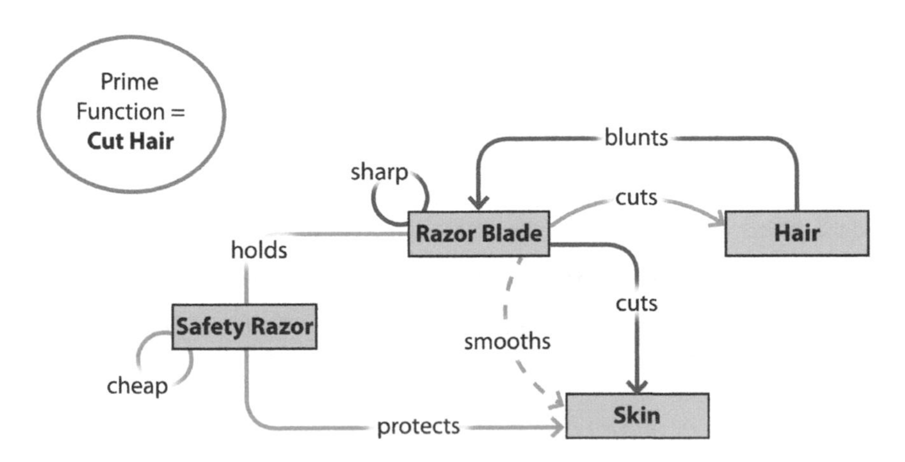

Simple Rules of Function Analysis

We want benefits which are delivered by functions and a *subject action object* delivers a *function.*

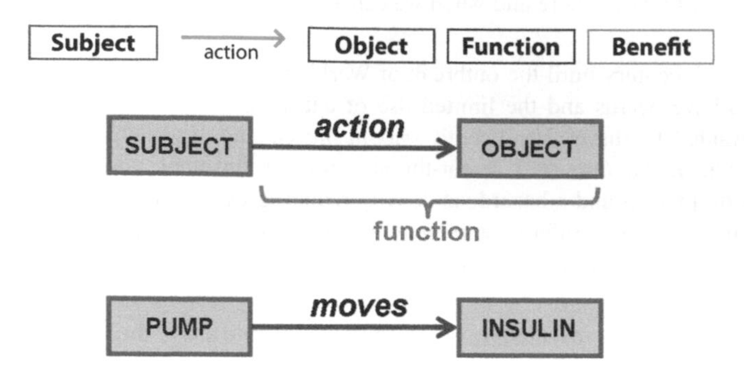

It is important not to confuse requirements/benefits and functions – benefits have no solutions.

S-a-Os deliver solutions and show in detail how we get functions (the physics of an S-a-O) and the structure of systems.

Finding Your S-a-O is Not Always Easy

S-a-Os are useful in problem solving because they are simple understandable units. A problem within one S-a-O can be solved with the Oxford Standard Solutions – a list of problem S-a-Os can be tackled one at a time. Most complex technical systems can be reduced to S-a-Os which provide a clear starting point for problem solving. Defining and describing each S-a-O takes some thought and the careful choice of words. There are simple rules for defining which is the subject and which is the object. Something has to be done to the *object* and this is done by the *action* and the *subject.*

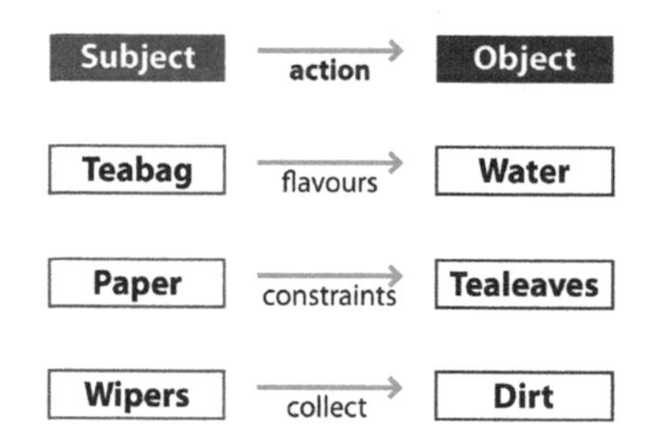

Defining the right S-a-O means we must ensure that the object is changed. This change occurs by the action and the subject working together on the object.

I *LOVE* **YOU** Is this a Subject action Object?

No! The object is not necessarily changed. **I** (subject) **love** (action) **YOU** (subject) may not have any effect on the object (**YOU**) – especially if you don't like me or even know me. The correct S-a-O in this case is:

YOU *INSPIRE LOVE IN* **ME** (ME the object is changed!)

Define the Correct S-a-O for a Thermometer

Which is the subject and which is the object when taking someone's temperature with a thermometer? *Thermometer* measures *Temperature* on the body is not an **S-a-O.** The thermometer does not change the temperature in any way but the body changes the temperature of the thermometer.

The correct S-a-O is:

Systems Are Made Up of S-a-Os

Once we have all the functions accurately mapped we join them together to see the whole system through Function Mapping. For a complete function to take place we need to ensure that each **Subject action Object** is present. By themselves the action or subject can produce no effect on the object – they need to act together on the object. Subject and object can be components or whole complex systems or simple components:

Teabag and fingers
Submarine and navy
River and Dam
Knife and apple
Nut and bolt
Hammer and nail
Laptop and information
TRIZ and engineer

Many popular systems have obvious problems yet to be solved

The effectiveness of an S-a-O depends on all three elements, and problems can be solved by changing one, two or all three elements.

The action or field provides the energy force that guarantees the *effect* of the subject on the object.

To change the object requires both an instrument or tool (the subject) and an action or energy field. To cut the apple needs the subject (the knife) and the action of moving the knife through the apple. As the object the apple is the passive receiver of the knife cutting it. Certain features of the two components, the apple and the knife, could influence how effective or easy is the delivery of the function 'cut apple'. Either a blunt knife and/or a very hard apple could make the function insufficient. The features of the components influence the effectiveness of the function.

The Oxford Standard Solutions which suggest ways of improving insufficient functions instruct us to look at ways of changing the subject and/or the object as well as the action itself.

Defining the Action Takes Careful Thought

Describing, finding the right active verb for the correct action can be a slightly tricky task. The subject and the object are nouns and the action itself is a verb. To check if an action is correct it should be possible to specify exactly how the action working with the subject changes the object it is working on. If actions are too loosely or inaccurately described then we will reduce the number of solutions we can potentially find as we will not understand the exact nature of the function we are seeking.

This way of correctly defining how the action changes the object is not difficult but may be counter-intuitive to our normal way of describing simple systems. So it needs careful thought and practice, as Function Analysis language is much more precise than ordinary language, for example:

System	Function Analysis Speak	Normal Speak
Plastic water bottle	Cap Constrains Liquid	Cap seals bottle
Piston in cylinder	Oil constrains piston	Oil lubricates cylinder
Lightning conductor	Wire conducts current	Wire conducts current
Measuring instrument	Signal informs instrument	Instrument measures signal

Function Maps Contain All the System and Relevant Environmental Elements

The main elements to consider are:

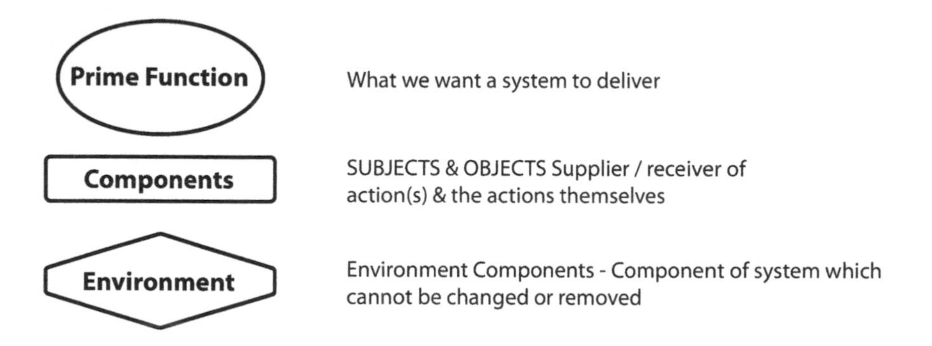

An important part of any TRIZ analysis is the context of the system and how it influences the system and possible solutions. *Environment components* are those outside our control which interact with our system. Unlike other components we haven't provided them, we have no control over them and we can't change them.

Toothbrush components. = bristles, handle

Environment components. = gums, teeth, saliva

They interact with the toothbrush but are not parts of that system and we can't usually change them.

System Development Through Extra Functions

For system improvement we must ensure that we understand the benefits we want to be delivered by that system and understand which functions deliver them and how. We must not forget the main purpose of our system and which function delivers this (Prime Function). For example if our system is a toothbrush and the Prime Benefit we want is clean teeth (to achieve an even higher level benefit – so they don't decay) then we have to understand *how* our system delivers this. The Prime Function is often the clue of how the system works. The Prime Function delivered by the toothbrush is the removal of plaque – this gives the result/benefit of cleaned teeth. Other benefits may include sparkling teeth, no damage to teeth or gums, fresh breath, no visible bits lodged on or between teeth, whitening achieved, smooth feel to teeth etc. The Prime Function of the toothbrush is to remove plaque and any improvements to a toothbrush involve making it better at removing plaque – it may also be to match the system to all the other benefits we want. This matching of benefits gives us better systems for the future (see the TRIZ 8 Trends of Evolution as part of the Ideal System capture stage of problem solving).

The simple systems below have simple Prime Functions which deliver their Prime Benefit and Ultimate Goal.

System	**Prime Function**	**Prime Benefit**	**Other benefits**	**Ultimate Goal**
Toothbrush	Remove plaque	Clean teeth	Good gums	Undecayed teeth
Vacuum Cleaner	Collect dust and dirt	Clean floor	No damage to carpet	Hygiene
Pen	Stain paper	Writing	Unscratched paper	Communication

For a toothbrush the problems include solving contradictions of brushing hard to clean teeth but brushing gently to prevent damage to gums which can give us an insufficient action (brushing gently). One way of overcoming insufficient actions suggested by the Oxford Standard Solutions is to add another field to the insufficient action (mechanical brushing).

Simple solutions suggested by the TRIZ Standard Solutions such as adding vibration improve many systems big and small

Problem Solving from the Function Analysis Problem List

We use the Function Analysis map to draw up a subsequent list of the problems (harmful, missing and insufficient functions) and can then use the Oxford Standard Solutions to helps us solve problems and the TRIZ 40 Principles to resolve the revealed contradictions. Standard Solutions are very simple lists of all the ways to solve problems recorded by science and technology, particularly in patents. They have been reduced to such simple and general language that they help with any problem not just technical ones. The Standard Solutions begin with the Trimming Rules, and show the ways of removing components or simplifying systems. This is a very important part of the TRIZ problem-solving process and the Trimming Rules are simple and show us how to remove components without losing useful functionality. Once we have constructed a clear and accurate Function Map of a management, structural or people problem we apply the relevant triggers in the Oxford Standard Solutions to each problem identified on the Function Map.

Standard Solutions begin by suggesting all the ways of simplifying the system by taking out problematic components and then offer all the ways of:

- dealing with harms
- improving any insufficiencies
- measuring and detecting.

Other Popular Problem-Solving Toolkits Need the Oxford Standard Solutions

Problems are solved with these simple to use Oxford Standard Solutions. TRIZ Function Analysis is powerful and links with other popular problem-solving toolkits like LEAN, TQM, Value Engineering, Systems Engineering and Six Sigma, all of which (like TRIZ) have their roots in the middle of the last century. These help identify problems but leave it to experience and brainstorming to solve the problems. TRIZ additionally offers the Standard Solutions. Engineers using any of the 'lean' problem solving kits find the Trimming Rules an invaluable aid to their problem solving as it contains all the ways of removing components without losing functions.

We use the Oxford Standard Solutions once we have identified the problem type – such as if we have something harmful. We can then step through all the ways to stop harm:

- trim
- stop/block, or prevent the harm so it no longer causes a problem
- transform it so it is no longer harmful (perhaps even making it useful)
- correct afterwards.

Oxford Standard Solutions for Solving Problems Mapped in Function Analysis

The simple TRIZ code for identifying problems in Function Analysis leads to matching problems to solutions. Function Analysis will normally reveal insufficiencies and harms and the Oxford Standard Solutions then gives us all the ways the world knows to deal with these, including how to simplify and reduce cost of systems and how to deal with harm and improve insufficiency, often pointing to the routes to the next generation system. This matching of problems to solutions is unique to TRIZ and so powerful it can help us solve any problem. When we identify a harmful action, we then check all the ways the world knows to deal with harmful actions using the simple TRIZ list of all the different 24 ways of dealing with harmful actions. Similarly there are 35 ways of overcoming insufficient actions. Using these method we will be given simple solution triggers to turn into real solutions. This is like using the 40 Principles or 8 Trends but with more detail because the Standard Solutions contain the most useful of the distilled TRIZ knowledge from contradiction and trends plus frequently used knowledge in the patent database.

Oxford Standard Solutions provide a comprehensive set of solution triggers to help solve the problems identified in a system through Function Analysis. When we have a complex situation and problems are hard to define, we overcome this by identifying the problems individually with a Function Map of the problem, to reveal the problems singly and showing their problem interactions. We then have a problem list and each of those problems can be solved one at a time. There are three main types of problem that Function Analysis can reveal:

- harmful actions
- useful actions, but which are insufficient
- measurement and detection.

We can correct any such identified problem, by adding or altering

- the subject ,
- the object
- the action
- or two, or all three, of them.

This is achieved by directly changing them, adding something to them or affecting them through the environment around them. We use the Oxford Standard Solutions after we have identified the problem type.

When we have something *insufficient* (the problem is that the delivered function is too weak, too slow, too uncontrolled etc.) then the two basic strategies to overcome the insufficiency are to improve, change or enhance the:

the components (subject and/or object)

the action/field.

When we have something *harmful*, we can then step through the four basic strategies of all the ways to deal with harm:

trim/eliminate – trim out the harm – Trimming Rules

stop – block the harm

transform into good – turn harm into good

correct afterwards – put right the harm.

Insufficiencies → 35 Standard Solutions for Improving a Function

Insufficient actions occur when there is no harm but we still have a problem. All three of the subject, action and object are present and are needed to deliver a function but it is insufficient in some way – it is too weak, too slow, too uncontrolled etc. To overcome insufficiency there are 35 conceptual answers, which show us all the known ways to improve the solution we've got.

These are simple lists of solutions of how to improve useful functions which are insufficient. In these lists the suggested solutions prompt us to look at ways of changing the subject and/or object and/or action in order to overcome the insufficiency or to introduce a new function but with minimal complications or disadvantages. There are three basic types of solutions.

To improve the components (subject and/or object) or their surroundings to enhance the delivered functions:

- *add* something to the components to overcome insufficiency (7 ways)
- *change/evolve* the components to overcome insufficiency (10 ways)
- *get a better or another action* – achieve more effective actions (18 ways).

Insufficiency of communication and understanding can be a big problem in many areas

Detection and Measurement → 17 Solutions

These 17 Standard Solutions are used for solving measuring or detection problems in engineering systems. There are 5 distinct types of solution:

1.	Replace detection or measurement with system change, or use a copy, or use two consecutive detections	3 Solutions.
2.	Create or build a measurement system. Some elements or fields must be added to the existing system	4 Solutions.
3.	Enhance the measurement system	3 Solutions.
4.	Measure a ferromagnetic field	5 Solutions
5.	Evolve the measurement systems	2 Solutions

Measurement problems need clever TRIZ solutions

Harmful Actions → 24 Standard Solutions for Overcoming Harm

There are four basic strategies for dealing with *harm*:

1	Trim/eliminate – Trim out the Harm	6 solutions
2	Stop – Block the Harm	11 solutions
3	Transform into good – Turn Harm into Good	4 solutions
4	Correct afterwards – Put right the harm	3 solutions

These 24 conceptual solutions for harm are not only applicable to technical engineering systems but can also be used to solve any problem with harmful actions – management, business, theoretical, scientific, or even personal. The 24 solutions are simple to understand and apply, and they offer all the conceptual answers the world has uncovered to deal with harm.

Many ways of solving a problem

The first category *trimming* is for removing harm by simplifying the system. This is by cutting out the subject which delivers harm and/or the object which receives it as well as expensive, complicated and/or troublesome components particularly if they contribute little to the main purpose of the system. In these lists for dealing with harms the suggested solutions prompt us to look at ways of changing the subject and/or object in order to eliminate the harmful action or to introduce a new function but with minimal complications or disadvantages.

Trimming – for Simplifying Systems, Reducing Costs and Removing Harms

System Development by Trimming

Once we have a Function Map which shows problems (harms and insufficiencies) then the first step for finding simple, elegant solutions by removing parts and simplifying the system is called 'trimming' in TRIZ. This TRIZ term describes how to simplify systems by removing components, or parts of components. Trimming is in some ways an inappropriate name and sounds like a gentle process. 'Pruning' or 'amputation' would be better terms to describe the process of cutting out as much as possible from a system while keeping all or as much functionality as possible.

Why Trim?

Trimming increases Ideality (same or better benefits, less costs, less harms) and helps us to solve problems in engineering systems. It helps us eliminate troublesome components by reducing complexity and part count. It is useful for most problem circumstances (it is an important step for TRIZ patent circumvention and patent strengthening) and to generally improve a system by removing harms, expense and complexity.

Trimming – One of the Most Powerful and Effective Steps in TRIZ Problem Solving

Trimming is a hard reality check, a moment for breaking psychological inertia and inappropriate attachments to the current system. It makes us question whether we need all the components, and forces us to confront the reality of slashing out whole sections of a system in order to improve it. If a Function Map shows components with lots of red lines we should think about trimming the components associated with them, especially if they are not directly involved in delivering the Prime Function. But trimming is not just a way of dealing with harm, excess or expense: it is something which should be considered for almost every component, and by applying the Trimming Rules to each component can create some imaginative, innovative and resourceful solutions.

Compact, Effective and Economic Systems

The clever engineering solutioneering part of trimming is the focused search for available resources to deliver the useful functions of components which are candidates to be trimmed by systematically applying the Trimming Rules. We ask what are the useful functions we lose if this component is removed from the system. What is already there and available to provide these functions? This cuts to the heart of problem solving – what do we really want and what is the best and most cost-effective ways of providing it? Applying the Trimming Rules is the systematic route to looking for available resources and covers all possibilities. It is simple and effective. Lean practitioners and proponents of Lean manufacturing or Lean production strive to minimize waste and use resources only to create value. Those who also learn TRIZ claim that TRIZ trimming adds a systematic step for matching resources to functions. Most Lean processes, although systematic throughout identifying what to trim, then rely on experience and brainstorming to make this important match to resources, rather than the application of the simple, systematic TRIZ rules.

Rules for Trimming

The purpose of trimming is to eliminate problems, so we normally look to trim something that has low ideality. We trim components which have high harm, expense and complexity for the benefits they deliver.

First trim or remove components that are expensive, have problems and/or are far from the Prime Function of the system (i.e. are probably non-productive in that they do not directly contribute to the useful output of system such as corrective components). Check to see if existing components and available resources could replace/substitute them and think about the value each component adds to the system and assess its Ideality (value plus consideration of harms) – trim if its Ideality is low.

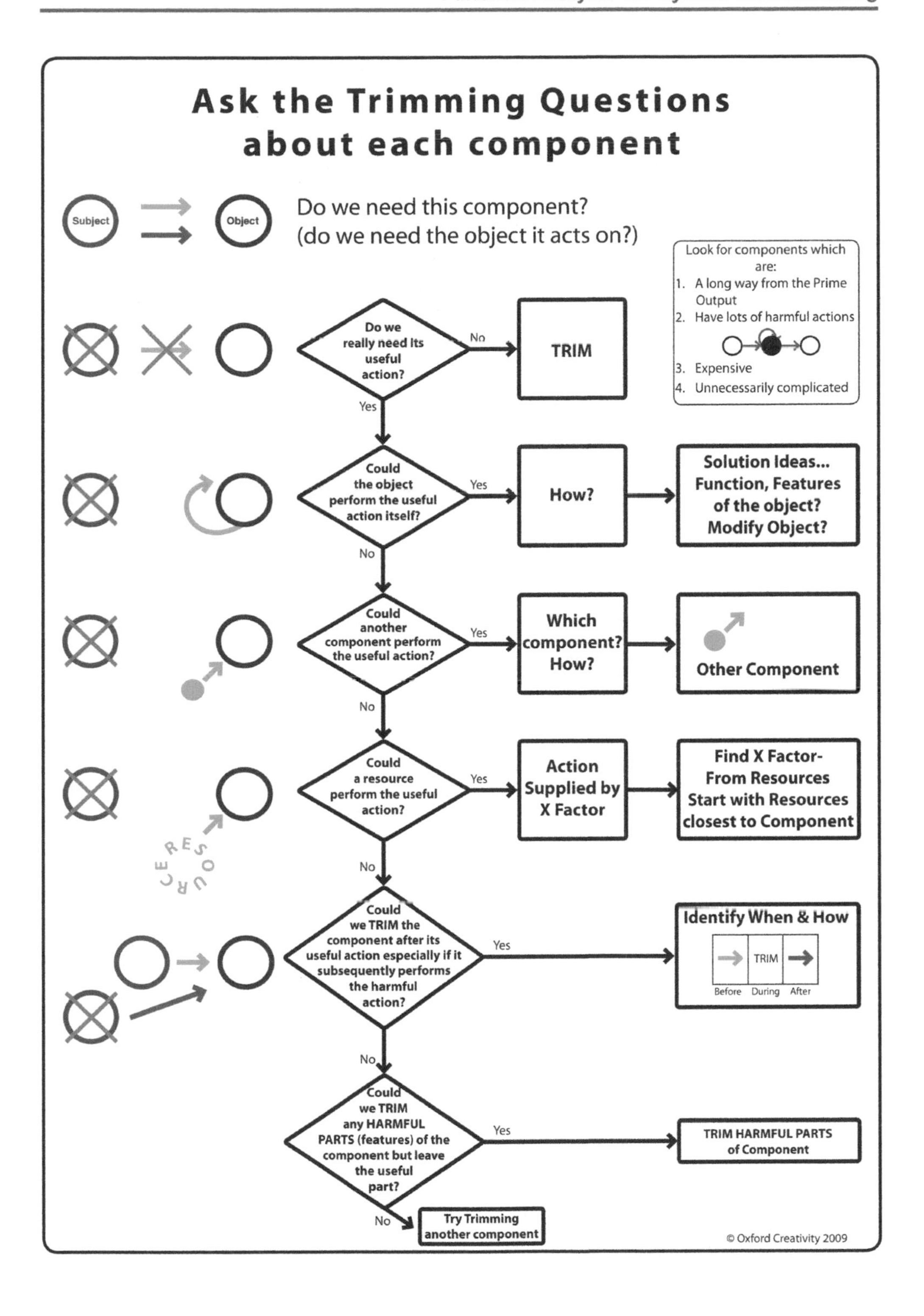
Ask the Trimming Questions about each component
Subject
Object
Do we need this component?
(do we need the object it acts on?)
Look for components which are:
1. A long way from the Prime Output
2. Have lots of harmful actions
3. Expensive
4. Unnecessarily complicated
Do we really need its useful action?
No
TRIM
Yes
Could the object perform the useful action itself?
Yes
How?
Solution Ideas... Function, Features of the object? Modify Object?
No
Could another component perform the useful action?
Yes
Which component? How?
Other Component
No
Could a resource perform the useful action?
Yes
Action Supplied by X Factor
Find X Factor- From Resources Start with Resources closest to Component
RESOURCE
No
Could we TRIM the component after its useful action especially if it subsequently performs the harmful action?
Yes
Identify When & How
TRIM
Before During After
No
Could we TRIM any HARMFUL PARTS (features) of the component but leave the useful part?
Yes
TRIM HARMFUL PARTS of Component
No
Try Trimming another component
© Oxford Creativity 2009

Guidelines for Transferring Responsibility for Useful Action

(e.g. Trimming out heaters from a small office)

1. Performs the same function on the same object (use the air-conditioning system).
2. Performs same function on a different object (obtain heat from ovens or water heaters).
3. Performs a different function on same object (fans which move air or special solar heat windows which store and release heat on demand).
4. Has a relevant resource (walls or floors or ceilings all have a large air contact area).

Remember that trimming creates problems (but that's fine – we are problem solvers!). Be prepared for outrageous trimming suggestions and never dismiss them out of hand. The other TRIZ tools are there to solve problems raised by trimming – use the entire TRIZ toolkit to improve and simplify systems.

Trimming Harmful Permanent Marker Pens from a Training Environment

I spend a lot of time lecturing and teaching TRIZ and problem solving in conference rooms and decided to use TRIZ Trimming to solve one ongoing problem that has annoyed and irritated me. I ruin/stain whiteboards by using the wrong (permanent marker) pens on them. I used the whole Function Analysis process to solve this.

There are usually two kinds of presentation media – flipcharts with flipchart pens and paper, and whiteboards with whiteboard pens. The flipchart pens permanently stain the whiteboard; the whiteboard pens are very faint when used on the flipchart and also stain the projector screen. At a glance the pens are very similar, same size, same caps, same colours – some small writing tells you which is which. For a busy presenter in the middle of a problem-solving session, with high energy in the room, stopping for careful examination of the pens before using them is not a practical option and just doesn't happen.

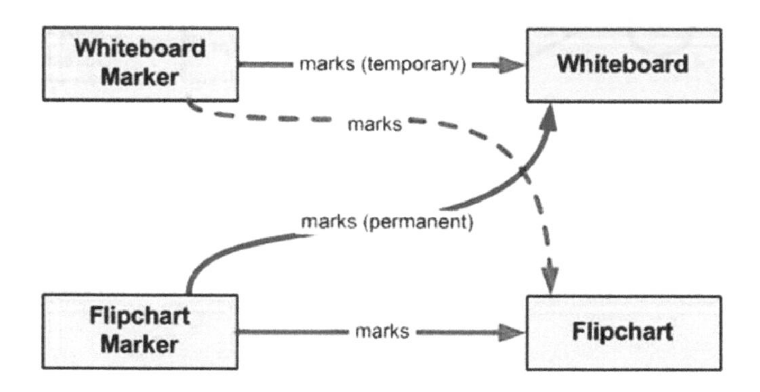

This is a simple problem but only when I drew a Function Map of the situation and applied the Trimming Rules did I see what was required. A simple Function Map of the situation showed the two problems (the harmful action and the insufficient action) and made the solutions obvious if I follow the rules of Function Analysis. Normally we begin by solving the harmful actions as they usually deliver the biggest improvement. Start with components with the most red lines. My solution was first to trim/remove from the room all the flipchart markers to stop ruining whiteboards, and solve the consequent, second problem of the whiteboard markers being too inadequate to read on flipchart paper. This was achieved by changing the flipchart paper. Further analysis of the flipchart had showed the problem was caused by the flipchart paper surface which was too rough and that if replaced by shiny, smooth flipchart paper then writing with a whiteboard pen was just as clear as with a flipchart pen. We could remove all the flipchart pens, stop all whiteboard damage, and have only one kind of pen available. Whiteboard pens are used on both flipchart and whiteboards and there were no more problems, confusion or damage.

Trimming a Component to Remove Harms Solves Problems, but It May Mean Losing Useful Functions Too

Components (subjects and/or objects) are within a system to act together to provide useful functions, when they also produce harm then if we trim either or both of them to remove/reduce problems, we also trim their useful functions. Useful functions can be lost or provided by other elements in or around the system.

For example imagine trimming out the mini-bar in a hotel room – for the hotel the mini-bar is a source of profit and provides a good service to the guests, but is expensive to audit and maintain. Spurious problems such as questions about mini-bar use are an irritation to busy guests on checkout and a source of friction between customers and the hotel. Mini-bar refilling and audit requires a great deal of careful work and administration, which is expensive. A simple Function Map may reveal some of the problems.

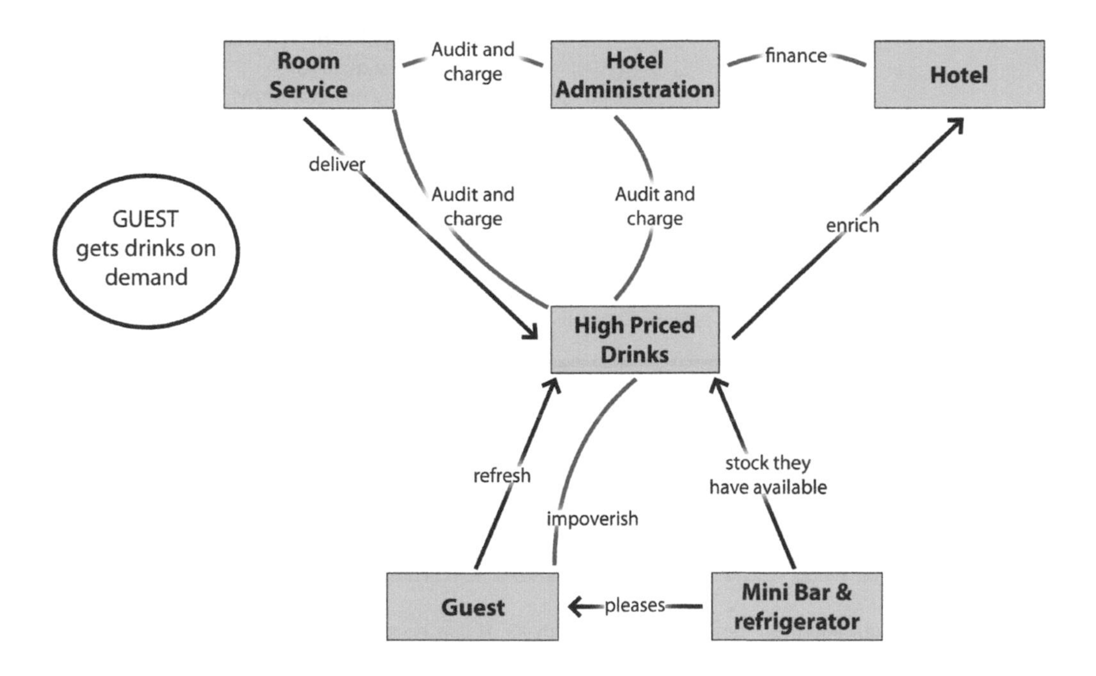

Recent solutions in innovative hotels have shown interesting solutions, including free mini-bars in every room modestly stocked each day or with nothing in them, no mini-bars but free ice and vending machines and/or good room service. There are many solutions to problems and all the interesting solutions to this problem can be found by using the Trimming Rules. The Trimming Rules suggest how to trim out components without losing anything useful.

When we need to simplify systems or get rid of harms the useful functions can be lost or provided by something else – to locate all the possibilities we need to apply the Trimming Rules.

Après Trim – Don't Stop Now!

Once a component has been trimmed out, don't stop – there can be more trimming.

After trimming, the system has now changed – it is now a new system, so we need to update the functional analysis and try trimming further. The trimming of one component often opens up opportunities for further trimming

Trimming requires an understanding of system resources to obtain our trimmed functions with what we've already got.

For this we need to understand and be aware of all our resources. This means we need to consciously make best use of all resources available and look for unused resources. Don't forget the harms and bad things/the causes of the problem – they are a resource too and may be mobilized. Try to get the functions of the trimmed components by only using the resources already there (rather than adding more inputs or throwing more money at it).

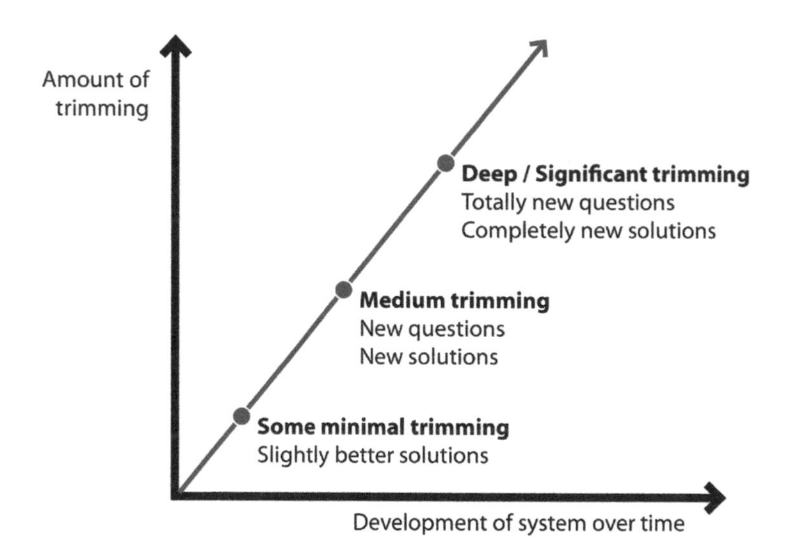

Function Analysis at Every Stage and for Every Kind of Difficult Problem

Most TRIZ problem solving requires Function Analysis. It can be a very small process with a simple Function Map for one component or quite a detailed process with a large Function Map for a large and whole system, very simple or very complicated or somewhere in between. Function Analysis is used at the highest level for examination of the whole system, and also used for detailed examination of the interactions of all the components. A Function Map is a snapshot of the system or parts of it for one moment in time. All the stages of a system process can be represented by several, separate Function Maps – one for each stage.

Function Analysis can be used to map systems at any of the stages of problem solving – it can certainly cover many of the different steps in creating and using systems to meet needs as shown in the diagram below.

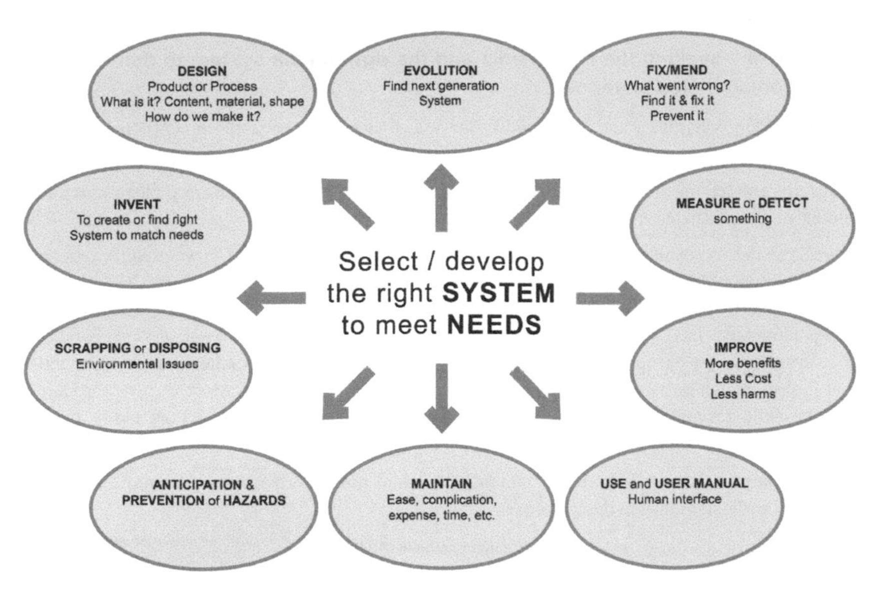

System could be a product or process, technical or management

Using Function Analysis on Real and Difficult Problems

Nuclear clean-up is one of the important and unresolved problems/challenges facing us all and the full brain power, experience and problem solving experience of our engineering and scientific communities could be usefully applied to this terrifying problem area. TRIZ is used and applied in the UK, and Function Analysis has been used to great effect especially when initially understanding the problems. A simple very general but real example is illustrated here.

The picture below is from a process to remove nuclear activity from contaminated water. The simple Function Analysis diagram shows the problems, the harmful actions and the insufficient but also good actions, and highlights contradictions (a red line and a green line from the same components). The picture below helped the designers define and understand the many problems with the initial system, and how to significantly improve their initial very flawed and problematic design.

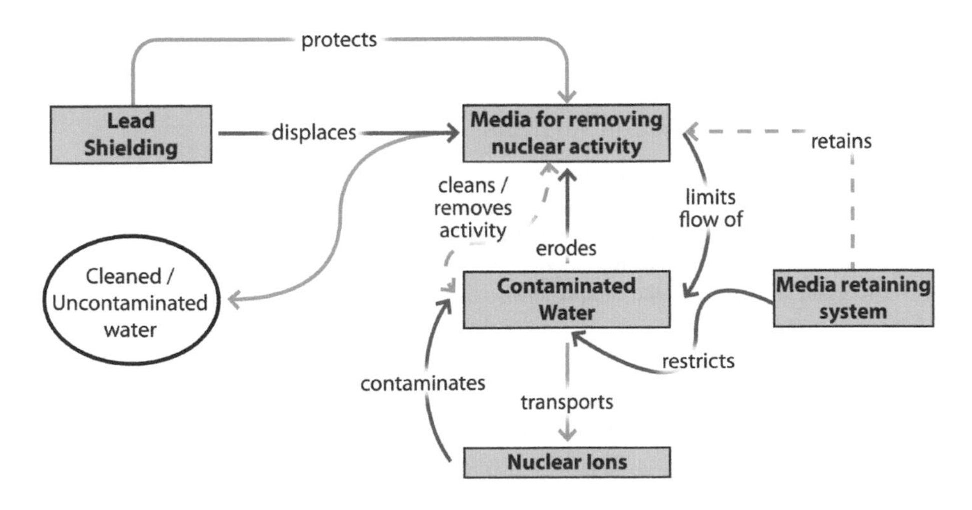

Nuclear problem – seeking the right media and the right media system to deliver nuclear clean-up of contaminated water

In this Function Map 'what we want' is cleaned water and is shown by the prime function (oblong shape).

The Ideal Outcome would be clean water – the Ideal System might be that a very small amount of media instantly absorbs and retains all the nuclear contaminates washed through it – delivering clean water, for a very small amount of contaminated material.

How we achieve the removal of nuclear contaminates involves some form of sieve/media/ material which collects nuclear contaminants from water as it flows through it (media in media retaining system). The media is expensive to purchase and dispose, and we want to use as little media as possible to minimize the used contaminated media for nuclear long-term storage. Therefore we want a very good, effective media in a system which ensures that the media is used up (collects contaminants) efficiently and evenly. The initial design challenge is a contradiction that the media must allow the water to flow through both as quickly as possible to maximize clean up and as slowly as possible to collect as many nuclear ions as possible.

The Function Analysis shows all the basic contradictions and problems with such a simple system. This is helpful at a very early stage of the design process.

Function Analysis Identifies All Significant Problems

Problems exist when seeking ways to close the Ideality gap for systems or processes, when they give us less than what we want, or give us what we want plus things we don't want, or cost too much in terms of time, energy and physical components. Problem solving often involves matching systems and Ideality and removing the gaps between them.

TRIZ Function Analysis provides simple visual map of our system which helps us see at a glance all the things which are right and wrong – anything we don't want or is inadequate or excessive or superfluous or harmful.

To draw a Function Map we have to understand how our simple or complex system works, to be able to highlight the problems, and also be able to separate all the problems from each other and deal with them one at a time. A Function Analysis can take us from a problem situation which is too complex for us to easily understand and communicate, to a simple problem list. This list of problems can then be solved one at a time with TRIZ tools.

System Analysis and Function Analysis

TRIZ Function Analysis was designed for engineering problems but works well for general, management, technical, scientific, detailed, complex or simple problems. When used with the TRIZ Solutions it is universally useful for understanding and solving all problems. Function Analysis is the system map, the route in, the key to systematically solving problems by asking the right questions and then locating the relevant answers from the comprehensive lists of all the world's answers which are contained in TRIZ.

Simple Answers to Simple Questions

To build a Function Map we need to map the functional interactions between the components of the system. A functional interaction is when one thing acts on another (heat melts wax, wife annoys husband, instrument measures signal etc.). Building a function diagram is useful for problem solving in two ways. First it helps us capture and visualize how the systems works (and see its problems at a glance) and second it enables us to apply the relevant TRIZ systematic problem solving tools such as the Oxford Standard Solutions and the 40 Principles. These give us simple answers or solution triggers to the now simple questions which the Function Analysis gives us as a problem list. The power of being able to build a list of all the problems and then be guided to all the possible solutions to each problem is the most useful part of TRIZ for problem solving.

Example of Function Analysis of a Single Item – a Coffee Cup

Function Analysis works on complex processes and can also help us see how we use products, and the functions they provide, both good and bad.

Even Simple systems may have many complicated problems

The starting point of any TRIZ problem is an Ideality Audit which can help us define both where we are and what we want (where we want to be). In TRIZ we can do this by defining the Ideality of everything we've got, everything we want and don't want, how much costs/ inputs we are prepared to make and what harms we will tolerate. Normally a system has one main benefit – its Prime Benefit (its purpose) delivered by one main function, its Prime Function, plus other functions. Once a system exists it often also has things we don't want or want to minimize such as harms and costs. For Function Analysis we can start with our Ideality Audit and need to record systematically everything we want (what it exists to deliver). The steps for creating a Function Analysis of a coffee cup include:

1. Ideality Audit (high level descriptions and simple drawings of the system we've got)
2. Function List of all components and their interactions (Prime System and its components as S-a-Os)
3. Draw Function Maps
4. Problem List
5. Trim (if appropriate)
6. Apply the rest of the Standard Solutions to solve problems one at a time
7. Define new system solutions

1. Ideality Audit

IDEALITY AUDIT	(fill in relevant parts according to the problem)
Ideal Outcome – Ultimate Goal + Prime Benefit + all other benefits (no costs and harms) =	
A perfect coffee cup, exactly the size and style we want?	
SYSTEM – Benefits we want	
Prime Benefit	Holds and enhances coffee enabling us to drink coffee
Other Benefits	easy to clean
	keeps coffee warm
	holds right amount coffee
	comfortable to hold and easy to carry
	pleasing to drink from
	pleasing design / style / material
	re-useable
	etc.
Acceptable costs / inputs	reasonable costs to buy or make – needs material, firing, design input etc.
	takes up space / storage
	cup needs to be located, chosen & decided on etc.
	kept clean / hygienic
Unacceptable costs / inputs	Too expensive or has great rarity value
	Hand washed
Acceptable harms/problems	needs washing up / messy
	cup stained by coffee
Unacceptable harms/problems	too big or too small
	coffee wrong temperature
	cup too fragile – chips/damages easily

Include other relevant factors for problem understanding and solving include such as:-

Constraints (Imposed conditions, rules or regulations) Domestic use, re-useable, safe and stable vessel to be stored in kitchen cupboard and washed in dishwasher.

Provide Simple Drawings of System if helpful

2. Function List of All Components and Their Interactions

Each Benefit is provided by a function. To gather benefits we can complete all or part of the Ideality Audit (how much we need to complete depends on our problem type) and consciously list them. In order to capture the functions, related to benefits, we draw the Function Maps. A cup provides the function of holding liquid so we can fulfil our purpose of drinking from it. How does a simple TRIZ Function Map show all the functions done by and to the individual components, and for parts or for the entire system? It simply links all the components with their relevant actions (good and bad) and this gives us a clear and simple insight of how the system works. It is always worth starting with a few relevant and single, individual S-a-Os. Remember that the subject does something to the object – cup holds coffee is an S-a-O which enables its usual use of drinking from it.

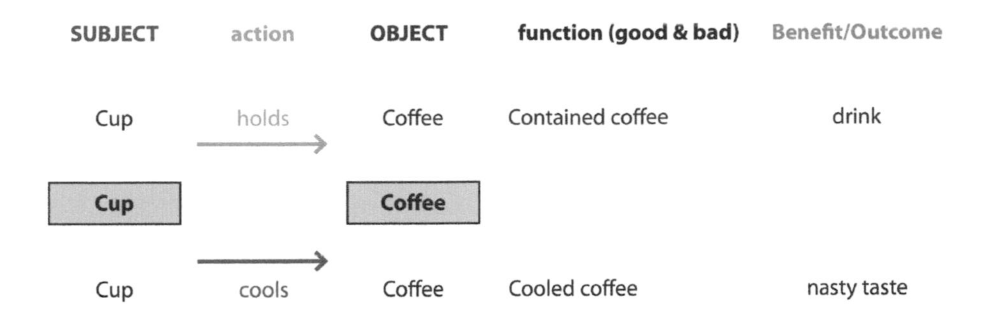

This helps us identify the problems associated with such a simple system such as Cup-cools-Coffee. We might then ask why/how does the cup cool the coffee? Answers may include: because of the cup shape; because of the cup material which is likely to absorb heat; because it might be made of a simple design with thick walls. As soon as we introduce components acting together to create a system it will have problems – as soon as we try and fulfil one function we will probably make some other function worse. In this case we can add the component features such as a thick/robust cup holds the coffee (good), is robust and hard to chip (good) but also cools coffee (bad).

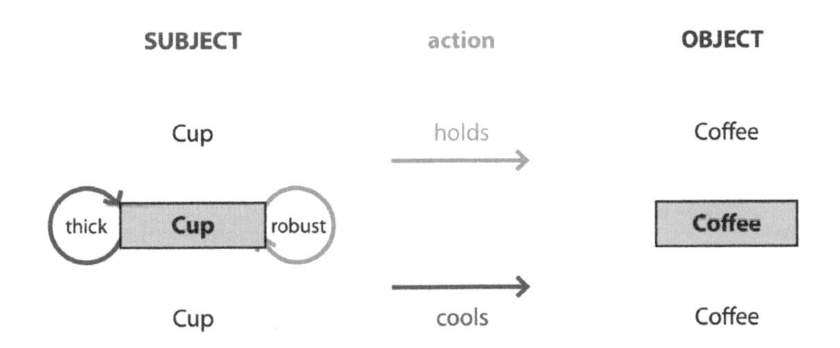

This reveals a contradiction and we can see at a glance that we have a contradiction because we have a green and a red arrow at the cup and between the cup and the coffee. Contradictions emerge when we invent, develop, fix or improve systems. Improve one thing and something else gets worse. We make our cup more robust (less likely to break) and it may get heavier, more expensive and cool down hot drinks. Technical advances mean that contradictions emerge everywhere and require innovative methods to solve them. TRIZ offers 40 Principles (40 simple triggers) to solve contradictions or we can just tackle the each harmful and insufficient actions separately and directly.

Function Maps give us simple visual models, a precise description and an understanding of the links between components and features. By highlighting all harmful, unwanted and redundant functions we can simplify cause–effect relationships and reveal the causes of problems, failure or faults in a product and see how to improve it. Describing a simple system by S-a-Os and the functions they deliver helps our understanding to check how close we are getting to our defined 'Ideality we want' (no problems) and how best to achieve it.

Each Subject-action-Object can be listed as a function statement if required or found to be useful. When we include benefits in this table this can remind us of what we want and help us remember any gaps between what we want and what the system delivers. A Function Statement table can be built by listing all the components and then noting how and if they act on other components – in good and bad ways.

A component can be anything that acts on something else. So components will be the obvious things like the cup, the coffee drinker, the table etc. but also a component could be less obvious – something like gravity or ambient temperature.

Prime Function of cup = holds coffee

Subject	action	Object	Function	Outcome/Benefits
Cup	contains/holds	Coffee	Held coffee (good)	Enables coffee drinking
Coffee	feeds energizes	Coffee-drinker	Revived coffee drinker	Pleasure/energy
Ambient air	cools	Coffee	Cooled coffee (harm)	Unsatisfactory experience
Coffee cup	burns	Table	Burn marks on table (harm)	Spoilt table

3. Draw TRIZ Function Maps

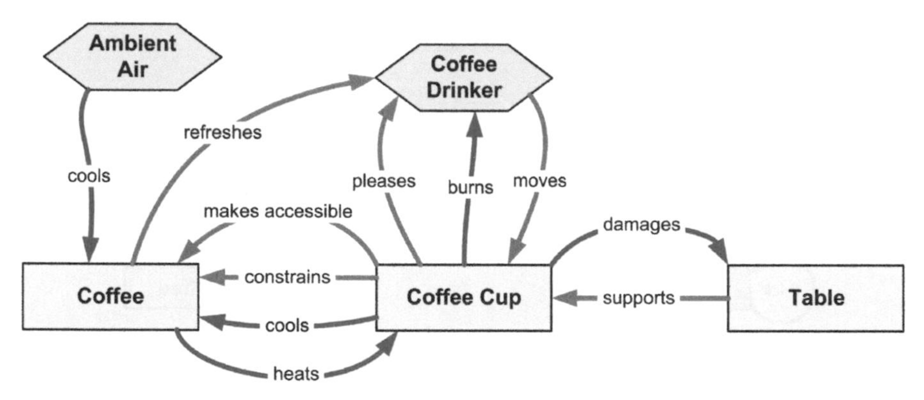

Try and identify the right verbs to describe some of the harmful actions in the above Function Analysis from the cartoon of the man holding the coffee cup

The Function Map as illustrated above shows the useful and harmful functions of a simple cup of coffee. We can also capture the interactions of the system with the environment. The environment may cool the coffee (or cover it in dust), the coffee's interaction with you (an environmental element) may be both beneficial (comforts, energizes) and/or harmful (keeps awake/raises blood pressure).

Function analysis, when done properly and accurately, delivers a problem list with all problems identified. We can then understand and deal with one problem at a time. We can solve problems of two linked S-a-Os by identifying and resolving contradictions (if one of the S-a-Os has a problem) with the 40 Principles. We can deal with a problem, an individual S-a-O, by solving problems with both component features and actions by applying the Oxford Standard Solutions. Each problem on our list should link to a certain type of Standard Solution. Once we have selected the appropriate type then this will help us systematically view all the proven, known, general solutions to the problem. These help trigger our brains to make connections to available resources and locate detailed and relevant solutions using our experience, knowledge and judgement.

Problems at a Glance

We can create Function Maps relevant to our product and process. We can focus in on particular areas of the system or a particular problematic moment of the process or use of the system. By highlighting all the unwanted and redundant results we can simplify cause–effect relationships and reveal at a glance all the problems – all excesses, insufficiencies and causes of failure or faults in products.

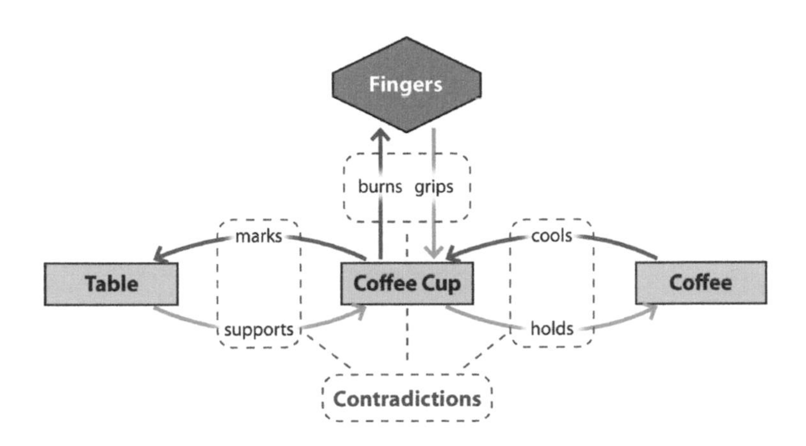

The final stages of problem solving for our coffee cup are:

4. Problem List – create the problem list from the various Function Maps ready to move onto steps 5–7 which Involve the actual Problem-Solving.
5. Trim/reduce parts/make simpler (if appropriate).
6. Apply the rest of the Oxford Standard Solutions (and 40 Principles if relevant) to solve problems one at a time.
7. Define new system solutions and solve the problems.

These last three steps prompt engineers to systematically explore the TRIZ Standard Solutions. This begins with Trimming (step 5) – simplifying the system by removing harmful excess and unnecessary parts. Step 6 prompts the application of the rest of the Standard Solutions for removing all other harms and overcoming any insufficiencies, and to finally and additionally apply the 40 Principles if contradictions have been uncovered.

All the work of steps 1–3 are about converging our analysis to give understanding to how the system works and defining all the problem functions within our system.

Step 4 is the summary in a problem list of problem functions. The last three stages of problem solving are the divergent stages when the relevant Standard Solutions to the problem are identified and the simple conceptual prompts they offer for solving the problems are starting points for divergence to real solutions. With relevant design expertise and engineering knowledge, available resources and ingenuity applied to each prompt, we can create a complete range of practical solutions to each problem.

*****

Below I show examples of how to problem solve for a system or product by following all the steps of Function Analysis.

Problem Solving with Function Analysis

OXFORD CREATIVITY

Start Here!

Problem Desciption

Constraints

Ideality Audit

	System we've got	System we want
Benefits		
Costs		
Harms		

Task 1 - Describe System

List all system components and their costs (or relative costs):

Component	Cost

Task 2 - Function Analysis

Draw a Function Analysis Diagram showing the interactions between all the components. Include negative or harmful interactions, and any insufficient or excessive interactions.
Using the list below may help you establish the relationships between the components. Once you are happy with your list, connect the components on the diagram to the right with arrows showing the type of action, and label each arrow to show what the action is. One example is shown to help you get started.

Subject	Action	Object	Type of action (effective, harmful or insufficient)

Overall Arrangement

Draw Your System here...

Function Analysis Diagram

Types of action

Effective

Harmful

Insufficient (or absent)

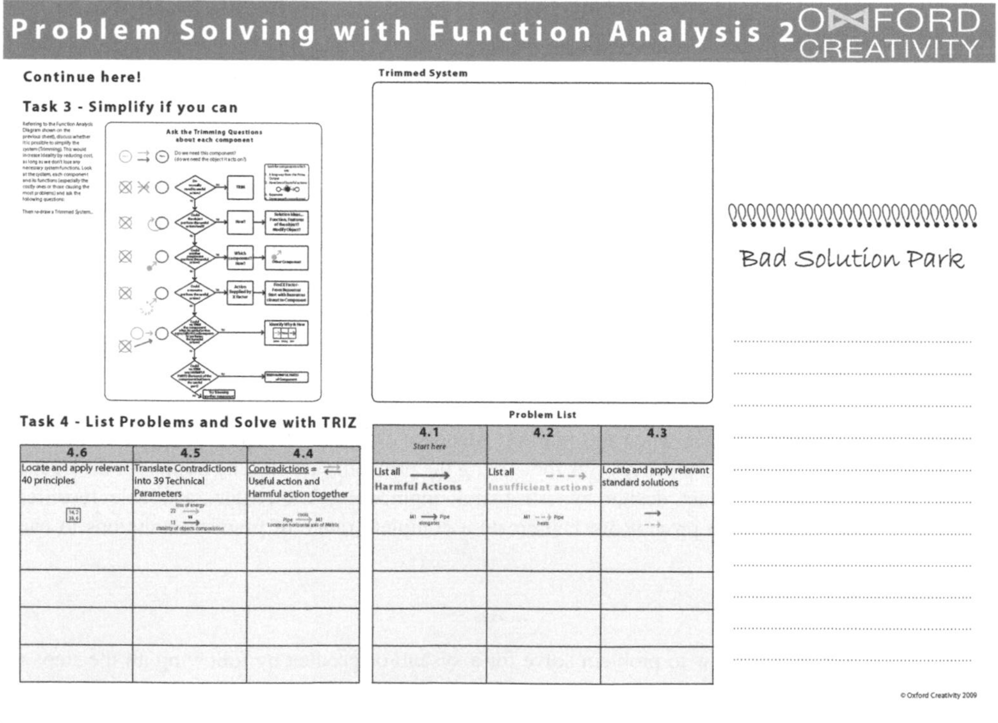

When I teach groups of engineers to problem solve with Function Analysis I use two A0 sheets with all the steps for Function Analysis set out as tasks 1–4 which when followed will help groups of engineers solve their problems. This process requires the Oxford Standard Solutions and the 40 Principles to be used the essential addition of their own intellect and engineering experience to complete the problem and find solutions.

Function Analysis for Locating and Dealing with the Causes of Problems – Roadside Bombs

TRIZ Tools are most effective when used combined together, which tool combination depends on the various problem types and the context of the problem. Function Analysis Tools will help us solve many problems and are particularly enhanced when used with the TRIZ Tool Thinking in Time and Scale (the 9-Boxes) to map both possible causes and possible solutions and focus on problem areas with an appropriate Function Map to show the problems and their causal links in detail.

A Function Map is a snapshot of one moment in time, showing all the good and bad functions – all the problems at that moment can be mapped and captured. Most problems do not occur at just one moment in time nor do they usually have just one cause. One simple way of fully understanding problems and the causes of problems is to expand the problem situation to understand its context in time and scale. One problem I have worked on with various organizations is the problem of roadside bombs which blow up passing vehicles killing and maiming the occupants. This is a horrific and very complex problem and one which urgently needs the best solutions our engineers can locate and apply. I use this problem to teach the relationship between causes of problems (hazards) to show how to locate many possible solutions, and show this problem solving process below. The causes and solutions suggested here are simply some that delegates have come up with during the exercises and are not supposed to suggest a finished solution. This is to show how TRIZ helps us approach and solve problems.

Problem of Detecting Roadside Bombs

Bombs are placed at the roadside and subsequently detonated when vehicles (carrying people) pass by.

There are many levels, times and places to look at this problem which might include:

- How might we detect such bombs before they cause harm?
- How can we prevent people being harmed by roadside bombs?

TRIZ Tools applied to the problem of detecting roadside bombs:

- Thinking in Time and Scale (9-Boxes)
- Functional Analysis (plus X-Factor where relevant)
- Applying the Oxford Standard Solutions.

Detecting Bombs in Time and Scale

When we begin a problem and there are many causes of the problems and a multitude of possible solutions and many places to apply those solutions then it s worth mapping the problem in Time and Scale.

Begin by mapping the situation in time steps – what are relevant steps before, during and after? Next in scale – how far should we zoom in and zoom out in scale? Where should we concentrate? The road? The vehicle? The bomb? When we use the Time and Scale tool we do have not have to choose one – we cover them all but in a simple and systematic method which displays a lot of relevant detail making it easy to understand and relate the different parts and steps to each other in a meaningful method.

	Before bomb placed	**During** bomb being placed	**After** bomb placed
Road and surroundings			
Around Bomb			

Causes of 'Successful' Roadside Bombs

Bombers are able to select and use suitable places to plant bombs (and observe & check the bombs)	Insufficient surveillance of bomb-planting sites. Bombers undetected on roads Bombers plant bombs on roads	Detonation is successfully initiated in response to approaching vehicle The Bomb must be detonated at the right time – something successfully detects that the target is in the right place
The Bomb Planter arrives undetected at the planting site The Bomb arrives undetected at the planting site	Planted Bomb undisturbed The Bomb remains unobserved until detonation Bomber remain unobserved	The Road User is near the bomb when the bomb detonates Insufficient vehicle protection The Road User must be vulnerable to the effects of the explosion
The Bomb successfully procured / made Components for the bomb obtained and assembled	Detonation is enabled either remotely or triggered by vehicle The Bomb is functional when it detonates	The Bomb must still be in the correct place when it is detonated Insufficient body protection

Successful Outcome from the Bomber's Perspective

A 'successful' bomb from the Bomber's perspective is one which has been made, put in position and detonated and hurt its target. Below is a Function Analysis from the viewpoint of the bomber showing what needs to happen to blow up passing vehicles with roadside bombs. The harm from a bomber's perspective is being detected.

One of the bomber's main problems is to prevent detection at any stage of the operation.

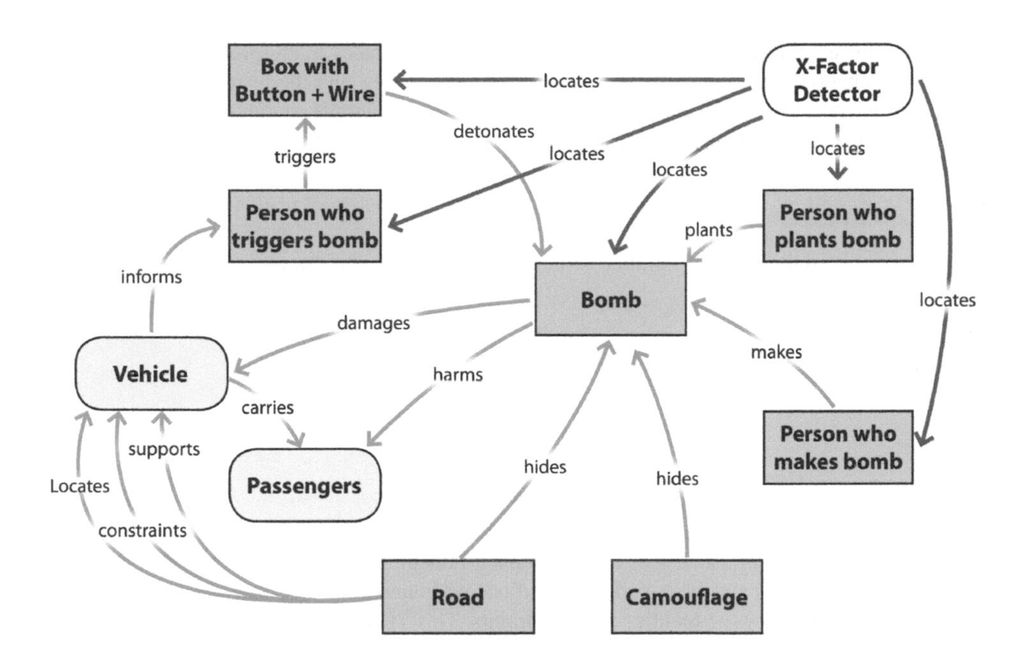

Outcome from the Bombee/Victim's Perspective

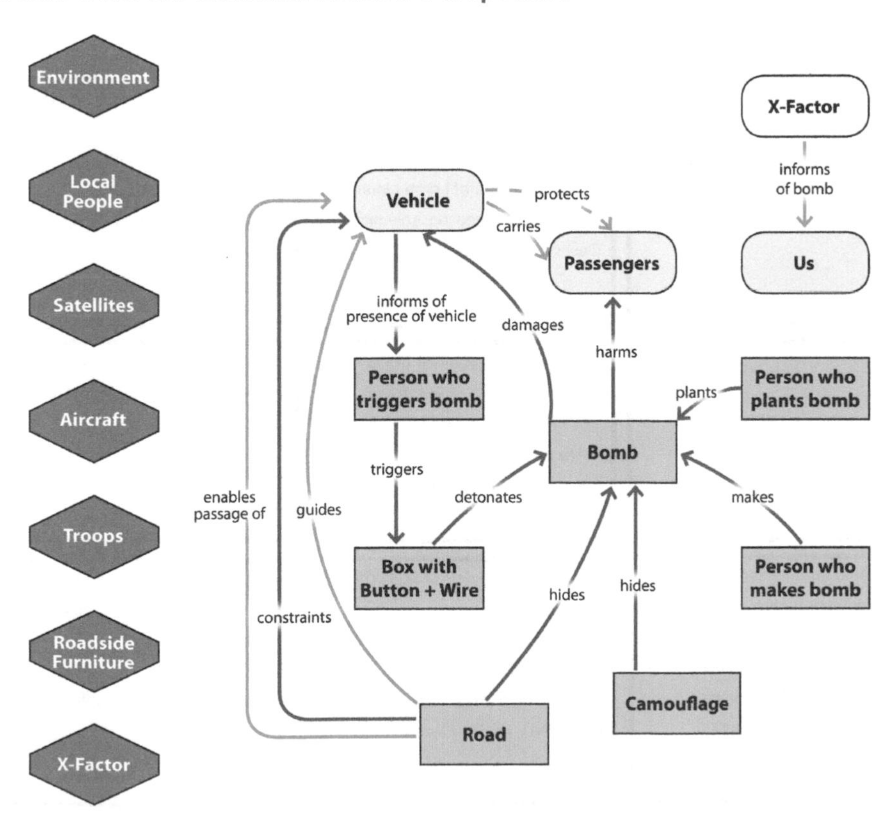

The problems of a vehicle under threat from a roadside bomb faces many problems, which can be mapped in a Function Analysis, showing the many harmful actions and the insufficient ones (such as vehicle insufficiently protects occupants). The diagram below attempts to capture the problems and gives a list of environmental components which could be brought into play to help the problem.

Finding Solutions to Roadside Bombs

This very serious problem has a number of conceptual triggers for answers in the TRIZ Standard Solutions:

1. Harmful Actions
2. Useful Actions, but which are Insufficient
3. Measurement and Detection.

Some Suggested Solutions

The answers below are some sample solutions from some TRIZ teams and in no way offered as the solutions to this terrible problem. This exercise of using Function Analysis on the problem of roadside bombs would yield the best and most workable solutions when undertaken by army teams who regularly face this problem and know the realities and constraints. This TRIZ exercise if thoroughly undertaken with the right people would discipline and structure the understanding of the problem, and offer all the possible conceptual answers when combined with the right experience. This would then yield many, if not most, of the possible answers to this problem.

Detect the intent to plant the bomb (intelligence) Detect bomb planter Surveillance +' spot the difference – something wrong'	Surveillance of possible bomb-planting sites Use locals as lookouts Detect signature of bomb planting in environment (acoustic, seismic etc) Observe behaviour of locals	Detect signature of effect of bomb on environment: Detect vapour (sniffing by machines or animals?) Magnetic field, thermal signature, other emission Routine scanning of road
Deter / remove bombers Detect activity of moving bomb to site. Detect bombs at checkpoints	Observe/detect planting of the bomb Observe patterns of behaviour Somehow get the bomb to tell us it is being planted The bomb planter himself somehow informs us	Regular sweeping of road Dummy (sacrificial) vehicles) Detonate any bombs
Detect bomb components during bomb making or transit Do something to bomb components (spike or dope with detectable agent) Create premature detonation or disabling of the bomb circuits Flood the market with inferior bomb-making materials.	Roadside informs us of bomb being planted Introduce something to significantly increase or decrease the stability of the bomb's explosive and/or detonator	Use sensors, locals, other road users, animals (rats), etc to locate planted bombs Somehow the bomb betrays its presence

Conclusion

We can use the problem lists from the Function Maps together with the Oxford Standard Solutions to find all the relevant solution triggers which would suggest areas to look for answers. By themselves the Standard Solutions to problems are theoretical – solutions become real, practical and useful when engineers with the relevant experience of the problem apply their knowledge and skills to each standard solution. In order to use the Standard Solutions we have to be able to list the problems in our system. We do this with Function Analysis to understand the system at one moment in time.

Problem and needs definition requires a clear an understanding of the system – how, where and when it is not meeting all our needs. TRIZ Function Analysis tools will help us understand the system, and map all its problems at one moment in time and will give us a picture of how functions are delivered and any shortfalls, or harms, or excesses – a problem list. The problems can then be prioritized and solved using the Oxford Standard Solutions. This process systematically checks that our system delivers all the things we want for the minimum cost and harm. This helps with problem solving but also for understanding how the new technology will fit in and replace existing systems. Generally speaking TRIZ is a great tool for solving technology communication issues, which can be at the root of many of the problems.

Case Study: Improving the Opening of the Bitesize Pouch at Mars

Mars Enjoys Immediate Success of New Pouch Packaging Concept

In 2003, Mars (at that time called Masterfoods) decided to launch a new packaging for its Bitesize product range. Bitesize simply means a product that can be eaten in one bite. Some of the products included in this category and made by Mars are M&M's, Maltesers, Minstrels, Revels and more recently, Mars Planets. The packaging change consisted in the move from a standard pillow bag (that was the main format used in this category at that time) to a standing Pouch.

The change delivered increased sales for Mars with features and benefits such as:

- horizontal pack gives better visibility on shelf;
- 'sharing experience' with a wide opening pouch on a table, making it easy to access together (or pass it around to friends);
- straight and easy opening that could be reclosed with a sticker (attached at the back of the pack);
- tear strip (upper part of the Pouch) is straight and easily removable so it does not block the hand getting inside the Pouch nor alter the aesthetic of the Pouch once open.

A multi-disciplinary team worked on the project and identified the best packaging machine and the right packaging material to deliver the concept. The straight opening was given by a micro-perforated line put on both sides of the Pouch next to the easy opening line printed on the design. The new packaging was launched and the expected result happened quickly: increased sales and a success in the market. This concept changed the way people were looking at the Bitesize segment and most of our competitors started to bring to the market their own version of the Pouch or even copied our product by buying the same packaging machine.

The whole market was affected and this change created a shift from standard pillow bags or poor standing bags to more premium packs.

The Pouch Problem

As it is often the case with new concepts and products the pouch had some problems. and some work had still to be done to move the concept along the S-Curve.

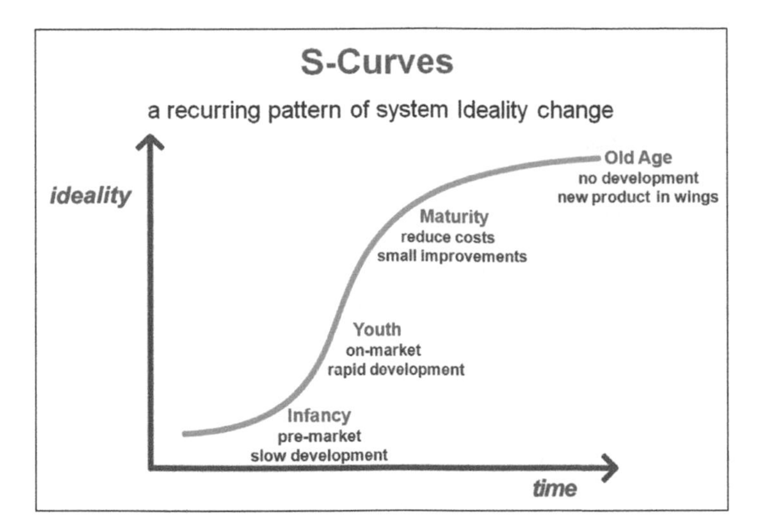

The pack that went out on the market was still in its infancy stage and was not perfect. One particular problem was the straightness of the opening – an important feature for consumers.

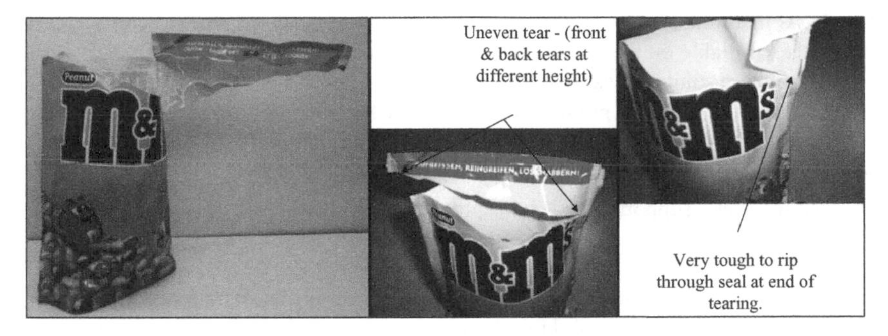

This problem was that when the consumer tears the pack open the tear lines were not always following the perforations but were going into erratic directions. This can create a big gap between the front and the back wall of the Pouch. As a result, it was nearly impossible to cleanly remove the tear strip, and the aesthetic of the Pouch itself (after opening) was awful.

In Picture 5 above you can see an illustration of the issue. In this case, the back tear followed the perforation line, but the front tear went astray and moved down towards the lower side of the Pouch. As the result, it is nearly impossible to remove the tear strip (it is attached on the side by at least a 2 centimetre seal) and the aesthetic is not nice at all. This problem was enhanced when the Pouch had a Euroslot on the top.

At that time we developed a method to check the opening performance and the results were far from satisfactory:

- straight opening: 5% (Euroslot: 0%)
- torn strip removed easily: 0% (Euroslot: 0%)
- average distance between front and back line: 15 mm (Euroslot: 30 mm)

We tried to solve the problem, as most people do naturally and instinctively, by using the standard trial and error method; i.e. you try one thing, then if it does not work, you try something else. Although this method is a result of the natural evolution of our mind through time, it is not perfect and could require an important amount of time and iterations before getting to the right solution.

As Genrich Altshuller quotes to the point in his book *The Innovation Algorithm*:

> During the process of evolution, our brain learns to find approximate solutions to simple problems. However, it does not develop mechanisms of slow and precise solutions to complex problems.

The issue that we had was that this Pouch opening problem was impacting one of our key consumer attributes and we had to solve it quickly. When we were in the trial and error phase, we involved our suppliers in the process, of course. But we were all stuck by the fact that we could do very little to the material itself. At that time (as it is still the case today), a lot of different film structures and combination of films were available on the market which had oriented-tear properties. The issue we had every time we moved to these solutions was that they affected the line efficiency (we were generating friction problems) or we lost the heat-seal properties we needed. Changing the film structure was not the right direction. The perforation we had on the film was not helping either: the tear did not follow it. We tried of course to play on the quality of this perforation, but without success.

All these attempts were supported by brainstorm sessions. But again, as Altshuller illustrates in his book:

> Brainstorming does not eliminate chaotic searching. In reality it makes searching even more chaotic. The absurdity of brainstorming as a searching process is compensated for by its quantitative factor – problems are attacked by a large team. Any gain here is achieved only through the reduction of inefficient attempts along the direction of the Inertia Vector.

The way we were tackling the problem was not succeeding or bringing a real step change. We were struggling. We needed to get away from the film and the perforation. To be quick, efficient and successful, we needed something different, something powerful, something able to direct our problem-solving search in a more heuristic way. TRIZ was this method.

Solving the Pouch Problem with TRIZ

In 2004, when we used TRIZ to solve the Pouch opening problem, it was neither unknown, nor new to Mars. Some people (including myself) were already trained by one of Altshuller's disciple (Victor Fey) and the method had been used to solve a problem in our coffee machine segment (Four square). But as often happens with much training initiatives in many companies, the competence tends to be lost because it is not used sufficiently. People work as they always did, in crisis mode using only trial and error and brainstorming to solve difficult problems. Sometimes they think it take too long to use their newly learned problem-solving methods – or would be too difficult. This was my feeling about

TRIZ at that time. The TRIZ training I experienced around 1999 was quite heavy going (five days in a row) and left me with a feeling that TRIZ was difficult and complex. It is only when I got the Oxford Creativity training in 2007 that I started to really understand TRIZ, its essence and how quick and easy the method was. The books of Altshuller I read in between were good contributors as well. But let's go back to 2004.

There were TRIZ people in Mars who were using the method and could help us find solutions to our problem. The time elapsed between my own training and 2004 was too big for me to lead this work. In fact we were lucky because the manager I had at the time was himself a TRIZ convert and TRIZ addict and took upon himself to lead the TRIZ work. He was trying to get the TRIZ culture and thinking process inside Mars (as I am doing now – which shows that TRIZ can be contagious once its essence is understood). So, in a rainy and dark day end of September 2004, we went to a meeting room and used the TRIZ method.

Finding Solution Concepts with TRIZ

The first thing we did (and that you do in the TRIZ approach) was to accurately define the problem and understand what we really wanted. Before we drew the function analysis map we defined our Ideal Outcome and the gap between what we wanted and the real situation:

1. We want the material to tear horizontally, but it tears either towards the top or towards the bottom of the Pouch.
2. We want to tear the material horizontally, but the tear stops when it reaches the Euroslot.
3. We want that the tear to follow the shape of the Euroslot, but it stops on the Euroslot.
4. We want the tear to follow the sealing line, but it stops on the Euroslot.
5. We want to two walls to be under tension, but only one side of the Pouch is under tension when we tear.
6. We want the tear on the two walls of the Pouch to mirror each other and arrive at the same position on the other side of the Pouch. This ensures that it looks good and is easy to remove the tear strip. But tear differently and the bigger the distance between the two tears, the more difficult it is to remove the tear strip.
7. We want the tears to go through the side seal of the pouch (to remove the tear strip), but when the tear lines are not at the same position, it does not work.
8. We want the tear to occur where two walls are close to each other (in the upper part of the Pouch), to ensure the tear is straight. As we go lower down in the Pouch height the walls move farther from each other (the product in the bottom of the bag makes the walls separate). The Euroslot creates problems by making the tear line lower on the Pouch, where the distance between walls is bigger and thus, the tear is worse. We could then define our Ideal Outcome for the Pouch opening as an easy straight tear, in the predicted position, with tear strip completely removed – more specifically:

   i The tear strip is easily removed as both front and back tears are aligned and finish at the same position (same height).

   ii The two walls are close to each other as the distance between the walls impacts the tear direction.

Ideal Outcome gave us the direction of our problem solving and the desired performance during the tearing action. Next step was to map our system within the TRIZ 9-Boxes time–scale matrix to get a clear overview of our system, the super-system, the sub-system and their relation to each other.

Time

Scale

Supplier manufacturing Process	Film on reels	Haguenau Pouch making machine
Each individual Pouch Wall	The two walls together + side notch	The Pouch open
Perforation Film molecules	Perforation Film molecules	Perforation Film molecules

Function maps revealed our problem as the need to deal with the harmful and insufficient actions during the tearing/opening of the pouch. To solve these issues we used the Oxford TRIZ Standard Solutions which offer 24 solution concepts for dealing with harms and 35 for dealing with insufficiencies.

We applied the relevant TRIZ Solution to different parts of the system to find ways to make the Pouch tear straight. Oxford TRIZ Standard Solutions suggest ways of dealing with both harms and insufficiencies and we tried those which seemed most relevant as shown below.

From the 'Block the Harm' section of the Oxford TRIZ Standard Solutions we applied:

> **H.2.2** Stop a harmful action being harmful: change the object so it is non-sensitive to the harmful action.

Applying this solution concept of changing the object so it becomes non-sensitive to the harm, one solution could be to move the Pouch two walls into one; then when the two walls are together, the tear would be straight.

To achieve this we had the following ideas at the System Level:

- Put cold-seal on each side of the wall. When they touch each other, they become one.
- Seal the area where you want the tear to happen (heat-seal or ultrasonic seal).
- Extract the air so both walls are close to each other.
- Use a zipper.
- Use a multi-layer laminate (Triplex).
- Use static electricity.
- Use Velcro.
- Void inside the Pouch to attract the walls together.

At Sub-System level:

- Use a non deformable material.
- Use more rigid material.
- Change the material orientation (molecule level).

From the 'Enhance the action' section of the Oxford TRIZ Standard Solutions we applied:

> **ia2** Add another action or extra field/supplement the action, *and*
>
> **ia3** Improve/evolve an action by finding a better one or add a new (second) field which is more easily controlled.

The idea here is to improve the efficiency of the perforation because its current action is insufficient: as the tear does not follow the line of the perforation.

At the Super-system Level (supplier process), the following solutions were generated:

- Fancy cut system (increase the number of perforation lines).
- Laser cut (new field).
- Change the shape of the cuts (seesaw type like on boxes).

At the System Level level:

- Cut completely one part of the laminate instead of just perforating it.
- Make half cuts like on cardboard.
- Change the position of the tearing notch.

From the 'Block the Harm' section of the Oxford TRIZ Standard Solutions we applied:

> **H.2.1.**Stop a harmful action being harmful: Counteract the harmful action with an opposing field

At the Super-System Level (Pouch-making machine):

- Create a thickness in the side walls to stop the tear getting down.
- Weaken the material at the side seal level so we can remove the tear strip even if the tears on each wall do not come to the same position on the opposite side.
- Add material or glue during Pouch-forming process.

At the System level level:

- Add something on the side walls to guide the opening.
- Use the reclose sticker to guide the opening.
- Remove the glue on partial areas to block the tear line.
- Use two different glues.
- Stress cracking: high or low temperature effect, with or without pressure.
- Change the shape of the Euroslot.

For this exercise, we used only 4 of the 76 Standards that we mixed with some of the 40 Principles of the Contradiction Matrix (segmentation in particular). The next step was then to meet our suppliers and check how we could put these concepts into industrial solutions.

Industrialization of the Concepts

When we met our suppliers, we explored the different concepts one by one and checked what we could do to solve our issue:

(a) Material

(i) I.1) Orientation (no bubble effect)

The idea was to check if we could block the bubble effect generated by the blown extrusion process to make one of the materials used in the structure. We supposed this effect made the tear go down (it looks like it is following a bubble shape).

(ii) I.2) Rigidity

Thicker film

More rigid film

More layers

(b) Make the two Pouch walls getting close together (walls glued after the pouch is made)

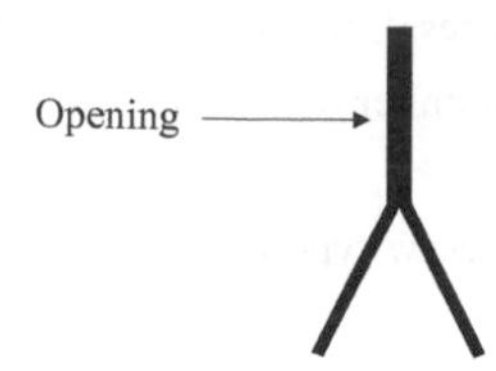

Using Cold Seal points.

(c) Guiding the opening better

(i) III.1) Improvement of micro-perforation

Cutting a different layer

Cutting different sides in an alternate way

Cutting all sides at the same place (total perforation after lamination)

Several parallel lines

Other knife width

Other knife shapes (easy opening system on boxes).

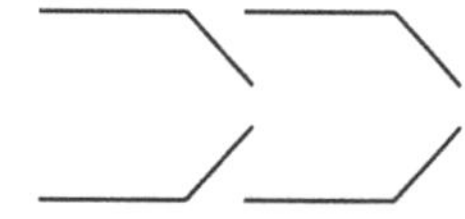

or

Segmentation of holes: multiplication of holes (like fancy cuts). This is coming back to multi-lines.

Vertical lines: multiplicity of vertical line to create tension differential.

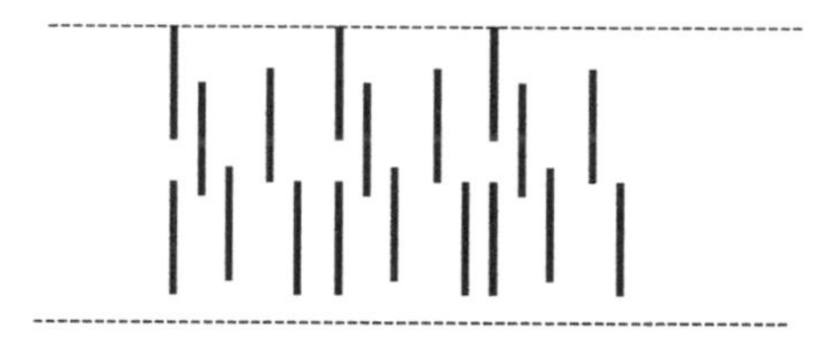

Vertical lines at the edge of the pouch to improve the tearing of the top part of the pouch.

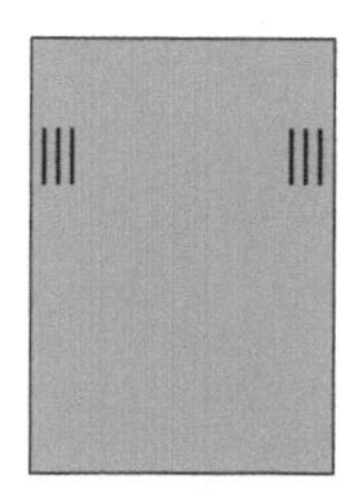

(ii) III.2) Material stress to get an opening orientation:

- Using the cutting reel on the line at the supplier to crush locally the film in the easy opening area.
- Heating the cutting reel (material strcss).

(d) Creating artificial thicker areas (to guide the opening like a ruler)

- Thicker lamination glue areas.

- Increase the glue weight to check impact on rigidity.
- Change glue type.
- No glue at all in the opening area.

- Hardening varnish.
- Combination of glue and varnish.

The Winning Idea and the Validation

A lot of work has been put into action to test and validate these ideas. Not all will be discussed here. The most original, clever and easy to implement idea was guiding the opening and blocking the tear from getting away from its intended path: the absence of glue in the laminate. This is the one that we implemented.

The most difficult in this development was to convince some of our suppliers to test it. First they rejected the idea (psychological inertia). Then they did a first timid trial, then a second one and then the solution was validated.

The final concept was to add three lines on each of the side walls (to cover for Pouch forming variation). The tear occurred in the middle of the area. Each time it started to get away from a straight line, it fell inside the glue free area and then was guided by it.

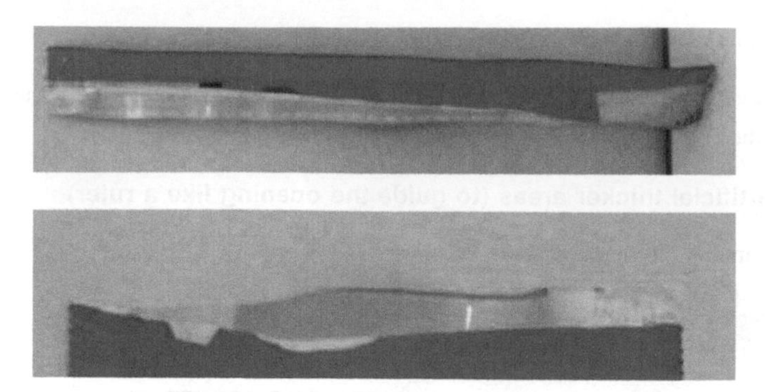

On picture 9, the effect of the Glue Free Lines is clearly visible.

Patenting the Idea

As the principle was quite unique and to ensure a proper protection, we decided to patent the idea:

(19) Europäisches Patentamt / European Patent Office / Office européen des brevets

(11) **EP 1 746 043 A1**

(12) **EUROPEAN PATENT APPLICATION**

(43) Date of publication:
24.01.2007 Bulletin 2007/04

(51) Int Cl:
B65D 65/40 (2005.01) B65D 75/58 (2005.01)
B32B 7/06 (2005.01)

(21) Application number: 05291526.1

(22) Date of filiing: 18.07.2005

(54) **Easy-open package made of two-or-more ply laminate including adhesive-free lines**

(57) The package comprises at least one wall of laminate having two plies bonded together by an adhesive wherein at least one adhesive-free lines is arranged between said two plies. This adhesive-free line(s) provide an easy-open feature. The package maybe a bag or a pouch having two main walls comprising each one or more adhesive-free lines (17, 18) arranged in correspondence to each others.

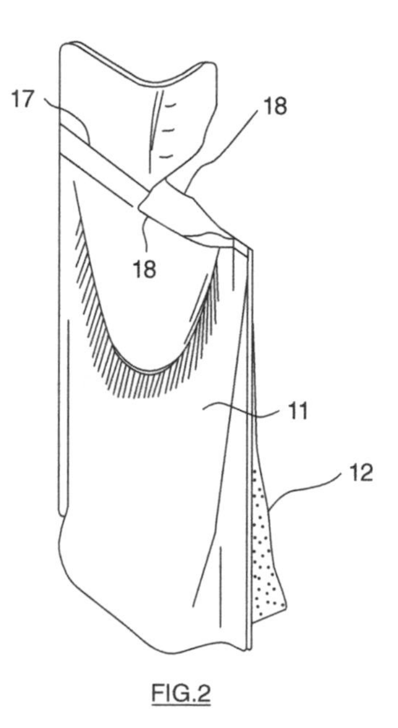

FIG.2

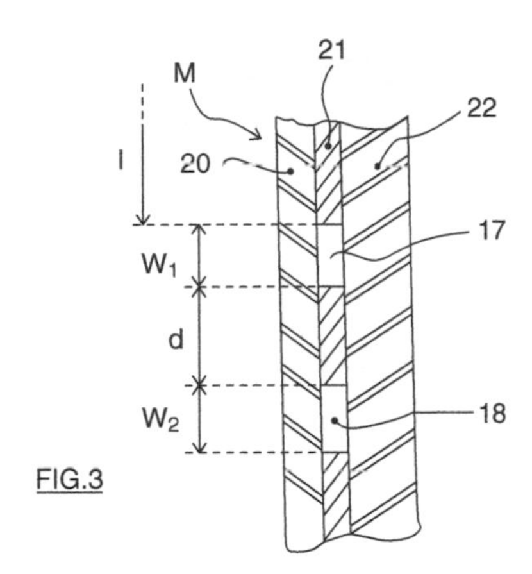

This system has been on the market for some time now on an M&M's Peanut Pouch.

The Future

As for all systems, invention and creativity are a journey along the S-curve and the adventure never finishes. When we wanted to move the system to another design (M&M's Crispy that has a blue background and M&M's Plain that has a brown background) we were faced with another aesthetic problem: under a dark colour, the glue free lines are visible and impact the visual aspect of the pouch. This was not acceptable for marketing. We then had to develop another solution that I cannot discuss here for confidentiality reasons. This new solution is on the market and has not only solved the aesthetic problem, but has also improved the opening efficiency.

Conclusion

I hope that this example has showed and convinced you that TRIZ is the powerful and innovative problem-solving method it claims to be. Since the development of the new Pouch and after new training with Oxford Creativity, I have made it one of my basic development, problem-solving and innovation tools. I used it to solve some other packaging problems, develop new packaging concepts and lead innovation in a new direction. The example above only illustrates some of the aspects of the method. TRIZ has some other tools that I did not describe: the Trends, the Ideality of a system, Functional Mapping, etc. The combination of all these tools is now helping us bring creativity to a new level.

Appendix 11.1
Oxford Standard Solutions
These are the Traditional TRIZ
76 Standard Solutions Re-Arranged
into Three Categories

Three Categories of Solution

Harm 24 Solutions

There are 4 basic strategies for dealing with harm, which can be achieved in 24 ways in total:

- Eliminate – trim out the harm – 6 ways
- Stop – block the harm – 11 ways
- Transform – turn harm into good – 4 ways
- Correct – put right the harm – 3 ways.

Insufficiency 35 Solutions

There are 2 basic strategies to improve, change or enhance functions by changing:

- the components
- the action/field, which acts between them.

Insufficiencies occur when there is no harm but we still have a problem. All three of the *Subject action Object* are present and together delivering a function, but it is insufficient in some way – too weak, too slow, too uncontrolled etc. To overcome insufficiency there are 35 conceptual answers, which show us all the known ways to improve the solution we've got:

- add something to the components – subject or object (7 ways);
- change/evolve the components – subject or object (10 ways) – using some of the 8 trends;
- achieve better/effective actions (18 ways).

Measurement 17 Solutions

Used for solving measuring or detection problems of engineering systems. The major recommendations are:

- try to change the system so that there is no need to measure or detect;
- measure a copy;
- introduce a substance that generates a field (introduce a mark internally or externally).

In this book there are two version of the Standard Solutions – the Oxford Standard Solutions and Altshuller's original 76 Standard solutions. The Oxford Standard solutions contain all the original 76 standard solutions but in a different order with some extensions and additions. The measurement sections is identical to the original (see Appendix 12.1 for details) and not repeated here. The numbers in brackets cross reference the Oxford Standard Solutions to the original 76 standard solutions which are detailed at the end of Chapter 12.

The numbers in brackets cross reference this version of the standard solutions to the original. Both versions of the Standard Solutions need relevant knowledge to yield practical solutions, see www. TRIZ4engineers.com.

Harms = H

H.1 Eliminate – trim out the harm – 6 ways

H.2 Stop – block the harm – 11 ways

H.3 Transform – turn harm into good – 4 ways

H.4 Correct – put right the harm – 3 ways

These 24 conceptual solutions for harm can be applied not just to technical engineering systems but can also be used to solve any problem with harmful actions – theoretical, scientific, management or even personal. They are simple to understand and apply. They are all the conceptual answers the world has uncovered to deal with harm. Here they are illustrated with examples to show how to remove, block, transform or correct harm.

The first category is for removing harm by simplifying the system. This is by cutting out expensive, complicated and/or troublesome components particularly if they contribute little to the main purpose of the system. There are many trimming examples in nature as shown below which shows how the first tadpoles 'trim' their smaller siblings by eating them.

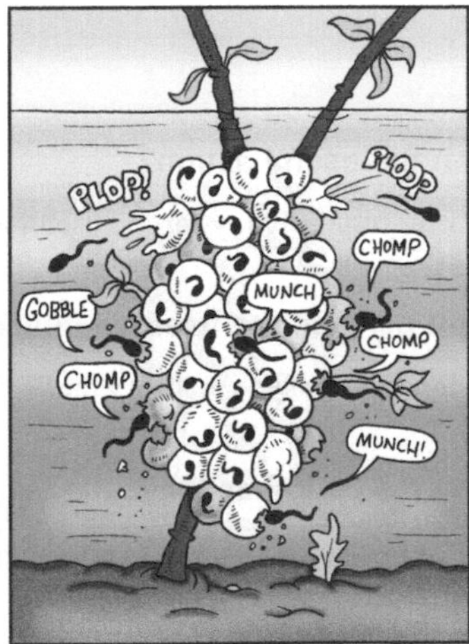

H.1 Trim/eliminate the Harm – 6 ways

Remove the components which have harm associated with them (including expense and complication). Concentrate on the essential functions of the system and look to remove (trim) anything which is a long way from the main purpose and has harm, high cost or needless difficulty associated with it. Trim components by integration of several components into one which still delivers all the functions (3.1.4.)

Trimming should be attempted with a ruthless eye – by asking *'do we really need this?'* about every element in the system and asking the trimming questions below.

Trim the subject but keep its useful action – obtain it from some other source.

H.1.1 For any component, do we need their useful action?

If not, then remove the component

H. 1.2 Could the object perform the useful action?

If yes, then trim the subject and transfer responsibility for the useful action to the object

H.1.3 Could another component perform the useful action?

If yes, then trim the subject and transfer responsibly for delivery the useful action to the other component

H.1.4 Could a resource perform the useful action?

If yes, then trim the subject and transfer responsibly for delivery the useful action to the resource

H.1.5 Could we trim the subject after it has performed its useful action?

If yes then trim the subject

H.1.6 Could we partially trim any harmful parts but leave any useful parts?

If yes then remove only that part of the Subject that is delivering the harmful action, or only that part of an object that is receiving the harmful action.

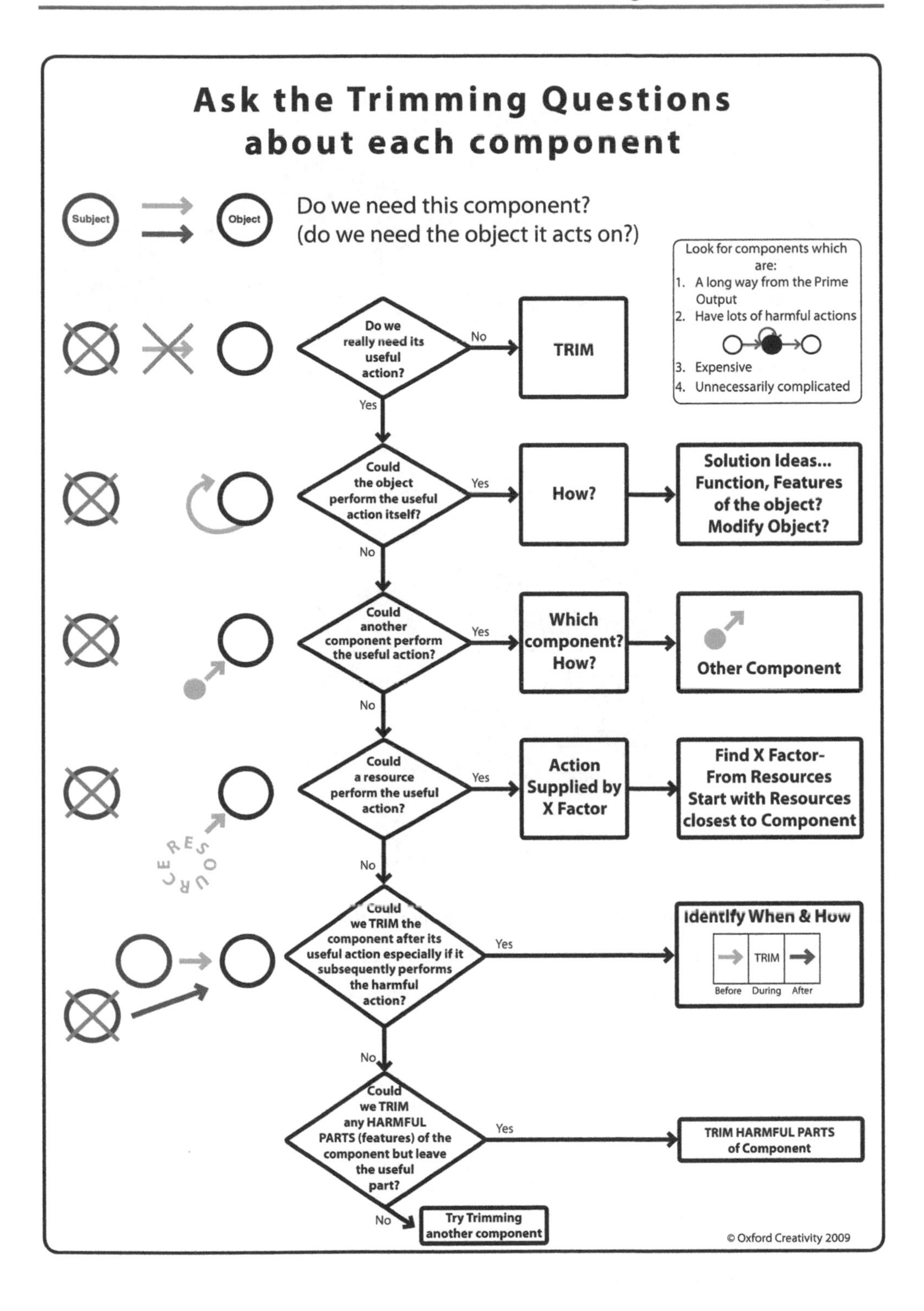
Ask the Trimming Questions about each component
Subject
Object
Do we need this component?
(do we need the object it acts on?)
Look for components which are:
1. A long way from the Prime Output
2. Have lots of harmful actions
3. Expensive
4. Unnecessarily complicated
Do we really need its useful action?
No
TRIM
Yes
Could the object perform the useful action itself?
Yes
How?
Solution Ideas... Function, Features of the object? Modify Object?
No
Could another component perform the useful action?
Yes
Which component? How?
Other Component
No
Could a resource perform the useful action?
Yes
Action Supplied by X Factor
Find X Factor- From Resources Start with Resources closest to Component
RESOURCE
No
Could we TRIM the component after its useful action especially if it subsequently performs the harmful action?
Yes
Identify When & How
TRIM
Before
During
After
No
Could we TRIM any HARMFUL PARTS (features) of the component but leave the useful part?
Yes
TRIM HARMFUL PARTS of Component
No
Try Trimming another component
© Oxford Creativity 2009

H.2 Block the Harm – 11 ways

H.2.1 Counteract the harmful action with an opposing field which neutralises the harm (1.2.4)

H.2.2 Change the object so that it is not sensitive to the harmful actions

H.2.3 Change the zone and/or duration of the harmful action to decrease its effects

H.2.4.1 Insulate from harmful action by introducing a new component/substance.

If it is not necessary for the two components to be in direct contact, then block the harm with a new component (1.2.1).

H.2.4.2 Insulate from harmful action by introducing a substance made from elements of the existing components

If it is not necessary for the two components to be in direct contact but we cannot introduce a new component/substance, then block the harm by introducing a substance which is made from the components, or parts of them, or by modifying them (includes using nothing – voids, bubbles, foam, vacuum, air etc.) (1.2.2).

H.2.5 Protect from harmful action with a sacrificial substance which attracts the harm to itself

Introduce a sacrificial substance to absorb the harm (1.2.3)

H.2.6 Protect from the harm of a necessary strong field by putting the full required force elsewhere

If the required field harms the system, then apply it indirectly, to another linked element (1.1.7).

H.2.7 Protect part of the system from harm

When a powerful, strong field/action is required which needs to be effective only in some parts (maximum/ present in some zones and minimum/absent effects in other zones), add a protective substance to the areas where the force may be weaker or is not required and will cause damage (1.1.8.1).

H.2.8 Reduce harm by using a weaker, minimum field/ action and enhancing it locally where required

If a powerful, strong field/action is required which needs to be stronger in some parts than others (maximum/strong/present in some zones and minimum/weak/ absent in other zones), the field should be minimal and the locations requiring the maximum effects can be made sensitive to the field, enhanced by adding a substance only where needed (1.1.8.2).

H.2.9 Use sub-systems to stop the harm

World Mothers (MMM) advocate that if we try to know the young people who live near us, and their families (as we used to in small village communities); then we can inhibit young people's bad behaviour on the street by telling their mothers.

H.2.10 Super-systems – use the environment to stop the harm

Cultural norms – direct/prevent/correct bad behaviours.

H.2.11 Switch off harm – harms may exist because of certain properties in a system.

Remove the harms by switching them off (e.g. remove magnetism by heating it above its Curie point, or by introducing an opposite magnetic field) (1.2.5).

H.3 Turn Harm into Good – 4 ways

H.3.1 Use the harm to deal with the harm

H.3.2 Use the harm for something good

H.3.3 Add another harm so that the combination of the two harms is no longer harmful

H.3.4 Amplify the harm until it delivers benefit

"Mr Frimley, sir, can I have a word about the motivational artwork..."

H.4 Correct the Harm – 3 ways

H.4.1 After a harmful action eliminate any of its harmful consequences

H.4.2 When there is an inevitable harmful effect of a harmful action, add an anti-action which controls/counteracts the harmful effect

Create prior or simultaneous anti-harm to counteract the harm.

H.4.3 Anticipate and design to eliminate future harms – eliminate a predicted harm

Use something that disappears after carrying out its work or becomes identical to substances already in the system or environment (5.1.3.)

Insufficiency = i

Insufficient functions – how to improve the solution we've got.

When all three of the *Subject action Object* are present and together delivering a function, but it is insufficient, the problem is that the delivered function is too weak, too slow, too uncontrolled etc.

The two basic strategies to overcome the insufficiency are to:

- **i** Improve the subject and/or object or their surroundings to enhance the delivered functions:
  - **i.1** Add something to the subject or object (7 ways).
  - **i.2** Change/evolve the subject and object (10 ways).
- **ia** Enhance the action – for when a field F (action) is missing or insufficient (18 ways).

i.1 Add Something to the Subject or Object – 7 ways

i.1.1 Add something to/inside the subject or object

Improvement is achieved by the introduction of internal additives, which can be permanent or temporary. This provides extra functions of enhancing/providing the required properties (1.1.2).

If no new substances can be permanently added introduce something which later disappears or decomposes (5.1.1.8).

i.1.2 Add something between the subject and the object which enhances/delivers the function

i.1.3 Use the environment

Use the external environment (or substances to enhance/provide the actions/functions (1.1.4).

i.1.4 Add something into the environment/the surroundings of the subject and object

When we need to provide some extra function(s) or change in some way, but it is not easy to change the components, then change the external environment – assuming that the external environment does not contain ready solutions as in **i.**1.3.

Replace the external environment with another one. Decompose it or add new substances into it (1.1.5).

i.1.5 Add something outside/around the subject or object

Introduce external additives into the present substances/components to improve features/properties or provide extra functions such as protection, enhanced controllability, prevention from melting etc. (1.1.3 and 5.1.1.3).

i.1.6 Add something from the environment/surroundings to enhance the function

i.1.7 Use/mbolise the environment/surroundings to enhance the subject and object.

Or if we want an improvement but can't add anything, use the deterioration or decomposition of anything in the system, the elements of the components or the environment to achieve an enhanced function (5.1.1.9).

i.2 Change/Evolve the Subject and Object – 10 ways

i.2.1 Change/segment the subject or object – increase the degree of fragmentation (2.2.2)

Segmentation – When an improvement is needed but we cannot replace components or add anything new, change the object by dividing the elements into smaller units (5.1.2).

i.2.2 Change the subject or object by introducing voids, fields, air, bubbles, foam

Increase the degree of fragmentation to a porous or capillary material that will allow gas or liquid to pass through (2.2.3 and 5.1.1.1).

When we need large amounts but we must use nothing, use voids, foams, inflatable structures etc.(5.1.4).

i.2.3 Improve systems by multiplying similar (or dissimilar) system elements or combining with another similar (or dissimilar) systems (3.1.1)

i.2.4 Increase the efficiency of the system by making it more flexible/adaptable/ dynamic (2.2.4)

i.2.5 Develop/improve the links between the system elements (links can be made more flexible or more rigid) (3.1.2)

i.2.6 System transition – increasing the differences between elements – up to opposites (3.1.3)

High heel shoes for parties – flat shoes for walking home.

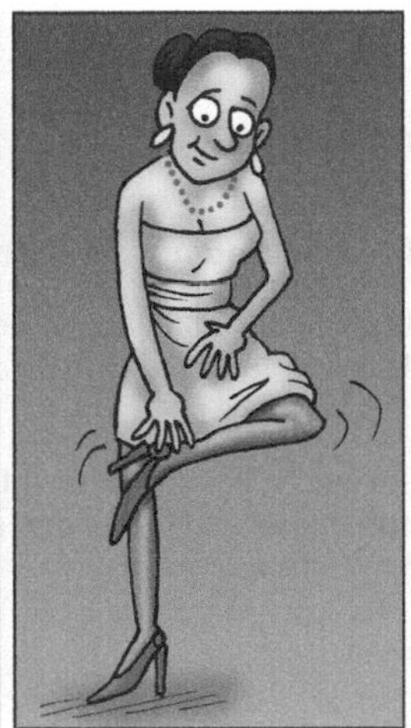

i.2.7 System Improvement by delivering opposite incompatible functions at different system levels (3.1.5)

One feature/function at a sub-system level and the opposite feature/function at the super-system level.

i.2.8 System Improvement: transition function delivery to the micro-Level (3.2.1)

i.2.9 Improve a system's effectiveness and controllability by developing one part of the system or one component to deliver its own, extra functions. (2.1.1)

i.2.10 Improve a system by changing the components/substances to deliver exactly what is needed in time and/or space.

Certain functions are only needed in certain areas or at particular times. Substances which produce them are placed only in the required places and/or for the required times – in place beforehand but may disappear after delivering their useful effect. Changes in space include from a uniform substance or uncontrolled substance, or to a non-uniform substance with a predetermined spatial structure – changes in time could be permanent or temporary (2.2.6)

ia Enhance the Action When a Field F (Action) is Missing or Insufficient – 18 ways

(Getting the right actions and getting actions right.)

To find the right field (action) when there is an object which needs to provide some extra function(s) or change in some way, we need a subject with a field (action) to deliver/complete the function.

Most of the other Standard Solutions are how to improve a solution we've already got, or how to keep the useful functions while dealing with the harmful solutions. This group of Standard Solutions helps us find a solution, particularly when used with EFFECTS – see www.TRIZ4engineers.com.

ia.1 Get an action – if a field F (action) is missing then add an appropriate action/ field (1.1.1)

Add a field to act on the object – When there is an object which needs to provide some extra function(s) or change in some way, but it is not easy to get the required change, then add a field/ action (also add a subject if required) to provide the extra functions. The object is subjected to the field which provides the change or the required extra function(s).

ia.2 Add another action or extra field – if you cannot change the existing system elements (2.1.2)

ia.3 Improve/evolve an action/field by finding a better one

Replace an inefficient or poorly controlled field with a better one (2.2.1). Replace a gravitational field with a mechanical field, a mechanical field with an electric field etc. Or use a field instead of a substance (5.1.1.2).

ia.4 Improve the effectiveness of an action by changing from a uniform action/ field (or uncontrolled field) to an action/field with predetermined patterns that may be permanent or temporary (2.2.5)

ia.5 Improve the effectiveness of the actions within a system by matching or mismatching the natural frequency of the actions (fields) with the natural frequency of the subject (tool) or the object it acts on (2.3.1)

ia.6 Improve the effectiveness of the actions within a system by matching (or mismatching) the frequencies of the different actions/fields being used (2.3.2)

ia.7 To achieve two incompatible actions – perform one action in the downtime of the other (2.3.3)

ia.8 Use actions/fields present in the system to create another field (5.2.1)

e.g. Separate the gas from the liquid in a flow of liquid oxygen by using centrifugal forces to move the liquid to the walls while the gas collects near the centre of the pipe.

ia.9 Use actions/fields that are present in the environment (gravity, ambient temperature, pressure, sunlight) (5.2.2)

ia.10 Achieve an extra action by using something already present in the system, or in the environment, as the source/provider of the extra fields/actions which can act as media or sources (5.2.3)

ia.11 Use excessive action

If a small amount of something is required but it is not easy to get the right degree of an action or exact amount of a substance change, use an excess of what is required and remove the surplus. Surplus substance is removed by an action/field and surplus action/field is removed by a substance (1.1.6).

ia.12 Use a small amount of a very active additive (5.1.1.4.)

ia.13 Concentrate the additive at a specific location (5.1.1.5)

ia.14 Introduce the additive temporarily (5.1.1.6.)

ia.15 Use a copy or model of the object in which additives can be used, instead of the original object, if additives are not permitted in the original (5.1.1.7)

ia.16 Improve an action/field by changing the phase of the existing field/substance (5.3.1)

ia.17 Achieve an action by using the phenomena which accompany phase change (5.3.3)

Water placed in holes in rocks- cracks it when it turns to ice.

ia.18 Achieve an action with dual properties by using substances capable of converting from one phase state to another (5.3.2)

Replace a single-phase state with dual-phase state (5.3.4).

Induce the dual-phase state by creating interaction between the parts of the system (5.3.5).

Classical TRIZ: Substance-Field Analysis and ARIZ

The earlier chapters covered many of the classical TRIZ tools including the contradiction toolkit, trends of evolution, ideal outcome, ideality, resources, creativity tools, knowledge tools, standard solutions, time and scale, all the TRIZ creativity and brain prompting tools each with their own processes and algorithms to step through problems. There are two further important parts of classical TRIZ: Substance–Field Analysis and ARIZ (Algorithm of Inventive Problem Solving). These are used mostly by traditional TRIZ practitioners to solve particularly challenging problems. They are rigorous and useful but not as widely used as the other TRIZ tools as many engineers often find them quite hard to understand, master and apply to their problems.

Both Substance–Fields and ARIZ may be partly responsible for TRIZ being seen as a difficult toolkit, and recent developments in TRIZ have resulted in the replacement of Substance Field Analysis by TRIZ Function Analysis by many TRIZ users, and ARIZ by simpler TRIZ algorithms. However both Substance–Field Analysis and ARIZ have an important role in tackling certain problems and no one could claim to understand all of TRIZ without appreciating, understanding and occasionally using these original and powerful TRIZ tools. When faced with really tough problems which have not been solved by the other TRIZ Tools the TRIZ engineer has these two rigorous approaches to apply to the problem.

A **Substance–Field model** (also known as Su–Field) is used to describe an exact problem, and give us detailed analysis of problems within existing technical systems, and the 76 Standard Solutions are then applied to correct problems with functions.

TRIZ for Engineers: Enabling Inventive Problem Solving, First Edition. Karen Gadd.

ARIZ is a step-by-step process which helps narrow problem information to an accurate problem definition. It then guides us through resource hunts and a focused final problem-solving stage. (ARIZ has many versions and one of the more popular and recent ones is explained below to show its logic and usefulness.)

ARIZ and Substance–Fields in Altshuller's Development of TRIZ Tools

In the 1950s as Altshuller uncovered the TRIZ problem-solving tools, it was the 40 Principles, with their unique power to solve contradictions, which were first widely used to solve problems within systems, and they have been popular and appreciated since that time. Other TRIZ tools were developed for problem understanding and solving including Ideal Outcome, Standard Solutions, resources and trimming. Development of where and when to use these tools became the first versions of ARIZ. This includes identifying the essential contradictions in a system as an early problem-solving step, which some find difficult and are uncertain how to get this right. This led to the searches for TRIZ solutions without contradictions.

By the mid-1970s Altshuller had developed the alternative (and complementary) approach to solving problems with Substance–Field Analysis and the 76 Standard Solutions, which do not necessarily require the uncovering of contradictions to solve problems, but offer solution directions for incomplete or insufficient systems, harms, measurement/detection and system evolution. Whether using the Substance–Field or the Contradiction toolkit it was found in practice that both approaches work well, and as they overlap they often take us to similar solutions. Altshuller's original Substance-Field Analysis is powerful for solving system problems as it focuses in on the problem area, and helps us locate and concentrate initially on just one problem function. This is by building a problem model which can then be corrected by applying the relevant 76 Standard Solutions. This method forces a zooming in right to the essential problem function (which takes us away from all irrelevant and distracting detail) and the Standard Solution will then suggest solutions to the revealed problems.

As Altshuller and his teams of engineers worked on the development of the TRIZ tools and processes there was a fairly smooth evolution except for ARIZ, which has had many versions and is still disputed and worked on by elements of the TRIZ community.

The tools were uncovered and developed roughly in the following sequence:

- 40 Principles 1946–1971
- ARIZ – Algorithm of Inventive Problem Solving, 1959–1985
- Separation Principles 1973–1985
- Substance–Field Analysis 1973–1985
- Standard Solutions 1975–1985
- Natural Effects (Scientific Effects) 1970–1980
- Patterns of Evolution 1975–1980

Substance–Field Analysis

Altshuller created this precise system for classifying problems within systems and their solutions, which is variously called Substance Field, S-Field or Su-Field Analysis. It uses simple triangles to focus in on each problematic function and the 76 Standard Solutions to solve them. Each Substance–Field is a problem model and denoted as a triangle representing the minimal technological system, and consists of one substance S_2 acting on another substance S_1 through a field F. Revealed problems can be solved by the relevant standard solution which in turn helps us consider ideas from the world's knowledge bases.

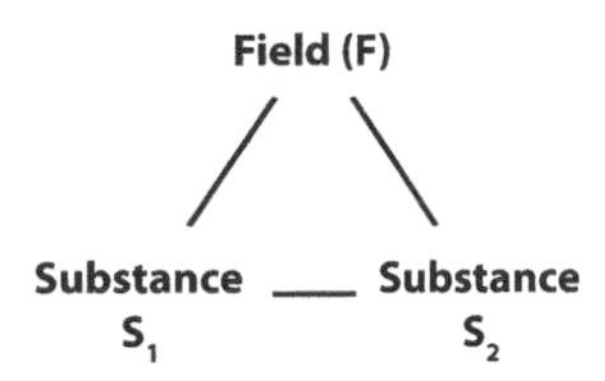

In his books *Creativity as an Exact Science* and *ASAIA* Altshuller (pp. 49–51) refers us to problems with contradictions but shows that problems within systems can be denoted by the interaction of two substances. He calls the solutions 'inventions' rather than the everyday engineering 'problem solutions', but shows how to solve various problems by getting substances and fields acting together to give us everything we need for solutions, by focusing on the exact moment and place where the problem occurs.

Substance–Field Model analysis (also Function Analysis) is very powerful for engineers because it takes us to an exact problem concept model with no confusing or irrelevant details. With small amounts of practice it can become a quick and very exact modelling tool for problems and helps with invention, problematic systems and process problems.

The system analysis and problem-solving tools of Function (or Substance–Field) Analysis and the Standard Solutions are seen by some as an evolution and replacement of the original Contradiction Toolkit and the Trends of Evolution. For many TRIZ users the 40 Principles and 8 Trends are initially easier to learn and share with others (the trends are a set of tools which deliver a conceptual vision for future, perfect products and are less concerned with problem solving as such). They are a very good way of starting to learn and use TRIZ and offer a viable way of dealing with some system problems. To others the uncovering of contradictions can prove difficult as there is a need to make bigger mental leaps to get to the solutions; these engineers often prefer the TRIZ system analysis methods which offer focused approaches for solving problems by analysing the system or a part of it which lead us straight to conceptual answers in the Standard Solutions. This takes us through a systematic process with smaller steps, and is therefore easier to follow and more certain of a result.

TRIZ Function Analysis and the Oxford Standard Solutions are a simpler substitute for Substance–Field Analysis and the 76 Standard Solutions. The logic of the two is similar and it takes us to the same powerful answers in the Standard Solutions. Obviously everyone should be encouraged to learn a wide range of tools to be able to choose and develop the TRIZ toolkit which best suits their needs and problem-solving style. Most who learn TRIZ and master Function Analysis are able to use them on real problems (see BAE Systems SRES Case Study) and those who master Substance–Fields and ARIZ additionally use them for the occasional more difficult company problems.

Building Substance–Field Models

The minimum technical system is represented in the relationship between the elements and their field by the mathematically elegant triangle.

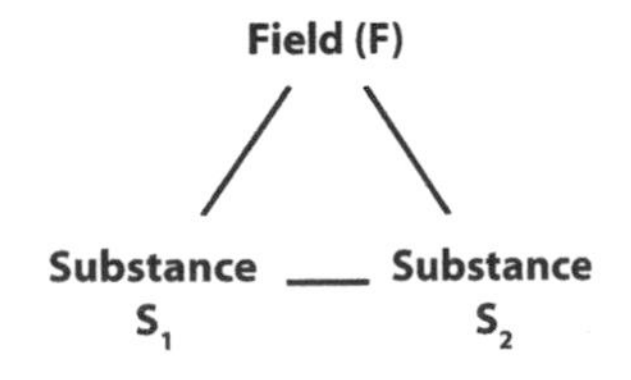

One substance acts on another substance, to provide a function. Functions can provide benefits or harms. Functions can be good/sufficient, insufficient, missing or harmful. The function is modelled as triangles with problems represented graphically as different types of lines or even missing lines to show precisely what is right and what is wrong with it – once the problem type is identified then the relevant Standard Solution can be located and applied to correct any problems, by changing, removing or adding substances or fields. Substance-Field Models take us to the focused analysis of functions, and getting the functions right is the key to problem solving and increasing Ideality.

$$\text{Ideality} = \frac{\text{Benefits}}{\text{Costs \& Harms}} = \frac{\text{Functions}}{}$$

Benefits
Good functions
Insufficient Function
Missing Function
Harmful Function
Harms
Costs

Substance–Field Models are used to identify and solve inventive problems and help us to define the problem in precise physical terms, by identifying the two essential objects/substances S_2, S_1 and how they interact by a field F. (S_2 acts on S_1 via a Field F). This enables us to draw a Substance–Field triangle of the localized problem zone which is the simplest model of any technical system. When more complex models are needed they can be made by linking more triangles together (called nodes). These triangles have a coded language to convey detailed problem information: a single complete Substance–Field (effective – no problems) has three straight lines; we denote any insufficiency by dashed lines; harms are shown as wavy (and/or red) lines; there are grey arrows for directionality.

The Transformation is denoted by an Operator to show when a Substance–Field Model has been made complete or has used, generated, absorbed the energy between them and has thus been improved to remove any problems or insufficiencies. The diagram below shows how the operator marks the transformation from an incomplete to a complete Substance–Field Model.

Fields/actions – The two elements interact, react and communicate through a field which could be anything such as:

- **mechanical** – vibration, leverage, agitation
- **thermal heating** – melting, boiling, friction
- **chemical** – glues, oils, solvents
- **electrical** – charges, currents, discharge
- **magnetic** – induction
- **electro-magnetic** light, radiation, or combinations such as
- **exothermic** (chemical and thermal).

(They are sometimes called METHCHEMEM = Reminder of the many fields and types.)

In TRIZ the term 'Field' has a wider definition than that of traditional physics and covers not only fields like mechanical, thermal, electric, magnetic, and chemical, electromagnetic, gravitational and nuclear fields but also human interactions to substances like smell, vision, touch (pain) and emotion (especially in management problem solving). Note that the field is *not* the same as the action as described in the Function Analysis chapter. The action is the plain-language description of the interaction between S_1 and S_2 – the sort of word(s) that we would see on the action arrow in an S-a-O diagram (e.g. cleans, informs, moves etc), whereas the field (F) is the energy that's associated with the interaction (e.g. mechanical, chemical, light of wavelength 500nm etc). Field = energy required or otherwise associated with the interaction between the two substances.

Definitions for Substance–Field

- Substance = any object no matter how complex referred to as S_1, S_2, S_3 etc.
- A Substance can be for whole systems, subsystems or single objects, tools or articles, e.g. submarines, screws, cables, engineers – all are substances.
- Substance S_1 is changed, processed, converted, discovered, inspected etc.
- Required action is accomplished by Substance S_2
- Field F provides the energy, force that guarantees the reaction of S_2 on S_1 (or their mutual interaction).
- These three active agents are necessary and sufficient for the result required of the problem.
- By themselves the field or substances can produce no effect. Substance S_1 needs an instrument S_2 and energy Field F.

Four Basic Substance-Field Model Types

1. Effective complete system (shown here)
2. Incomplete system
3. Ineffective/ insufficient complete system
4. Harmful complete system

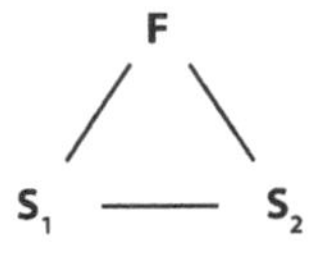

Incomplete Substance–Field Model (no field/ interaction between elements) transformed to a complete Substance–Field Model by adding a field

Insufficient Substance–Field Model – Transformed to a Complete Substance–Field Model by adding Another Substance

Harmful Substance–Field Model

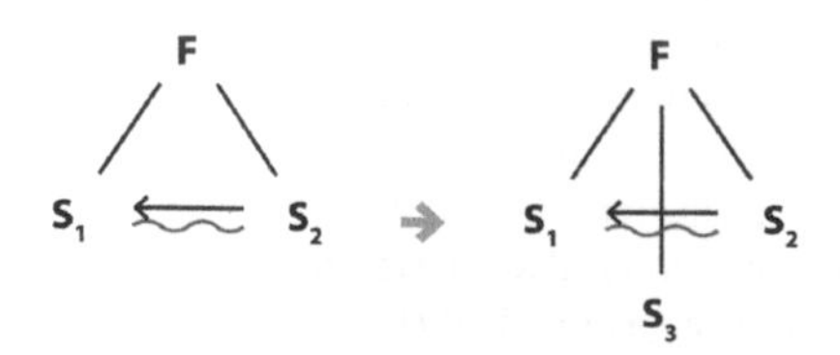

Substance–Field Models show us the exact problem with the function and help us identify which of the Standard Solutions will guide us to solve our problem, whether something is missing, insufficient, harmful, in need of development or requiring measurement or detection. When we build a Substance–Field Model of our problem (or system) it is to help us to understand the essential problem, and after identifying our problem type, it then shows us which of the 76 Standard Solutions will help us and for this we need to understand the Classification of 76 Standard Solutions. The 76 Standard Solutions are divided into five classes and their various sub-classes according to the type of typical engineering problems they solve as shown below.

76 Standard Solutions and Accessing Them with Substance–Field Models

Substance–Field and the whole method with ARIZ were built up from TRIZzing TRIZ. Altshuller noticed that certain problem types were solved by powerful combinations of the 40 Principles and some of the effects (such as ferromagnetism). Altshuller then set out to combine these solutions with the 40 Principles and the 8 Trends into one list – the 76 Standard Solutions. He divided them into five classes of Standard Inventive Solutions:

- Class 1: Building and Destruction of Substance–Field Models
- Class 2: Development of Substance–Field Models
- Class 3: Transition to Super-system and Micro level
- Class 4: Standards for Detection and Measuring
- Class 5: Standards on Application of Standards.

Class 1: Building and Destruction of Substance–Field Models

How to do something we want to do, completing an incomplete system and eliminating harmful actions.

Two Subclasses Containing 13 Standard Inventive Solutions

This class of Standard Solutions solves problems by constructing or destroying the Substance–Field Model, if it is incomplete or has Harmful Functions.

1.1 **Synthesis or building a Substance–Field.** Something is missing – add a substance or field.

1.2 **Destruction of a Substance–Field.** Removing/eliminating harm by blocking it, switching it off, counteracting or drawing it off.

Class 2: Development of Substance–Field Models

Development of Substance–Field Models for improving the efficiency of Systems by introducing minor modifications.

Four Subclasses and 23 Standard Inventive Solutions

The major recommendations from this class are: use chain Substance–Fields, use double Substance–Fields, segmentation and dynamization, rhythm coordination and using magnetic substances.

2.1 **Add new substances or fields to improving insufficient actions**

2.2 **Change the substances to enhance the effectiveness of the action**

2.3 **Enforcing matching rhythms**

2.4 **Ferromagnetic field models**

Class 3: Transition to Super-system and Micro Level

For finding solutions in the super-system or sub-systems, and for improving insufficiency, suggests ways to apply the trends of evolution to the system.

Two Subclasses Containing 6 Standard Inventive Solutions

3.1 **Simplicity–Complexity–Simplicity (Mono–Bi–Poly)** – Transition to super-system and to bi and poly systems.

3.2 **Transition to the sub-systems**

Class 4: Standards for Detection and Measuring

Solutions for problems involving measurement and detection

Five Subclasses and 17 Standard Inventive Solutions

The major recommendations of this class are to change a system so that there is no need to measure/detect by the following strategies:

- Measure a copy
- Introduce a substance that generates a field (introduce a mark internally or externally).

4.1 **Indirect methods**

4.2 **Create a measurement system by adding elements or fields**

4.3 **Enhance a measurement system**

4.4 **Measure ferro-magnetic field**

4.5 **Direction of evolution of measurement systems**

Class 5: Standards on Application of Standards

These are sophisticated approaches to system simplification which recommend how to introduce new Substances or Fields or scientific effects more effectively than in the four previous classes

Five Subclasses and 17 Standard Inventive Solutions

The basic strategies recommended are:

- Instead of a substance introduce a field.
- Instead of a substance introduce a void.

- Introduce a substance for a limited time.
- Introduce a little bit of a substance, but in a very concentrated way.

5.1 **Introducing Substances** – indirect methods for introducing substances under restricted conditions

5.2 **Introducing Fields under restricted conditions**

5.3 **Phase transitions**

5.4 **Clever use of natural phenomena**

5.5 **Generating higher or lower forms of Substances**

Guide to which Class of Standard Solutions to apply to problem

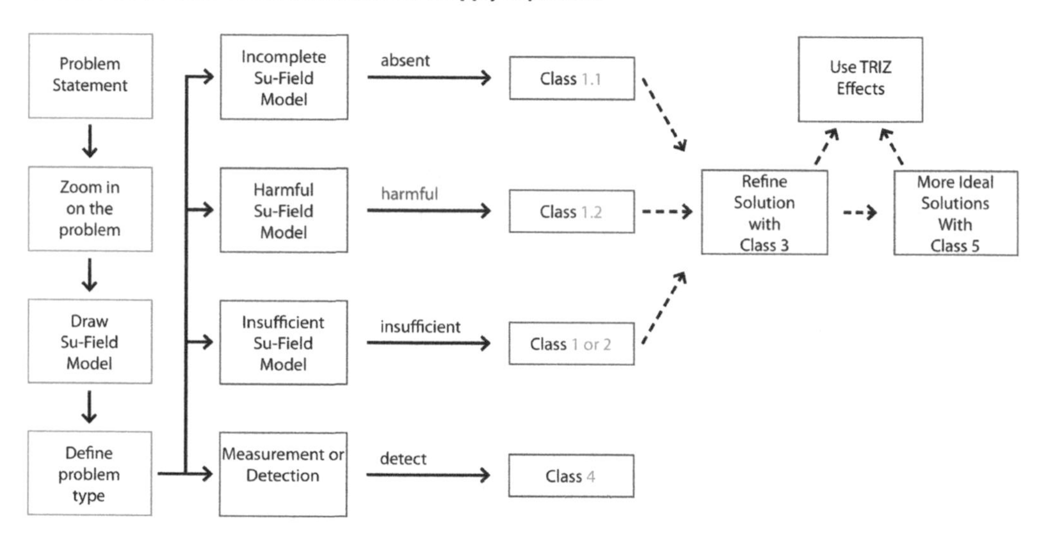

Simple Steps for Applying Substance–Field Model Analysis to Problems

1. **Problem statement** – describe the system and the problematic part of the system.
2. **Zoom in on the problem area** and identify the problem components (substances) and any relevant interactions(fields) including with the environment.
3. **Draw a Substance–Field Model of the problem**
4. **Define the problem type:**
   - missing elements? (incomplete Substance–Field Model? Class 1.1)
   - harmful elements? (do we need to destroy the Substance–Field Model we've got? Class1.2)
   - insufficient elements? (incomplete Substance–Field Model? Classes 1 or 2)
   - Do we wish to predict/activate the evolution of the substances? (Class 3)
   - Is it a detection or measurement problem? (Class 4)

5. **Select the appropriate Standard Solution(s) from the suggested class**
6. **Add in relevant knowledge and experience to transform the conceptual solution to a practical one** (Use the problem-solving individual's and team's own knowledge, company knowledge, industry knowledge and beyond that to patent databases etc. available on the internet – searching for analogous and relevant solutions to the problem functions.)

Simple Example of Substance–Field Analysis Using the Standard Solutions

A **'Jet of Water' from a hose is used to clean the Patio stones.** There could be a number of problems with this system – the water could clean insufficiently or it could create harmful effects (destroying the surface). Cleaning could be excessive or absent. This system has many potential problems

1. **Problem statement** – describe the system and the problematic part of the system.

   The water jet does not clean the patio stones.

2. **Zoom in on the problem area** and identify the problem components.

   Water does not move the dirt from the patio stones.

3. **Draw a Substance–Field Model of the problem**

   S_2 = *water*, S_1 = *dirt*, *Field* = *moves*

   The interaction between the water and the dirt is insufficient.

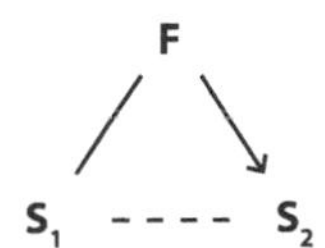

4. **Define the problem type**

   Problem Type = Insufficient – looking at the 76 Standard Solutions and at the Standard Solutions Chart below it suggests Classes 1 or 2.

5. **Select the appropriate Standard Solution(s) from the suggested class**

   Standard Solution 2.1.2. Add another field – if you cannot change the system elements to improve it then add another field – Double Substance–Field Model – Improve a system's effectiveness and controllability without changing the elements of the existing system by adding a second field to S_2.

6. **Add in relevant knowledge and experience to transform the conceptual solution to a practical one**

   From the use of various vibrations, water-based cleaning systems which have been around for many decades such as electric toothbrushes we might suggest a jet washer which adds vibration to the ordinary jet of water from a hose pipe and makes it a more effective cleaner of the stones. This could also have been prompted by Solution 2.2.5 'A Substance–Field Model can be enhanced by replacing homogenous or unstructured fields with either heterogeneous fields or fields of permanent or variable spatial structure', i.e. we have put a time-based 'structure' into the field by pulsing it.

   Or we could also look at a double Substance–Field by adding a chemical field as well as the mechanical field of the water – e.g. by the inclusion of bleach, detergent or some other chemical cleaning agent as suggested by jewellery cleaners which have also been available long before jet washers were invented.

ARIZ – An Algorithm for Inventive Problem Solving

Using ARIZ on Challenging Problems

ARIZ is a step-by-step problem-solving process which mandates certain TRIZ tools for certain steps in its process. This addresses one of the difficulties in using TRIZ of knowing which tools to use for a particular problem, and where and when to use them. This helps TRIZ beginners facing particularly complex problems, as ARIZ helps tackle any problem of any type and difficulty. Developing ARIZ was itself problematic, as it set out to prescribe one process for using TRIZ tools which will cover any problem, at any stage of development or use, of any level of difficulty or of any problem type. When Altshuller recognized the need for ARIZ his aim was that it could be used to guide any TRIZ user to the solution of any problem. He called this the 'Algorithm of Inventive Problem Solving' (Algoritm Reshenia Izobretatelskih) or ARIZ. Not unsurprisingly for such an ambitious aim, ARIZ developed slowly and through a series of different versions. These were named after the year in which they were published, such as ARIZ-69 or ARIZ-71.

The usefulness of ARIZ comes from its power as a universal 'one size fits all' algorithm but this universal fit has been achieved by going for something big enough to fit around even the largest, most complicated and most intractable problem. Thus for big, difficult or complex problems ARIZ may be a good fit; but for some smaller or simpler problems it can be too rigorous, too complicated, too long winded, and far from ideal in that it is an excessive approach and much more than is needed, as it involves a large number of small steps, and can have us inching towards a solution, when we could do it with bigger strides. (ARIZ can be too much – a sledgehammer to crack a nut.) The strength of ARIZ is that it segments the problem understanding into many small but highly directed steps, which leaves little room for the user to be led astray by irrelevant detail, psychological inertia or be distracted by blind alleys.

TRIZ usually offers fast, effective problem solving and this contrasts it to other toolkits which prescribe many steps. ARIZ is closer to other toolkits and evokes many different reactions from users. Some find its highly systematic and methodical approach suits their style and approach, while others can feel daunted (or even bored) by its sheer size and complexity. Many engineers are just too busy to try something with a steep learning curve that takes such a long time to master and follow. After successful problem solving with ARIZ, some may feel that it is not for them, but will still keep it in reserve as a useful addition to the TRIZ toolbox, to be used on difficult problems which have resisted all other attempts to solve them. Thus ARIZ can be kept for those 'once a year' problems while TRIZ thinking and the many straightforward tools can be used everyday by engineers to deliver powerful insights on simple problems and situations.

The version of ARIZ that is presented here is based upon ARIZ-85C and is a simple lesson with its essential steps and logic. There are many, more comprehensive excellent guides to ARIZ which give much more detail and many examples and TRIZ Master Victor Fey offers particularly good guides to ARIZ in his innovation and TRIZ books, and the website TRIZ.co.kr offers a free guide to ARIZ.

Overall Structure of the ARIZ Algorithm

The ARIZ algorithm can be broken down into a sequence of five distinct steps, shown below:

Step 1 Problem Definition

Step 2 Uncovering of System Contradictions

Step 3 Analysis of System Contradictions and Formulation of Mini-Problem

Step 4 Analysis of Resources

Step 5 Development of Conceptual Solutions

begin
1. Problem Definition
2. Uncovering of System Contradictions
3. Analysis of System Contradictions and Formulation of Mini-Problem
4. Analysis of Resources
5. Development of Conceptual Solutions
end

Looking at the flow diagram, it is immediately apparent that most of the activity in ARIZ is directed towards understanding the nature of the problem and the resources that are available to solve it. Only at the final stage does the algorithm call for the generation of solutions. In practice, however, the user will find solution ideas ('Bad Solutions') emerging at any stage of the process and this is always useful and to be encouraged. As we have seen in earlier chapters, all ideas which arise during problem understanding contain powerful insights of both needs and solutions to fulfil them, and should be captured and recorded in a Bad Solution Park. When using ARIZ, such useful solutions naturally arise during the detailed understanding of the problem progresses as we work through the ARIZ algorithm.

Step 1 ARIZ – Problem Definition

In this first stage of ARIZ we refine and clarify our understanding of the system in which the problem occurs and the objective of the problem solving.

(i) Identify the Primary Function and Output of the System

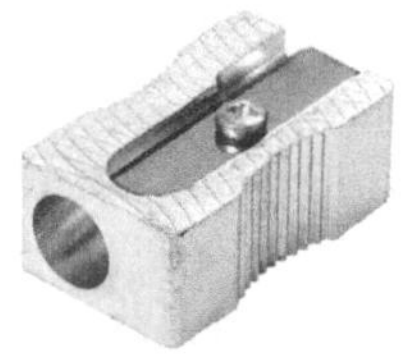

All systems exist for a reason, to fulfil a purpose, to perform some *primary function* which delivers some *primary benefit/output* of the system.

ARIZ begins by identifying the primary function and output of our system. For example, for a pencil sharpener:

1. PROBLEM DEFINITION

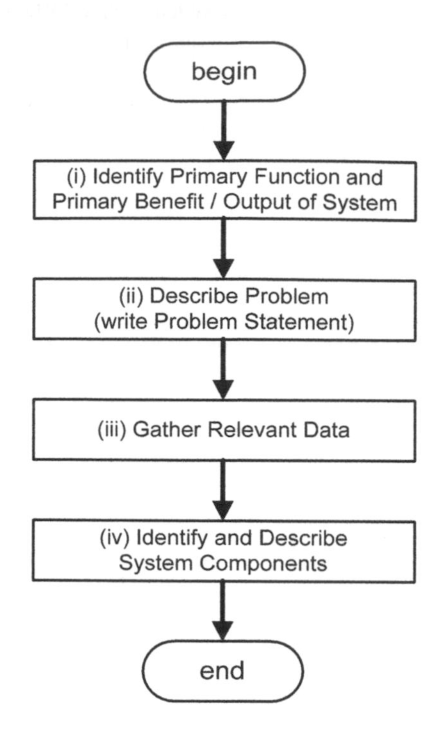

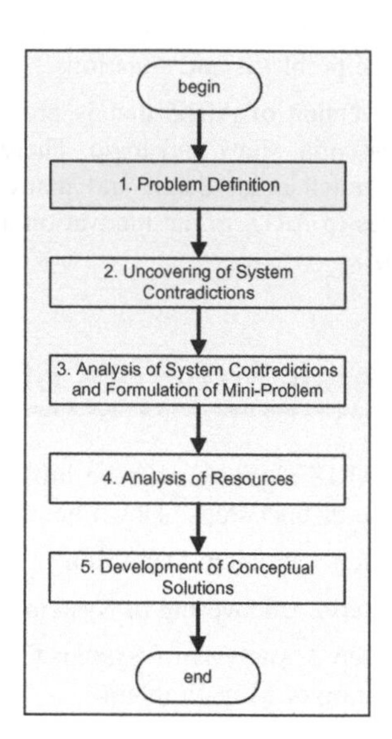

Name of System:	*Pencil Sharpener*
Primary Function of System	*To transform blunt pencils into sharp pencils that are suitable for writing or drawing*
Primary Benefit/Output of System:	*Sharpened Pencils*

(ii) Describe the Problem – Write Problem Statement

This is to generate a simple, clear, short and concise statement of the problem (agreed and understood by all interested parties). It is important that the problem statement does imply any particular solution or impose any unnecessary constraints. For example:

For our pencil sharpener with the *Primary Benefit/Output* = sharpened pencils. Our problem solving will be aimed at a particular problem such as the messiness from small pieces of wood

Problem Statement = Mess from wood shavings

For other simple problems such as a leaking flange joint:

The *Primary Benefit/Output* of a pipe flange joint = to allow easy assembly or disassembly of the pipes while maintaining leak proof flow of the fluid in the pipe

Problem Statement = Pipe flange is leaking

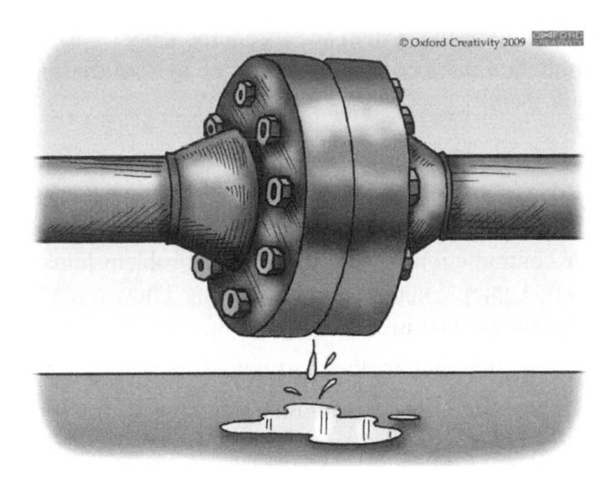

(iii) Gather Relevant Data

We need to collect together all the relevant data on the problem. Often there will be a 'problem owner' – a person or a team who need to have the problem solved. A discussion with the problem owner is often the best way of collecting together the relevant data. During this process it may become apparent that the problem statement needs to be updated or refined.

(iv) Identify and Describe System Components

The output from this step will be a list of the components of the system together with a brief description and drawings of the system and its function or purpose of relevant components. The Time and Scale (9-Boxes) and Functional Analysis tools are both useful to help understanding at this stage especially for large or complex systems.

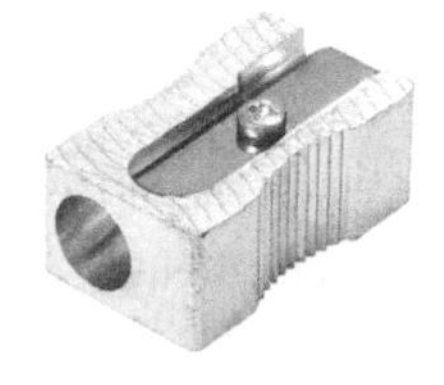

The TRIZ rule for simple language is important, and we need to describe each of the system components using the clear, simple accurate language (as non-technical as possible and with no acronyms or abbreviations). This is to avoid the traps of psychological inertia that can be inherent even in the simple names of systems. For example, the term 'Ice-Breaker' evokes an image of a ship that is able to move through ice covered water by breaking up the ice in its path. However, the simple purpose of the system is to move between two points separated by ice-covered water, and then it becomes immediately apparent that 'breaking-ice' is merely one possible *solution* but that the name 'ice breaker' has a solution contained in its name. Also when we are describing what a component delivers we should remember that it usually delivers more than one function. For example, in the case of our simple pencil sharpener we might have:

Component	Function
Blade	Removes unwanted wood and lead from the pencil
Body	Holds the pencil in the correct position relative to the blade. Allows the pencil to rotate Enables the user to hold the pencil sharpener, resisting the rotation force resulting from the rotation of the pencil Helps to hold the blade in the correct position Protects user from blade
Screw	Holds the blade securely to the body Allow the blade to be removed for sharpening or replacement

Once we have some understanding of the system and its components we can move to the next stage of ARIZ to look for the contradiction pairs contained within the system design or the problem and deal with the problems by solving them.

Step 2 ARIZ – Uncovering of System Contradictions

Solving Conflicts to Form Contradiction Pairs

In ARIZ we are looking for contradictions at the heart of the problem (close to the Primary Output), which we need to express in two complementary opposite forms, known as a 'contradiction pair'. There are two kinds of contradictions described in TRIZ:

- we improve part of the system but another part gets worse (*technical contradiction*)
- we want functions/features and the opposite functions/features (*physical contradiction*)

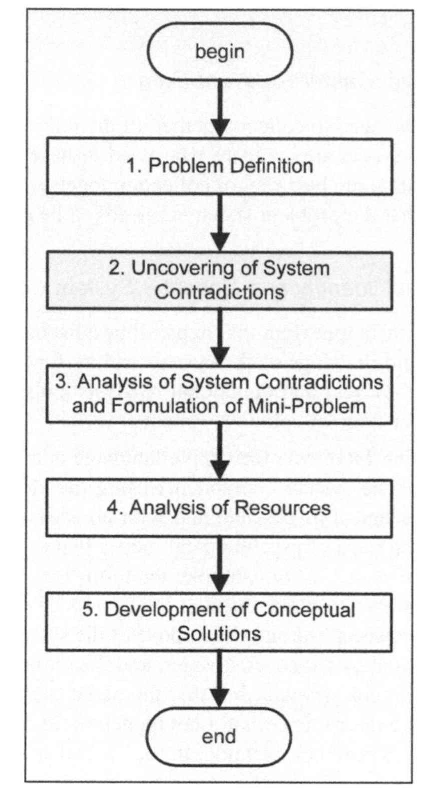

A contradiction pair is a physical contradiction made up of two technical contradictions (two opposite outcomes which each offer solutions delivering more benefit or less harm and less cost – but each solution makes something else get worse).

Typical contradiction pairs are opposite outcomes:

- We achieve the primary benefit, but we get harm x or incur cost y.
- We avoid harm x or cost y, but we don't get the primary benefit.

or

- Component delivers benefit, but with unwanted harms or costs.
- Component is absent (as is its benefit), but we avoid its associated harm or cost.

or

Component has useful property, but with unwanted harms or costs.

Component has the opposite useful property, but with unwanted harms or costs.

Conflicts need to be expressed in these two opposite ways (a contradiction pair).

At the end of this stage we will have identified one or more conflict/contradiction pairs. For example:

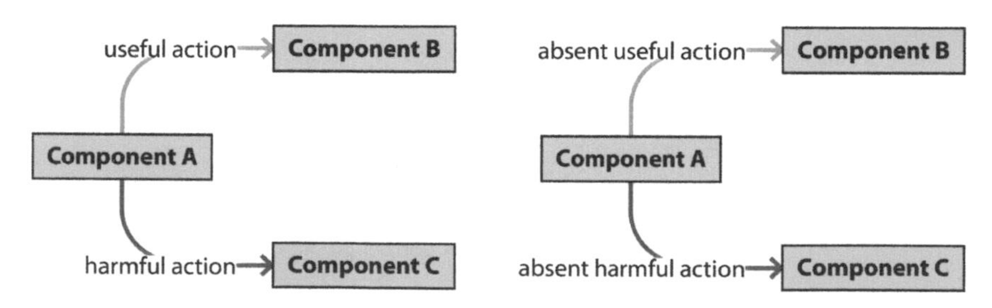

This is only an example – there can be many other forms, such as Component A present and Component B absent – there is a very wide range of possible contradiction pairs. Note that there could be a number of these contradictions, but in each case the contradiction is expressed in just two ways (a conflict or contradiction pair) which represent the 'extremes' or opposites of the situation, i.e. to formulate these we describe a situation which gives us what we want, which may well create problems. We keep well away from compromise – we push *away* from the compromise point in the middle to get the most extreme form of the contradiction which gives us a good outcome. We often make the system 'ideal' (maximum benefit *or* zero harm *or* zero cost) in a deliberate way to create the contradiction pair which both give us what we want.

Note also that the contradictions are expressed in functional terms relating to the components and they are not described as parameters; for example, Hot Oven Burns Pie and Hot Oven Cooks Pie rather than High Temperature Burns Pie.

When we have Opposite functions/features already as a contradiction pair they are fairly easy to formulate as opposites or a contradiction pair:

- Component A needs a feature which delivers functions or reduces harms or costs but which has consequent unwanted harms or costs.
- Component A also needs the opposite feature to deliver the functions or reduces a harm or cost but which also causes unwanted harms or costs.

For example a Ha-ha solves the problem of a barrier with a contradiction pair which needs *to be there* to keep the sheep out but it spoils the view *and* needs *not to be there* – doesn't spoil the view but it doesn't keep the sheep out.

Age old solutions to deliver opposite benefits (a barrier that is there and NOT there)

Technical Contradictions when we improve one part of the system and another part gets worse, takes a little more thought to put into a contradiction pair of opposites. **Achievement of Primary Benefit creates problems**

To achieve the primary benefit (e.g. easy to carry/light), we get harm x or incur cost y (e.g. weak).

We avoid harm x or cost y (e.g. weak) but then we don't get the primary benefit (e.g. easy to carry/light).

We want our table LIGHT and NOT LIGHT.

or

The presence of useful functions causes downsides.

We often think of solutions which have serious downsides

Taking to extremes helps us see the good and bad about solutions

We achieve the primary benefit *safety* but we get harm x or incur cost y.

We avoid harm x or cost y, but we don't get the primary benefit *safety*.

We want safety to be present, and we want safety to be absent – as it has many downsides.

We choose a contradiction at the end of this stage.

Step 3 ARIZ – Analysis of Systems Contradictions and Formulation of Mini-Problem

This point contains a major decision branch in the algorithm – the choice between two strategies of how to deal with the contradiction that was selected during Stage 2. Whichever route is now chosen, by the end of this stage we will have formulated a 'mini-problem', and we will have the definition of a new problem that, if solved, will result in the solution of the original problem. If the contradiction involves an 'auxiliary tool', i.e. a system component that is not directly concerned with delivering the prime benefit of the system, then the chosen strategy for eliminating the contradiction will be to trim the 'auxiliary tool' and our mini-problem will be a trimming proposal. Otherwise, the strategy will be to tackle the contradiction and, in this case, our mini-problem will take the form of a Standard Solution.

Analysis of System Contradictions and Formulation of Mini-Problem

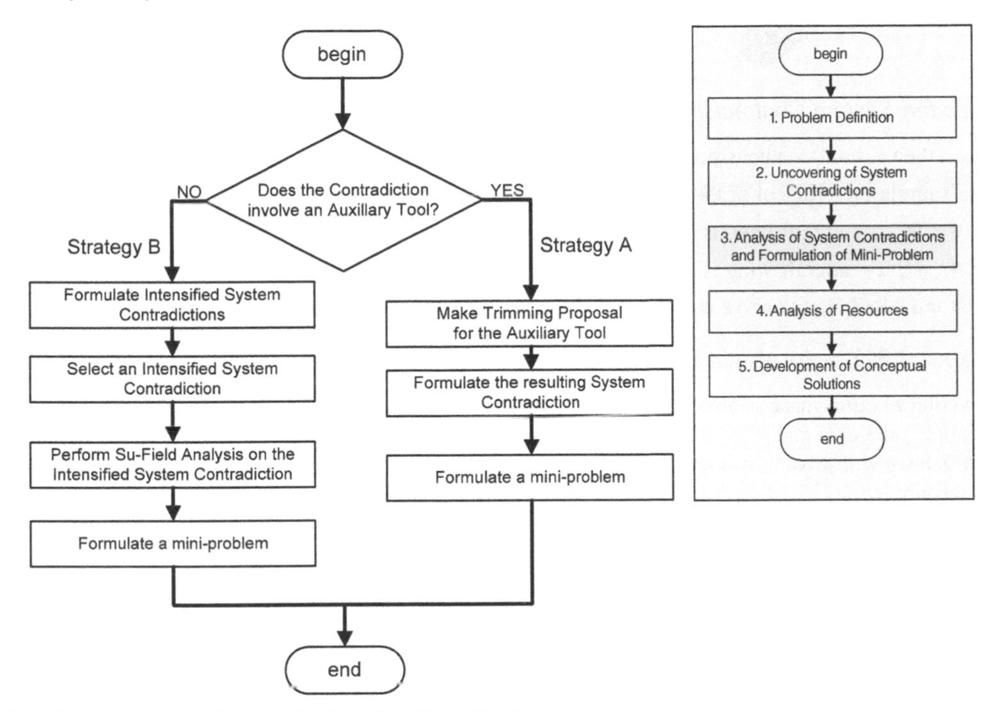

Testing for the Involvement of an Auxiliary Tool

The term 'auxiliary tool' means a component that is not directly concerned with the delivery of the primary benefit/output of the system. In complex systems it is not unusual for a large proportion of system components to be 'auxiliary'.

Does the Contradiction Involve an Auxiliary Tool?

If yes – then trim auxiliary tool and formulate the resulting system contradiction and formulate a mini contradiction. For example, an ice-cream scoop has an extra hinged spoon inside the scoop to help eject the ice cream. This has the contradiction of being useful to eject ice-cream but harmful as it adds complication and cost and is hard to clean and therefore unhygienic. This is an auxiliary tool that could be trimmed. This would come from a contradiction pair such as:

- TC 1 – a deep scoop provides a perfect, large ball of ice-cream but needs something such as an extra hinged spoon to remove the ice-cream from the scoop.
- TC 2 – a shallow scoop provides less ice-cream in a blob, but is easy to remove from the scoop.

What ARIZ should solve is the perfect shape, delivering the correct amount of ice-cream in a simple scoop with easy removal of ice-cream (i.e. give us the primary output of the correct quantity of ice-cream easily served in a single scoop).

This would lead to finding ways of the deep scoop itself being modified so that it ejects ice-cream easily – for example being made flexible.

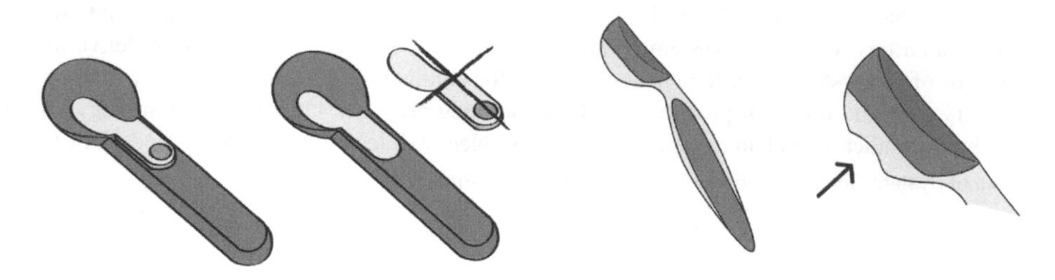

Does the System Contradiction Involve an Auxiliary Tool?

If no– then formulate intensified system and select an intensified system contradiction.

For example the problem of the perfect size of aircraft wings:

- TC 1 – aircraft wing is big for high lift is but is heavy.
- TC 2 – aircraft wing is small and has little weight but provides low lift.

What the ARIZ should solve is a high lift /low lift wing or a big/small wing.

Step 4 ARIZ – Analysis of Resources

The contradiction space is also known as the 'operational space' in ARIZ. Essentially we are defining the zone – the time and place where the contradiction occurs, and hence the most likely space and time in which we will solve it. It is important to zoom in on the exact place and the moment in time to search for resources.

Analysis of Resources

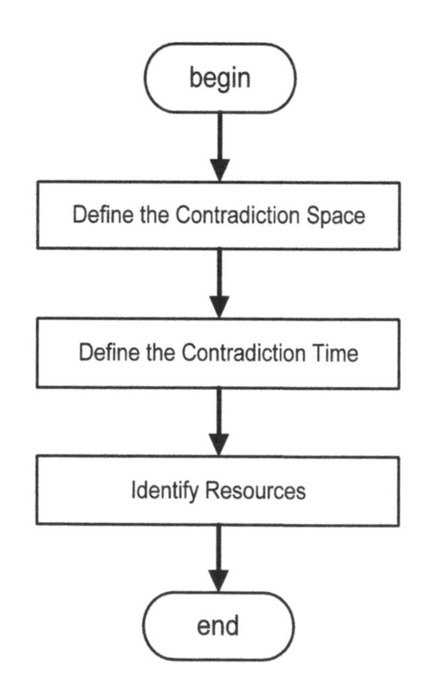

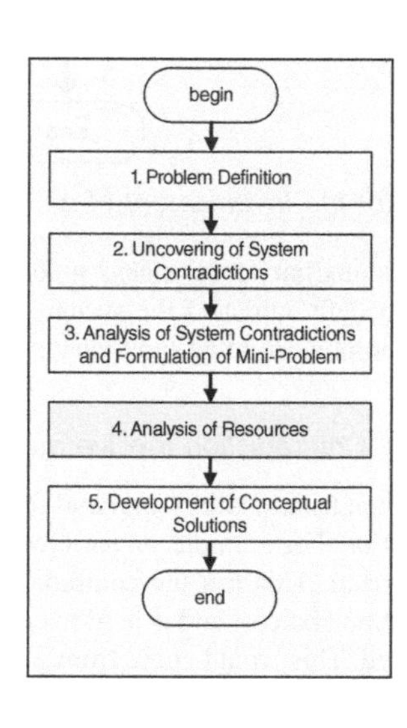

We are seeking available and present resources in the operational space, and in the operational time (includes the actual conflict time as well as the time immediately preceding and following the conflict). We need to locate all the resources closest to the problem area and problem occurrence.

Step 5 ARIZ – Development of Conceptual Solutions

We look for conceptual solutions which are made up of available resources. We examine each resource in turn as the solution provider – the element that is going to solve the mini-problem identified earlier. In each case we try the Ideal Outcome to locate opposite benefits and force the problem into a search for opposite functions/benefits – a physical contradiction situation and then solve using the separation principles.

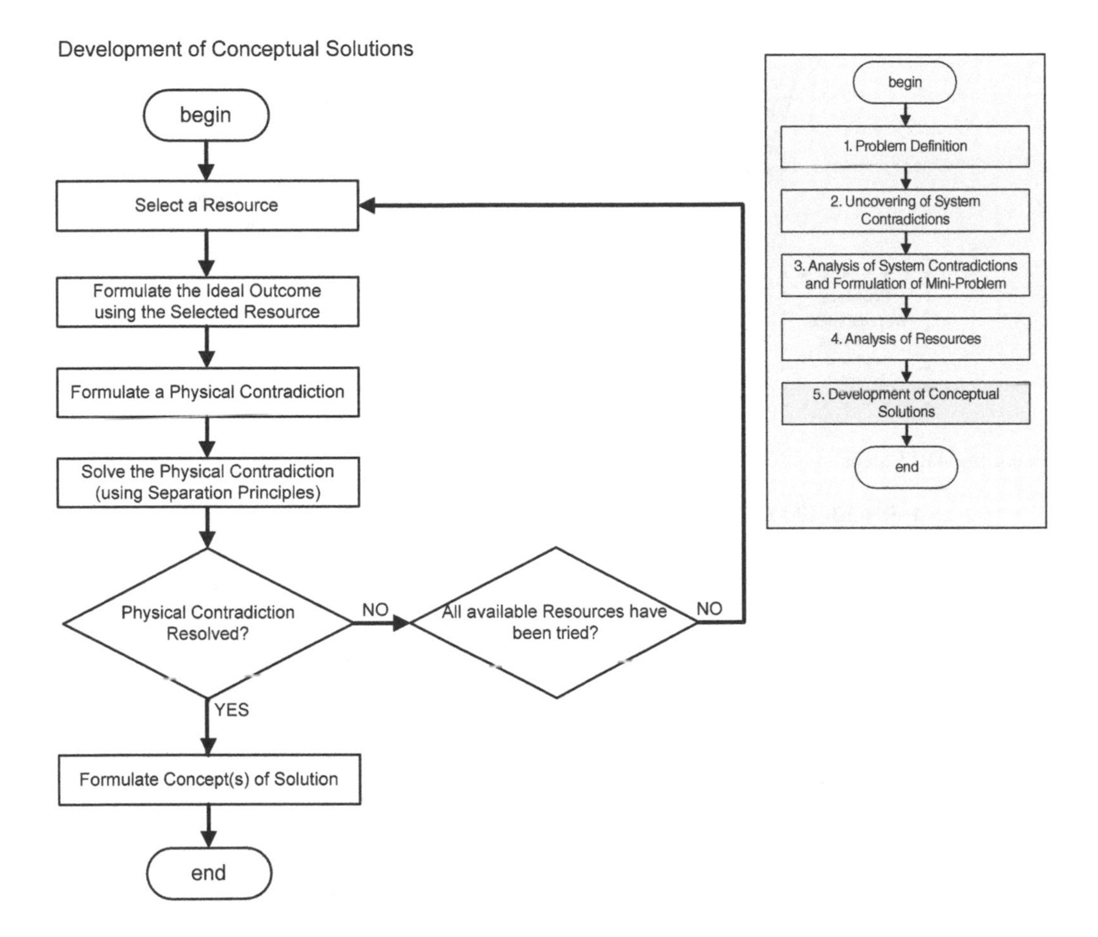

ARIZ Summary

There are many versions of ARIZ each with many steps – sometimes up to 80 – but in general there are these five broad areas of problem understanding and problem solving. In the TRIZ literature there are some simple examples of engineering problems solved by the TRIZ tools and processes. In *ASTIA* Altshuller describes a problem which he says occurs everywhere – it is a problem of a pipe blocked with wet coal in a power station. ARIZ is applied here to this problem to help illustrate its basic steps.

Using ARIZ to Solve a Problem with Coal Blocking a Pipe

Many electric power stations use coal which is delivered and loaded into large silos. The coal is often delivered wet to the power station and is stored outside. It is moved from the silos through a hopper to pipes which feed it into a ball mill which crushes the coal. The crushed coal is then moved to a cyclone separator which sorts the fine coal to be burnt and returns the coarse coal to the ball mill to be crushed again. The system works well when the coal is dry enough, but when the coal gets more than about 12% saturated with water then the coal starts to stick in the pipes and blockages appear. This is a difficult situation and solutions which involve drying the coal are problematic as the fine coal can self-ignite causing fires and explosions.

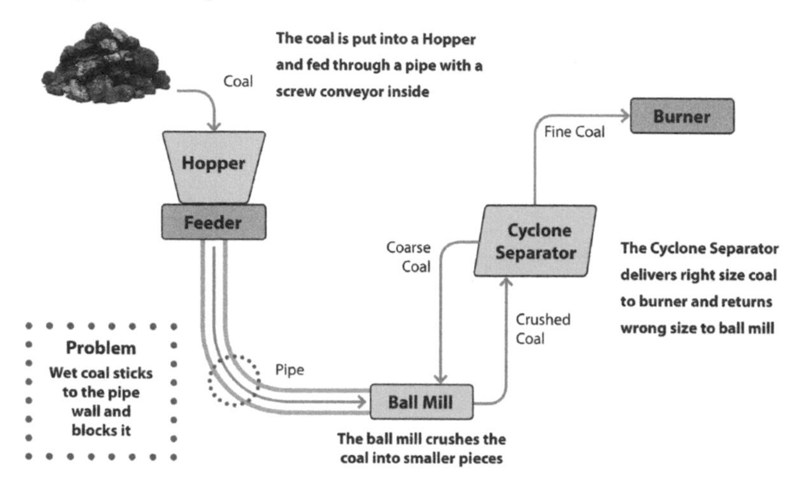

Using the ARIZ steps:

1. PROBLEM DEFINITION

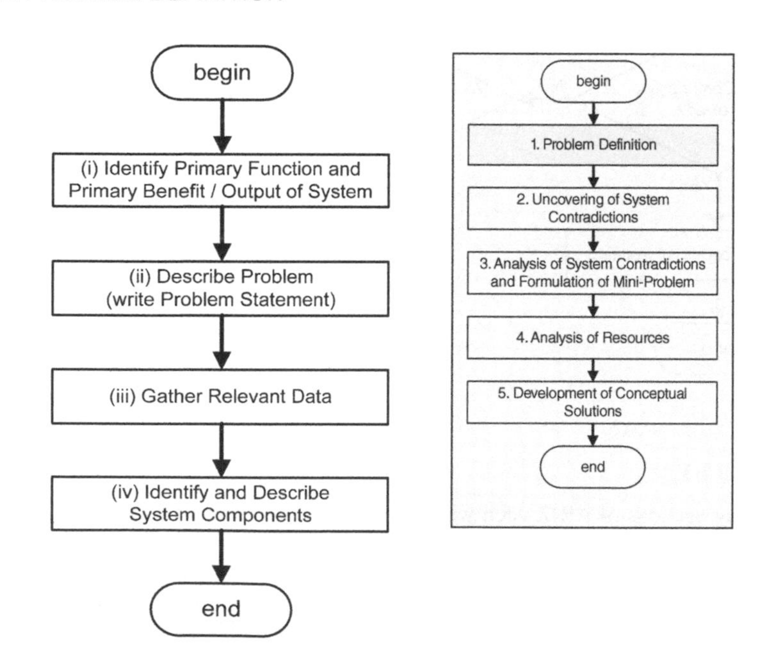

Step 1 ARIZ – Problem Definition

ARIZ on Coal Problem – Preparation and Problem Understanding

i. Identify the Primary Function/Output of the system

Coal needs to be moved from hopper to ball mill.

ii. Problem Formulation and description

Wet Coal blocks the pipe – It is sticking inside the Pipe preventing movement of coal to the ball mill.

iii. Background information and relevant data

Wet Coal is supplied with 22–25% water present-this is part of the problem as the pipe blocks – sticking occurs if the coal has more than 12% water. Dry coal is available, but it is more expensive.

iv. Identify the main components of the system, and describe them in very simple (non-technical) terms (This can be a simple list or it can be expanded using 9-Boxes and / or functional analysis)

Take out acronyms, technical terms or jargon

Hopper = metal container

Feeder = rotating table

Wet Coal = coal with more than 20% water present

Fine Coal = powdered coal

Pipe Wall

Ball Mill = rotating drum with metal balls

Cyclone Separator = device that separates coal by size

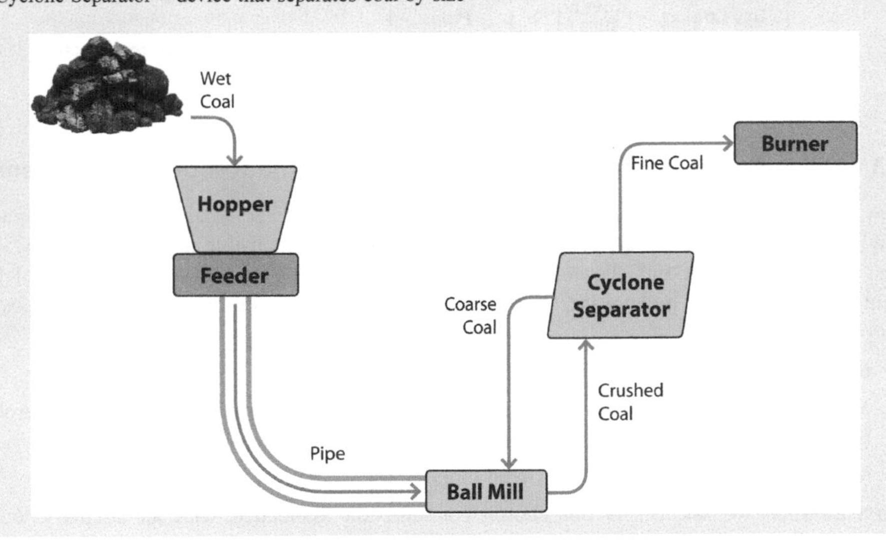

Step 2 ARIZ – Uncovering of System Contradictions

Identify the System Conflicts/Contradictions that Lie in or Behind the Problem

This is formulated by identifying the bad/harmful output or cost that is associated with delivering the good or primary function/output (and/or getting the primary benefit) it can be caused by the need for

- opposite functions/features, *or*
- we improve part of the system and another part gets worse.

ARIZ on Coal Problem – Formulation of System Conflicts

Contradiction 1

Use wet coal as it is cheap but it has a problem - it blocks the pipes.

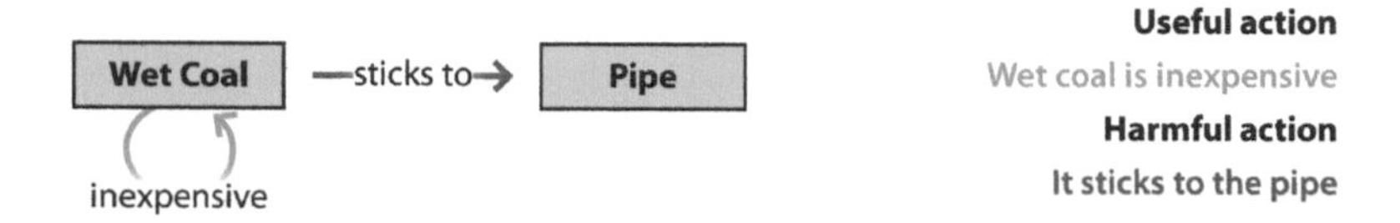

Contradiction 2

Use dry coal but it is expensive but it does not block the pipes.

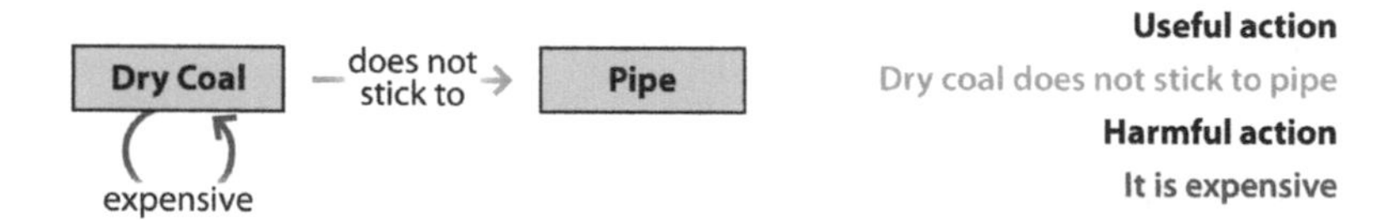

Step3 ARIZ – Analysis of Contradictions and Formulation of Mini-Problem

We now need to select one of the contradictions and analyse it to see what to do next (in practice we may want to work through all of the contradictions). It seems sensible to select the worst – the 'biggest and baddest'. For our chosen contradiction pair we will decide to go down one of three different routes – in effect picking the 'mini-problem' that we are going to solve. The three possible mini-problems are:

- Mini a) – eliminate (i.e. trim) the auxiliary tool/the article.
- Mini b) – no tool is eliminated (useful and harmful actions are retained) – so the mini-problem is to find an X-resource that can eliminate or neutralize the harmful action.
- Mini c) – otherwise, eliminate the main tool.

The first question we ask is: 'Is our chosen contradiction associated with an auxiliary tool/the article, or is it associated with the thing that delivers the primary function of our system (the main tool)?'

If it's the auxiliary tool/the article (AT) then we go down the mini-problem a) route, eliminate (i.e. trim) the auxiliary tool/the article.

If it's the main tool then we go for mini problem b) or c).

In the case of b) or c), we should use the Standard Solutions to suggest a model for the solution.

ARIZ on Coal Problem – Contradictions and Formulation of the Mini-Problem

Looking at the three options to formulate the mini-problem: we probably need to solve Mini b

Mini a) eliminate (i.e. trim) the auxiliary tool/the article.

We cannot trim the coal – this is not an option.

Mini b) no tool is eliminated (useful and harmful actions are retained) – so the mini-problem is to find an X-Factor in resources that can eliminate or neutralize the harmful action. This mini-problem to find an X-Factor resource to stop the coal blocking the pipe – this is an option.

Mini c) otherwise, eliminate the main tool (the pipe).

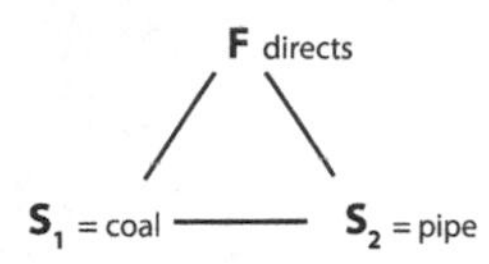

or as Function Analysis

Could we TRIM the pipe? This may be an option, but we would need more information about the system.

Analyse the System in More Detail to Formulate the Mini-Problem

This requires looking to define an X-Factor which will solve the problem as in option b) above (locate an X-Factor resource to stop the coal blocking the pipe).

We now draw the conflict zone in detail so we can understand exactly what is happening. Harmful Substance-Model – we are seeking S3 which will stop harm

S3 = X-Factor

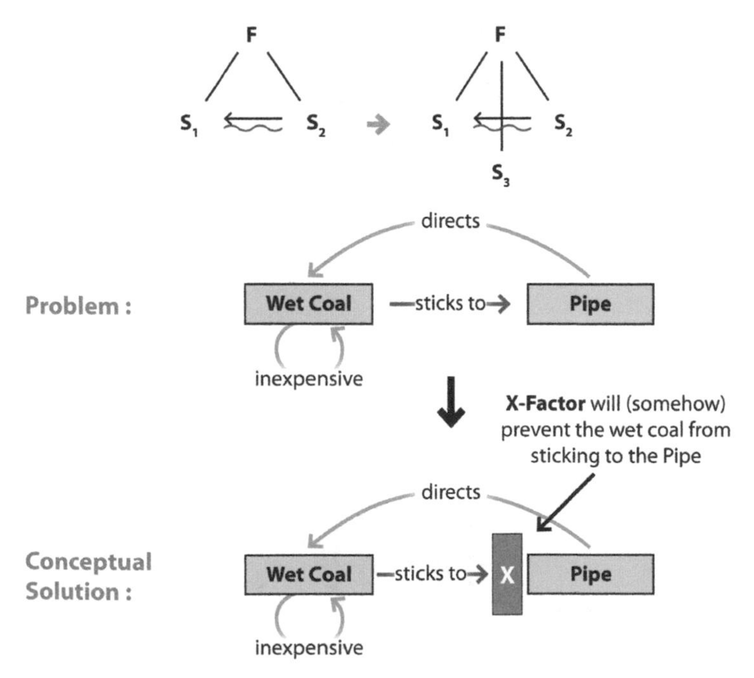

Zoom in to the Conflict Zone

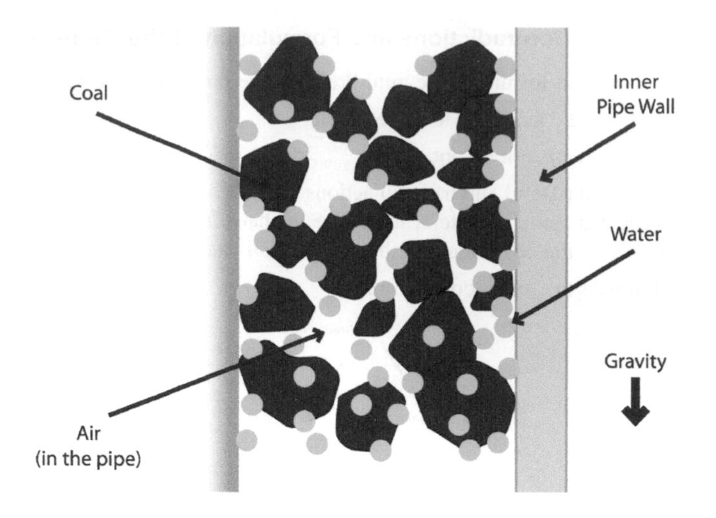

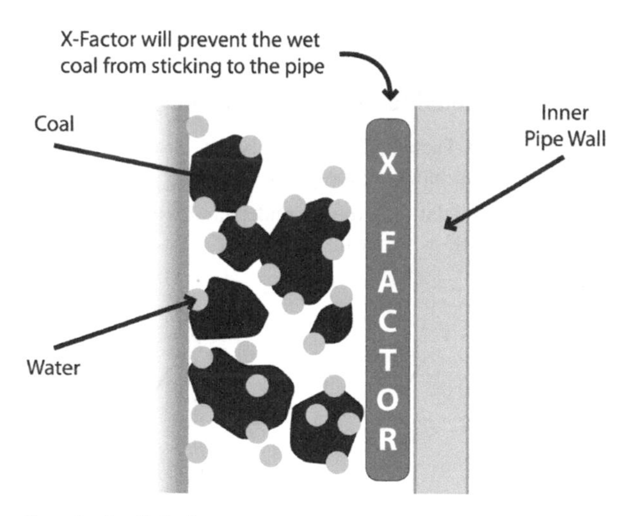

Describe the X-Factor

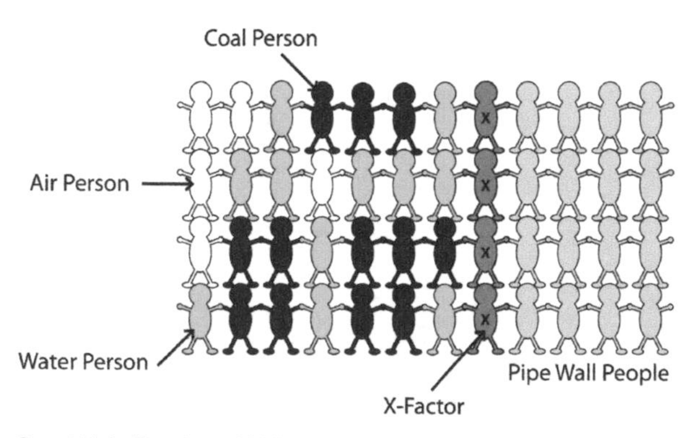

Smart Little People and X-Factor

This Substance–Field model would suggest we look at the Standard Solutions for dealing with harm and choose one such as *Standard Solution 1.2.2. – Useful and harmful effects exist in a system. It is not necessary for* S_1 *and* S_2 *to be in direct contact but we cannot introduce a new substance. Block the harm by introducing a substance S3 which is made from* S_1 *and* S_2 *(or parts* of them) or by modifying them.

Alternatively we can define the X-Factor in a Function Analysis Diagram.

Step 4 ARIZ – Search for Available Resources to Provide X-Factor

Analysis of Resources

begin

Define the Contradiction Space

Define the Contradiction Time

Identify Resources

end

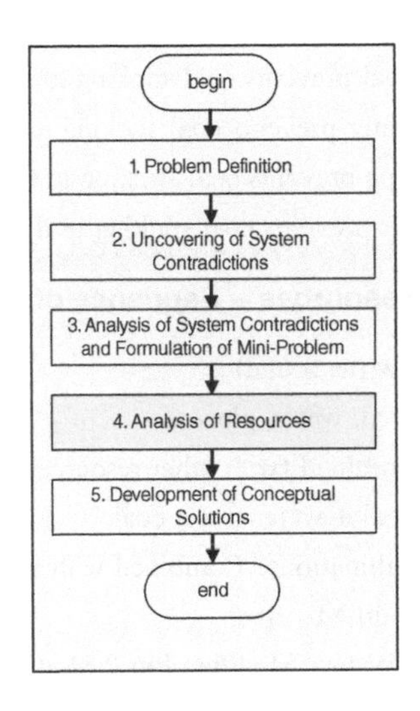

Contradiction Time – When is the Conflict /Problem?

Time-based resources:

- before coal blocks the pipe
- during/while coal is blocking the pipe
- after coal has blocked the pipe.

Contradiction Space – Where is the Conflict /Problem?

Space-based resources

- environment/super-system – heat, space etc.
- system/coal, water, wall
- components/sub-systems coal dust (powder)

Identify Resources

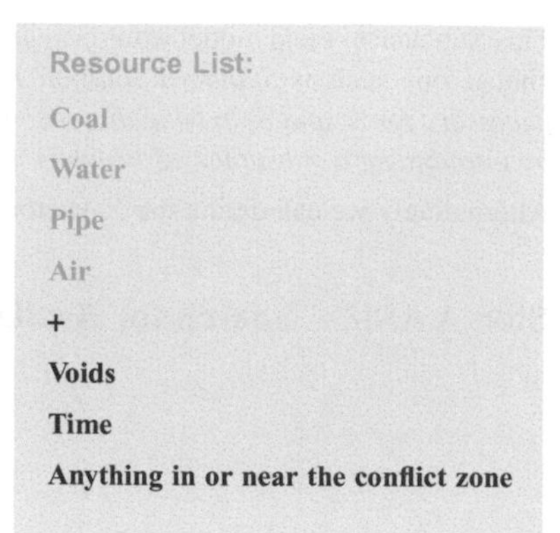

Resource List:
Coal
Water
Pipe
Air
+
Voids
Time
Anything in or near the conflict zone

A resource hunt, starting at the zone of conflict (in time and in space) and working outwards, with the aim of identifying resources that might help solve the problem, have responsibility transferred to them and/or be the X-Factor.

The Ideal Outcome states: that there should be no additional components and that we should make the X-Factor out of an available resource.

X-Factor should ideally be made from: coal, water, pipe, air

or some combination/modification of them

Examine each element in turn:

- Coal prevents coal sticking to the pipe wall
- Water prevents coal sticking to the pipe wall
- Pipe prevents coal sticking to the pipe wall
- Air prevents coal sticking to the pipe wall

Using Resources – Sequence of Preference

a. Raw (as found)

Coal, water, pipe, air

b. Combined (with other resources)?

Coal + water = wet coal

Other combinations? Combined with voids or time?

c. Modified

Modified Water? Modified Pipe? Modified Coal?

Wet Coal, Dry Coal?

Coal dust?

Step 5 ARIZ – Development of Conceptual Solutions

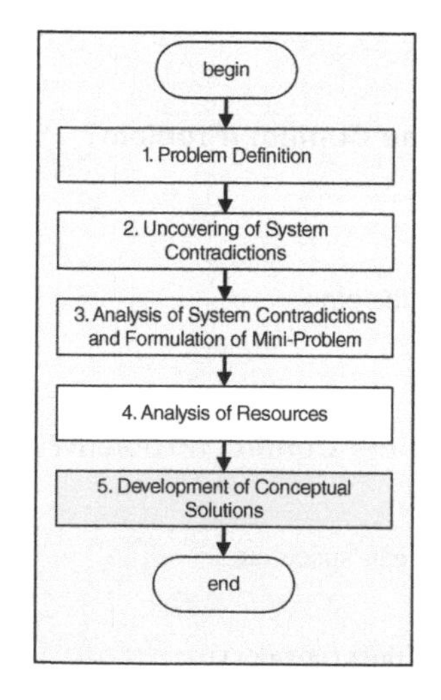

This is to find solutions – particularly to select the resource to provide the X-factor which will solve the problem. This depends upon the mini-problem formulated back in Step 2 and the analysis of the system conflicts in Step 3

Looking at the suggested options of Mini a), b) and c) – we decided that our option is only b) and we are looking for an *X-Factor to stop the coal blocking the pipe.*

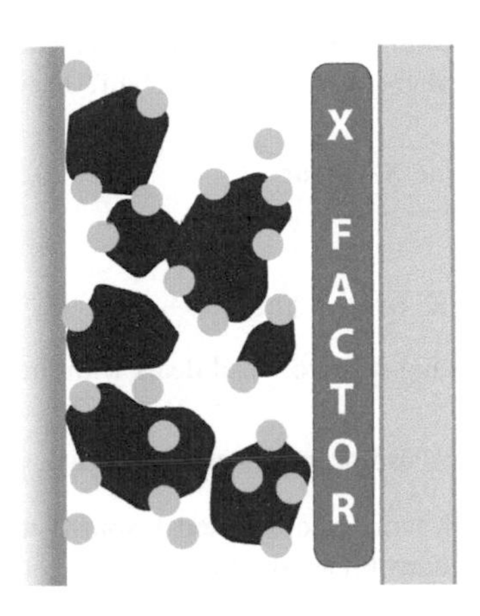

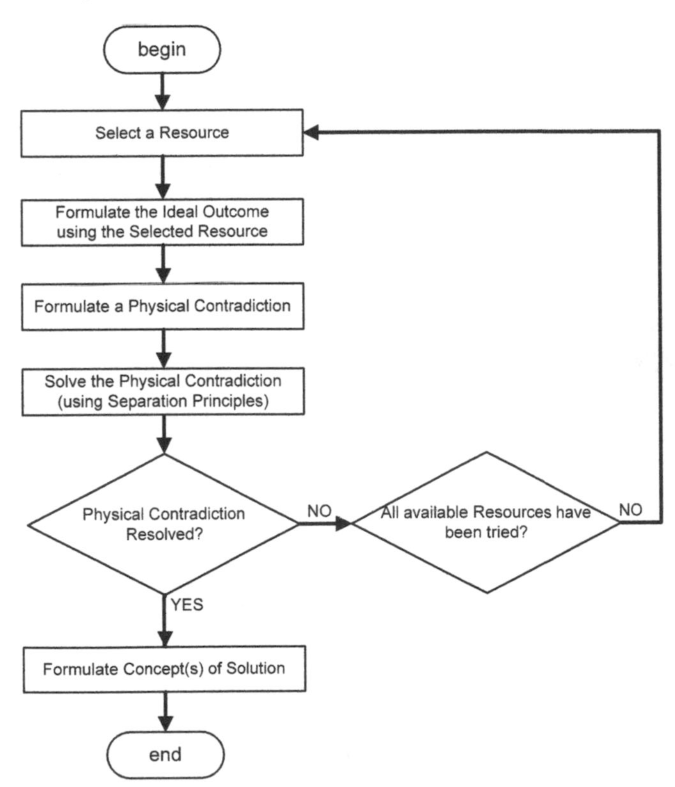

ARIZ on Coal Problem – Selecting a Resource

This mysterious X-factor stops the coal blocking the pipe – we need to select a resource which does this. If the right resource has not been found then we need to look wider beyond the conflict zone for resources within the system which could solve the problem. The problem of wetness of coal causes the coal to stick to pipe wall. We want the coal to be wet (wet coal is cheap but sticks) and we want the coal to be dry (dry coal is expensive but does not stick). Where do we want dry coal? Everywhere? Or just next to the pipe wall? Are there any analogies of systems with sticky wetness being a problem? (e.g. Making pastry or bread – we add more dry flour). Does dry coal (or coal dust, like flour) exist anywhere in the wider system? If so could we add it to the wet coal just before it enters the pipe?

We are selecting **COAL DUST** as the **X-Factor.**

Frame an Ideal Outcome/IFR of X-Factor

X-factor has to be made out of existing resources and deals with the harmful action – blocks pipe. No obvious solution?

Frame physical contradictions for the use of the resource:

> to provide the required useful action the selected resource must have property P *and* to eliminate/neutralize unwanted harmful action it must have property anti-P.

To prevent a blockage:

- the coal in the pipe must be dry, but to remain cheap it has to be wet.
- the dry coal powder has to be *present* between the wet coal and the wall but it should be *not present* until it is needed (the dryness will not last long so it should only appear when it is needed).

If we want dry coal as the X-Factor next to the pipe wall where can we find dry coal,

does it already exist in the system? Yes … it is an output of the cyclone separator.

Dry coal dust – is an available resource and one good solution is to add this to the hopper from the cyclone separator.

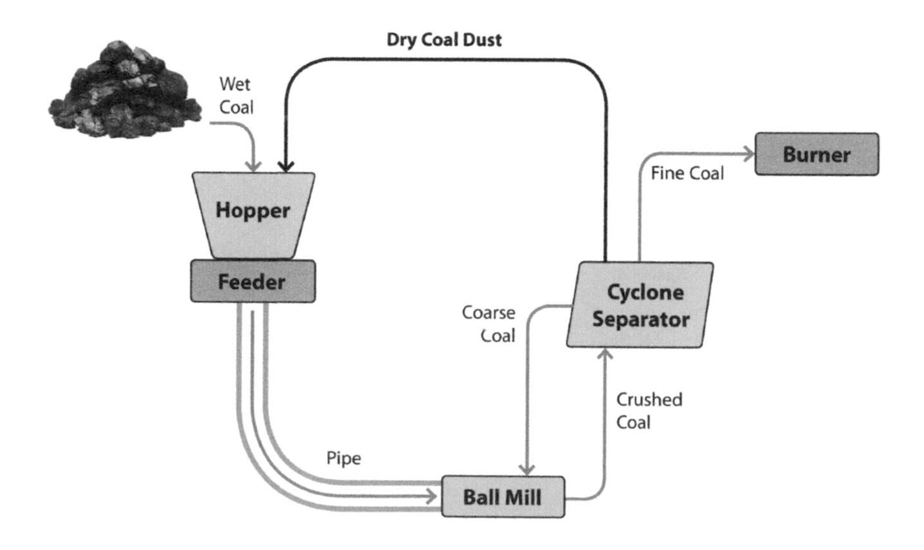

Conclusion

The two classic TRIZ tools Substance–Field Analysis and ARIZ are important for defining and solving challenging problems. They are both rigorous and powerful and worth mastering, yet in many companies they are not widely applied, with only a handful of TRIZ experts using them only for particularly difficult problems.

Substance–Field Analysis

Altshuller created Substance–Field analysis as an alternative to the Contradiction Matrix and the 40 Principles for classifying inventive problems and their solutions. Substance–Field triangles help to define problems, and the 76 Standard Solutions to solve them. Substance–Field Models are useful for considering ideas from knowledge bases after using the 76 Standard Solutions; they require greater understanding of physics and the effects than the other TRIZ tools. They also work well to help us structure the problems (and can reveal contradictions) but with a few basic rules and the 76 Standard Solutions, they give us accurate modelling of problem situations and direct us to solutions. Each Substance–Field Model is made up of at least one triangle. The triangle represents the minimal technological system and consists of a tool, Substance 2 (S_2), and an object which it acts on, Substance 1 (S_1), via a field (F) providing the means of their interaction.

TRIZ Function Analysis and the Oxford Standard Solutions are an easier to use substitute and are often taught for simpler but practical problem solving instead of Substance–Field Analysis and the 76 Standard Solutions. The logic of the two approaches is similar and it takes us to the same answers in the Standard Solutions. When faced with a choice between Contradictions or System Analysis (either Substance–Field or Function), I have found that engineers will frequently initially favour just one of these approaches, and become confident and competent in their chosen TRIZ toolkit. But later they switch to the other and eventually use both equally. As we can't predict which one they will choose first, learning both approaches and all the TRIZ Tools is important to encourage confidence in TRIZ. Also TRIZ users, particularly engineers, like to feel they have understood the entire toolkit even if they only use a part at any one time, depending on the problem. The different approaches will take us to the same TRIZ solution triggers and guide TRIZ problem solvers to solutions for problems and system development.

Simple Steps for Applying Substance-Field Analysis to Problems

1. **Problem Statement** – describe the system & the problematic part of the system
2. **Zoom in on the problem area** and identify the problem components (substances) and any relevant interactions (fields) including with the environment
3. **Draw a Substance-Field Model of the problem**
4. **Define the problem type:**

Missing elements?	Class 1.1
Harmful elements?	Class1.2
Insufficient elements?	Classes 1 or 2
Evolution of the substances?	Class 3
Detection or measurement problem?	Class 4
Extra improvements / simplification?	Class 5

5. **Select the appropriate Standard Solution(s) from the suggested class**

6. **Add in relevant knowledge and experience to transform the conceptual solution to a practical one** – (use the problem solving individual's and team's own knowledge, company knowledge, industry knowledge and beyond that to patent databases etc. available on the internet – searching for analogous and relevant solutions to the problem functions.

Using ARIZ

ARIZ guides us to solve difficult and perplexing problems and helps us focus on the real problem. It covers all problem types, contains many steps and is thorough but not quick in its approach. (see page 388)

Most of the activity in ARIZ is directed towards understanding the nature of the problem and then locating the best resources available to solve it. Only at the final stage does the algorithm call for the generation of solutions. ARIZ is the TRIZ ultimate problem-solving process and directs us through all the steps needed to solve any problem, it has great power and rigour to help locate clever solutions.

Appendix 12.1
Traditional TRIZ 76 Standard Solutions

The 76 Standard Solutions are in five classes, with various sub-classes, which are used according to the type of engineering problems they solve. The five classes are:

- Class 1: Building and Destruction of Substance–Field Models 13 solutions
- Class 2: Development of Substance–Field Models 23 solutions
- Class 3 System Transitions and Evolution 6 solutions
- Class 4: Detection and Measuring 17 solutions
- Class 5: Extra Helpers 17 solutions.

Class 1: Building and Destruction of Su–Field/Substance–Field Models

Class 1 helps us solve problems by building or destroying the Su–Field Models if they are incomplete or have harmful functions. Class 1 has two sub-classes containing 13 Standard Inventive Solutions:

1.1 Building of Su–Fields (if incomplete) 8 solutions

The major recommendations from this sub-class are:

- Make Su–Field complete.
- Make it minimally workable by introducing an internal additive.
- Make it minimally workable by introducing an external additive.
- Use minimal–maximal mode (add more and remove the extras; add less and enhance locally).

1.2 Destruction of Su–Field (harms) 5 solutions.

The major recommendations from this sub-class are:

- Introduce the third substance between the given two substances.
- Introduce the third substance from the super-system.
- Introduce the third substance that is a modification of one of the given two substances.
- Introduce a sacrificial substance.
- Introduce a field that counteracts the harmful field.

Sub-class 1.1: Building/completing of a Su–Field (if it is incomplete)

The desired functions/effects are absent. Seeing what is missing and then looking for a simple solution from the lists provided in the standard solutions is systematic and surprisingly powerful.

1.1.1 Complete the Su–Field

A minimum technical system consists of two substances (S_1 and S_2) and a field (F). Any system that does not contain at least these three elements cannot deliver a useful function. Thus, where we do not have a complete Su–Field, typically because either S_2 or F (or both) is missing, then the missing elements must be introduced to complete the Su–Field:

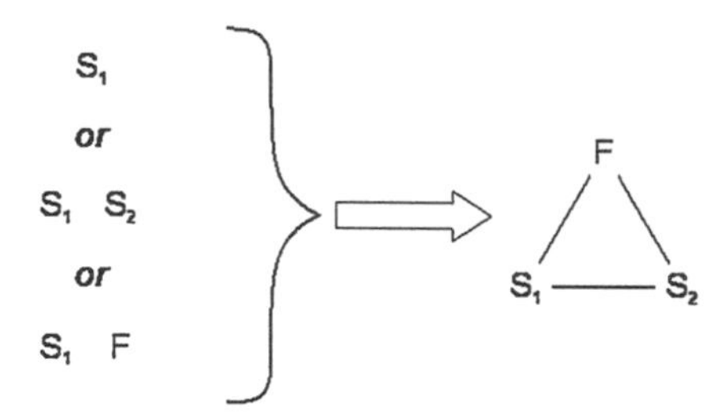

The completed Su–Field will typically take one of the following forms:

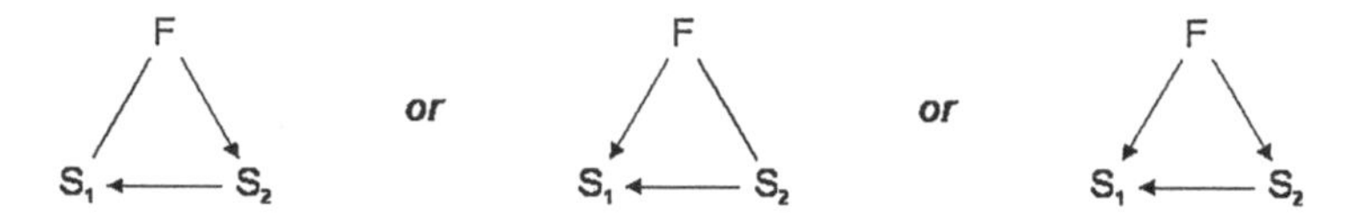

1.1.2 Transition to internal complex Su–Field

If we have a Su–Field which needs to be improved but the required change is not straightforward and if there are no restrictions on the use of additives, then the problem can be solved by the addition of additives into S_1 or S_2. The additives may be introduced permanently or temporarily.

The additive is denoted by S_3. The round brackets indicate that the additive S_3 is internal to S_1 or S_2.

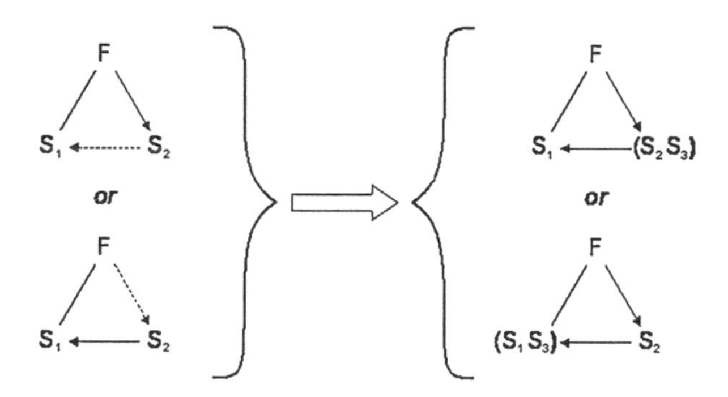

1.1.3 Add something outside the substance

When we have a Su–Field which needs to provide some extra function(s) or change in some way, but it is not easy to change as required then transition to an external complex Su–Field. This is achieved by the introduction of external additives into the present substances. This is to provide the extra functions of enhancing controllability or providing required features or properties.

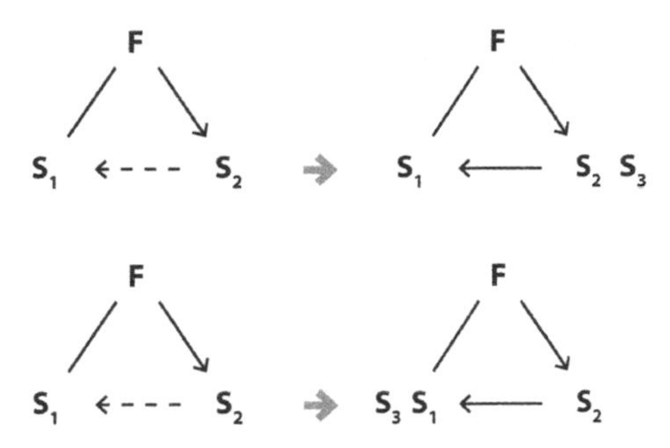

1.1.4 Use the environment

When we have a Su–Field which needs to provide some extra function(s) or change in some way, but it is not easy to change as required then transition the Su–Field by using the external environment as the substance.

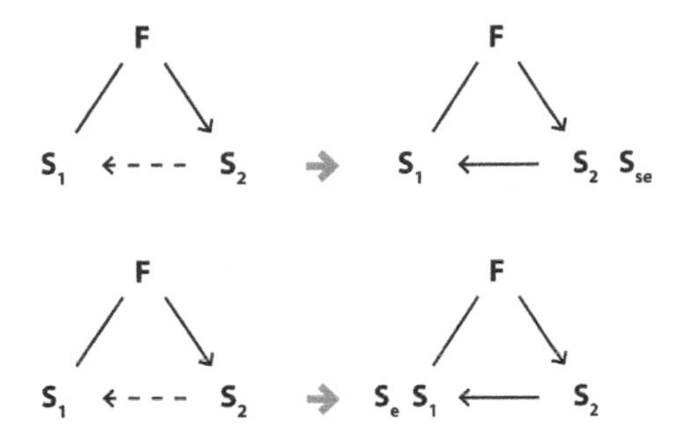

1.1.5 Add something to the environment

When we have a Su–Field which needs to provide some extra function(s) or change in some way, but it is not easy to change as required then transition the Su–Field by changing the external environment (if the external environment does not contain ready solutions as in 1.1.4). Replace the external environment with another one. Decompose it or add new substances into it.

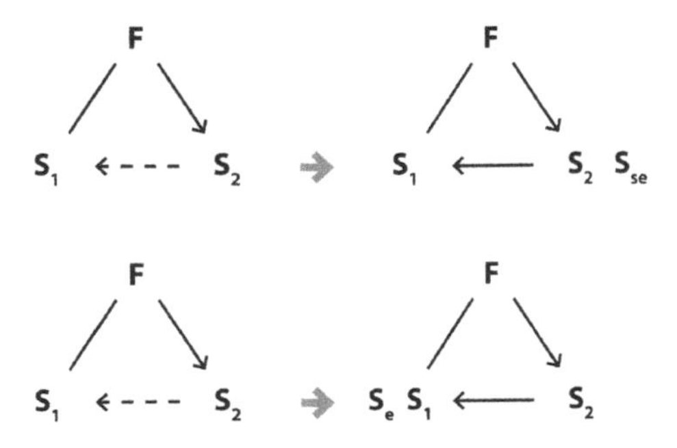

1.1.6 Be excessive

If a small amount of something is required but it is not easy to get the right degree of an action or exact amount of an substance change, use an excess of what is required and remove the surplus. Surplus substance is removed by a field and surplus field is removed by a substance.

1.1.7 Protect the substance by putting the full required force elsewhere

If a moderate field can be applied but this is insufficient but a greater field will damage the system, apply the larger field to another element linked to the original. A substance which cannot take the full action directly can achieve the desired effect through linkage to another substance.

1.1.8.1 Maximum field but not everywhere

If a powerful, strong field/action is required which needs to be more effective in some parts than others (maximum/ large/strong in some zones and minimum/small/weak effects in other zones), add a protective substance S_3 to the places where the weaker effects are required.

1.1.8.2 Minimum field enhanced where required

If a powerful, strong field/action is required which needs to be stronger in some parts than others (maximum/ large/strong in some zones and minimum/small/weak effects in other zones), the field should be minimal and the locations requiring the maximum effects can be enhanced by a substance S_3.

Sub-class 1.2: Eliminating, blocking or reducing harms

1.2.1 Useful and harmful effects exist in a system.

It is not necessary for S_1 and S_2 to be in direct contact. Block the harm by introducing a new substance S_3.

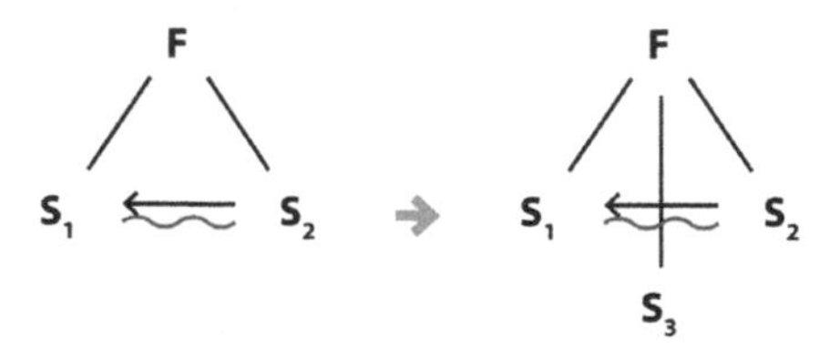

1.2.2 Useful and harmful effects exist in a system

It is not necessary for S_1 and S_2 to be in direct contact but we cannot introduce a new substance. Block the harm by introducing a substance S_3 which is made from S_1 and S_2 (or parts of them) or by modifying them (includes using nothing – voids, bubbles, foam, vacuum, air etc.).

1.2.3 Sacrificial substance

A field causes harm on a substance. Introduce a sacrificial substance S_3 to absorb the harm.

1.2.4 Counteract harm

Useful and harmful effects exist in a system in which the elements S_1 and S_2 must be in contact. Counteract the harmful effect by creating a dual Su–Field in which the useful effect is provided by the existing field F_1, while a new field F_2 neutralises the harm (or turns it into good)

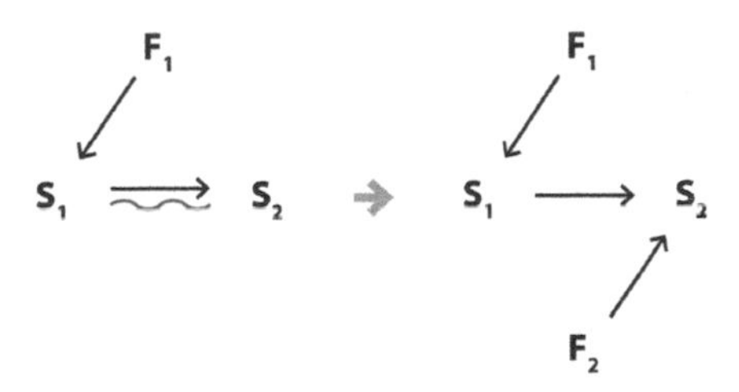

1.2.5 Magnetic harm

A harm may exist because of magnetic properties in a system. Remove the harm by switching off the magnetism (by heating it above its Curie point, or by introducing an opposite magnetic field).

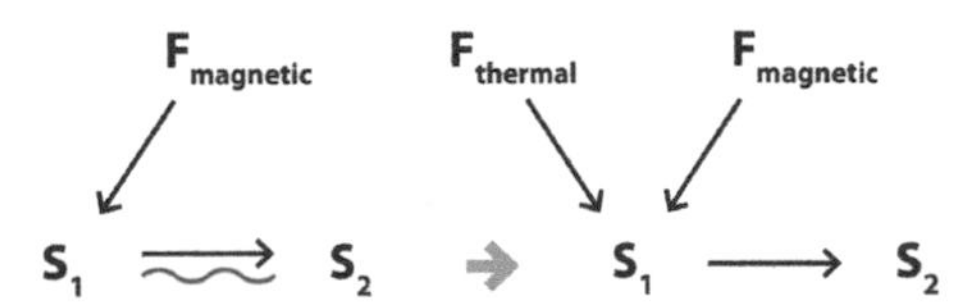

Class 2 Development of Substance–Field Models

This class is used for improving the efficiency of engineering systems by introducing minor modifications. It offers concept solutions of how to improve and evolve systems. The major recommendations from this class are:

- use of chain Su–Fields
- use of double Su–Fields
- segmentation (including porosity increase)
- dynamisation
- rhythm coordination
- use of magnetic substances.

Class 2 contains 4 subclasses and 23 Standard Solutions:

2.1 **Transition to complex Su–Field Models** 2 solutions

2.2 **Evolution of Su–Fields Models** 6 solutions

2.3 **Evolution of rhythms** 3 solutions

2.4 **Complex forced Su–Field Models** 12 solutions.

Sub-class 2.1: Transition to Complex Substance–Field Models

2.1.1 Improve a system's effectiveness and controllability by developing one component (to deliver its own functions)

Transform one of the parts of a Su–Field to its own independently controllable Su–Field. Convert the single Su–Field model to a chained model by having S_2 with F_1 applied to S_3 which applies F_2 to S_4. The two models can be independently controlled.

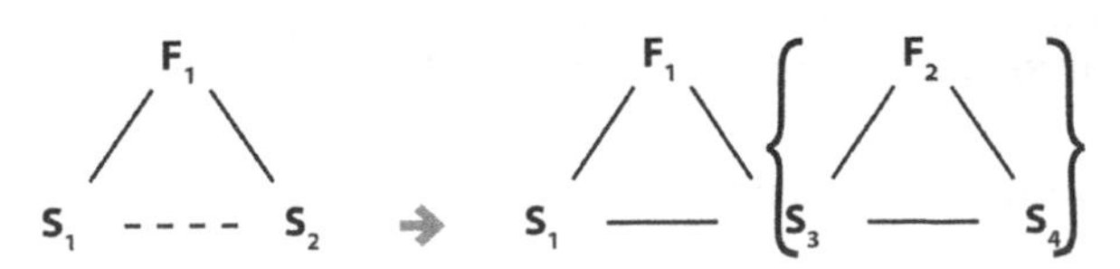

2.1.2 Add another field

If you cannot change the system elements to improve it then add another field. The double Su–Field Model improves a system's effectiveness and controllability without changing the elements of the existing system by adding a second field to S_2.

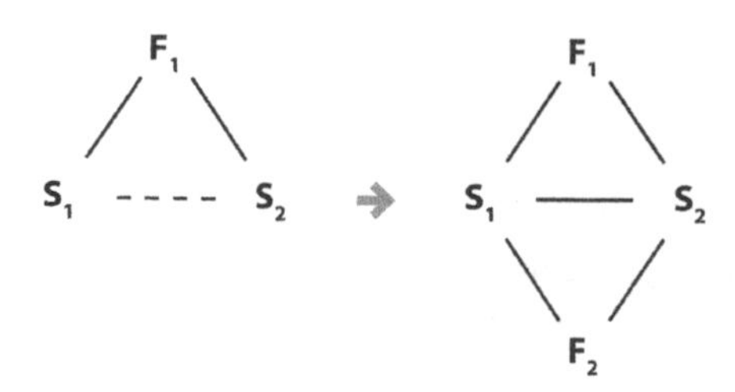

Sub-class 2.2: Evolution of Su–Field Models

2.2.1 Replace an uncontrolled or poorly controlled field with an easily controlled field.

Replace a gravitational field with a mechanical field, a mechanical field with an electrical field etc.

2.2.2 Increase the efficiency of Su–Field Models by segmenting the object – change/segment S_2

2.2.3 Increase the efficiency of Su–Field Models by changing the object from a solid to a porous or capillary material that will allow gas or liquid to pass through

2.2.4 Increase the efficiency of Su–Field Models by making the system more flexible/adaptable/dynamic

2.2.5 Increase the efficiency of Su–Field Models by changing from a uniform field (or uncontrolled field) to a field with predetermined patterns that may be permanent or temporary

2.2.6 Efficiency of a Su–Field Model can be improved by changing the substances from a uniform substance or uncontrolled substance to a non-uniform substance with a predetermined spatial structure (permanent or temporary)

Sub-class 2.3: Evolution of Matching Rhythms

2.3.1 Efficiency of a Su–Field Model can be improved by matching or mismatching the natural frequency of the field (action) with the natural frequency of the subject (tool) or the object it acts on

2.3.2 Efficiency of a complex Su–Field Model can be improved by matching (or mismatching) the frequencies of the different fields/actions being used

2.3.3 When we have two incompatible actions, perform one action in the downtime of the other

Sub-class 2.4: Complex Forced Su–Field Models

This sub-class uses ferromagnetism etc. to add functions and control. The use of ferromagnetic material (S_F) and magnetic fields (F_{Fe}) can be an effective way to improve system performance.

2.4.1 Use ferromagnetic materials (S_F) and/or a magnetic field (F_{Fe}) to add functions

2.4.2 Instead of just using a solid ferromagnetic substance, segment that ferromagnetic substance into particles

Use ferromagnetic substances in the form of granules, powder, very fine powder etc.

2.4.3 Use magnetic ferro-fluids, which are very small colloidal ferromagnetic particles suspended in kerosene, silicone or water

When activated, magnetism occurs in milliseconds and the fluid appears solid. The fluid can be moved/positioned with a magnetic field.

2.4.4 Change the efficiency of the structure by using capillary structures that contain magnetic particles or liquid

2.4.5 Introduce magnetic additives (such as a coating) into one of the components or substances

Give a non-magnetic object magnetic properties, which may be temporary or permanent.

2.4.6 Introduce ferromagnetic materials into the external environment of the system

This can be used when it is not possible to make components or a system magnetic.

2.4.7 Control a ferromagnetic system with physical effects

Heating a substance above the Curie point to lose its ferromagnetism is one such example.

2.4.8 Increase the efficiency of a magnetic field by segmenting it

This can make it self-adjusting, more flexible, dynamic, variable etc.

2.4.9 Improve a ferromagnetic field by transitioning from unstructured to structured fields, or vice versa

Introduce ferromagnetic particles, and then apply a magnetic field to move the particles.

2.4.10 Match the rhythms in the ferromagnetic fields

For example, use mechanical vibration.

2.4.11 Use an electric current to create magnetic fields

Used when we cannot introduce ferromagnetic particles.

2.4.12 Use electro-rheological fluids where viscosity is controlled by an electric field

Class 3: System Transitions and Evolution – Transition to Super-system and Sub-system

This class is used for solving problems by developing solutions at different levels in the system (super-system or sub-system). The major recommendations from this class are how to improve systems by combining elements or combining with other systems.

Class 3 contains 2 sub-classes containing 6 Standard Solutions:

3.1 Simplicity – complexity – simplicity (mono – bi – poly) and increasing flexibility and dynamization (no links, rigid links, flexible links, "field" links) 5 solutions

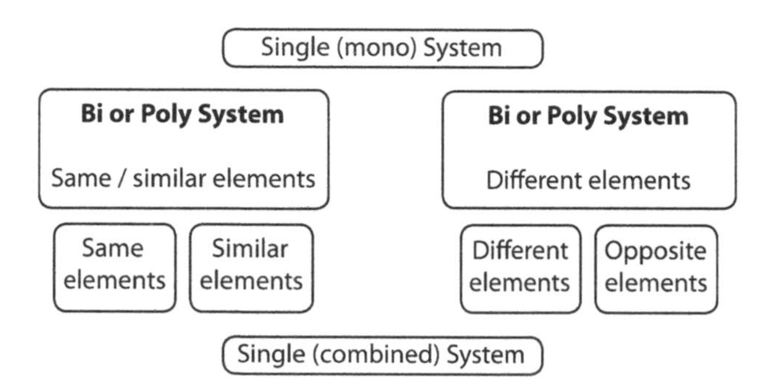

3.2 Transition to micro-level (smart substances) 1 solution

Sub-class 3.1: Transition to the System – Bi-systems and Poly-systems – Improving Systems by Adding Elements

3.1.1 System transition – improve systems by combining with another system or multiply/copy system elements

3.1.2 Develop/improve the links between the system elements

Links can be made more flexible or more rigid.

3.1.3 System transition – increasing the differences between elements – to opposites

3.1.4 System simplification

Achieve all functions but reduce/trim components. Integrate several components into one but still deliver all the functions.

3.1.5 System improvement by delivering opposite incompatible functions at different system levels

One feature/function at a sub-system level and the opposite feature /function at the super-system level.

Sub-class 3.2: System Transition to the Micro-Level

3.2.1 System improvement: transition function delivery to the micro-level

Class 4: Solutions for Detection and Measurement

This class is used for solving measuring or detection problems in engineering systems. These solutions have many distinguishing features, especially the use of indirect methods and the use of copies.

Detection and measurement are typically for control. Detection is binary (something either happens or doesn't happen) and measurement has some level of quantification and precision. For example, a length measurement might be 2.15 m ± 0.01 m. Often the most innovative solution is automatic control which removes formal detection/measurement by taking advantage of physical, chemical or geometrical effects.

The major recommendations of this class are:

- Try to change the system so that there is no need to measure/detect.
- Measure a copy.
- Introduce a substance that generates a field (introduce a mark internally or externally).

Class 4 contains 5 sub-classes and 17 Standard Solutions:

4.1 Indirect Methods	3 Solutions
4.2 Create or Build a Measurement System	4 Solutions
4.3 Enhancing the Measurement System	3 solutions
4.4 Measure Ferromagnetic-field	5 solutions
4.5 Direction of Evolution of the Measuring Systems	2 solutions

Sub-class 4.1: Indirect methods

4.1.1 Change the problem so there is no need for measurement or detection

Instead of measuring something to subsequently add control, stopping or adjustment use closed cycle or feedback mechanisms (applicable to economic and financial situations as well as technical systems)

4.1.2 Measure a copy or an image

Used if 4.1.1 can't be used.

4.1.3 Transform the problem into detection of consecutive, successive changes

Used when we need to make a measurement but can't remove the need for measurement or measure a copy.

Sub-class 4.2: Create or build a measurement system

4.2.1 We need to measure something, an element S_1 – but can't do it directly

We add something to create a measurement system. Instead of measuring some feature of S_1, we add S_2 and measure some feature of S_2 which is connected to S_1 – this is to convey information/measurement/some change in S_1.

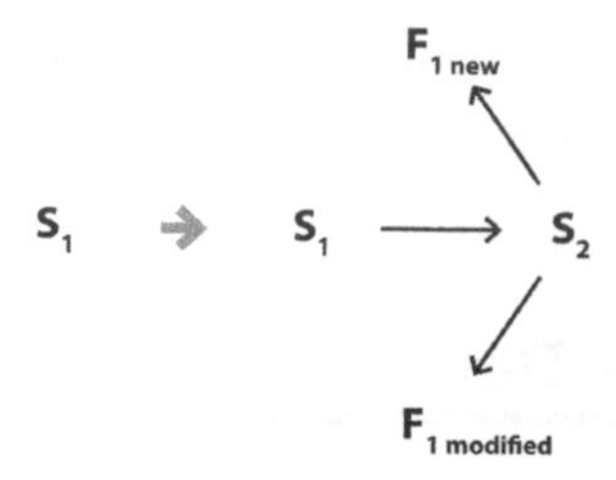

4.2.2 We need to measure something S_1 – but can't do it directly

We add something to create a measurement system which is made of an extra/new component acting on a field which conveys the information we want. If an incomplete S-a-O cannot be detected or measured, complete the S-a-O with a field as an output

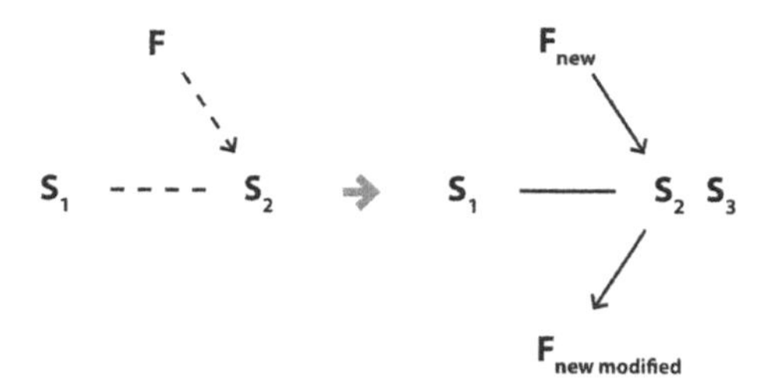

4.2.3 If a system or substance is difficult to measure and we cannot add anything into the system, then introduce an additive externally into the surrounding environment, which reacts to a change in the original system

Change the state of the environment to reveal /measure any changes in the additive.

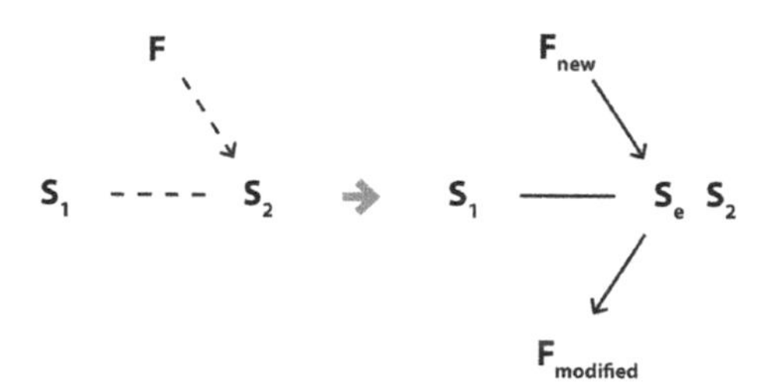

4.2.4 If additives cannot be introduced into the system environment as in 4.2.3, then create them by decomposing or changing the state of something that is already in the environment

Measure the effect of the system on these created additives.

Sub-class 4.3: Improving/enhancing measurement systems through fields

4.3.1 Use natural phenomena

Use scientific/physical effects that are known to occur in the substances/system, and measure/determine the state of the system by observing changes in the effects.

4.3.2 Use system resonance

If changes in a system cannot be measured or detected directly and no field can be passed through it, measure the excited resonant frequency of the entire system, or individual parts of the system.

4.3.3 Use resonance of joined object

If 4.3.2 is not possible (cannot excite an object to resonance), join it to another object and measure its resonant frequency (or measure the resonance in its environment).

Sub-class 4.4: Use Extra Substances and Fields to Help Measurements

Adding ferromagnetic materials for measurement was popular before the development of remote sensing, miniature devices, fiber optics, etc.

4.4.1 Use a suitable detectable substance (such as ferromagnetism) to improve measurement

Add or make use of a substance (such as ferromagnetic) and a field (such as magnetic) in a system.

4.4.2 Add easily detectable particles to a system (such as ferromagnetic particles S_{fm})

This facilitates measurement by detection of the resulting appropriate field

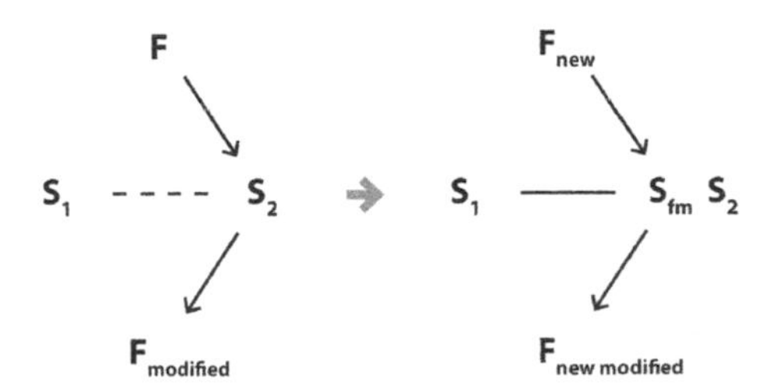

4.4.3 Put detectable additives into the substance

If detectable particles (such as ferromagnetic) cannot be added directly to the system or a substance cannot be replaced by detectable particles (such as ferromagnetic S_{fm}), construct a complex system, by putting detectable additives (such as ferromagnetic) into (or attaching to) the substance.

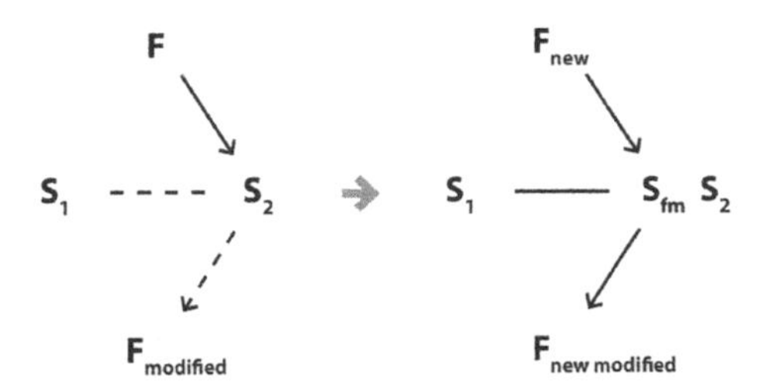

4.4.4 Put detectable additives into the environment

If detectable particles (such as ferromagnetic) cannot be added directly to the system or a substance cannot be replaced by detectable particles (such as ferromagnetic), construct a complex system by putting detectable additives (such as ferromagnetic) into the environment if they cannot be added to the system.

4.4.5 To improve measurement systems use effects such as Curie point, Hopkins and Barkhausen, etc.

Sub-class 4.5: Direction of Evolution of the Measuring Systems

4.5.1 Use more than one measurement system to get a better/more accurate result

If a single measurement system does not give sufficient accuracy, use two or more measuring systems, or make multiple measurements.

(Measuring systems transition to form bi- and poly-systems.)

4.5.2 Measurement systems evolve towards indirect measurement of features or derivatives of the function being measured

Instead of a direct measurement of a phenomenon; measure the first and then the second derivatives in time or in space.

Class 5: Extra Helpers

After using the other four classes of the Standard Solutions, Class 5 is additionally helpful for further general improvements and simplification of systems. These Standard Solutions give recommendations of how to introduce new substances or fields or use scientific effects more effectively after applying the relevant Standard Solutions in the four previous classes.

Class 5 solutions help when simplifying or trimming the system to remove components or to reduce the strength of the relevant interaction. The first four classes of Standard Solutions above often lead to solutions which increase complexity because we are often adding something to the system to solve the problem. This fifth class shows how to get something extra through simplification but without introducing anything new.

The useful recommendations from this class are:

- Instead of a substance, introduce a field.
- Instead of a substance, introduce a void.
- Introduce a substance for a limited time.
- Introduce a little bit of a substance, but in a very concentrated way.
- Use phase changes.
- Get the substance or environment to change themselves to solve the problem.
- Use segmentation.

Class 5 contains 5 sub-classes and 17 Standard Solutions:

5.1 Indirect methods for introducing substances under restricted conditions	4 Solutions
5.2 Introducing fields under restricted conditions	3 Solutions
5.3 Phase transitions	5 Solutions
5.4 Clever use of natural phenomena	2 Solutions
5.5. Generating higher or lower forms of substances	3 Solutions

Sub-class 5.1: Indirect Methods for Introducing Substances under Restricted Conditions

5.1.1 Indirect ways

Solutions for when it is necessary to introduce a substance but it is not allowed.

5.1.1.1 Achieve what you want indirectly by introducing voids, fields, air, bubbles, foam

5.1.1.2 Use a field instead of a substance

5.1.1.3 Use an external additive instead of an internal one

5.1.1.4 Use a small amount of a very active additive.

5.1.1.5 Concentrate the additive at a specific location.

5.1.1.6 Introduce the additive temporarily

5.1.1.7 Use a copy or model of the object in which additives can be used, instead of the original object, if additives are not permitted in the original.

5.1.1.8 When no new substances can be permanently added for a system improvement, introduce a chemical compound which can be later decomposed.

5.1.1.9 When no new substances can be permanently added – obtain the required effect by decomposition of either the environment or the object itself

5.1.2 Segmentation

When an improvement is needed but we cannot replace components or add anything new, change the object by dividing the elements into smaller units

5.1.3 Introduce a substance that disappears after carrying out its work or becomes identical to substances already in the system or environment

5.1.4 When we need large amounts but we must use nothing

Use voids, foams, inflatable structures etc.

Sub-class 5.2: Introducing Fields under Restricted Conditions

5.2.1 Use fields present in the system to cause the creation of another field

5.2.2 Use fields that are present in the environment

For example, use gravity, ambient temperature, pressure or sunlight.

5.2.3 Use substances that are the sources of fields

If we cannot introduce another field by using the fields present in the system or in the environment, then use the fields for which the substances present can act as media or sources.

Sub-class 5.3: Use of Phase Transitions

5.3.1 Change its Phase

Obtain performance improvement (without introducing another substance) by phase transition of an existing substance.

5.3.2 Dual properties are obtained by using substances capable of converting from one phase to another

5.3.3 Use the phenomena which accompany phase change

5.3.4 Dual properties are achieved by replacing a single-phase state with dual-phase state

5.3.5 Dual properties are achieved by replacing a single-phase state with dual-phase state, which can be improved by creating an interaction between the parts (phases) of the system

Sub-class 5.4: Applying the Natural Phenomena

(also called "Using Physical Effects")

5.4.1 Self-controlled transitions

If an object is alternating between physical different states, it should transition from one state to the other by itself using reversible transformations.

5.4.2 From a weak input field produce a strong output field

This is done by keeping the transformer substance in a condition close to a critical condition, near a phase transition point, so that the energy stored in the substance is released by the weak input signal and gives a strong output signal.

Sub-class 5.5: Generating Higher or Lower Forms of Substances

5.5.1 Obtaining the substance particles (ions, atoms, etc.) by decomposing a substance at a higher structural level (e.g. molecules)

5.5.2 Obtaining the substance particles (e.g. molecules) by combining particles of a lower structural level (e.g. ions)

5.5.3 Applying the Standard Solutions 5.5.1 and 5.5.2

If a substance of a high structural level has to be decomposed, the easiest way is to decompose the nearest highest element. When combining particles of a lower structural level, the easiest way is to complete the nearest lower element.

Part Six
How to Problem Solve with TRIZ – the Problem Solving Maps

TRIZ Problem-solving Maps and Algorithms

TRIZ for the Right Functions at the Right Time in the Right Places

TRIZ helps engineers create or improve systems, which are minimal and elegant but have all the necessary functions to deliver all benefits. All the TRIZ tools are there to help get the right functions and all the functions absolutely right, for the least possible costs (all inputs) and for the least possible harms (no problems and absolute minimum harmful outputs including to the environment). In TRIZ the Ideal System has only the benefits with no costs and harms – and as benefits are provided by functions the Ideal System has only good functions provided free from available resources.

All the TRIZ tools help us focus on understanding the functions we really want, and how best to get them, using available technologies and solutions. TRIZ thus helps us recognize and locate the existing answers to our problems. Solving problems with the TRIZ toolkit can involve just a few tools or the entire toolkit depending on the complexity of the problem. The TRIZ algorithms to help us can be simple flowcharts for solving contradictions, looking for the next evolutionary step for a component or locating resources, or very much more detailed algorithms requiring several of the tools to be used. This chapter contains the simple algorithms plus a master problem-solving algorithm which puts all the smaller, simple ones together to solve any problem – in addition there is ARIZ (see Chapter 12).

The simplest TRIZ flowchart is the Prism of TRIZ which is a simple convergent /divergent tool for finding existing solutions, converging our real-life problems to a conceptual problem with conceptual answers and diverging the conceptual solutions back to real-life answers by adding back the relevant knowledge and experience.

TRIZ for Engineers: Enabling Inventive Problem Solving, First Edition. Karen Gadd.

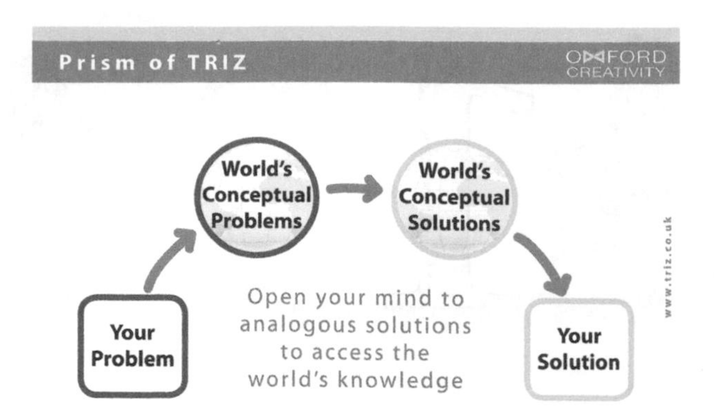

Where Do We Start with TRIZ? Which Tools When?

Problem solving can be hard without TRIZ

During TRIZ workshops these are the most frequently asked questions, and the first answer is that the detailed problem-solving route depends on the type and difficulty of problem. However there are simple stages which we can apply to any problem:

- Defining the Ideal for setting the direction of the problem solving and the scope for stakeholders.
- Problem Understanding – mapping the problem context in Time and Scale and capturing the relevant facts for both requirements and the system.
- Finding, creating and combining all the previous, spontaneous and known Solutions and Concepts. Filling up a 'Bad Solution' Park using everyone involved in the problem.
- Choosing/developing/transforming bad solutions into better solutions using TRIZ. Map all TRIZ transformed solutions in 9 boxes.

The general steps outlined above can be roughly mapped to show their sequence. The first step, the Ideal Outcome, is shown to the right of the 'after' stage as in solution terms it shows us what we would ideally like in the future.

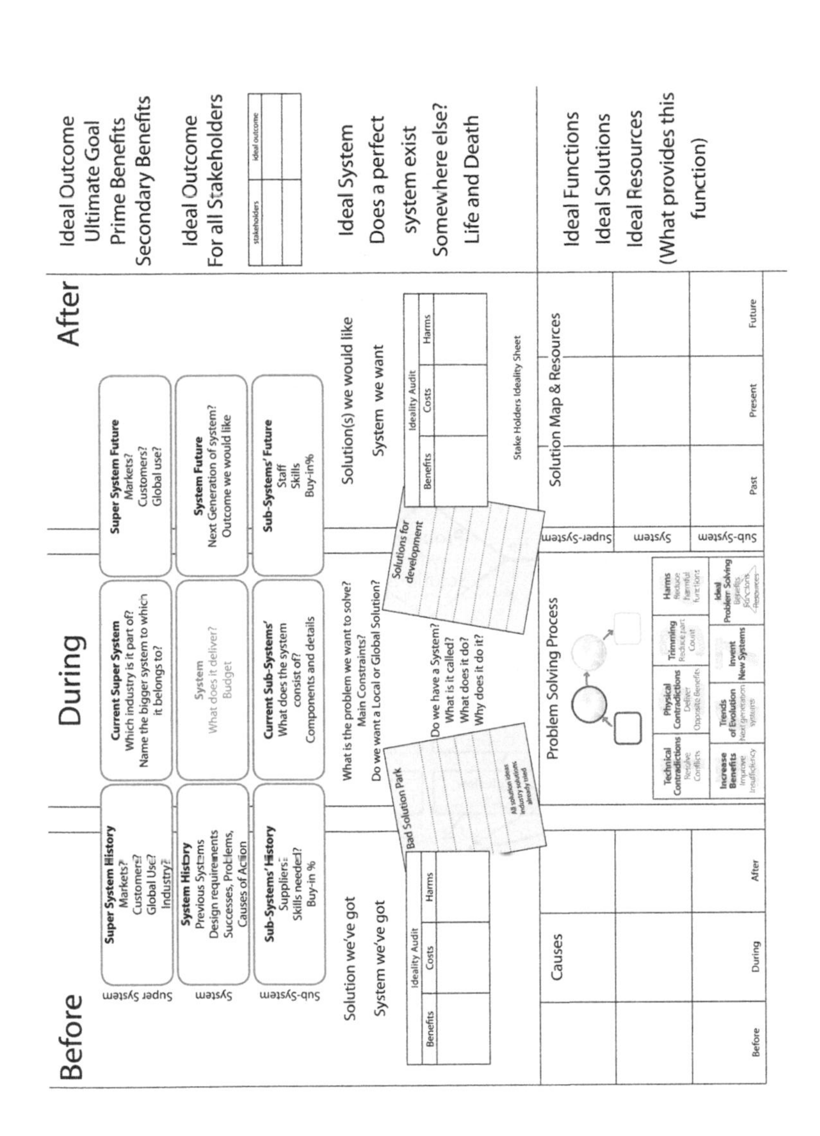
Before
During
After
Super System
System
Sub-System
Super System History
Markets?
Customers?
Global Use?
Industry?
System History
Previous Systems
Design requirements
Successes, Problems,
Causes of Action
Sub-Systems' History
Suppliers?
Skills needed?
Buy-in %
Current Super System
Which industry is it part of?
Name the bigger system to which it belongs to?
System
What does it deliver?
Budget
Current Sub-Systems'
What does the system consist of?
Components and details
Super System Future
Markets?
Customers?
Global use?
System Future
Next Generation of system?
Outcome we would like
Sub-Systems' Future
Staff
Skills
Buy-in%
Ideal Outcome
Ultimate Goal
Prime Benefits
Secondary Benefits
Ideal Outcome
For all Stakeholders
stakeholders
ideal outcome
Solution we've got
System we've got
Ideality Audit
Benefits
Costs
Harms
Bad Solution Park
What is the problem we want to solve?
Main Constraints?
Do we want a Local or Global Solution?
Do we have a System?
What is it called?
What does it do?
Why does it do it?
Solutions for development
Solution(s) we would like
System we want
Stake Holders Ideality Sheet
Ideal System
Does a perfect
system exist
Somewhere else?
Life and Death
Causes
Before
During
After
Problem Solving Process
Technical Contradictions
Physical Contradictions
Trimming
Harms
Increase Benefits
Trends of Evolution
Invent New Systems
Ideal Problem Solving
Solution Map & Resources
Super-System
System
Sub-System
Past
Present
Future
Ideal Functions
Ideal Solutions
Ideal Resources
(What provides this
function)

The diagram on the previous page is not a flowchart but a map of the tools we use at each stage – the Ideal and Before represent problem understanding – the During is the actual problem-solving process when we apply the TRIZ problem solving tools such as the 40 Principles or the TRIZ Standard Solutions to solve the problems. The After is the mapping and selection of the solutions we have created, uncovered, combined etc.

The TRIZ Flowchart of how we might step through the map might take many forms such as the one below:

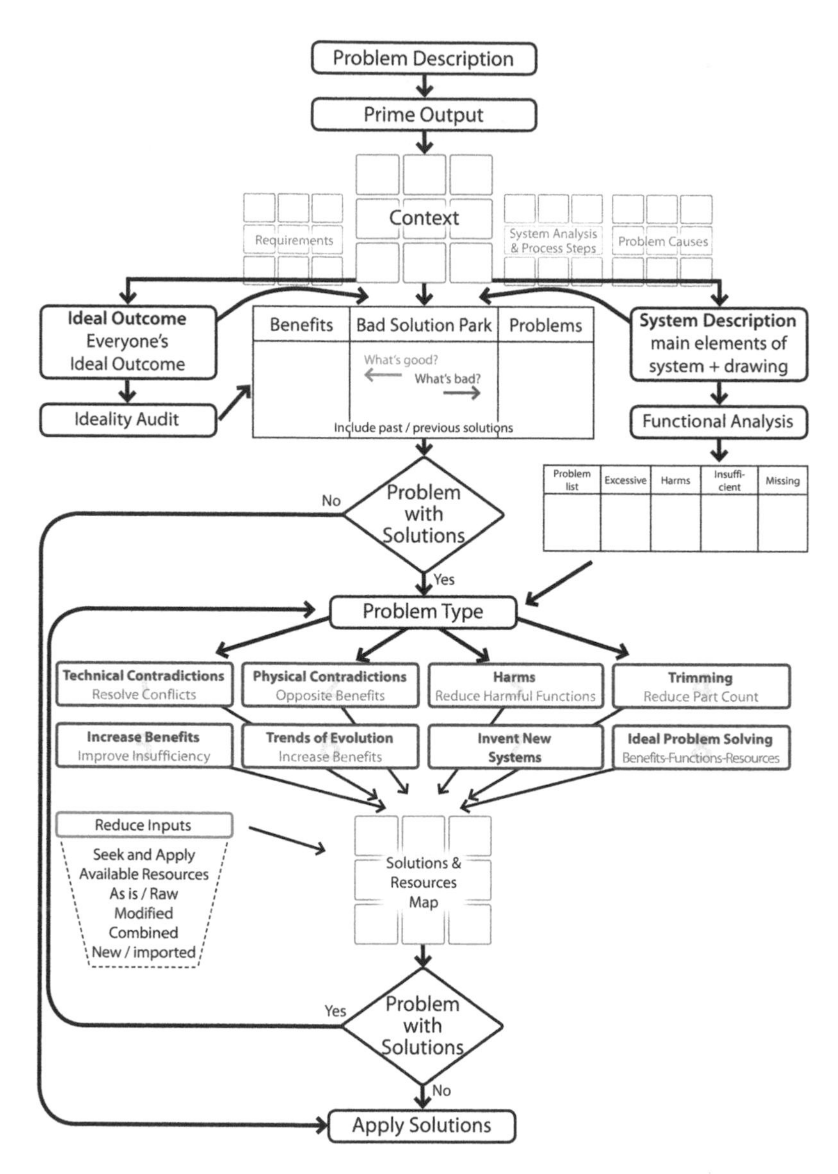

Although the flow-chart above looks complicated it is fairly straightforward to apply and TRIZ problem solving is thorough and fast. The ultimate speed to success depends on many factors but in particular it is the level of complexity and difficulty of each unique problem, the team involved, the extent of their undivided time and attention, and which tools (TRIZ and non-TRIZ) the team knows and uses and are applied to the process.

The powerful TRIZ tools are primarily there for problem solving, and solutions to problems are to be found by closing the gaps between 'what we want' (requirements) and 'what we have' (the current system). Additionally there are TRIZ tools to precede problem solving with definitions of the Ideal Outcome and TRIZ Tools for analysis of both the requirements (Ideality Audit, and the 9-Box requirements map for all stakeholders) and system analysis (function map and 9-Boxes). Additional help for both these stages is offered by the 13 TRIZ creativity triggers to jolt the brain into understanding the essential problem (they also help us with solutions). In addition there is also a separate toolset of the TRIZ Trends for next generation products (although these can be applied to problem solving as well).

Using the TRIZ toolkit helps us with focused problem understanding and then to uncover most, if not all, good, possible solutions; many of the tools overlap in their functionality for both understanding and solving. To tackle a particular problem we would very rarely use all of the TRIZ tools but generally a selection of the appropriate tools, and these will always include the ones we like best for fast problem understanding and solving.

In TRIZ the selection of the right tools for a particular problem is nearly always more than one option. Whatever we use depends on both what is appropriate and our own preferences – and this can only come with experience and practice. So how do we start? In this chapter there is a series of various small algorithms which combine subsets of the TRIZ tools and are helpful for particular problems, and more importantly always seem to work in live, industrial problem solving. Additionally there are two all-encompassing algorithms – one is ARIZ, explained in Chapter 12, and the simpler version shown here which retains much of ARIZ logic but is easier to understand and apply. However if you are the kind of person who doesn't want to read the manual, but likes to open the box and start playing with the new purchase – you can certainly do this with TRIZ by just trying one or more of these smaller algorithms. Useful insights for solutions will be found by trying all the tools on various problems, because whatever the problem, all of the TRIZ tools will help in some way, and even if it is the least perfect tool for the task it will still be useful in some way.

TRIZ is Immediately Useful but Understanding Takes Time and Practice

When after their first day of TRIZ someone says – I can't use it all – why not? I try and explain that one day of teaching is a great start but that's what it is … a beginning. Most of us are not fluent in French after our first lessons, and mathematics cannot be taught in a day. TRIZ use and understanding is built by learning and applying TRIZ tools to problems, and this can be very effective from day one, but effectiveness varies from person to person. Some people seem to grasp the range and power of TRIZ very quickly, others take more time.

My own first experience with TRIZ was excitement, and some confusion, and I cautiously applied one or two tools, then tried others as I got to know them better, until my confidence was built and I could see the whole picture more clearly. After thirteen years of TRIZ teaching I still learn something new every day about TRIZ. Everyone is different and each of us learns in different ways and at different speeds. The wide ranging, overlapping TRIZ toolkit appeals to everyone in a slightly different ways so it is not always wise to dictate one best way of using it.

My advice to the engineers I teach is to begin by building familiarity and confidence with your own initial favourite TRIZ tools and explore their power. Initially everyone has at least one favourite, and other TRIZ tools which they initially distrust or under-rate; therefore adhering to the ones you both like and believe in is probably the right initial approach (one man's meat is another man's poison). Also for everyone their own personal list of TRIZ favourites changes with time and use, and periodically each of us will embrace a less familiar TRIZ tool and try it on everything, until eventually, the whole toolkit is understood liked and appreciated.

There are TRIZ tools for you, whatever kind of problem solver you are

One aspect when first learning TRIZ which puzzled me was that some practitioners only taught about a third/half of the tools, asserting that the rest were unnecessary, and that amongst many TRIZ practitioners very different subsets of TRIZ tools are taught. Only years of experience made me realize that they each taught only their initial favourite TRIZ tools, and always only focussed on these. This focus and enthusiastic championing of initial favourite TRIZ tools has good and bad aspects, but can be a particular impediment within companies, when a new TRIZ enthusiast introduces only their own favourite subset of the TRIZ toolkit to their company. In one of our leading engineering companies one new, senior TRIZ champion asserted his will and insisted that they only needed his favourites – the 40 Principles, 8 Trends, Function Analysis and Smart Little People as a complete toolkit. Ten years later this company still only teaches these TRIZ tools to be used by its thousands of engineers.

My own approach has been to teach/offer all the TRIZ tools. This is because different people have different favourites and everyone starts in a slightly different place with TRIZ. Amongst my colleagues we all have current favourites and have to resist the temptation to sell these hard to TRIZ novices. Favourite tools do have patterns in time for each individual on their TRIZ journey, and at a TRIZ conference recently one new practitioner was extolling the wonders of ARIZ and an old TRIZ hand commented loudly that they must be in their third year of using TRIZ (the third year is often the stage when everyone finally gets to use and appreciate the power of ARIZ for a while). However it doesn't matter which tools we use most frequently and which we currently like the best, as long as we use the toolkit, and try it out on a wide range of problems.

Problems vary according to the stage of development of the system, the type of problem and the means to solve it, and as the TRIZ tools overlap there can be no hard and fast rules about which tool to use when.

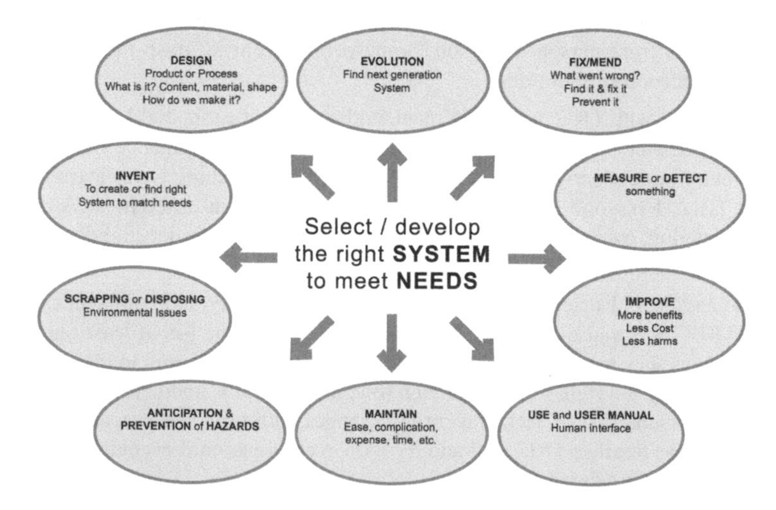

The important factor is that TRIZ has a wide range of tools to tackle almost all aspects and all stages of problem solving The power and fun of TRIZ problem solving is phenomenal when a group of confident, clever TRIZ practitioners work together to solve difficult problems. The good team is far more effective faster and get a lot further than one 'hero' coming up with their one solution to tackle the problem.

The TRIZ toolkit is rigorous and wide-ranging and different tools are needed for different problems of various levels of difficulty. There is a wide range of activities in problem solving and there is a TRIZ tool for all of them. TRIZ uniquely offers solution lists for problem solving but offers other powerful tools to help us think clearly and to understand the problem.

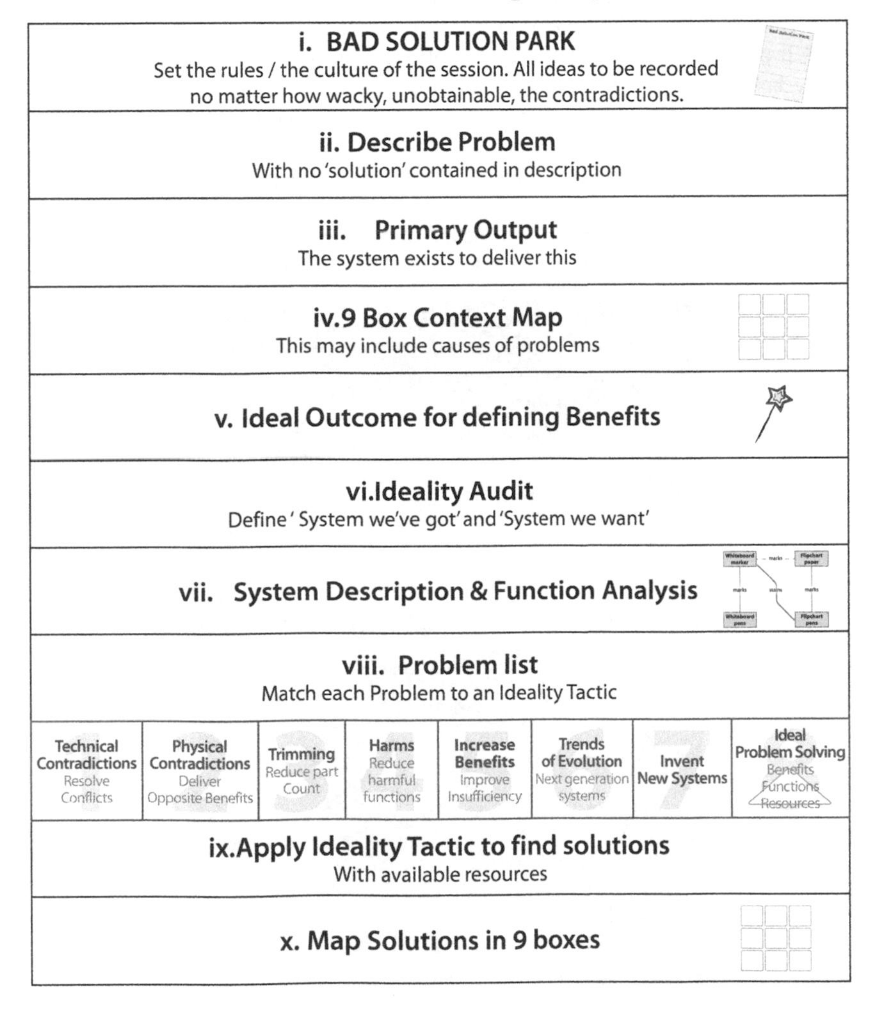

Problem Solving Steps
i. BAD SOLUTION PARK
Set the rules / the culture of the session. All ideas to be recorded no matter how wacky, unobtainable, the contradictions.
ii. Describe Problem
With no 'solution' contained in description
iii. Primary Output
The system exists to deliver this
iv.9 Box Context Map
This may include causes of problems
v. Ideal Outcome for defining Benefits
vi.Ideality Audit
Define ' System we've got' and 'System we want'
vii. System Description & Function Analysis
viii. Problem list
Match each Problem to an Ideality Tactic
Technical Contradictions
Resolve Conflicts
Physical Contradictions
Deliver Opposite Benefits
Trimming
Reduce part Count
Harms
Reduce harmful functions
Increase Benefits
Improve Insufficiency
Trends of Evolution
Next generation systems
Invent New Systems
Ideal Problem Solving
Benefits
Functions
Resources
ix.Apply Ideality Tactic to find solutions
With available resources
x. Map Solutions in 9 boxes

Two Fundamental Areas in Practical Technical Problem Solving

New systems – practical invention of new systems and next generation systems.
Improving the systems we've got by solving its problems without changing the system (more benefits, less problems, lower costs).

Both involve following processes to deliver innovation, creativity, clever solutions and new concepts, and all need the TRIZ problem understanding and tools.

All routes need some understanding of what we want and the TRIZ Ideal Outcome Tool is a very good starting place for every kind of problem solving – it is an essential step route in for many of TRIZ processes.

New Systems – Invention and Next Generation Systems

Inventing or creating a new product for market acceptance means finding a new system to match and meet some unfulfilled needs – or often means combining a selection of technologies to form a new system in order to meet new requirements. Here the journey from the first ideas for invention to launch onto the market means it must be developed, checked, made useable and able to be cost-effectively manufactured. It must pass the first test of getting off the ground when its Ideality Balance reaches a point when the benefits it fulfils are worth all its investments, all its costs and exceed its harms and problems. In some practical way the system must become useful and in market terms it must begin to fly. During its journey to and once it has reached that point all the TRIZ tools are helpful to improve it and the TRIZ Trends can help predict its journey (through several generations of systems) to perfection. Practical inventing of systems for market acceptance and next generation systems TRIZ tools include Ideal Outcome, Prism of TRIZ, Effects database, 8 TRIZ Trends, Creativity Triggers, Function Analysis and Thinking in Time and Scale (see www.TRIZ4engineers.com).

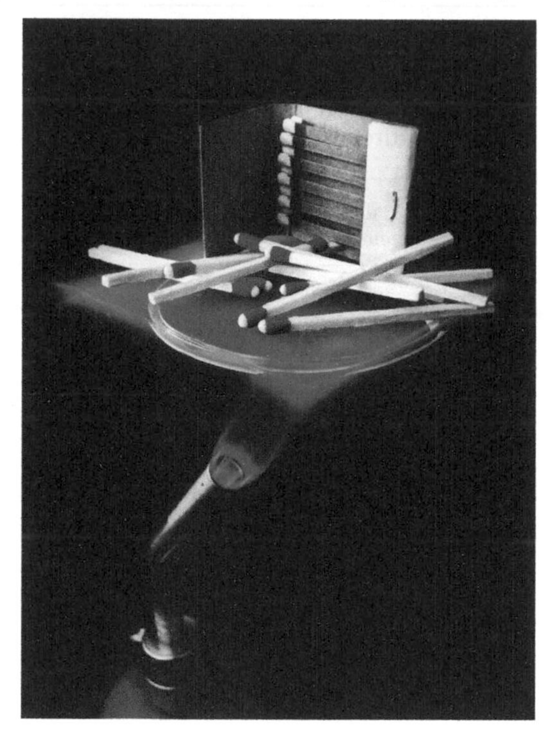

New materials such as aerogel offer many opportunities for new systems. Reproduced from http://stardust.jpl.nasa.gov/photo/aerogel.html

With invention we have to be willing to look beyond our own knowledge and ask the 'how to' questions and delve into the powerful TRIZ effects and new knowledge on the internet. To locate the next generation needs a willingness to look at new technologies which may as yet be unproved in the industry it is being invented for, but may well be accepted and useful in another completely different industry.

Improving Systems, Delivering More Benefits, Reducing Inputs and Dealing with Any Problems

TRIZ Tools – 40 Principles, Standard Solutions (Insufficiencies, Harms and Trimming), Creativity Tools Thinking in Time and Scale and the 8 TRIZ Trends.

Once a system has been invented, improving a system means increasing Ideality (getting benefits up and/or costs and harms down) and involves dealing with functions in some way – either providing them, improving them to give better or more benefits (outputs we want) and/or reducing their costs (all inputs) and/or dealing with or eliminating their harms (all unwanted outputs).

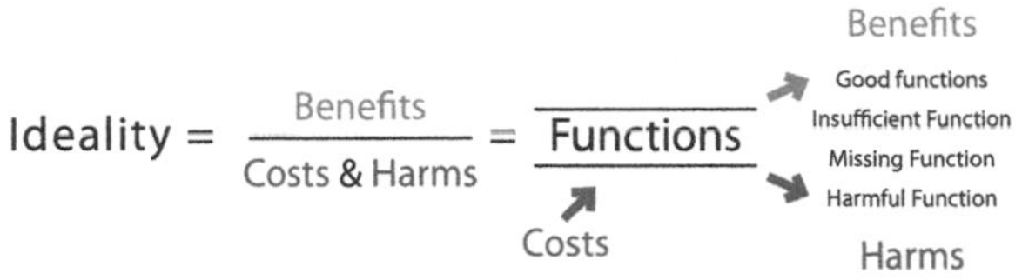

$$\text{Ideality} = \frac{\text{Benefits}}{\text{Costs \& Harms}} = \text{Functions}$$

Benefits
Good functions
Insufficient Function
Missing Function
Harmful Function
Harms
Costs

The TRIZ ways of changing functions is by application of the relevant TRIZ solution concepts from the TRIZ overlapping lists; 40 Inventive Principles for solving contradictions (ways of getting opposite benefits or increasing benefits while decreasing costs and harms), 8 Trends of Evolution for perfecting systems and the 76 Standard Solutions which cover a whole range of solutions. All 100 or so answers help us tackle any problems with functions, and solve those problems to improve a whole system or parts of the system. Functions are provided by components and systems and Ideality can be ascribed to a part or a whole system – we can improve any system by raising its ideality – improving at least two of the three options of more benefits, less costs, less harms. A system with more benefits and proportionately more costs is not a better system, just a more expensive one; while a system with less costs and less benefits is just a cheaper system. A better system is one which has an improved ratio of benefits, costs and/or harms. Ideality increase can also be from replacing the system with a better one. Technology advances often are achieved with higher Ideality from better/different systems.

Ultimately we are trying to maximize benefits for the least costs and harms. Maximizing Ideality means finding exactly the right functions to deliver benefits/needs for the minimum inputs – and as functions cannot exist without inputs (a system) then problem solving means matching needs to the right system, under whatever circumstances the problem occurs, and getting the ideality of that chosen system as high as possible. There are eight principal routes to achieving this – the Ideality Tactics.

Ideality Tactics – Improve Outcomes, Reduce Harms and Reduce Costs

$$\textbf{Ideality for any system} = \frac{\Sigma\ \text{Benefits produced by the system (outputs we want)}}{\Sigma\ \text{Costs (inputs)} + \Sigma\ \text{Harms (outputs we don't want)}}$$

There are a total of eight basic problem-solving routes to increasing Ideality – the Ideality Tactics:

1. Solve the Technical Contradictions – start with a bad solution and resolve its conflicts (40 Principles)
2. Solve Physical Contradictions – separate and achieve the opposite benefits (40 Principles)
3. Trimming – simplify system, reduce costs and complexity using resources – (Trimming Rules)
4. Deal with Harmful Actions – (24 ways of dealing with harm)
5. Increase Inadequate Benefits – (35 ways for dealing with insufficiency)
6. Evolution of current system with TRIZ Trends to improve the system towards perfection (also useful for IP and patents)
7. Invention and ways of providing unfulfilled requirements (Ideal Outcome and resources)
8. Ideal Fast Problem Solving – benefits-functions-resources.

Problem Understanding and Solving Routes and Applying the Ideality Tactics

Problem Understanding Steps

Begin with the general background and problem understanding steps:

i. Create and constantly fill in a Bad Solution Park – capture all and everyone's solutions at every stage. Include previous solutions tried. Capture what's good (in benefits) and what's bad (in problems) about each solution. Allow all solutions no matter how wacky or weird.

ii. Problem – rough description of the problem and general problem situation and constraints.

iii. Primary output – what is the main output the system exists to deliver?

iv. 9-Box Context Map – background information and relevant data and chosen scale of solution also map causes of problems if this is relevant or under question.

v. Define Ideal Outcome (to capture benefits).

vi. Ideality Audit – define the gaps between what we have and what we want.

vii. System description – identify the main components of the system, and describe them in very simple (non-technical) terms – draw picture of system. Draw function analysis maps to identify problem types.

viii. Select problem type for appropriate Ideality Tactic

ix. Follow Ideality Tactic algorithm for solution concepts

x. Map each solution in the Solution Map and match to available resources for solutions

i) Bad Solution Park – Capture All Solutions at every Stage of the Problem Understanding

This sets the culture of the session for everyone's participation and sharing solutions. We should include previous solutions tried. Capture what's good (in benefits) and what's bad (in problems) about each solution. The logic of TRIZ problem solving often uses our Bad Solutions – the solutions we can't help thinking up while describing the problem (all contradiction problem solving starts with a Bad Solution). Our Bad Solution (also called our 'ugly baby') was generated with our top brain power and usually contains good things and bad things. All problem understanding and solving should be in the presence of a Bad Solution Park (large sheet of paper) and everyone should have stickies to capture all their Bad Solutions (initially none should be lost or discouraged – no negativity tokens should be issued).

TRIZ helps us break out of discouraging behaviour

We can further analyse the Bad Solutions by recording what's good about them – what benefits do they fulfil – in the 'What do we want? 'column, and what's bad about them in the 'Problem with Solution?' column. If we have both then we have a technical contradiction. If we have conflicting (opposite) benefits then we have a physical contradiction and can apply the relevant Ideality Tactics.

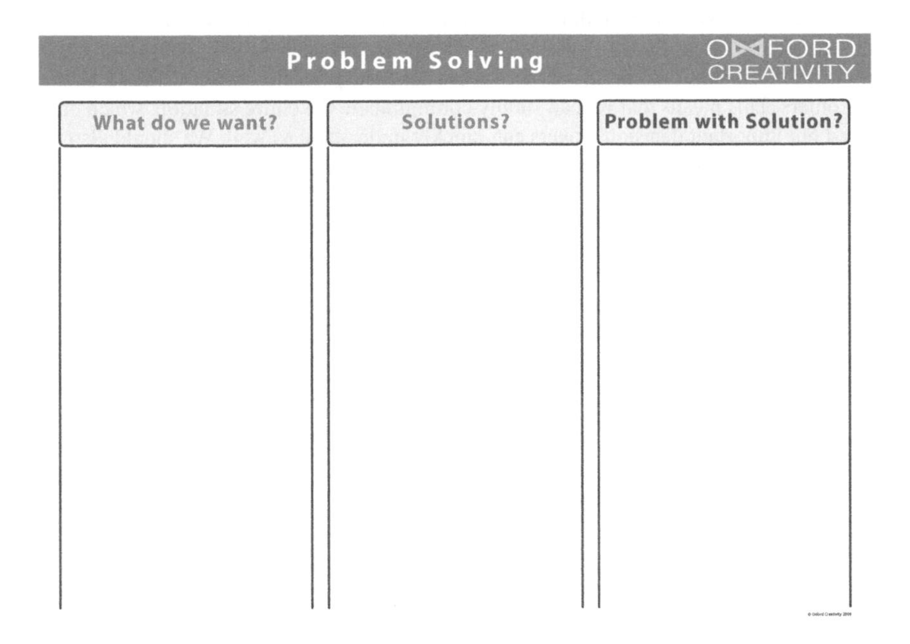

ii) Problem – Describe the General Problem Situation and Constraints

Don't offer any solutions while describing the problem. All problem solving begins with a problem, so starting with some kind of problem statement is essential (even if it is inaccurate, incomplete and will probably change). Essentially a top-of-the-head, rough description of the problem in a couple of sentences, a quick summary, is all that is required here (for example for a heat exchanger, problem statement could be to increase efficiency and decrease its size and constraints could be inlet and outlet conditions cannot change and the fluids cannot mix).

Problem situations vary enormously and the problem may be clear and well defined, or we may have a rather messy problem situation, from which an understood problem or problem list will emerge after some analysis. All we need here is a small statement of the problem reality and the main constraints.

iii) Identify the Primary Output of the System

What is the main output the system exists to deliver?

This is an important step and often forgotten – when we start problem solving we can get lost in problem details almost immediately. The statement of the Primary Outcome reminds everyone the main thing we are trying to achieve – what the system is there to deliver. There are many other benefits which we do not consider here – we just want to record the main one, its Primary Benefit which delivers its Primary Function. For example for a heat exchanger, the primary output/benefit is 'heating or cooling' (the Primary Function is exchange heat between fluids). Primary Benefit for a toothbrush is 'clean teeth' (the Primary Function is remove plaque). When problem solving it is easy to lose sight of the essentials and the simple statement of this main output is an obvious but important step.

iv) 9-Box Context Map – Background Information and Relevant Data and Scale of Solution

Fill in the 9-Box Context Map and add any relevant information about the particular problem. This can be done in a number of ways, but to quickly complete a simple context map delivers a simple summary and helps with an essential understanding of the situation (any missing facts can be added or revised later). Mapping in the 9-Boxes the important factors in Time and Scale is a way of capturing the whole problem context. It also helps us summarize and understand the problem environment which is valuable to us and to others. This means that we can simply communicate the entire situation, which ensures that less informed but important decision makers can quickly understand as well. We should also record the Scale of Solution – are we trying to create a new system? Or fix the one we've got to solve the problem?

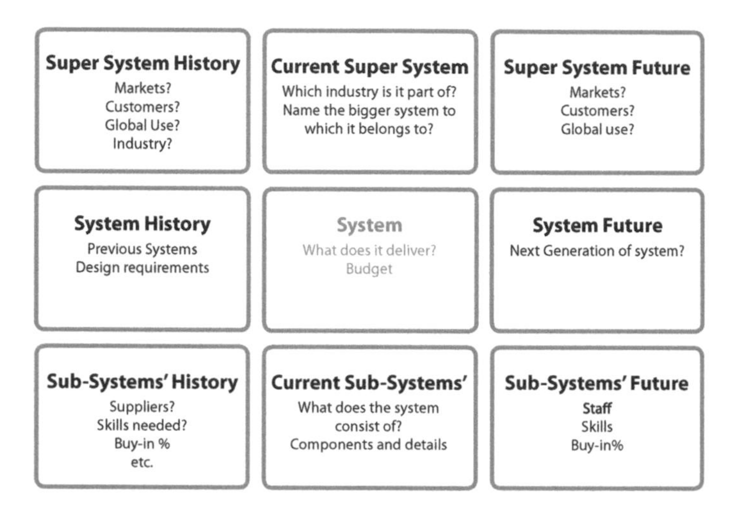

Super System History
Markets?
Customers?
Global Use?
Industry?
Current Super System
Which industry is it part of?
Name the bigger system to which it belongs to?
Super System Future
Markets?
Customers?
Global use?
System History
Previous Systems
Design requirements
System
What does it deliver?
Budget
System Future
Next Generation of system?
Sub-Systems' History
Suppliers?
Skills needed?
Buy-in %
etc.
Current Sub-Systems'
What does the system consist of?
Components and details
Sub-Systems' Future
Staff
Skills
Buy-in%

Put the problem system in the middle box and the end result in the future middle box. After an initial description of the problem, we start by understanding the context of the problem in Time and Scale.

In Scale we map the system environment, its bigger picture (what is it part of?). We also map the details of the system.

In Time we map the future (where are we going, what are we hoping to achieve?) and the past – the relevant general history, covering as many days/ months/years/decades as appropriate.

Start by quickly answering the following questions (we can revise the answers later if we need to):

- Simple and general problem system description (middle box).
- What is the end result we want? What are we trying to do? Outcomes?
- Why do we want this? Think about this answer in the past, present and future.
- For us – define us.
- For others – define who is relevant.

Map relevant essential issues, changing constraints, trends from the past which may be pertinent at any scale level such as changing regulations (Super-system boxes) or changing workforce skills (Sub-system boxes) or what budget issues are relevant, such as timing of money etc.?

We can use the Thinking in Time and Scale tool and draw as many different 9-Boxes as are needed to map the essential elements of the problem. Use more than 9 boxes if necessary but capture essential, distilled data which may be needed for the problem understanding and solving. Map causes of the problems especially if there are many linked causes creating the problem situation. If there are unknown causes then this is a particularly valuable addition to problem understanding. (See Chapter 4: Thinking in Time and Scale for more details.)

v) Define Ideal Outcome (to Capture Requirements)

The Ideal Outcome is fairly easy to roughly define, but for a detailed, exact description, there are simple routes to follow. Defining the ultimate goal needs clarity of thought but is important to record. Ultimate goal is at a higher level and beyond the Primary Benefit. It is the higher purpose of the system (for a toothbrush the Primary Benefit is clean teeth – the ultimate goal is probably un-decayed teeth) and defines why the system exists and was created. This may influence any expectations that may affect the way the problem is tackled – especially from different stakeholders.

Ideal Outcome = Ultimate Goal
Primary Benefit
Benefits Benefits Benefits
Benefits Benefits Benefits

The perfect system is our final theoretical aim, and an interesting part of initial problem understanding is to set the direction towards perfect solutions, by defining the Ideal Outcome. For this we begin by imagining the Ideal and use this and the other TRIZ creativity tools to stimulate solutions which deliver everything we want (delivered by available resources). This simple thinking asks us to imagine some Ideal Outcome – when the Ideal System solves the problem itself without any costs or inputs and without any harms or problems.

vi) Ideality Audit – Define the Gaps Between What We Have and What We Want

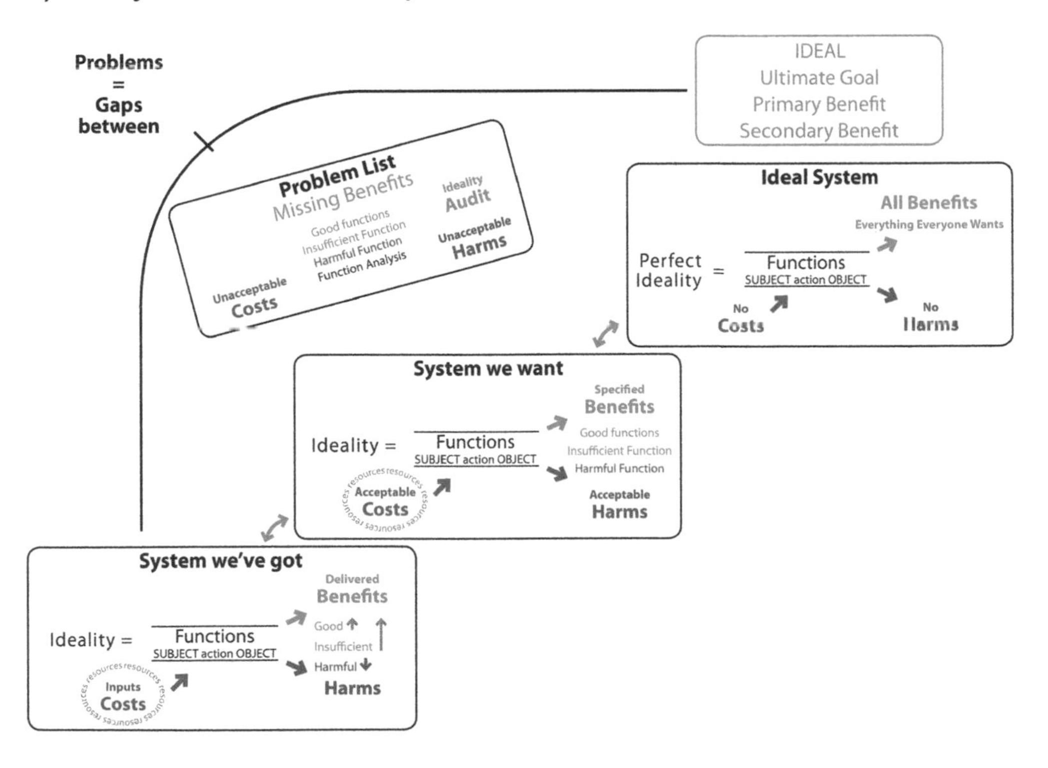

The previous steps are designed to bring us to this point. This final problem definition of the Problem Gap may be very different from the problem description in step one. The Problem Gap should define the main difference(s) in benefits between the system we've got and the system we want – problem solving then should remove the gap. When the system is further analysed with function analysis (the next step in the process) we produce a detailed problem list revealing the many individual problems. These are problems at the function level and need solving one at a time.

IDEALITY AUDIT First roughly define Ideal Outcome – Ultimate Goal and everything we want then define the following:

IDEALITY	**SYSTEM we've got**	**SYSTEM we want**	**Ideality Gap**
Prime Benefit			
Other Benefits			
Prime Output / Function			
Acceptable costs / inputs			
Unacceptable costs / inputs			
Acceptable harms/problems			
Unacceptable harms/problems			

Include other relevant factors for problem understanding and solving include such as:

Constraints (Imposed conditions, rules or regulations)

Simple Drawings of System (if relevant)

This Ideality Audit can be more detailed by filling in the chart below for the 'system we've got' giving all the relevant data. To accurately complete this Ideality Audit chart it needs to be done in conjunction with the next step – Function Analysis, to accurately reveal any problems with functions.

Having understood our benefits, harms and costs, identify them according to the criteria in the tables below. This identifies our problem list

Benefit How? →	Function ← Why?	Sufficient?	Insufficient?	Missing?	Excessive?

Costs/Inputs	Acceptable	Unacceptable

Harm/Problems	Acceptable	Unacceptable

vii) System Description – Identify the Main Components of the System

Describe all components in very simple (non-technical) terms – draw a picture of system and map it in Function Analysis to identify problem types. (This can be a simple component list, or it can be expanded using 9-Boxes and/or a functional analysis.)

Take Out Acronyms, Technical Terms or Jargon

This is an important step and helps break psychological inertia by taking out all technical jargon. By describing all components in basic language we ensure that we understand them and see what they do and why they are part of the system. Simple drawings are often useful at this stage.

Then follow the steps Tasks 1 and 2 in the diagram below to create a function map of the system.

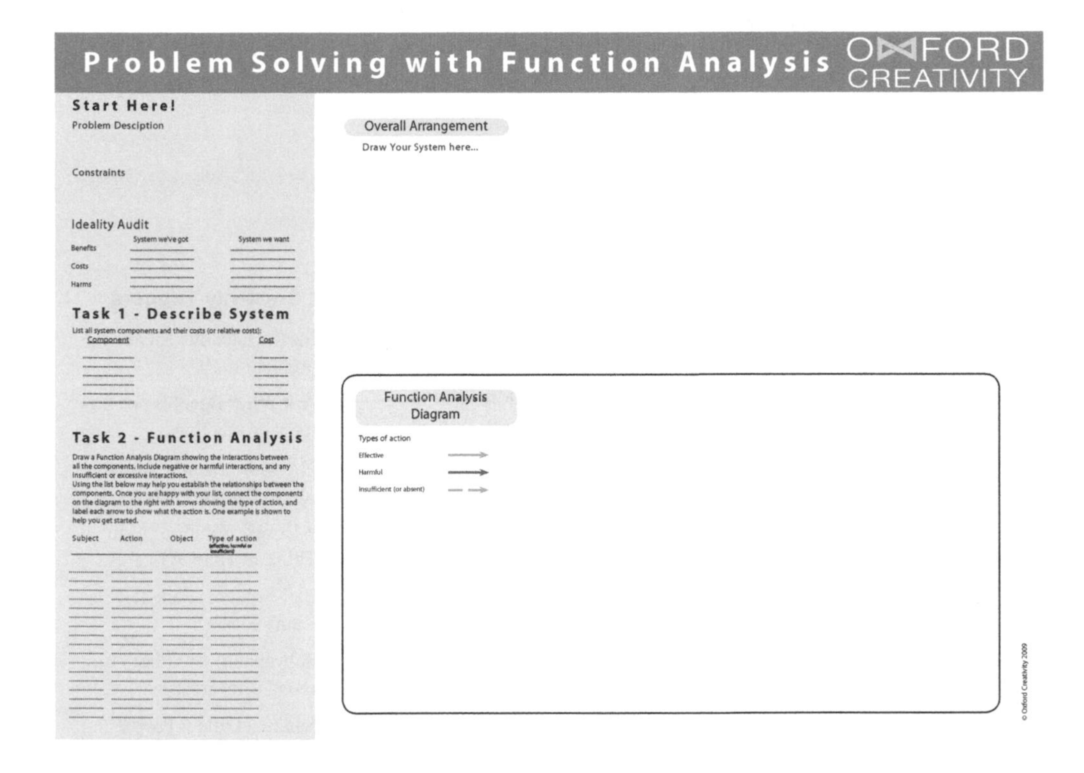

viii) Select Problem Type for Appropriate Ideality Tactic

ix) Follow Ideality Tactic Algorithm to Reveal all Relevant Solution Concepts

Choose the right Problem Type. After defining our problem, we solve by applying one or more of the Ideality Tactics.

x) Fill in the 9-Box Solution Park

Final stage after the Ideality Tactics is to map all the solutions in the 9-Box Solution Map and match to available resources for clever solutions. We want resource recognition for cost-effective, smart solutions – looking for available raw resources, or resources we can modify, or resources we can combine together and only finally bringing in new resources.

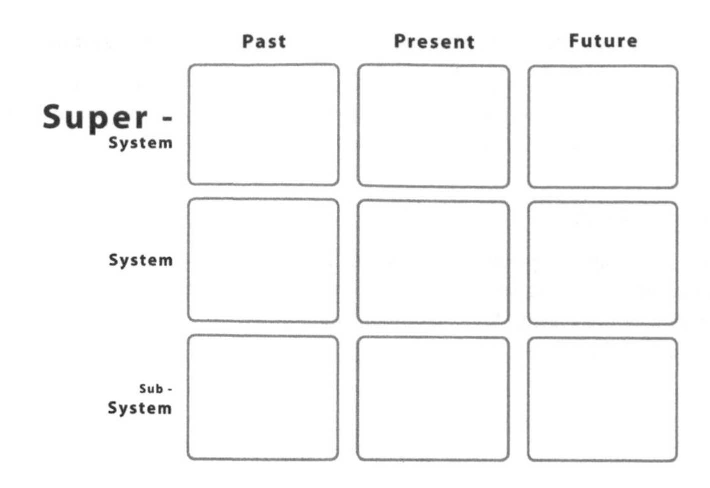

Applying the Ideality Tactics

At any stage of the problem understanding and the problem solving process we can try one or more of the TRIZ Creativity Triggers. The first one The Ideal is powerful and defining the Ideal will point us in the right direction for the right solutions and we can move from the Ideal back towards reality. The search for solutions can is helped by defining the Ideal Solution which just somehow solves the problem itself. When defining the Ideal many solutions will occur and all solutions should be captured.

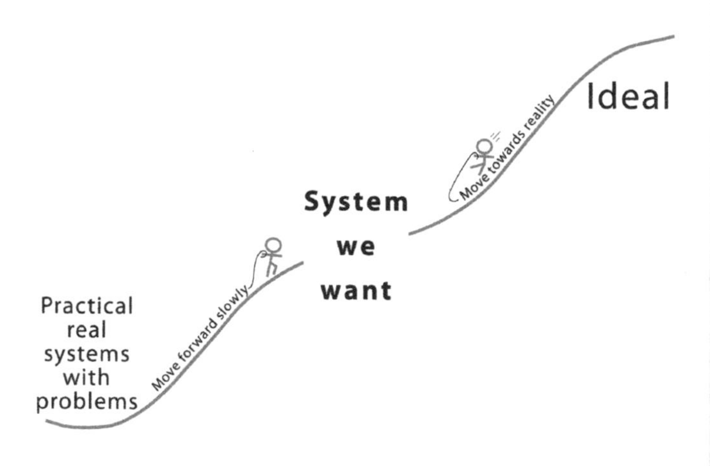

TRIZ Creativity Triggers

1. Ideal Outcome solves the problem itself
2. Ask Why? Benefits not features or functions
3. X-Factor (some resource solves the problem)
4. 9-Boxes Solution Map – Think in Time & Scale
5. Bad Solution Park (all the team's ideas)
6. Subversion – Other way round – Invert it
7. Smart Little People – model the problem
8. Size-Time-Cost – exaggerate the problem
9. Simple language (no technical jargon)
10. Idea/Concept – concepts behind each idea
11. Prism of TRIZ – locate existing solution concepts
12. Life and death analogies (others' critical solutions)
13. Combine good solutions (carrot/cabbage)

1. Increase Ideality by Solving the Technical Contradiction

Technical Contradictions occur when we improve something and something else gets worse (e.g. strength vs. weight). In the Ideality Equation when we add to the benefits of one element, then the costs and/or harms may increase. We need to find a way to increase benefits without making costs or harms any worse. We do this by solving Technical Contradictions and unlinking the components so that we can change one without changing the other.

Technical Contradictions are solved with the Contradiction Matrix and the 40 Principles.

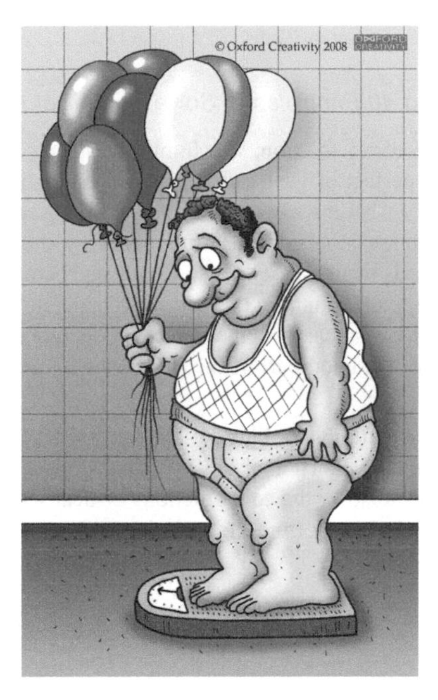

Anitweight is one of the 40 principles for solving contradictions

Solving Technical Contradictions

Increasing Ideality means improving some benefits without making any other benefit get worse or increasing benefits without making costs and/or harms worse, or reducing costs and/or harms without making benefits decrease.

Increase benefits without increasing costs or harms.

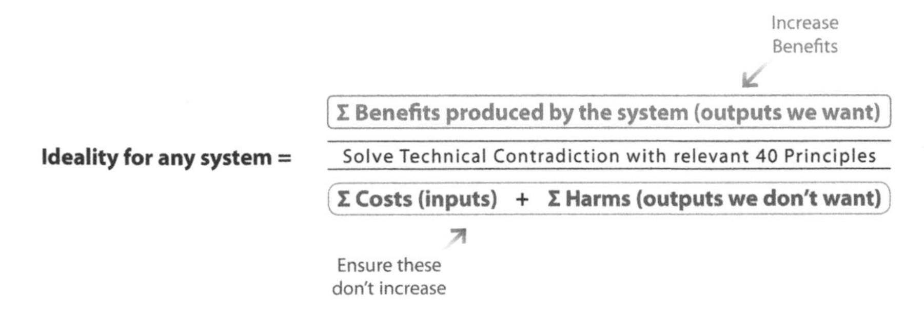

Alternatively reduce costs or removing harms without a decrease in benefits.

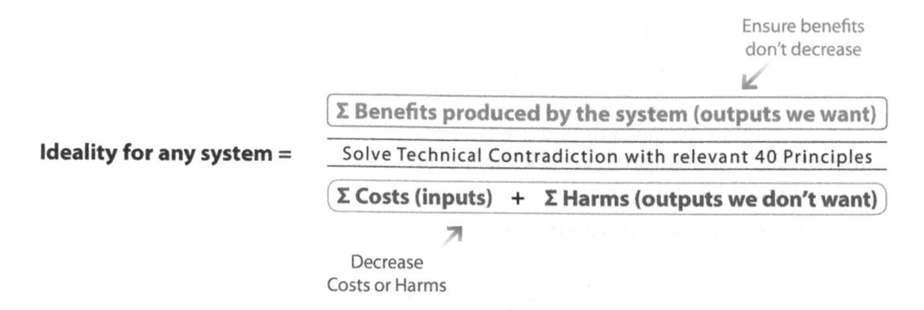

Bad Solutions to Uncover and 40 Principles to Solve Technical (Chained) Contradictions

When we have a problem, we picture a solution, which improves, fixes, makes something better about the problem (no matter how imperfect or what subsequent problems it creates). We call this a 'Bad Solution' and we need to identify what it makes better and what subsequently gets worse. We map all the problems with the 'bad solution' noting what is improved by the solution (in the what's good? column) and what deteriorates or is made worse (in the what's bad? column). This uncovers contradictions. We then use the TRIZ Contradiction matrix, and the 40 Principles to locate better solutions than the original 'Bad Solution'. We have 'unchained' the improving feature or function from the deteriorating one. This ensures that we get the good part of the solution without the bad part, and have solved the contradiction.

When initially looking for solutions to the problem, it can be assisted by the TRIZ creativity tools to first stimulate solution ideas, which we then analyse for 'What's good? and What's bad? We can then apply the TRIZ solution concepts to solve the contradictions or deal with any harms, which help us to develop a new solution – often a long way from first solution.

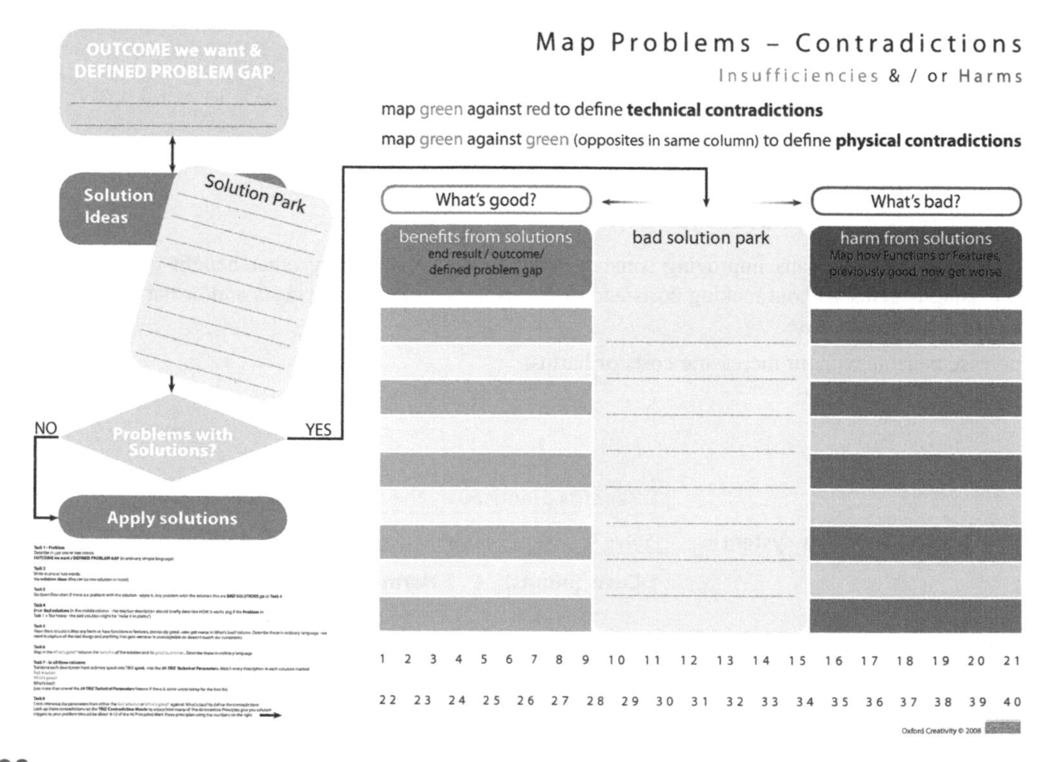

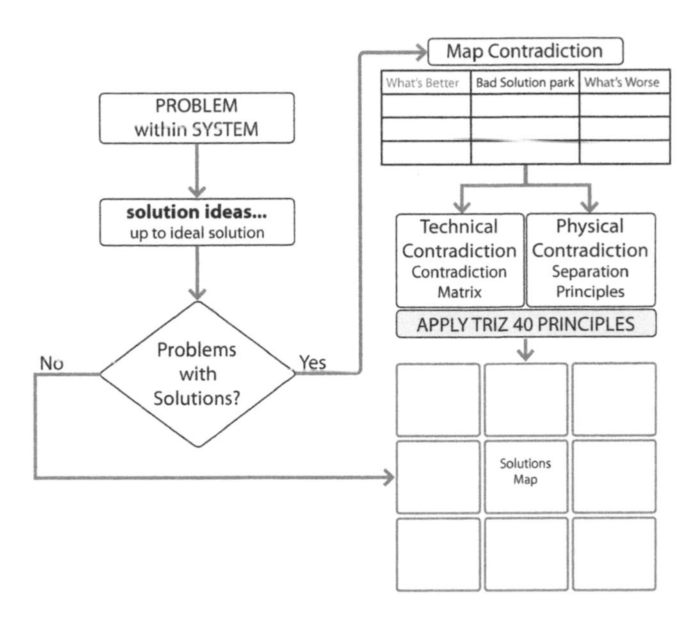

2. Solve Physical Contradictions

Separate and achieve the opposite benefits (use the Separation Principles and the relevant 40 Principles).

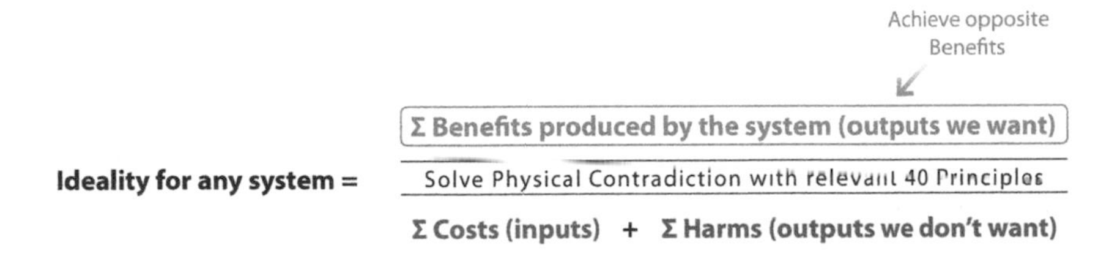

Physical Contradictions are solved using the Separation Principles and the 40 Principles.

Use Ideal Solutions to Find and Solve Physical (Opposite) Contradictions

After defining a problem situation, describe an Ideal Outcome to that problem. Try and imagine how that Ideal Outcome might be realized. Look for known ways to picture how to deliver everything we want.

When we define the Ideal it usually includes opposites – mutually exclusive benefits (fattening chocolates which make you thin?), i.e. when we have a problem and picture an Ideal Outcome which gives us all benefits, it is everything we want – we describe something apparently impossible with some of those benefits conflicting and opposite to each other (such as something big and something small). Identify the Opposite Contradiction(s) of these benefits, and solve by *separating* them and then using the TRIZ Separation Principles and the TRIZ 40 Principles – to suggest a new system which delivers all the benefits we want as shown in the sequence below (see Chapter 5 for more details).

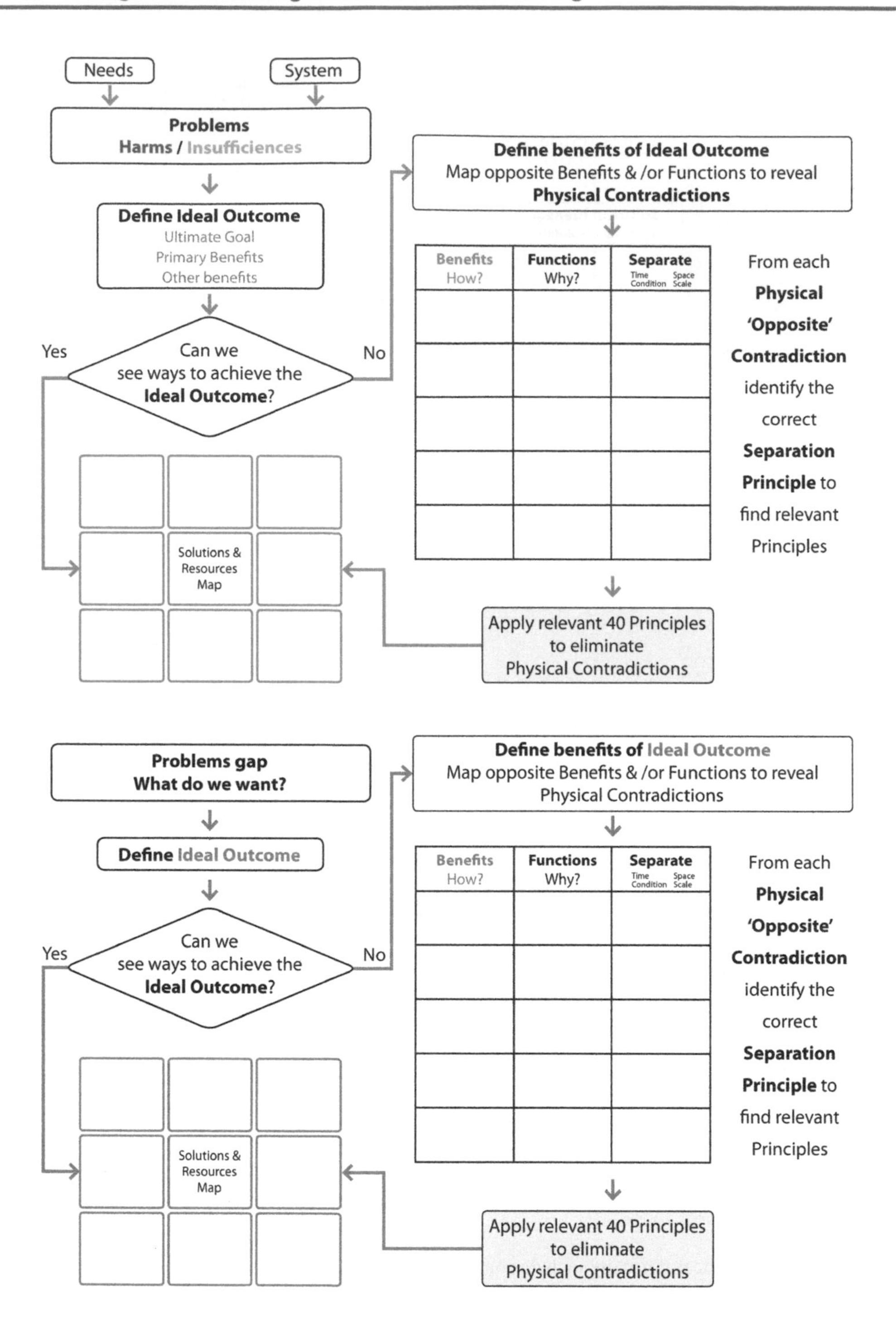

3. Trimming – Reduce Costs and Harms, Then Apply Resources

When we want to simplify systems (reduce costs, harms and complexity) we can look to trim out components by applying the Trimming Rules. This means we are looking to eliminate or trim out components to reduce harm, excess and complexity and then replace any of their essential useful function by locating and mobilising useful resources from other components, environmental resources or any other resource available within or around the system.

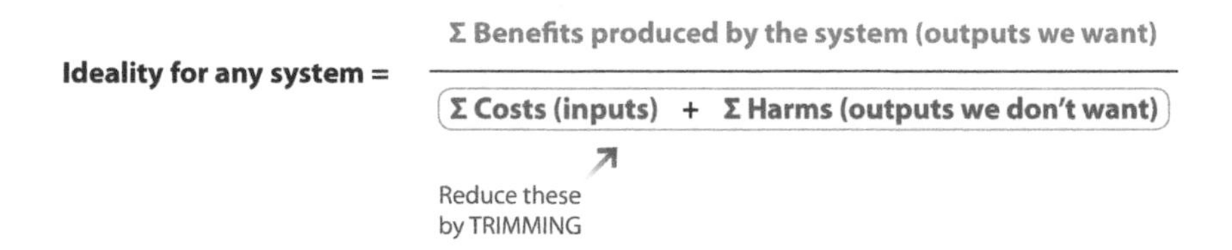

Σ Benefits produced by the system (outputs we want)
Ideality for any system =
Σ Costs (inputs) + Σ Harms (outputs we don't want)
Reduce these
by TRIMMING

These strategies are applied to decrease any excessive inputs. We need to identify where components have harms and/or may be over-designed (excessive actions) and reduce them accordingly. Trimming is simplifying systems by removing parts (especially the most complex and expensive and those with harms or problems) and ensuring their useful functions are still provided – ideally provided by available resources. When components are selected to be trimmed, they have been recognized and defined from a TRIZ Function Analysis. Once identified TRIZ has tools for how to trim without losing any benefits. Apply the Trimming Rules as shown below:

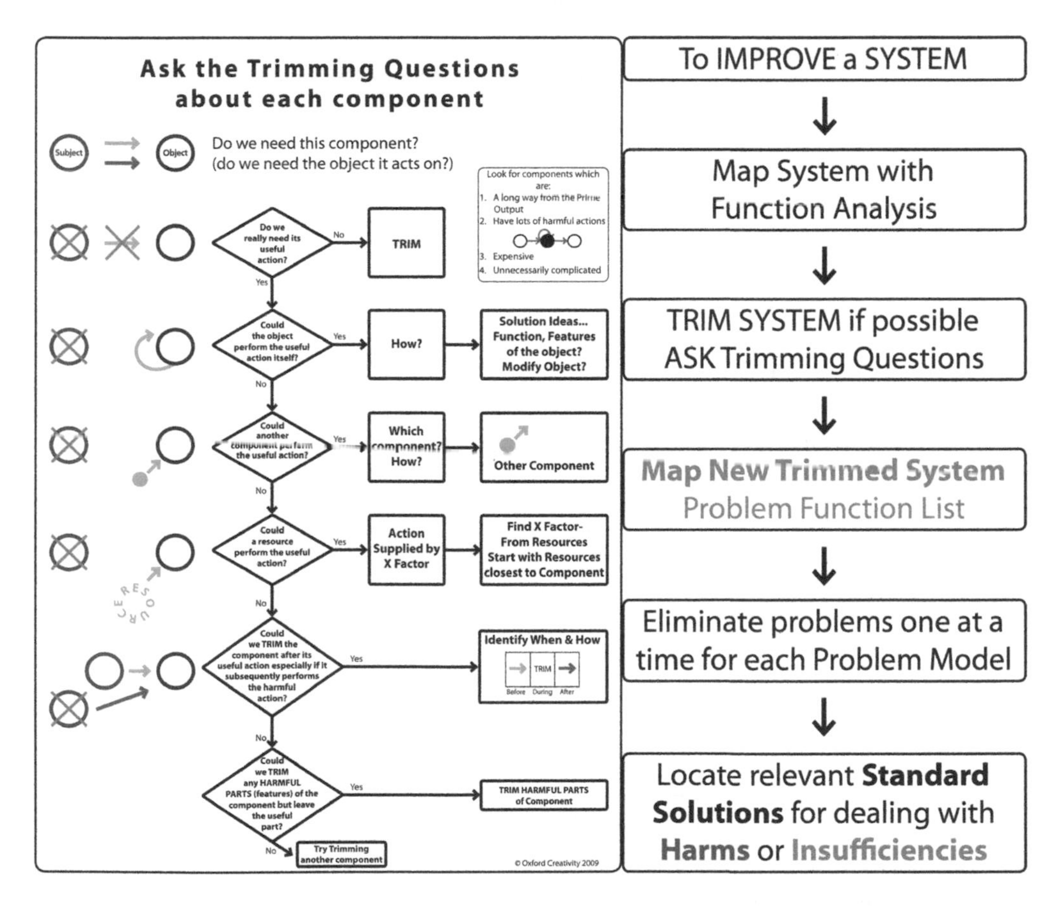

Map New Trimmed System Problem Function List
Eliminate problems one at a time for each Problem Model

1. ______________________________

2. ______________________________

4. Remove or Reduce Harmful Actions

24 ways of dealing with harm (see Oxford Standard Solutions page 359).

The system has outputs we don't want which are harmful. Harmful actions are normally recognized and defined by doing a TRIZ Function Analysis. Once identified apply the TRIZ tools for dealing with harms:

- **Eliminate** – Trim out the Harm – 6 ways
- **Stop** – Block the Harm 11 Ways
- **Transform** – Turn Harm into Good – 4 ways
- **Correct afterwards** – Put right the harm – 3 ways

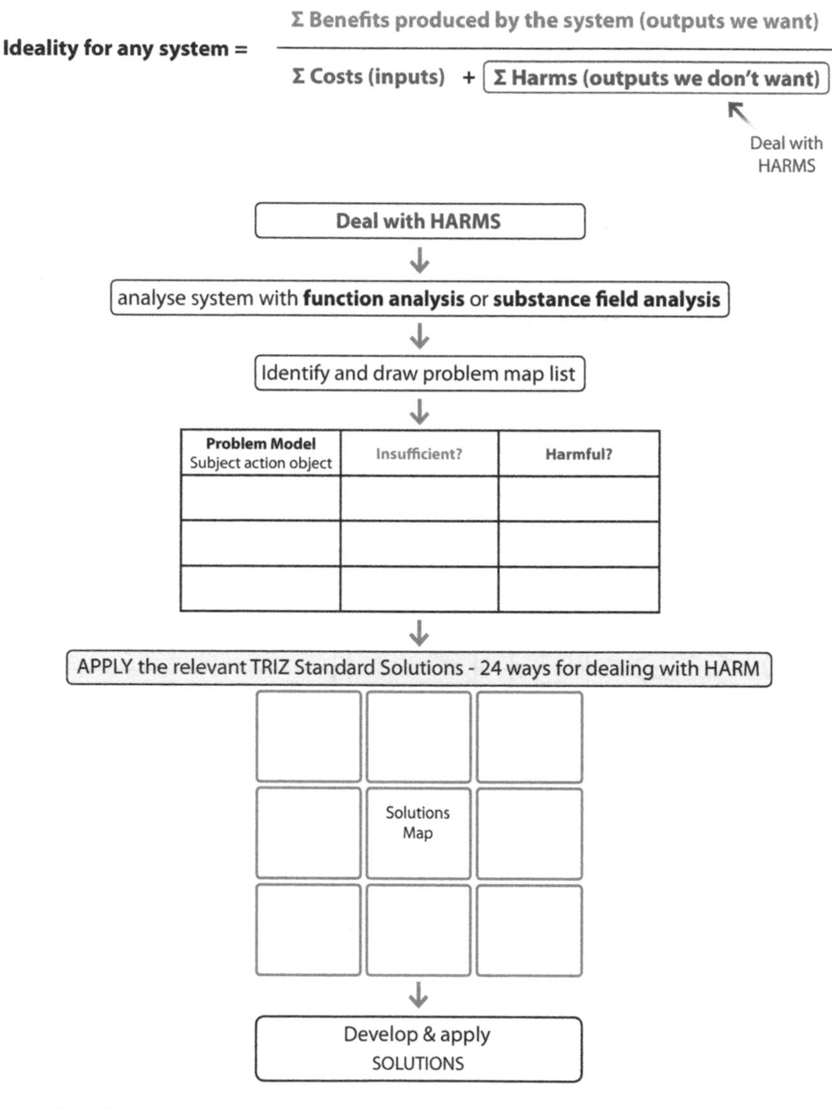
Ideality for any system =
Σ Benefits produced by the system (outputs we want)
Σ Costs (inputs) + Σ Harms (outputs we don't want)
Deal with HARMS
Deal with HARMS
analyse system with function analysis or substance field analysis
Identify and draw problem map list
APPLY the relevant TRIZ Standard Solutions - 24 ways for dealing with HARM
Solutions Map
Develop & apply SOLUTIONS

Problem Model Subject action object	Insufficient?	Harmful?

5. Increase Inadequate Benefits

Inadequate Benefits – when there is an insufficiency in an action there are 35 solutions for insufficiencies.

The system has useful actions but these are in some way insufficient. These can be recognized and defined from a TRIZ Function Analysis. Once identified apply the TRIZ for increasing benefits.

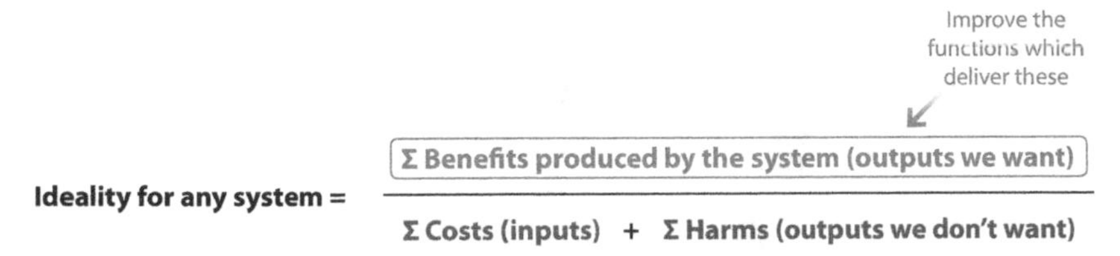
Improve the functions which deliver these
Ideality for any system =
Σ Benefits produced by the system (outputs we want)
Σ Costs (inputs) + Σ Harms (outputs we don't want)

Insufficient actions can be dealt with using the Oxford Standard Solutions, page 367.

There are two basic strategies to improve change or enhance functions by changing the:

- **components**
- **action/field** which act between them.

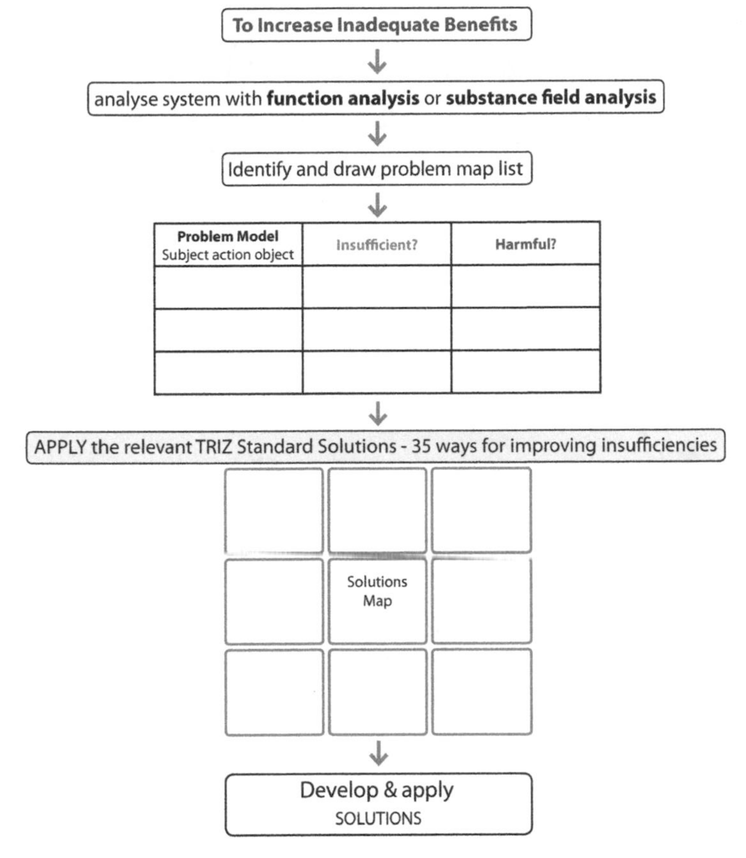
To Increase Inadequate Benefits
analyse system with function analysis or substance field analysis
Identify and draw problem map list
Problem Model
Subject action object
Insufficient?
Harmful?
APPLY the relevant TRIZ Standard Solutions - 35 ways for improving insufficiencies
Solutions Map
Develop & apply
SOLUTIONS

6. Follow the Trends to Improve the System Towards Perfection

Next generation – we can evolve the current system by applying TRIZ Trends by examining the current system and mark on each of the TRENDS below:

- where we are
- where we think we should be
- highlight the trends gaps.

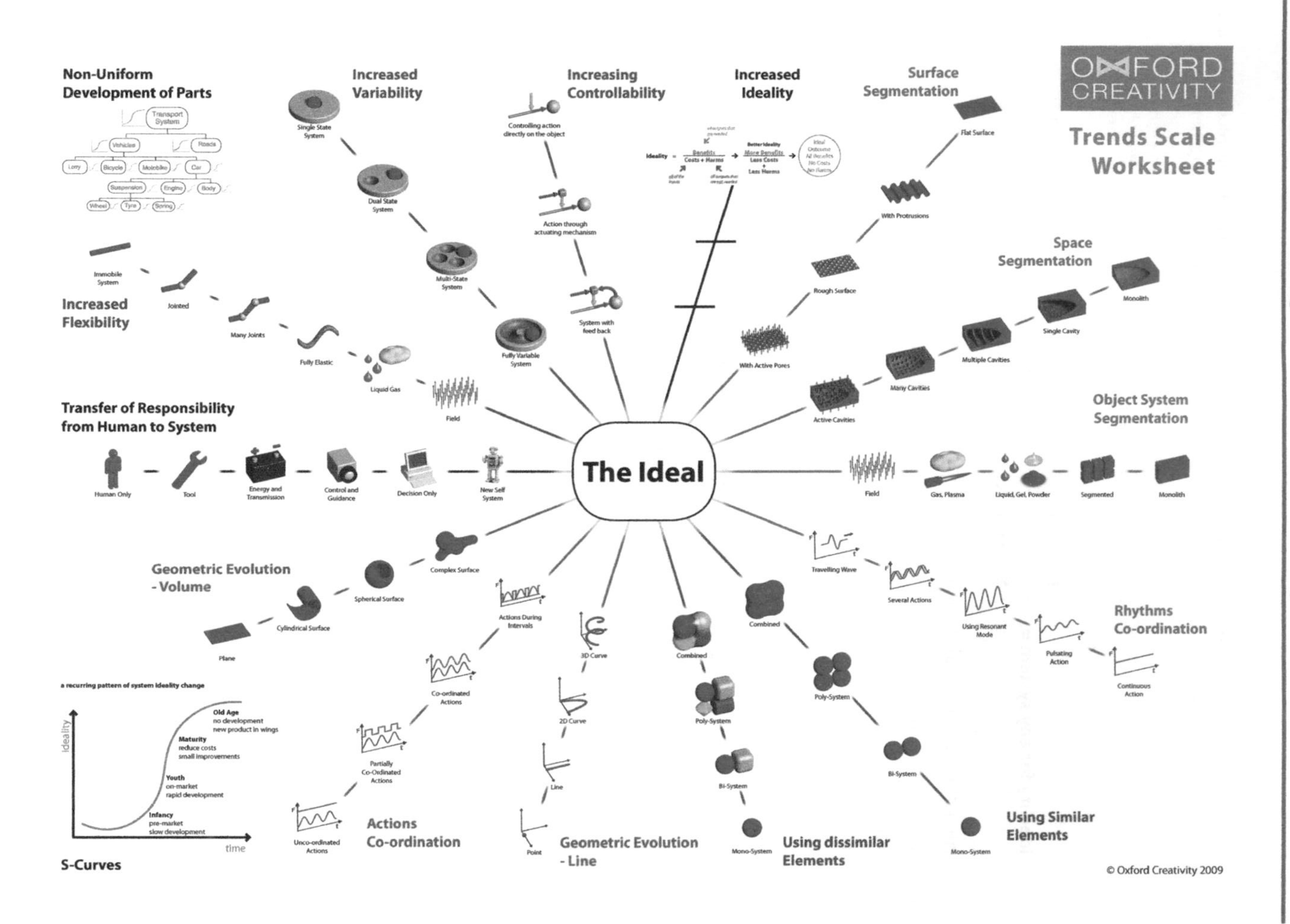
OXFORD CREATIVITY
Trends Scale Worksheet
The Ideal
Non-Uniform Development of Parts
Transport System
Vehicles
Roads
Lorry
Bicycle
Motorbike
Car
Suspension
Engine
Body
Wheel
Tyre
Spring
Increased Variability
Single State System
Dual State System
Multi-State System
Fully Variable System
Increasing Controllability
Controlling action directly on the object
Action through actuating mechanism
System with feed back
Increased Ideality
Surface Segmentation
Flat Surface
With Protrusions
Rough Surface
With Active Pores
Space Segmentation
Monolith
Single Cavity
Multiple Cavities
Many Cavities
Active Cavities
Object System Segmentation
Field
Gas, Plasma
Liquid, Gel, Powder
Segmented
Monolith
Increased Flexibility
Immobile System
Jointed
Many Joints
Fully Elastic
Liquid Gas
Field
Transfer of Responsibility from Human to System
Human Only
Tool
Energy and Transmission
Control and Guidance
Decision Only
New Self System
Geometric Evolution - Volume
Plane
Cylindrical Surface
Spherical Surface
Complex Surface
Actions Co-ordination
Unco-ordinated Actions
Partially Co-Ordinated Actions
Co-ordinated Actions
Actions During Intervals
Geometric Evolution - Line
Point
Line
2D Curve
3D Curve
Using dissimilar Elements
Mono-System
Bi-System
Poly-System
Combined
Using Similar Elements
Mono-System
Bi-System
Poly-System
Combined
Rhythms Co-ordination
Travelling Wave
Several Actions
Using Resonant Mode
Pulsating Action
Continuous Action
a recurring pattern of system ideality change
Ideality
time
Old Age
no development
new product in wings
Maturity
reduce costs
small improvements
Youth
on-market
rapid development
Infancy
pre-market
slow development
S-Curves
© Oxford Creativity 2009

Next Generation Systems

When the system is hard to improve – it is probably (not always) time to locate a new, next generation system (move from improve) by combining existing new and emerging technologies. The TRIZ Trends will help us see how to perfect the systems we've got and look for characteristics of the next generation.

Successful products can predict their own future with TRIZ

Next generation thermometer – Trends suggest it will smaller, flexible, cheaper, easier to use etc.

An old-fashioned thermometer comprised of a sealed glass tube containing mercury with a scale printed on the side. When the thermometer was placed in a person's/child's mouth it warmed the thermometer and the mercury expanded within the tube, and reading off the scale on the glass where the mercury reached gave the temperature. Ideality increases were achieved by making it as small as possible, as robust as possible (unlikely to be bitten in half when in the mouth, and surviving being dropped on the floor) and as easy to read as possible. All the Ideality improvements were taking the glass thermometer up its S-curve (shown in green below). When a new technology comes along, shown in the blue curve, such as the thermo-chromic plastics which change colour according to the temperature they immediately replaced the mercury thermometers for most domestic use. This was especially true for use with children, as they were safe, easy to use, easy to read, cheap and non invasive (as they could be placed on the forehead).

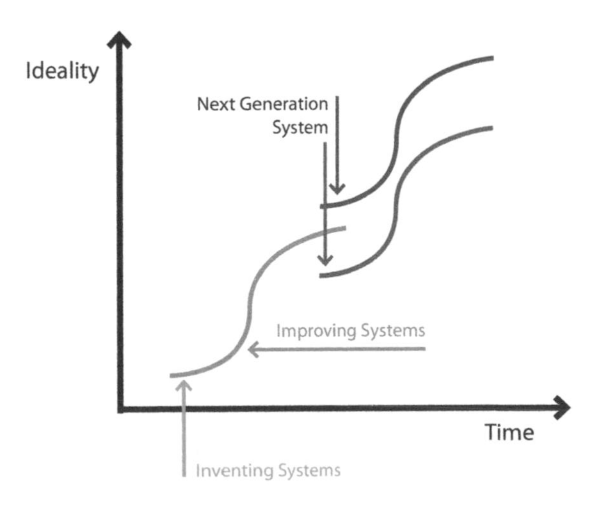

7. Invention and Ways of Providing Unfulfilled Requirements

Use the Ideal Outcome and resources.

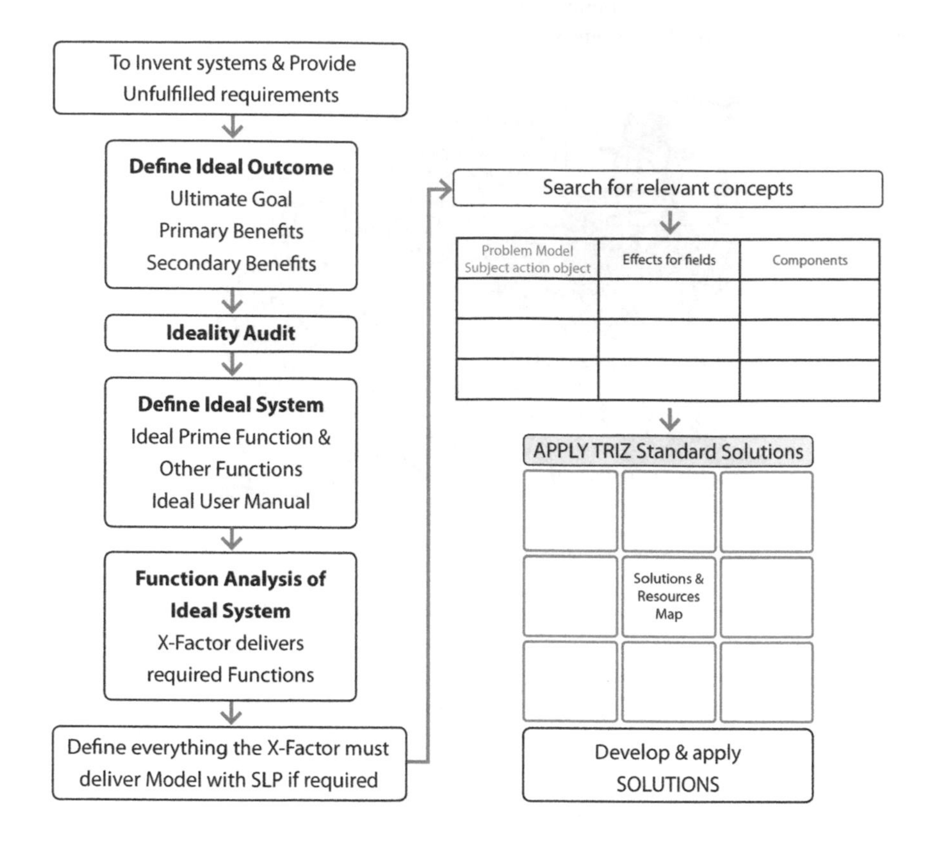

8. Ideal Problem Solving – Looking for Ideal Solutions from Resources

Ideal-Functions-Resources

Use readily available, already present, under-used resources to create clever solutions.

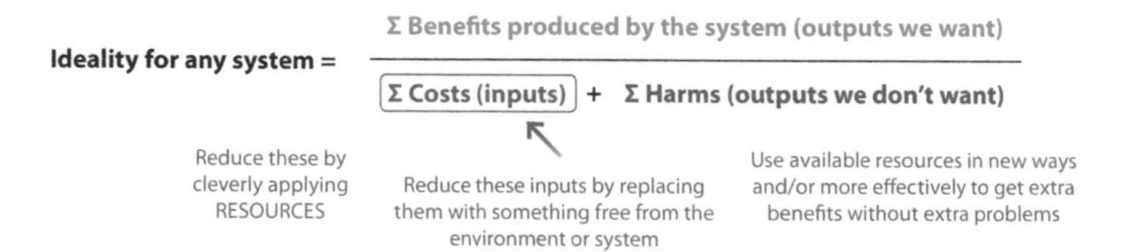

If costs remain the same, we can increase benefits and/or decrease harms by using the inputs in additional ways or in different ways by using them more efficiently. In TRIZ this is referred to as 'using the resources you've got'. Altshuller called this the basis for invention, but it is highly effective for clever solutions and reducing costs and harms. Awareness of the full potential of all resources in the inputs (including harmful ones), in the features of the system and in converting harmful outputs to useful outputs is all part of the conscious, deliberate and systematic search for resources.

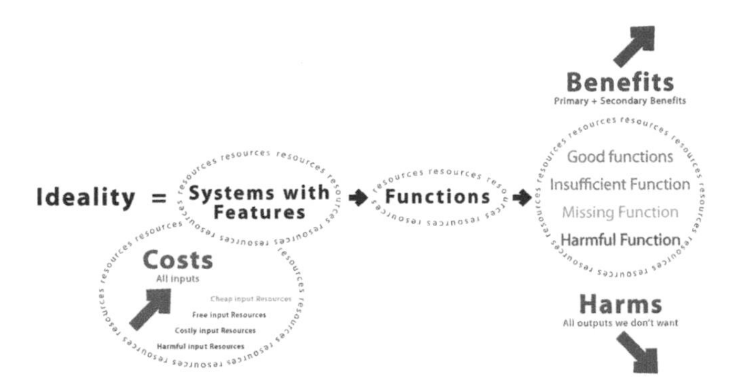

9. Ideal – Benefits – Functions – Resources

This offers to shortcut the problem solving by starting with the Ideal and then cut straight to a resource hunt. Defining the Ideal Outcome is a very powerful, very quick way of summarizing top-level requirements, and inevitably sets us thinking about solutions and ways of providing the Ideal. By consciously searching for those solutions in the available resources gives us a simple short cut of the problem process and can be very quick and effective.

RESOURCE HUNT to mobilise the resources we've got

SOMETHING (already there) WILL PROVIDE THE FUNCTIONS WE ARE SEEKING (including X-FACTOR)
We list all the resources in and around our system particularly those near the problem and needed only when the problem occurs. We try and use resources unchanged but otherwise in the order :-

- **UNCHANGED / RAW - as found**
- **COMBINED - with other available resources**
- **MODIFIED - changed to provide the needed function**

Assuming we have multiple resources - Resource A, Resource B and Resource C
and we also need a combination of functions - FUNCTIONS 1 and 2 - try each in turn
(Define exactly where and when the function is needed)

Ask the simple question of each resource in turn

Can Resource A provide Function 1?
Can Resource A provide Function 2?
Can the component (which Resource A needs to act on) provide the needed functions itself?

Can Resource B provide Function 1?
Can Resource B provide Function 2?
Can the component (which Resource B needs to act on) provide the needed functions itself?

Can Resource C provide Function 1?
Can Resource C provide Function 2?
Can the component (which Resource C needs to act on) provide the needed functions itself?

Do this with a team of experts/engineers with the relevant experience.
This simple exercise solves many problems elegantly, cheaply and effectively.

Four Simple Steps to Define the Ideal, Uncover Required Functions and Deliver Them from Resources

1. What is the *problem*?
2. Define the *Ideal* to list the essential requirements/ needs to define benefits.
3. Uncover/define the *functions* which deliver those benefits.
4. Search for *resources* around the system to deliver those essential functions.

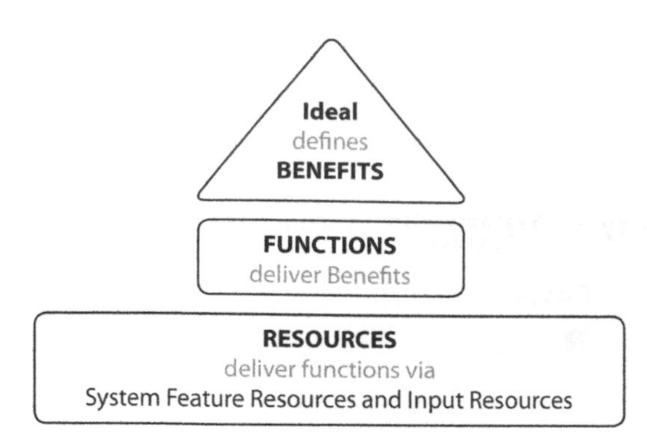

In essence TRIZ shows us how to solve problems by applying available resources in ways suggested by simply understanding the essential functions we are seeking, or by accessing the appropriate the TRIZ solution triggers and matching them to resources. For example in a ship sinking in cold seas the Ideal Outcome is that everyone survives – benefits are to survive, stay alive, and be found and taken to safety. The essential functions are stay afloat, stay warm, stay together with other survivors, be found /spotted by rescue and collected and taken home. The only way to deliver all these functions is with the available resources, shown in the diagram above – if it is a closed system with no new inputs available, the functions can only be delivered from resources already there in the system. The discipline of fast awareness of all resources is helpful in all situations.

Summary on Resources and the Ideality Tactics

Clever solutions use the correct and available resources. Where to find resources? We search amongst available inputs, the components of the system, harms etc. Focusing the search by starting with the Ideal and being aware of what we want is a quick powerful approach, matching what we want by a clever consciousness of what is already available with some clear engineering solution thinking creates very cost-effective answers.

The Power of TRIZ Problem Solving

TRIZ problem-solving success is based on keeping it as *simple* as possible. The reason TRIZ helps us solve any problem is that we get out of any unnecessary detail and move into the zone of the exact problem area; we define the exact functions we have to provide or fix and this helps us both understand and solve the very precise problem. Solving problems has simple fundamental routes in TRIZ shown as the eight Ideality Tactics in this chapter. TRIZ is the only and totally unique solution toolkit and offers all recorded conceptual answers to conceptual problems – all the fixes for problem functions. To successfully problem solve with TRIZ we ensure that our problem is reduced and categorized as a simple conceptual problem (using the Prism of TRIZ which helps us use the world's knowledge to find all known solutions – see TRIZ EFFECTS on www.TRIZ4engineers.com).

All technical or business problems are gaps between our requirements (benefits we want provided by functions) and our system (which provides the functions). Good problem solving delivers the system with the exact required functions (no more, no less) for the inputs we are prepared to make, and with only the harms (problems) we are prepared to tolerate. Problem solving is fixing or providing the right functions (with Standard Solutions) or providing conflicting (opposite) functions (with 40 Principles) to give us all our benefits. Extreme problem solving delivers perfection of the Ideal Solutions which give us all benefits with no costs and harms.

There is a slight extra challenge to the above in two areas of problem solving:

- Invention with its big gap of no system, which means we have to select or create the right system – after first checking to see if it already exists.
- Problems which have unknown or very complex causes – work is first needed to locate the root causes of problems before we can solve them.

What is the problem? How can we find the gaps?

Define requirements/benefits – what we want using the TRIZ tools of Ideality Audit, Ideal Outcome and the 9-Box requirements map for all stakeholders. All help us define the Problem Gaps we need to close.

Define system problems at the function level with system understanding using the TRIZ tools of 9-Boxes and Function Analysis. This delivers a Problem List which we can solve one at a time.

Solve the problems – Only TRIZ offers conceptual solutions for all Conceptual Problems = insufficiencies, harms or contradictions. These conceptual solutions are transformed into real answers by experienced engineers applying their knowledge to mobilize available resources to deliver solutions with the most benefits, least costs and least harms and also applying the world's knowledge filtered by the TRIZ EFFECTS:

- Conflicts between benefits (contradictions) – TRIZ offers 40 solutions.
- Insufficiencies – TRIZ offers 35 solutions.
- Harmful – TRIZ offers 24 solutions.
- TRIZ also offers 13 Creativity Triggers to help problem understanding and solution generation.

The Ideality Tactics guide us to the logic and power of TRIZ and offer simple approaches to master the individual tools, the TRIZ processes, its unique solution concepts and mental stimuli.

All these show how TRIZ helps both appropriately broaden and narrow our thinking, so that we can problem solve like great engineers and famous inventors. The TRIZ tools are like ladders we can put in place to reach good solutions – the genius may always leap straight to these solutions (and we may sometimes) but for everyday, time-tabled inspiration TRIZ offers small but sure links to the best answers.

Daily TRIZ Thinking for mental agility

The investment in mastering TRIZ is worthwhile for those who like solving problems because TRIZ guides us to systematically solve any problem. Daily use helps us quickly find the best solutions to easy problems, and gives us mental agility for harder problems, and eventually the ability to solve seemingly impossible problems – 'innovative problems' – which are defined as those we wouldn't otherwise know how to solve. People familiar with TRIZ can quickly offer very practical and seemingly ingenious and obscure answers to problems. Practice at TRIZ thinking is a simple and guaranteed way of staying mentally fit and realising our full potential, and can make us all think and look like a genius.

© Oxford Creativity 2009
Finish this sentence:
"IF AT FIRST YOU DON'T SUCCEED..."
TRY, TRY AND TRY AGAIN
RE-DEFINE SUCCESS
DESTROY ALL EVIDENCE OF YOUR FIRST ATTEMPT
BLAME YOUR COLLEAGUES
GIVE UP
DELEGATE AND MAKE IT SOMEONE ELSE'S PROBLEM

© Oxford Creativity 2010
"IF AT FIRST YOU DON'T SUCCEED..."
USE TRIZ. IT NEVER FAILS!
AND USE MY CARTOONS

GOOD LUCK!
IT'S FUNNY, BUT THE MORE I PRACTICE THE LUCKIER I GET!
© Oxford Creativity 2010

Case Study
BAE Systems 'SRES' Ducting Design

Problem Context

An air force has a requirement for the **'SRES'** to all aircraft (note: fictitious acronym, for security and confidentiality).

The SRES system consists of various units and antennae along with associated wiring and cooling provisions where necessary. The area for development is a new cooling duct to supply low pressure/ low temperature air to the SRES and PS (power supply) units installed within an equipment crate. This new duct 'taps' cooling air from the existing plenum chamber[1] attached to the rear of the crate (via a newly introduced cut-out). Earlier development activities and ECS (Environmental Control System) testing arrived at the development solution shown below (Figures 1 and 2) from which the design department have been tasked with creating a production solution.

It has to be noted that the retro modification is split into two stages:

- **Stage 1** (Provisioning), the SRES and PS units are not fitted – however, it is still necessary to fit a small section of the duct to the rear of the crate. The reason for this 'stub duct' is to maintain space provision for the continued use of existing systems whilst simultaneously negating the later removal of the equipment crate and subsequent refit/retests when SRES is finally installed. The stub duct requires sealing off at this stage in readiness for Stage 2.
- **Stage 2** (Installation), the existing units/racks and the stub duct blanking are removed to allow installation of the manifold duct part, new rack and SRES/PS units.

The problems identified with the current development solution are:

1. The duct parts comprise of many fabricated aluminium alloy parts welded together, with the welding dressed out at critical locations. Clearly, this is a very labour-intensive and costly method of manufacture that would also require a lot of tooling.
2. Both duct parts are rigid. Furthermore the stub duct is riveted to the equipment crate plenum chamber and the manifold duct is bolted to the SRES/PS rack assembly. Consequently, it is imperative that these duct parts are perfectly aligned (notably in X and Z) in order to prevent a build up of stresses when assembled. Due to aircraft build differences, design would need to impose controls such as close manufacturing tolerances on the duct parts, use of packing/shim and possibly further (costly) tooling for use at the installation stage.
3. At Stage 1, failure of the clamping force exerted by the clamp ring may lead to loss of the sealing cap and consequent loss of cooling air to vital aircraft equipment including the main computer. The development solution is reliant only upon the friction between mating surfaces to withstand the air pressure being applied from within the crate plenum chamber.

[1]Depending upon aircraft build block standard, this existing plenum-type chamber is manufactured from either SPF (super plastic formed) aluminium alloy or GRP (glass reinforced plastic) with slightly different profiles. Hence two versions of Stub Duct were designed at the development stage. For cost and kitting considerations, it has been envisaged that only one type of ducting shall be designed for production.

TRIZ for Engineers: Enabling Inventive Problem Solving, First Edition. Karen Gadd.

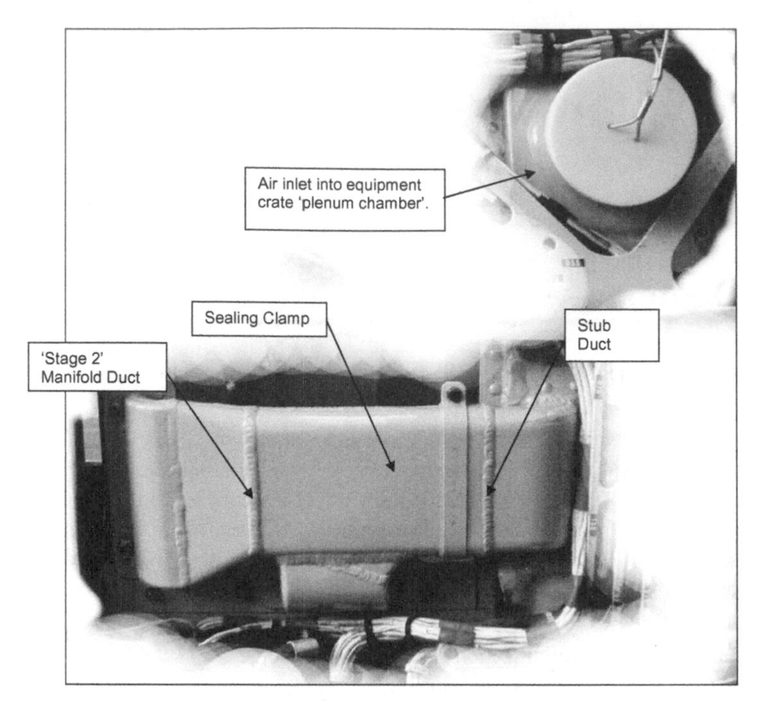

Figure 1 Looking forward on rear of crate – development version.

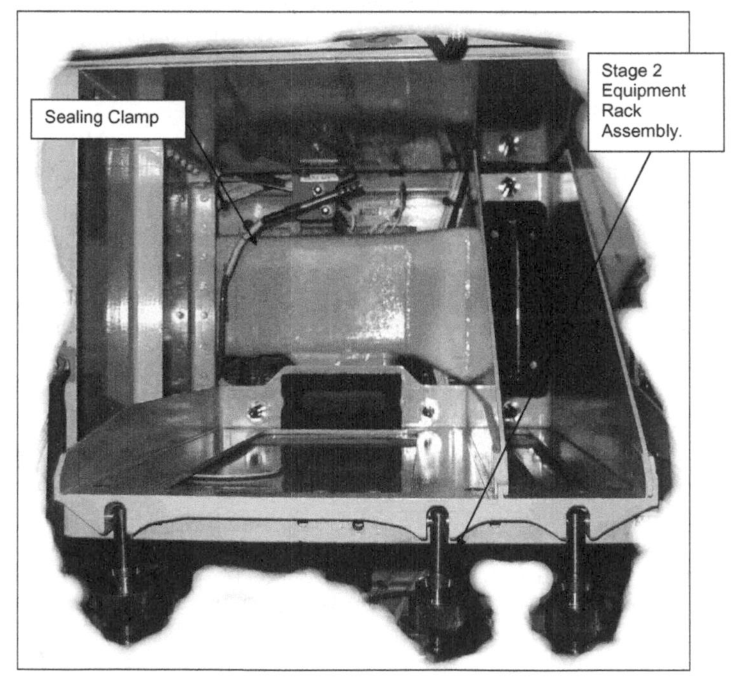

Figure 2 Looking aft at front of crate.

4. Disassembly/assembly of the clamp ring (bolt and nut) would be awkward at Stage 2 due to limited access as the Zone 12 crate would be fitted in the aircraft. Risk of FOD (foreign object damage).

Design constraints are:

1. Extensive ECS testing has previously been performed at the development stage in order to determine duct internal geometry and restrictor plate requirements so as to provide suitable 'balanced' cooling airflow to all required electrical units. To avoid further testing, design are unable to deviate too far from the duct internal geometry offered by the development solution.
2. The ECS department have specified that air loss from any of the duct interfaces/joints is not acceptable.
3. There is very little clearance between the back of the equipment crate and the aircraft frame structure.

System Modelling and Analysis

The functional analysis model of the SRES ducting arrangement for the development solution shows clearly how parts of the system interact with each other and more importantly whether the actions between them are *useful* or *harmful*. Both these problem types have conceptual solutions in the TRIZ Standard Solutions – 24 ways of dealing with *harms* and 35 ways of dealing with *insufficiency*.

Where there are both useful and harmful actions between two subjects or objects, we have some kind of contradiction. In this case, the ring clamp is shown to have two useful actions of securing and supplying air to the two duct parts but it is also has a harmful action of inducing stresses as a result (i.e. if, due to aircraft build differences or build up of manufacturing tolerances, they are not correctly aligned).

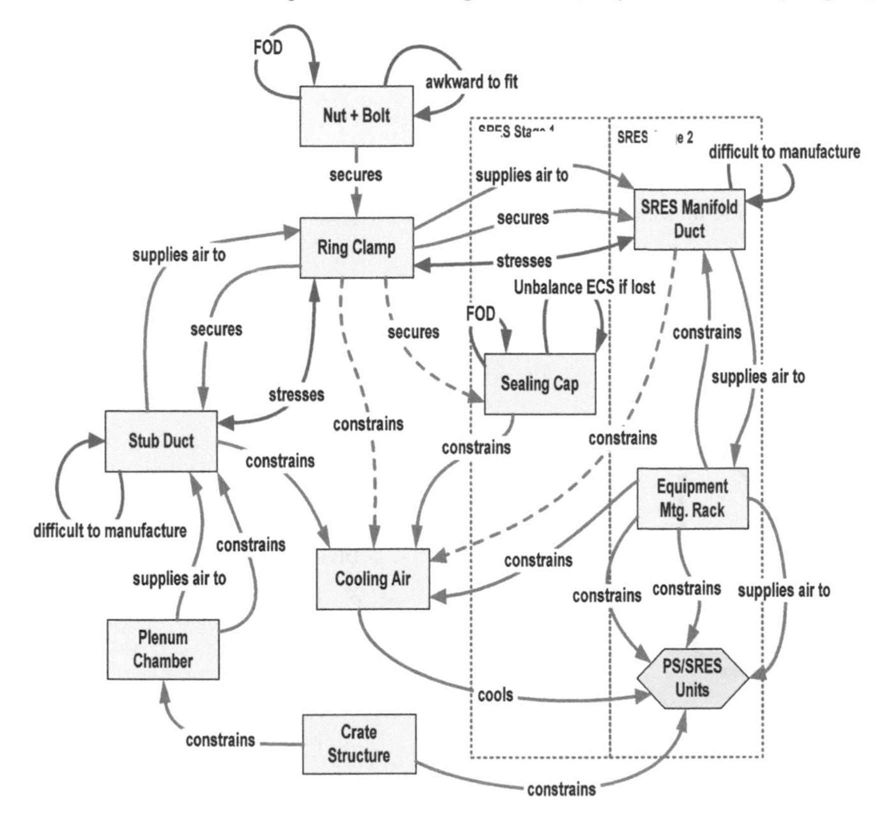

Uncovering and Solving Physical and Technical Contradictions

Step 1: Identify/uncover the contradictions.

Step 2: Solve the contradictions by applying the relevant selection of the 40 Principles.

Step 3: Generate possible solutions from the TRIZ conceptual prompts – the Oxford Standard Solutions.

Step 4: Look at each solution and identify what is bad and what is good – solve those contradictions if appropriate.

Solutions – Physical Contradictions

Physical Contradiction 1

The following physical contradiction has been revealed: The duct parts need to be *rigid* to maintain the correct airflow through the duct – i.e. the internal duct geometry must not be compromised. However, the duct also needs to be *flexible* to allow for aircraft build differences when joining the duct parts. When faced with a physical contradiction we ask: 'Do we need these opposites at the same time and in the same place?'. If the answer is no then we see if we can separate the opposite parameters to have them *both* either at different times or in different places/space. We need *rigid* and *flexible* at the same time but not in exactly the same place; because the majority of the duct must to be rigid (maintaining its shape) but the duct joint area could be flexible to allow for any duct misalignment, the Separation Principle 'Space' is thus used to solve this physical contradiction. The following of the 40 Principles are relevant for solving such physical contradictions like wanting something to be both rigid and flexible in different places: (1) Segmentation, (2) Taking Out, (3) Local Quality, (4) Asymmetry, (7) Nested Doll , (13) The Other Way Round, (14) Spheroidality/Curvature, (17) Another Dimension, (24) Intermediary, (26) Copying, (30) Flexible Membranes and Thin Films, (40) Composite Materials.

Solution Physical Contradiction 1a – (3)(30)

Fasten (e.g. rivet) a flexible rubber sleeve to the open end of the stub duct. The sealing cap/SRES manifold duct can slide into the sleeve and be secured using some kind of clamp, tie-wrap, Jubilee clip, etc.

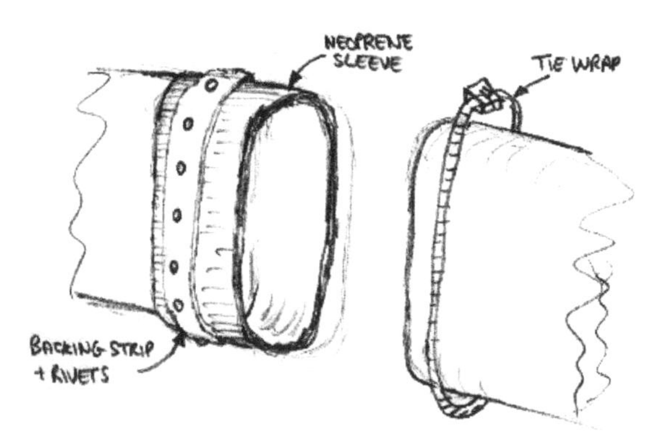

Physical Contradiction 1a

Advantages:

✓ Simple

✓ Relatively cheap

Disadvantages:

✗ Due to oval shape, tie-wrap may not apply enough pressure to seal sufficiently.

✗ Like the DOI version, the Stage 1 seal cap could 'blow-off' as its security relies upon friction only.

Solution Physical Contradiction1b – (7)(30)

Have sealing cap/manifold duct slide into the stub duct and use a wiper type seal to take up any duct misalignment.

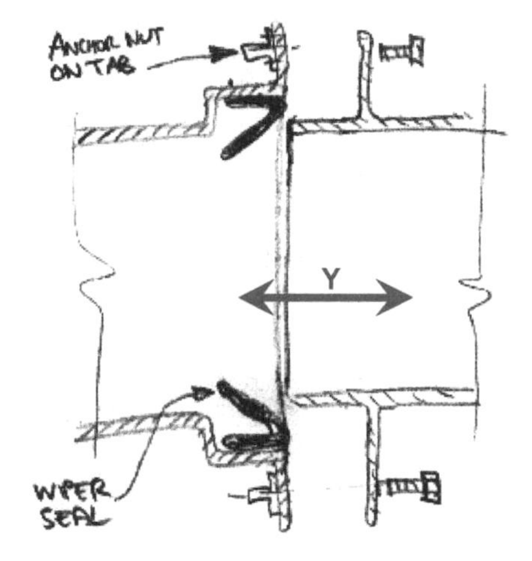

Physical Contradiction 1b

Advantages:

✓ Good seal

✓ Secure Stage 1 sealing cap.

Disadvantages:

✗ More complex to manufacture.

✗ More costly than Physical Contradiction1a.

✗ Orientation of the anchor-nut tabs still allows for the build up of stresses in the 'Y' direction.

Solution Physical Contradiction1c – (30)

Thin membranous flaps bonded on inside of stub duct. The intention is that the airflow through the duct will lift the flaps outwards to seal against the inside of the SRES manifold duct. Note: TRIZ encourages us to look for all *resources* available to us and make use of them when possible; here we are making use of the 'air-flow' resource.

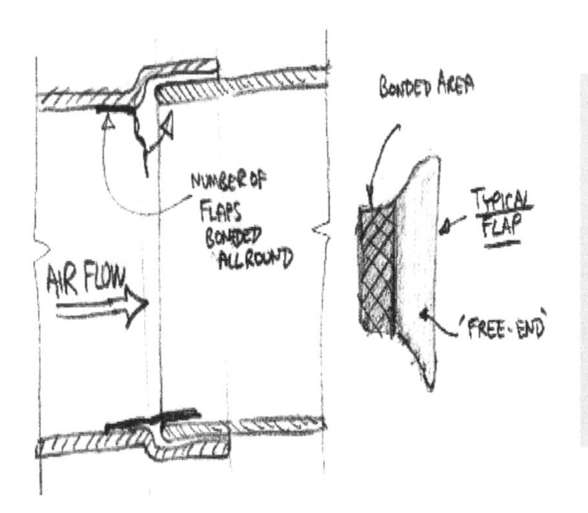

Physical Contradiction 1c

Advantages:

✓ Cheap.

Disadvantages:

✗ FOD risk if a flap becomes detached – potential damage to equipment.

✗ The system is rendered open to the environment when there is no airflow – potential ingress of water, fuel, hydraulic oil, etc.

✗ No means of securing Stage 1 sealing cap.

Solution Physical Contradiction 1d – (24)(30)

Fit an intermediary rubber, bellows-type sleeve between duct flanges.

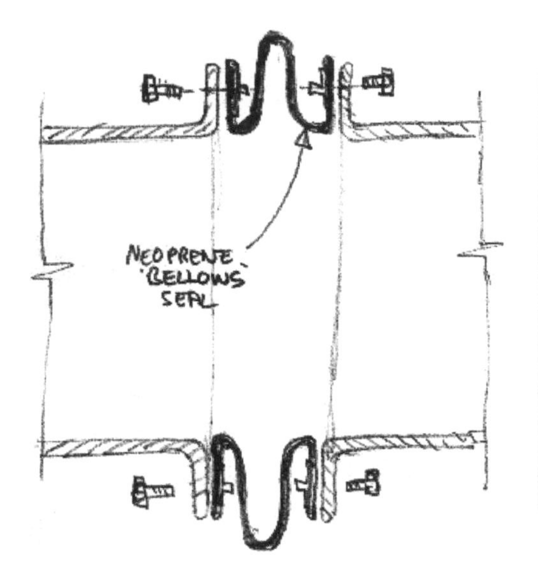

Physical Contradiction 1d

Advantages:

✓ Good seal.

✓ Duct parts secured effectively.

✓ No stresses in X, Y or Z.

Disadvantages:

✗ No space provision for the flange all around the duct periphery.

✗ No access to fit securing bolts at rear of duct when fitting Stage 2 manifold duct.

✗ Airflow turbulence may be created due to shape of bellow-type seal.

Solution Physical Contradiction 1e – (30)

Seal duct joint using standard adhesive 'duct-tape' wrapped around the joint area.

Physical Contradiction 1e

Advantages:

✓ Very cheap.

✓ No stresses in X, Y or Z.

Disadvantages:

✗ Environment (heat, hydraulic oil, water, etc) may deteriorate the adhesion of the tape allowing it to become detached.

✗ Very limited access in aircraft to properly fit tape at Stage 2.

Technical Contradictions

Technical Contradiction 1

The Function Analysis of the SRES ducting has highlighted that the clamp ring is used to secure the two duct parts together (to ensure no loss of air flow) which could also induce stress into the parts during assembly. Looking through the list of 39 Engineering Parameters (or features) we can select 'Loss of Substance' (i.e. the clamp ring is securing to prevent air loss) as an improving feature and 'Strength' or 'Stress' as the worsening feature. By cross referencing 'Loss of Substance' against 'Strength' on the Contradiction Matrix, it is suggested that we look at the following Inventive Principles:

(35) Parameter Change

(28) Replace Mechanical System

(31) Porous Materials

(40) Composite Materials

Note: see Technical Contradiction 2 below for 'Loss of Substance' versus 'Stress'.

Solution Technical Contradiction 1a – (35)(28)

Elasticated sleeve fitted over joint area rather than a stiff clamp ring.

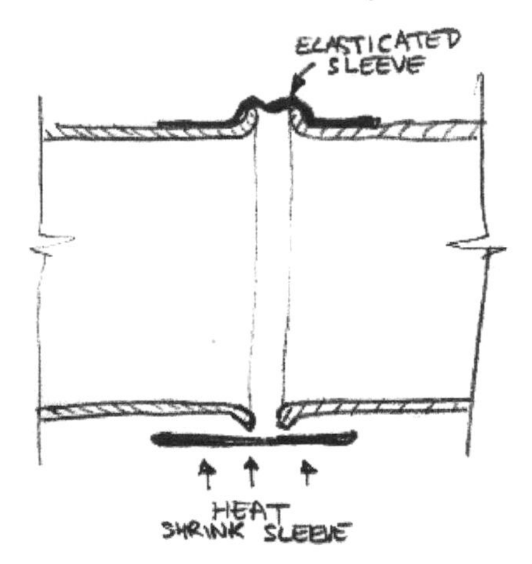

Technical Contradiction 1a

Advantages:

✓ Very cheap.

✓ Reduced part-count.

✓ Potentially a very good sealed joint.

✓ No stresses in X, Y or Z.

Disadvantages:

✗ Risk of heat-shrink splitting and very poor access to apply heat-shrink at Stage 2.

✗ Risk of sleeve sliding away from joint area.

✗ Sleeve exposed to mechanical damage.

Solution Technical Contradiction 1b – (31)(28)

Omit the clamp ring and compress a porous rubber section into the joint area of the ducts that seal on assembly. Any duct misalignment would still allow air to pass through with minimal loss.

Technical Contradiction 1b

Advantages:

✓ Potentially a very good sealed joint.

✓ No stresses in X, Y or Z.

Disadvantages:

✗ Too much restriction on airflow.

✗ Pores could eventually become blocked due to particles within the airflow.

✗ No means of securing Stage 1 sealing cap.

✗ Seal could dislodge leading to air loss at joint and partial blockage of the downstream duct (equipment overheating).

Solution Technical Contradiction 1c – (40)(28)

Replace the clamp ring for a neoprene impregnated nylon moulded sleeve that can be secured to each duct end by means of tie-wraps or Jubilee clips.

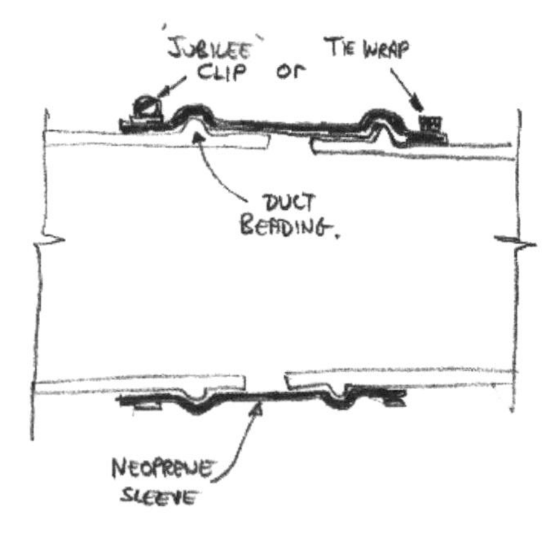

Technical Contradiction 1c

Advantages:

✓ Simple design.

✓ Beading helps to prevent sleeve from sliding and improves sealing.

✓ No stresses in X, Y or Z.

Disadvantages:

✗ Due to oval shape, tie-wrap may not apply enough pressure to seal sufficiently.

Technical Contradiction 2

By cross referencing 'Loss of Substance' against 'Stress' on the Contradiction Matrix, it is suggested that we look at the following Inventive Principles:

(3) Local Quality

(36) Phase Transition

(37) Thermal Expansion

(10) Prior Action

Solution Technical Contradiction 2a – (3)

Investigate additive layer manufacture (ALM) hard plastic duct body blending into flexible rubber sleeve in joint area. Secure with tie-wrap or Jubilee clip.

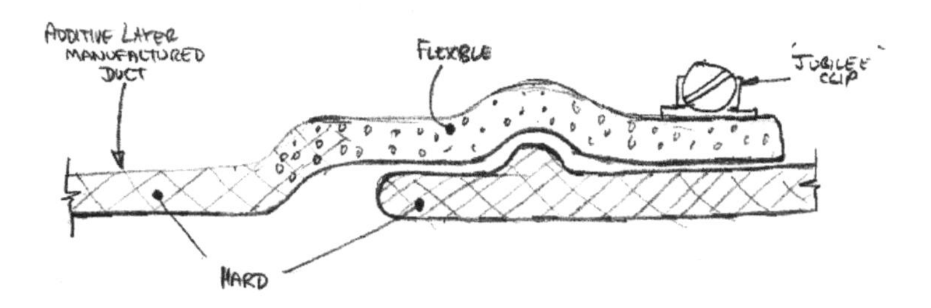

Technical Contradiction 2a

Advantages:

✓ No tooling required for manufacture.

✓ Low cost (dependent on quantity required).

✓ Consistent, high-quality parts produced.

Disadvantages:

✗ Manufacturing process still in development stage.

✗ Materials not yet qualified for aircraft use.

Solution Technical Contradiction 2b – (37)

Heat-shrink sleeve fitted over joint area. [See Technical Contradiction 1a.]

Solution Technical Contradiction 2c – (10)

Introduce use of packers, laminated shim, floating anchor nuts (at interface of SRES manifold duct with rack assembly) to eliminate any mismatch in duct alignment (and thus stress).

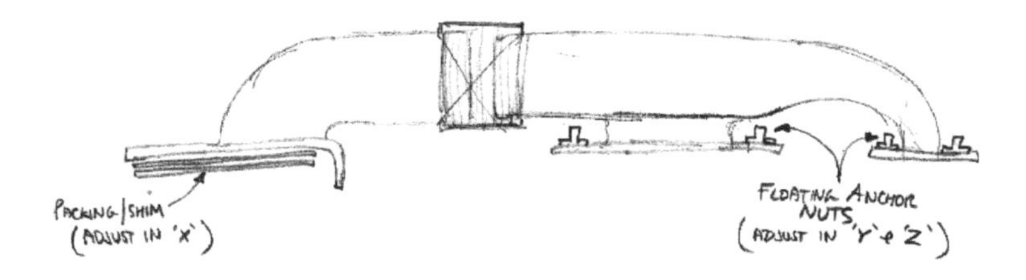

Technical Contradiction 2c

Advantages:

✓ Floating anchor-nuts and use of packing allow for very good duct alignment – hence low stress on Stage 2 assembly.

✓ DOI-type sealing ring can be fitted without the need to manufacture a dedicated rubber seal.

Disadvantages:

✗ Tedious method of assembly to establish correct thicknesses of shim/packing.

Solution Technical Contradiction 2d – (10)

Ensure SRES stub duct and manifold ducts are positioned accurately during fitment to the crate by making use of tooling fixtures.

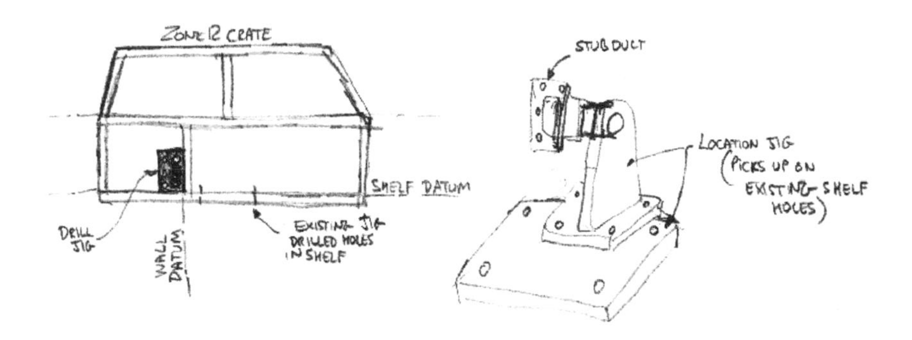

Technical Contradiction 2d

Advantages:

✓ Highly accurate assembly produced making for good interchangeability if parts ever need replacing.

✓ Tooling makes for quicker fitting of parts (e.g. no marking out and drilling guided by the drill jig).

Disadvantages:

✗ Tooling is expensive to produce.

Technical Contradiction 3

The Function Analysis also shows inherent technical contradictions within the SRES manifold and SRES stub duct parts. Although both are doing their intended function of containing the airflow, they are very difficult to *manufacture* and therefore costly (i.e. the ducting is of a complex *shape* and is made up of many intricately shaped/folded sheet metal components welded together). Extracting the first technical contradiction of 'Shape' versus 'Ease of Manufacture' we are guided into looking at the following Inventive Principles:

(17) Another Dimension
(32) Colour Change
(1) Segmentation
(28) Replace Mechanical System

Solution Technical Contradiction 3a – (1)

Manufacture the duct part in layers using glass-fibre or carbon-fibre rather than by fabrication.

Technical Contradiction 3a

Advantages:	Disadvantages:
✓ Composite fibre manufacture ideal for complex duct geometry.	✗ Initial expense of tooling.
✓ Lightweight components.	✗ Any important external surfaces (i.e. non-tool faces) require skilled manual finishing.
✓ Aircraft qualified materials.	✗ Labour intensive lay-up of plies and joining of duct parts.

Technical Contradiction 4

Following on from Technical Contradiction 3, the other technical contradiction we can explore is 'Shape' versus 'Productivity'. Note: the 'Productivity' parameter is the closest we can find that captures the 'cost' element we are interested in. The Contradiction Matrix reveals the following Inventive Principles:

(17) Another Dimension
(26) Copying
(34) Discarding and Recovering
(10) Prior Action

Solution Technical Contradiction 4a – (10)(34)

Manufacture duct parts using a lost-wax casting process.

Technical Contradiction 4a

Advantages:	Disadvantages:
✓ Ideal for complex duct geometry.	✗ Difficulty removing 'ceramic' layer from narrow duct interior.
✓ Relatively high strength parts produced.	✗ Costly process (energy and labour intensive).
✓ Aircraft qualified materials.	✗ Risk of warping/shrinkage of large or complex components during the cooling process.

Applying the Standard Solutions for Dealing with Harm in Problem Solving

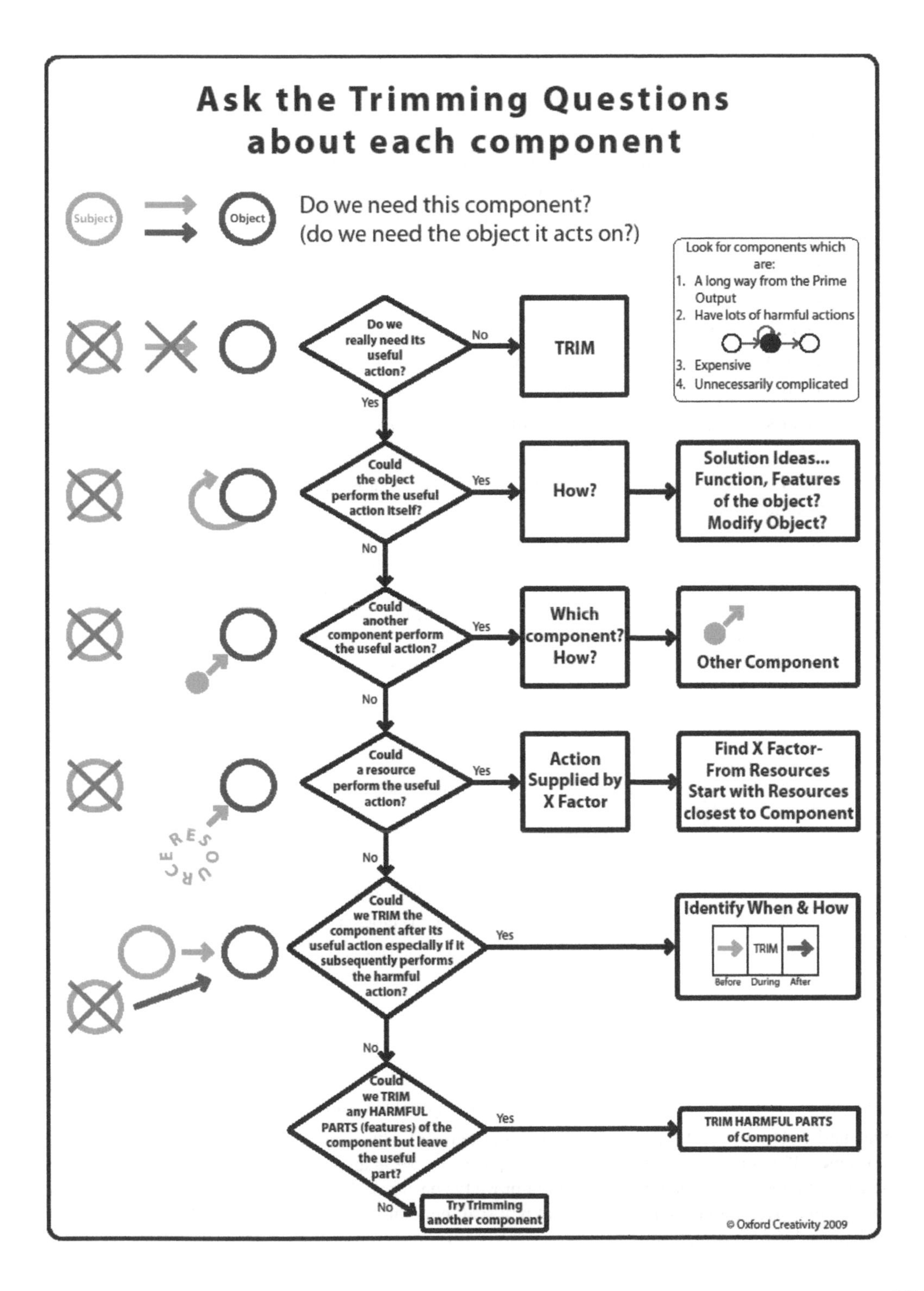
Ask the Trimming Questions about each component
Subject
Object
Do we need this component?
(do we need the object it acts on?)
Look for components which are:
1. A long way from the Prime Output
2. Have lots of harmful actions
3. Expensive
4. Unnecessarily complicated
Do we really need its useful action?
No
TRIM
Yes
Could the object perform the useful action itself?
Yes
How?
Solution Ideas... Function, Features of the object? Modify Object?
No
Could another component perform the useful action?
Yes
Which component? How?
Other Component
No
Could a resource perform the useful action?
Yes
Action Supplied by X Factor
Find X Factor- From Resources Start with Resources closest to Component
RESOURCE
No
Could we TRIM the component after its useful action especially if it subsequently performs the harmful action?
Yes
Identify When & How
TRIM
Before
During
After
No
Could we TRIM any HARMFUL PARTS (features) of the component but leave the useful part?
Yes
TRIM HARMFUL PARTS of Component
No
Try Trimming another component
© Oxford Creativity 2009

The Standard Solutions offer four associated groups for dealing with harm of which there are a number of methods available that allow us to apply them:

1. **Eliminate** – Trim out the harm – Trimming Rules – 6 ways
2. **Stop** – Block the Harm – 11 Ways
3. **Transform** -Turn Harm into Good – 4 ways
4. **Correct** – Put right the harm – 3 ways

It is often useful to begin by applying the Trimming Rules – 6 ways to eliminate the harm by trimming the subject, which has harm associated with it, but ensuring that we keep all its useful actions when necessary and obtain them from some other source – preferably already there and freely available.

- **Trim 1** – For any component – do we need their useful action? If not then trim the component.
- **Trim 2** – Could the *object* perform the useful action? If yes then trim the *subject.*
- **Trim 3** – Could another component perform the useful action? If yes then trim the *subject.*
- **Trim 4** – Could a *resource* perform the useful action? If yes then trim the *subject.*
- **Trim 5** – Could we trim the *subject* after it has performed its useful action? If yes then trim the *subject* at the appropriate time.
- **Trim 6** – Could we partially trim any harmful *parts* but leave any useful parts? If yes then just trim the *harm.*

With reference to the Function Analysis diagram, there is a problem with the clamp ring inducing stresses into the duct parts. Trimming Rules state 'If the subject's useful action cannot be eliminated, obtain it from some other source and eliminate the subject.' Expanding on this statement, the ring clamp's useful actions (securing and ensuring air supply) are still required but we can obtain these actions from the duct parts themselves by incorporating a flexible seal and securing device within them; thus, we can 'trim-out' the clamp ring component and hence its associated harmful effect.

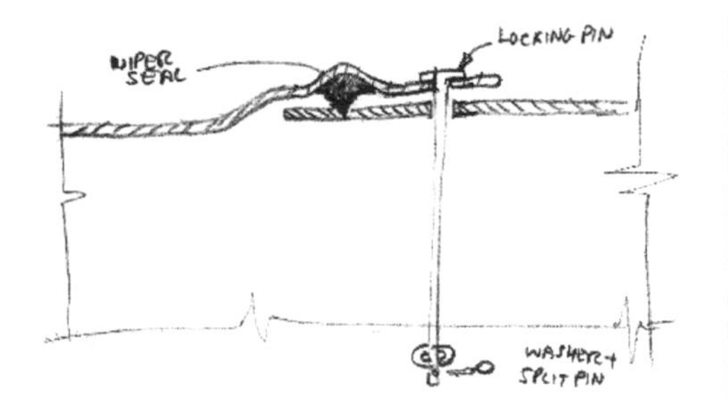

Standard Solutions – Trimming Rules

Advantages:

✓ Good seal.

✓ Positive locking of duct parts.

Disadvantages:

✗ Increased cost due to manufacture of a dedicated seal.

Looking at the other Standard Solutions for harm we have 11 ways to stop/block harmful actions.

Standard Solution H 2.4.1 states: 'Insulate from harmful action by introducing a new component / substance. If it is not necessary for the two components to be in direct contact then block the harm with a new component.' Applying this we could improve the design of the ring clamp by adding a soft rubber sponge-like layer onto its inner surface that will seal the joint and take into account duct misalignment without inducing any high stresses.

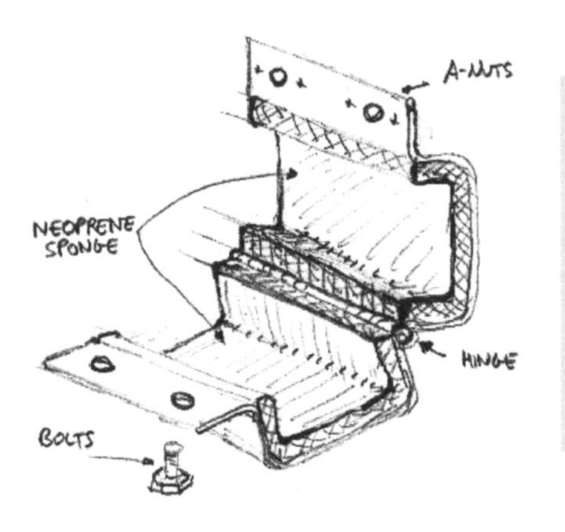

H 2.4.1 Insulate from harmful action

Advantages:

✓ Good seal.

✓ Easy access to fit 'clam-shell' locking bolts.

✓ No stresses in X, Y or Z.

Disadvantages:

✗ Possibly too bulky to fit in the restricted space available.

Apply Standard Solutions for Insufficiency (35 Ways)

There are two basic strategies to overcome the insufficiency – these are to improve, change or enhance the:

1. components (subject and/or object)
2. action/field.

The 35 strategies to overcome insufficiency are:

i.1.1 Improve the subject and/or object or their surroundings to enhance the delivered functions (7 ways).

i.1.2 Change/evolve subject and/or object (10 ways).

i.a.1. Enhance the action – for when a field F (action) is missing or insufficient (18 ways).

The current design of the sealing cap (SRES Stage 1) relies upon the frictional clamping force of the ring clamp only to hold it in place. Loss of the sealing cap in flight would unbalance the ECS system and could possibly lead to important equipment overheating. Applying Standard Solution i.1.1, which states: 'Add something to/or inside the subject or object, improvement is achieved by the introduction of internal additives, which can be permanent or temporary. This provides extra functions of enhancing/providing the required properties', for example, a 'quick-release' pin to mechanically lock the sealing cap onto the stub duct.

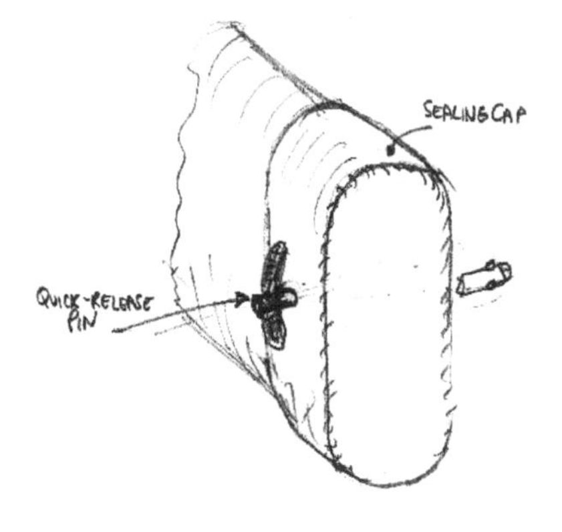

Standard Solution i.1.1

Advantages:

✓ Simple design.

✓ Effective neoprene face-seal incorporated in base of cap.

✓ Secure locking of cap.

Disadvantages:

✗ If cap is manufactured from a 'soft' material (e.g. GRP) there is a risk of wear around the hole – leading to loss of the pin.

Applying Standard Solution i.1.2, which states: 'Add something between the subject and the object which enhances/delivers the function', might suggest bonding the sealing cap onto the stub duct using PRC.

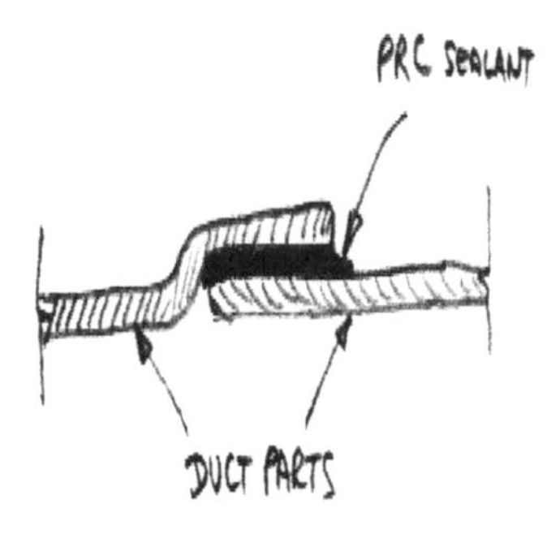

Standard Solution i.1.2

Advantages:

✓ Simple.

✓ Cheaper alternative to a dedicated seal.

✓ Secure locking of cap.

Dis-advantages:

✗ Difficult to ensure a satisfactory seal all around the duct part at Stage2 due to limited access.

✗ Risk of sealant dripping into the duct interior and affecting airflow.

Standard Solution **i**.1.5 states: 'Add something outside/around the subject or object – introduce external additives into the present substances/components to improve features/properties or provides extra functions.' For example, discard the ring clamp and add locking tabs onto the stub duct and sealing cap that allow the components to be bolted together.

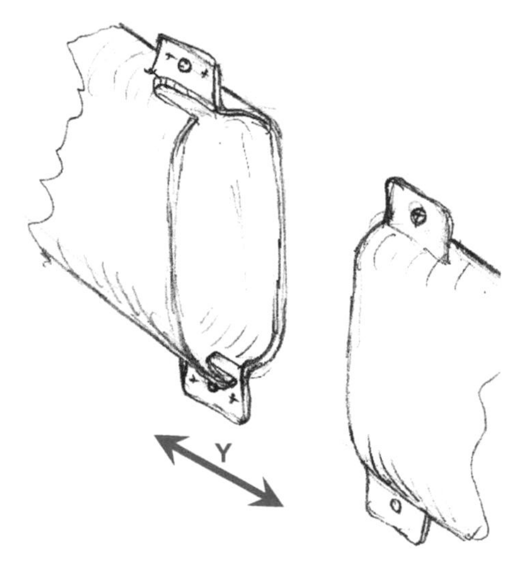

Standard Solution i.1.5

Advantages:

✓ Secure locking of duct parts.

✓ Orientation of tabs allows for misalignment in 'Y' – hence avoiding build up of stress.

Disadvantages:

✗ Tabs may be vulnerable to damage.

✗ Complex manufacture required to ensure correct location of tabs.

Final Solution

Combining the benefits of a selection of solution ideas, we arrived at the following final solution:

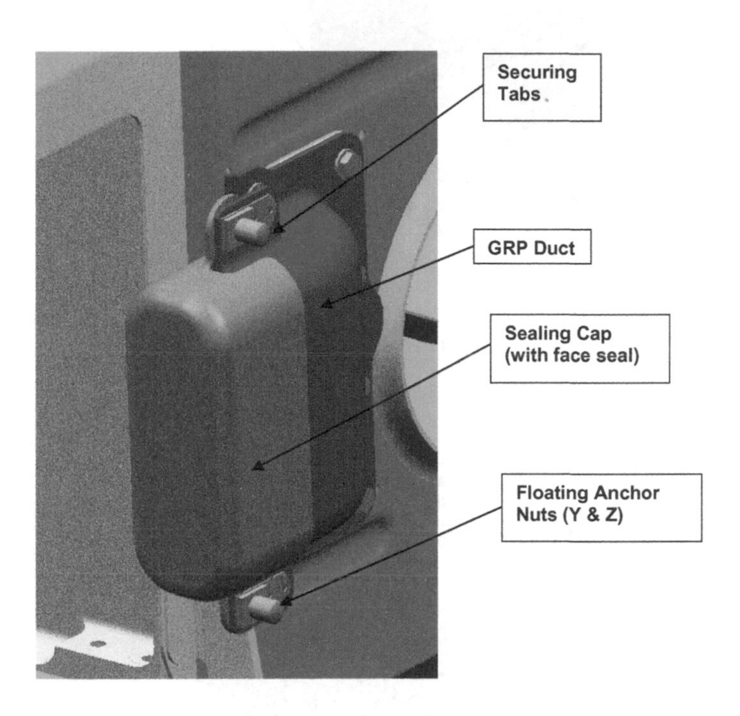

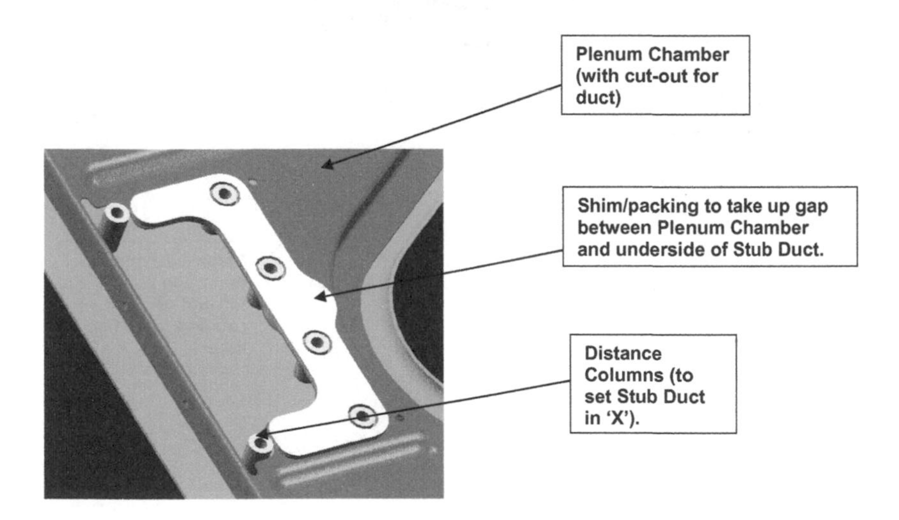

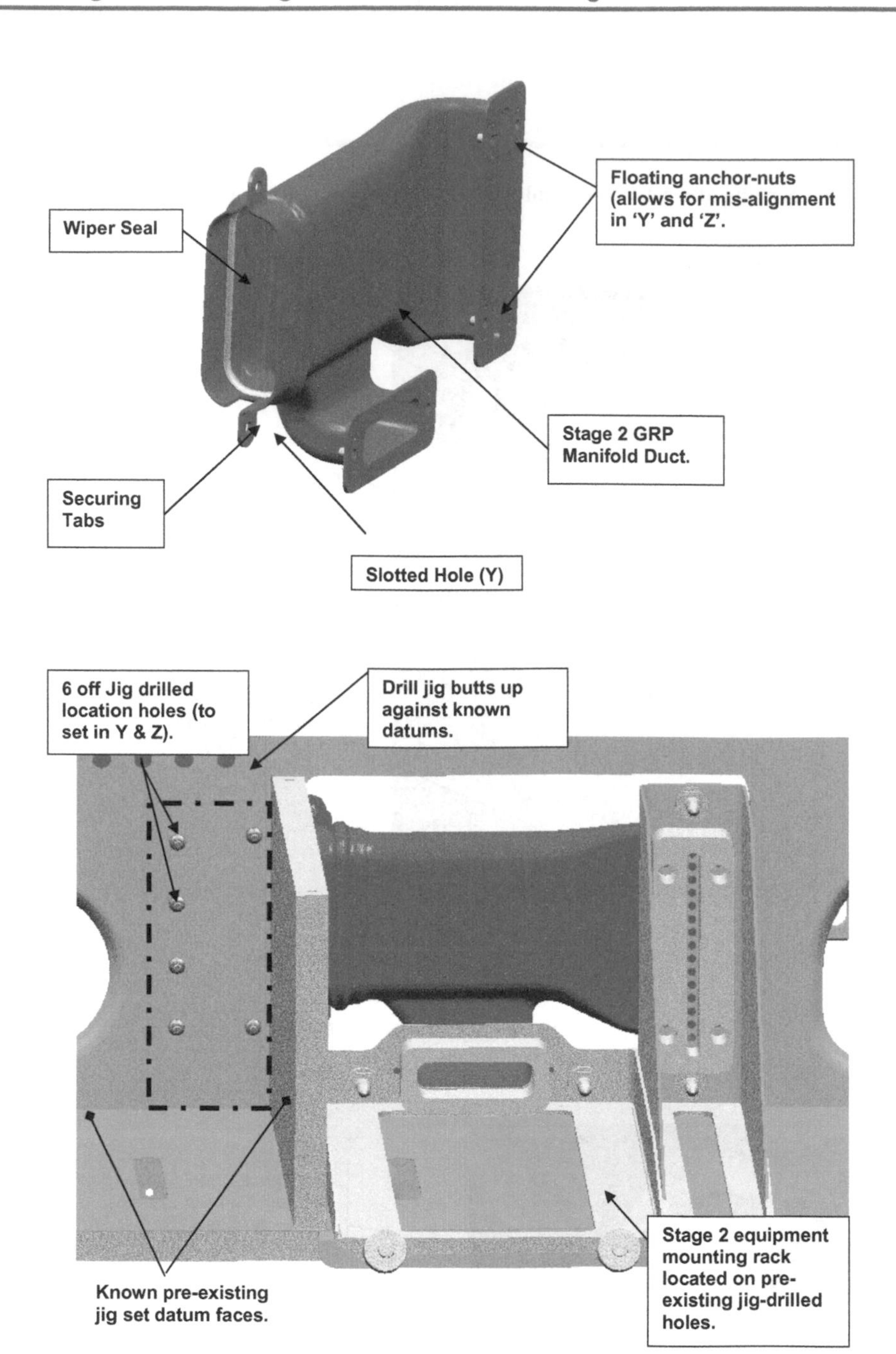
Wiper Seal
Floating anchor-nuts (allows for mis-alignment in 'Y' and 'Z'.
Stage 2 GRP Manifold Duct.
Securing Tabs
Slotted Hole (Y)
6 off Jig drilled location holes (to set in Y & Z).
Drill jig butts up against known datums.
Known pre-existing jig set datum faces.
Stage 2 equipment mounting rack located on pre-existing jig-drilled holes.

Summary and Conclusions

TRIZ provides very useful tools to help focus on the problems at hand and arrive at solutions far quicker and easier than the more traditional methods such as brainstorming. During the SRES case study, the Functional Analysis map was found to be extremely useful in providing a clear overview of the system whilst at the same time highlighting any problem areas from which to identify and solve any Contradictions via the matrix. The Standard Solutions method took slightly longer to use but offered a different perspective of looking at the problem. It must be noted that nearly all the solutions generated had advantages and disadvantages; thus (in time) we could probably carry out further work using the TRIZ tools in order to reduce and eliminate the disadvantages (either by solving the contradictions or eliminating the harms using the TRIZ solutions for both).

Benefits of TRIZ to the BAE Systems Team

The main benefits that our team found using TRIZ on this project were:

- Improved the initial definition of problem(s) – thus taking us in the right direction for better solutions.
- Guided our engineers to the generation of many creative, innovative solutions in a very efficient manner.
- Systematically helped us locate many/most of the possible improvements on the current design.
- Gave awareness of access to conceptual solutions, resources and knowledge bases (i.e. patent databases).
- Improved our confidence to get quality, reliability and safety (e.g. by getting it 'right-first-time').
- Greater engineer and customer satisfaction.
- Better designs and time saving. and cost savings.
- Helped us create a culture of innovative thinking – enhancing and complementing current practices.

Appendix I
39 Parameters of the Contradiction Matrix

The Contradiction Matrix is one of the easiest TRIZ tools to use when TRIZ problem solving. The matrix itself, and the 40 Principles it refers to are public domain and available to all engineers for solving contradictions. The matrix consists of 39 improving and worsening features. The following explanations were taken from the *TRIZ Journal* by Ellen Domb from translations of Altshuller's work.

No.	Title	Explanation
	Moving objects	Objects which can easily change position in space, either on their own, or as a result of external forces. Vehicles and objects designed to be portable are the basic members of this class.
	Stationary objects	Objects which do not change position in space, either on their own, or as a result of external forces. Consider the conditions under which the object is being used.
1	Weight of moving object	The mass of the object, in a gravitational field. The force that the body exerts on its support or suspension.
2	Weight of stationary object	The mass of the object, in a gravitational field. The force that the body exerts on its support or suspension, or on the surface on which it rests.
3	Length of moving object	Any one linear dimension, not necessarily the longest, is considered a length.
4	Length of stationary object	Same.
5	Area of moving object	A geometrical characteristic described by the part of a plane enclosed by a line. The part of a surface occupied by the object. OR the square measure of the surface, either internal or external, of an object.
6	Area of stationary object	Same.
7	Volume of moving object	The cubic measure of space occupied by the object. Length x width x height for a rectangular object, height x area for a cylinder, etc.

TRIZ for Engineers: Enabling Inventive Problem Solving, First Edition. Karen Gadd.

Table (*Continued*)

No.	Title	Explanation
8	Volume of stationary object	Same.
9	Speed	The velocity of an object; the rate of a process or action in time.
10	Force	Force measures the interaction between systems. In Newtonian physics, force = mass x acceleration. In TRIZ, force is any interaction that is intended to change an object's condition.
11	Stress or pressure	Force per unit area. Also, tension.
12	Shape	The external contours, appearance of a system.
13	Stability of the object's composition	The wholeness or integrity of the system; the relationship of the system's constituent elements. Wear, chemical decomposition, and disassembly are all decreases in stability. Increasing entropy is decreasing stability.
14	Strength	The extent to which the object is able to resist changing in response to force. Resistance to breaking.
15	Duration of action by a moving object	The time that the object can perform the action. Service life. Mean time between failure is a measure of the duration of action. Also, durability.
16	Duration of action by a stationary object	Same.
17	Temperature	The thermal condition of the object or system. Loosely includes other thermal parameters, such as heat capacity, that affect the rate of change of temperature.
18	Illumination intensity	Light flux per unit area, also any other illumination characteristics of the system such as brightness, light quality, etc.
19	Use of energy by moving object	The measure of the object's capacity for doing work. In classical mechanics, Energy is the product of force x distance. This includes the use of energy provided by the super-system (such as electrical energy or heat.) Energy required to do a particular job.
20	Use of energy by stationary object	Same.
21	Power	The time rate at which work is performed. The rate of use of energy.
22	Loss of Energy	Use of energy that does not contribute to the job being done. See 19. Reducing the loss of energy sometimes requires different techniques from improving the use of energy, which is why this is a separate category.
23	Loss of substance	Partial or complete, permanent or temporary, loss of some of a system's materials, substances, parts or subsystems.
24	Loss of Information	Partial or complete, permanent or temporary, loss of data or access to data in or by a system. Frequently includes sensory data such as aroma, texture, etc.
25	Loss of Time	Time is the duration of an activity. Improving the loss of time means reducing the time taken for the activity. 'Cycle time reduction' is a common term.

Table (*Continued*)

No.	Title	Explanation
26	Quantity of substance/the matter	The number or amount of a system's materials, substances, parts or subsystems which might be changed fully or partially, permanently or temporarily.
27	Reliability	A system's ability to perform its intended functions in predictable ways and conditions.
28	Measurement accuracy	The closeness of the measured value to the actual value of a property of a system. Reducing the error in a measurement increases the accuracy of the measurement.
29	Manufacturing precision	The extent to which the actual characteristics of the system or object match the specified or required characteristics.
30	External harm affects the object	Susceptibility of a system to externally generated (harmful) effects.
31	Object-generated harmful factors	A harmful effect is one that reduces the efficiency or quality of the functioning of the object or system. These harmful effects are generated by the object or system, as part of its operation.
32	Ease of manufacture	The degree of facility, comfort or effortlessness in manufacturing or fabricating the object/system.
33	Ease of operation	Simplicity: The process is not easy if it requires a large number of people, large number of steps in the operation, needs special tools, etc. 'Hard' processes have low yield and 'easy' process have high yield; they are easy to do right.
34	Ease of repair	Quality characteristics such as convenience, comfort, simplicity, and time to repair faults, failures or defects in a system.
35	Adaptability or versatility	The extent to which a system/object positively responds to external changes. Also, a system that can be used in multiple ways for under a variety of circumstances.
36	Device complexity	The number and diversity of elements and element interrelationships within a system. The user may be an element of the system that increases the complexity. The difficulty of mastering the system is a measure of its complexity.
37	Difficulty of detecting and measuring	Measuring or monitoring systems that are complex, costly, require much time and labour to set up and use, or that have complex relationships between components or components that interfere with each other all demonstrate 'difficulty of detecting and measuring.' Increasing cost of measuring to a satisfactory error is also a sign of increased difficulty of measuring.
38	Extent of automation	The extent to which a system or object performs its functions without human interface. The lowest level of automation is the use of a manually operated tool. For intermediate levels, humans program the tool, observe its operation, and interrupt or re-program as needed. For the highest level, the machine senses the operation needed, programs itself and monitors its own operations.
39	Productivity	The number of functions or operations performed by a system per unit time. The time for a unit function or operation. The output per unit time, or the cost per unit output.

Appendix II
Contradiction Matrix

TRIZ for Engineers: Enabling Inventive Problem Solving, First Edition. Karen Gadd.

OXFORD CREATIVITY

40 Inventive Principles

Separation Principles for Solving **Physical Contradictions**

No.	Inventive Principle	Space	Time	Condition	Scale
1	Segmentation	✓	✓		✓
2	Taking Out	✓			
3	Local Quality	✓			✓
4	Asymmetry	✓			
5	Merging				✓
6	Universality				✓
7	Nested Doll	✓	✓		
8	Anti-Weight				✓
9	Prior Counteraction		✓		
10	Prior Action		✓		
11	Cushion in Advance		✓		
12	Equipotentiality				✓
13	The Other Way Round	✓			✓
14	Spheroidality - Curvature	✓			
15	Dynamics		✓		
16	Partial or Excessive Action		✓		
17	Another Dimension	✓			
18	Mechanical Vibration		✓		
19	Periodic Action		✓		
20	Continuity of Useful Action		✓		
21	Rushing Through		✓		
22	Blessing in Disguise				✓
23	Feedback				✓
24	Intermediary	✓	✓		
25	Self-Service				✓
26	Copying	✓	✓		
27	Cheap Short-Living Objects		✓		✓
28	Replace Mechanical System			✓	
29	Pneumatics and Hydraulics		✓	✓	
30	Flexible Membranes / Thin Films	✓			
31	Porous Materials			✓	
32	Colour Change			✓	
33	Homogeneity				✓
34	Discarding and Recovering		✓		
35	Parameter Change			✓	
36	Phase Transition			✓	
37	Thermal Expansion		✓		
38	Accelerate Oxidation			✓	
39	Inert Environment			✓	
40	Composite Materials	✓			✓

39 Technical Parameters

Improve this one (rows) without making this one worse (columns)

Parameters (columns): 1 Weight of moving object; 2 Weight of Stationary Object; 3 Length of moving object; 4 Length of Stationary object; 5 Area of moving object; 6 Area of stationary object; 7 Volume of moving object; 8 Volume of stationary object; 9 Speed; 10 Force (Intensity); 11 Stress or pressure; 12 Shape; 13 Stability of the object's composition; 14 Strength; 15 Duration of action of a moving object; 16 Duration of action of a stationary object; 17 Temperature; 18 Illumination Intensity; 19 Use of energy by a moving object; 20 Use of energy by a stationary object; 21 Power; 22 Loss of energy; 23 Loss of substance; 24 Loss of information; 25 Loss of time; 26 Quantity of substance; 27 Reliability; 28 Measurement accuracy; 29 Manufacturing precision; 30 Object-affected harmful factors; 31 Object-generated harmful factors; 32 Ease of manufacture; 33 Convenience of Use; 34 Ease of repair; 35 Adaptability or versatility; 36 Device complexity; 37 Difficulty of detecting and measuring; 38 Extent of automation; 39 Productivity

Improve ↓ / Worsening →	1	2	3	4	5	6	7	8	9	10	11	12	13
1 Weight of moving object		-	15 8 29 34	-	29 17 38 34	-	29 2 40 28	-	2 8 15 38	8 10 18 37	10 36 37 40	10 14 35 40	1 35 19 39
2 Weight of stationary object	-		-	10 1 29 35	-	35 30 13 2	-	5 35 14 2	-	8 10 19 35	13 29 10 18	13 10 29 14	26 39 1 40
3 Length of moving object	8 15 29 34	-		-	15 17 4	-	7 17 4 35	-	13 4 8	17 10 4	1 8 35	1 8 10 29	1 8 15 34
4 Length of stationary object	-	35 28 40 29	-		-	17 7 10 40	-	35 8 2 14	-	28 10	1 14 35	13 14 15 7	39 37 35
5 Area of moving object	2 17 29 4	-	14 15 18 4	-		-	7 14 17 4	-	29 30 4 34	19 30 35 2	10 15 36 28	5 34 29 4	11 2 13 39
6 Area of stationary object	-	30 2 14 18	-	26 7 9 39	-		-		-	1 18 35 36	10 15 36 37		2 38
7 Volume of moving object	2 26 29 40	-	1 7 4 35	-	1 7 4 17	-		-	29 4 38 34	15 35 36 37	6 35 36 37	1 15 29 4	28 10 1 39
8 Volume of stationary object	-	35 10 19 14	19 14	35 8 2 14	-		-		-	2 18 37	24 35	7 2 35	34 28 35 40
9 Speed	8 28 13 38	-	13 14 8	-	29 30 34	-	7 29 34	-		13 28 15 19	6 18 38 40	35 15 18 34	28 33 1 18
10 Force (Intensity)	8 1 37 18	18 13 1 28	17 19 9 36	28 10	19 10 15	1 18 36 37	15 9 12 37	2 36 18 37	13 28 15 12		18 21 11	10 35 40 34	35 10 21
11 Stress or pressure	10 36 37 40	13 29 10 18	35 10 36	35 1 14 16	10 15 36 28	10 15 36 37	6 35 10	35 24	6 35 36	36 35 21		35 4 15 10	35 33 2 40
12 Shape	8 10 29 40	15 10 26 3	29 34 5 4	13 14 10 7	5 34 4 10		14 4 15 22	7 2 35	35 15 34 18	35 10 37 40	34 15 10 14		33 1 18 4
13 Stability of the object's composition	21 35 2 39	26 39 1 40	13 15 1 28	37	2 11 13	39	28 10 19 39	34 28 35 40	33 15 28 18	10 35 21 16	2 35 40	22 1 18 4	
14 Strength	1 8 40 15	40 26 27 1	1 15 8 35	15 14 28 26	3 34 40 29	9 40 28	10 15 14 7	9 14 17 15	8 13 26 14	10 18 3 14	10 3 18 40	10 30 35 40	13 17 35
15 Duration of action by moving object	19 5 34 31	-	2 19 9	-	3 17 19	-	10 2 19 30	-	3 35 5	19 2 16	19 3 27	14 26 28 25	13 3 35
16 Duration of action by Stationary object	-	6 27 19 16	-	1 40 35	-	-	-	35 34 38	-				39 35 3 23
17 Temperature	36 22 6 38	22 35 32	15 19 9	15 19 9	3 35 39 18	35 38	34 39 40 18	35 6 4	2 28 36 30	35 10 3 21	35 39 19 2	14 22 19 32	1 35 32
18 Illumination Intensity	19 1 32	2 35 32	19 32 16		19 32 26		2 13 10		13 19 10	26 19 6		32 30	32 3 27
19 Use of energy by moving object	12 18 28 31	-	12 28	-	15 19 25	-	35 13 18	-	8 15 35	16 26 21 2	23 14 25	12 2 29	19 13 17 24
20 Use of energy by stationary object	-	19 9 6 27								36 37			27 4 29 18
21 Power	8 36 38 31	19 26 17 27	1 10 35 37		19 38	17 32 13 38	35 6 38	30 6 25	15 35 2	26 2 36 35	22 10 35	29 14 2 40	35 32 15 31
22 Loss of energy	15 6 19 28	19 6 18 9	7 2 6 13	6 38 7	15 26 17 30	17 7 30 18	7 18 23	7	16 35 38	36 38			14 2 39 6
23 Loss of Substance	35 6 23 40	35 6 22 32	14 29 10 39	10 28 24	35 2 10 31	10 18 39 31	1 29 30 36	3 39 18 31	10 13 28 38	14 15 18 40	3 36 37 10	29 35 3 5	2 14 30 40
24 Loss of Information	10 24 35	10 35 5	1 26	26	30 26	30 16		2 22	26 32				
25 Loss of Time	10 20 37 35	10 20 26 5	15 2 29	30 24 14 5	26 4 5 16	10 35 17 4	2 5 34 10	35 16 32 18		10 37 36 5	37 36 4	4 10 34 17	35 3 22 5
26 Quantity of Substance	35 6 18 31	27 26 18 35	29 35 14 18		15 14 29	2 18 40 4	15 20 29		35 29 34 28	35 14 3	10 36 14 3	35 14	15 2 17 40
27 Reliability	3 8 10 40	3 10 8 28	15 9 14 4	15 29 28 11	17 10 14 16	32 35 40 4	3 10 14 24	2 35 24	21 35 11 28	8 28 10 3	10 24 35 19	35 1 16 11	
28 Measurement Accuracy	32 35 26 28	28 35 25 26	28 26 5 16	32 28 3 16	26 28 32 3	26 28 32 3	32 13 6		28 13 32 24	32 2	6 28 32	6 28 32	32 35 13
29 Manufacturing Precision	28 32 13 18	28 35 27 9	10 28 29 37	2 32 10	28 33 29 32	2 29 18 36	32 23 2	25 10 35	10 28 32	28 19 34 36	3 35	32 30 40	30 18
30 Object affected harmful factors	22 21 27 39	2 22 13 24	17 1 39 4	1 18	22 1 33 28	27 2 39 35	22 23 37 35	34 39 19 27	21 22 35 28	13 35 39 18	22 2 37	22 1 3 35	35 24 30 18
31 Object-generated harmful factors	19 22 15 39	35 22 1 39	17 15 16 22		17 2 18 39	22 1 40	17 2 40	30 18 35 4	35 28 3 23	35 28 1 40	2 33 27 18	35 1	35 40 27 39
32 Ease of manufacture	28 29 15 16	1 27 36 13	1 29 13 17	15 17 27	13 1 26 12	16 40	13 29 1 40	35	35 13 8 1	35 12	35 19 1 37	1 28 13 27	11 13 1
33 Convenience of Use	25 2 13 15	6 13 1 25	1 17 13 12		1 17 13 16	18 16 15 39	1 16 35 15	4 18 39 31	18 13 34	28 13 35	2 32 12	15 34 29 28	32 35 30
34 Ease of repair	2 27 35 11	2 27 35 11	1 28 10 25	3 18 31	15 13 32	16 25	25 2 35 11	1	34 9	1 11 10	13	1 13 2 4	2 35
35 Adaptability or versatility	1 6 15 8	19 15 29 16	35 1 29 2	1 35 16	35 30 29 7	15 16	15 35 29		35 10 14	15 17 20	35 16	15 37 1 8	35 30 14
36 Device complexity	26 30 34 36	2 26 35 39	1 19 26 24	26	14 1 13 16	6 36	34 26 6	1 16	34 10 28	26 16	19 1 35	29 13 28 15	2 22 17 19
37 Difficulty of detecting and measuring	27 26 28 13	6 13 28 1	16 17 26 24	26	2 13 18 17	2 39 30 16	29 1 4 16	2 18 26 31	3 4 16 35	36 28 40 19	35 36 37 32	27 13 1 39	11 22 39 30
38 Extent of automation	28 26 18 35	28 26 35 10	14 13 17 28	23	17 14 13		35 13 16		28 10	2 35	13 35	15 32 1 13	18 1
39 Productivity	35 26 24 37	28 27 15 3	18 4 28 38	30 7 14 26	10 26 34 31	10 35 17 7	2 6 34 10	35 37 10 2		28 15 10 36	10 37 14	14 10 34 40	35 3 22 39

Improve ↓ / Worsening →	14	15	16	17	18	19	20	21	22	23	24	25	26
1 Weight of moving object	28 27 18 40	5 34 31 35	-	6 29 4 38	19 1 32	35 12 34 31	-	12 36 18 31	6 2 34 19	5 35 3 31	10 24 35	10 35 20 28	3 26 18 31
2 Weight of stationary object	28 2 10 27	-	2 27 19 6	28 19 32 22	19 32 35	-	18 19 28 1	15 19 18	18 19 28 15	5 8 13 30	10 15 35	10 20 35 26	19 6 18 26
3 Length of moving object	8 35 29 34	19	-	10 15 19	32	8 35 24	-	1 35	7 2 35 39	4 29 23 10	1 24	15 2 29	29 35
4 Length of stationary object	15 14 28 26	-	1 40 35	3 35 38 18	3 25	-		12 8	6 28	10 28 24 35	24 26	30 29 14	
5 Area of moving object	3 15 40 14	6 3	-	2 15 16	15 32 19 13	19 32	-	19 10 32 18	15 17 30 26	10 35 2 39	30 26	26 4	29 30 6 13
6 Area of stationary object	40	-	2 10 19 30	35 39 38		-		17 32	17 7 30	10 14 18 39	30 16	10 35 4 18	2 18 40 4
7 Volume of moving object	9 14 15 7	6 35 4	-	34 39 10 18	2 13 10	35	-	35 6 13 18	7 15 13 16	36 39 34 10	2 22	2 6 34 10	29 30 7
8 Volume of stationary object	9 14 17 15	-	35 34 38	35 6 4		-		30 6		10 39 35 34		35 16 32 18	35 3
9 Speed	8 3 26 14	3 19 35 5	-	28 30 36 2	10 13 19	8 15 35 38	-	19 35 38 2	14 20 19 35	10 13 28 38	13 26		10 19 29 38
10 Force (Intensity)	35 10 14 27	19 2		35 10 21	-	19 17 10	1 16 36 37	19 35 18 37	14 15	8 35 40 5		10 37 36	14 29 18 36
11 Stress or pressure	9 18 3 40	19 3 27		35 39 19 2	-	14 24 10 37		10 35 14	2 36 25	10 35 3 37		37 36 4	10 14 36
12 Shape	30 14 10 40	14 26 9 25		22 14 19 32	13 15 32	2 6 34 14		4 6 2	14	35 29 3 5		14 10 34 17	36 22
13 Stability of the object's composition	17 9 15	13 27 10 35	39 3 35 23	35 1 32	32 3 27 15	13 19	27 4 29 18	32 35 27 31	14 2 39 6	2 14 30 40		35 27	15 32 35
14 Strength		27 3 26		30 10 40	35 19	19 35 10	35	10 26 35 28	35	35 28 31 40		29 3 28 10	29 10 27
15 Duration of action by moving object	27 3 10		-	19 35 39	2 19 4 35	28 6 35 18		19 10 35 38		28 27 3 18	10	20 10 28 18	3 35 10 40
16 Duration of action by Stationary object		-		19 18 36 40		-		16		27 16 18 38	10	28 20 10 16	3 35 31
17 Temperature	10 30 22 40	19 13 39	19 18 36 40		32 30 21 16	19 15 3 17		2 14 17 25	21 17 35 38	21 36 39 31		35 28 21 18	3 17 30 39
18 Illumination Intensity	35 19	2 19 6		32 35 19		32 1 19	32 35 1 15	32	13 16 1 6	1 31	1 6	19 1 26 17	1 19
19 Use of energy by moving object	5 19 9 35	28 35 6 18	-	19 24 3 14	2 15 19		-	6 19 37 18	12 22 15 24	35 24 18 5		35 38 19 18	34 23 16 18
20 Use of energy by stationary object	35				19 2 35 32	-				28 27 18 31			3 35 31
21 Power	26 10 28	19 35 10 38	16	2 14 17 25	16 6 19	16 6 19 37			10 35 38	28 27 18 38	10 19	35 20 10 6	4 34 19
22 Loss of energy	26			19 38 7	1 13 32 15			3 38		35 27 2 37	19 10	10 18 32 7	7 18 25
23 Loss of Substance	35 28 31 40	28 27 3 18	27 16 18 38	21 36 39 31	1 6 13	35 18 24 5	28 27 12 31	28 27 18 38	35 27 2 31			15 18 35 10	6 3 10 24
24 Loss of Information		10	10		19			10 19	19 10			24 26 28 32	24 28 35
25 Loss of Time	29 3 28 18	20 10 28 18	28 20 10 16	35 29 21 18	1 19 26 17	35 38 19 18	1	35 20 10 6	10 5 18 32	35 18 10 39	24 26 28 32		35 38 18 16
26 Quantity of Substance	14 35 34 10	3 35 10 40	3 35 31	3 17 39		34 29 16 18	3 35 31	35	7 18 25	6 3 10 24	24 28 35	35 38 18 16	
27 Reliability	11 28	2 35 3 25	34 27 6 40	3 35 10	11 32 13	21 11 27 19	36 23	21 11 26 31	10 11 35	10 35 29 39	10 28	10 30 4	21 28 40 3
28 Measurement Accuracy	28 6 32	28 6 32	10 26 24	6 19 28 24	6 1 32	3 6 32		3 6 32	26 32 27	10 16 31 28		24 34 28 32	2 6 32
29 Manufacturing Precision	3 27	3 27 40		19 26	3 32	32 2		32 2	13 32 2	35 31 10 24		32 26 28 18	32 30
30 Object affected harmful factors	18 35 37 1	22 15 33 28	17 1 40 33	22 33 35 2	1 19 32 13	1 24 6 27	10 2 22 37	19 22 31 2	21 22 35 2	33 22 19 40	22 10 2	35 18 34	35 33 29 31
31 Object-generated harmful factors	15 35 22 2	15 22 33 31	21 39 16 22	22 35 2 24	19 24 39 32	2 35 6	19 22 18	2 35 18	21 35 2 22	10 1 34	10 21 29	1 22	3 24 39 1
32 Ease of manufacture	1 3 10 32	27 1 4	35 16	27 26 18	28 24 27 1	28 26 27 1	1 4	27 1 12 24	19 35	15 34 33	32 24 18 16	35 28 34 4	35 23 1 24
33 Convenience of Use	32 40 3 28	29 3 8 25	1 16 25	26 27 13	13 17 1 24	1 13 24		35 34 2 10	2 19 13	28 32 2 24	4 10 27 22	4 28 10 34	12 35
34 Ease of repair	11 1 2 9	11 29 28 27	1	4 10	15 1 13	15 1 28 16		15 10 32 2	15 1 32 19	2 35 34 27		32 1 10 25	2 28 10 25
35 Adaptability or versatility	35 3 32 6	13 1 35	2 16	27 2 3 35	6 22 26 1	19 35 29 13		19 1 29	18 15 1	15 10 2 13		35 28	3 35 15
36 Device complexity	2 13 28	10 4 28 15		2 17 13	24 17 13	27 2 28 29		20 19 30 34	10 35 13 2	35 10 28 29		6 29	13 3 27 10
37 Difficulty of detecting and measuring	27 3 15 28	19 29 39 25	25 34 6 35	3 27 35 16	2 24 26	35 38	19 35 16	19 1 16 10	35 3 15 19	1 18 10 24	35 33 27 22	18 28 32 9	3 27 29 18
38 Extent of automation	25 13	6 9		26 2 19	8 32 19	2 32 13		28 2 27	23 28	35 10 18 5	35 33	24 28 35 30	35 13
39 Productivity	29 28 10 18	35 10 2 18	20 10 16 38	35 21 28 10	26 17 19 1	35 10 38 19	1	35 20 10	28 10 29 35	28 10 35 23	13 15 23		35 38

Improve ↓ / Worsening →	27	28	29	30	31	32	33	34	35	36	37	38	39
1 Weight of moving object	3 11 1 27	28 27 35 26	28 35 26 18	22 21 18 27	22 35 31 39	27 28 1 36	35 3 2 24	2 27 28 11	29 5 15 8	26 30 36 34	28 29 26 32	26 35 18 19	35 3 24 37
2 Weight of stationary object	10 28 8 3	18 26 28	10 1 35 27	2 19 22 37	35 22 1 39	28 1 9	6 13 1 32	2 27 28 11	19 15 29	1 10 26 39	25 28 17 15	2 26 35	1 28 15 35
3 Length of moving object	10 14 29 40	28 32 4	10 28 29 37	1 15 17 24	17 15	1 29 17	15 29 35 4	1 28 10	14 15 1 16	1 19 26 24	35 1 26 24	17 24 26 16	14 4 28 29
4 Length of stationary object	15 29 28	32 28 3	2 32 10	1 18		15 17 27	2 25	3	1 35	1 26	26		30 14 7 26
5 Area of moving object	29 9	26 28 32 3	2 32	22 33 28 1	17 2 18 39	13 1 26 24	15 17 13 16	15 13 10 1	15 30	14 1 13	2 36 26 18	14 30 28 23	10 26 34 2
6 Area of stationary object	32 35 40 4	26 28 32 3	2 29 18 36	27 2 39 35	22 1 40	40 16	16 4	16	15 16	1 18 36	2 35 30 18	23	10 15 17 7
7 Volume of moving object	14 1 40 11	26 28	25 28 2 16	22 21 27 35	17 2 40 1	29 1 40	15 13 30 12	10	15 29	26 1	29 26 4	35 34 16 24	10 6 2 34
8 Volume of stationary object	2 35 16		35 10 25	34 39 19 27	30 18 35 4	35		1		1 31	2 17 26		35 37 10 2
9 Speed	11 35 27 28	28 32 1 24	10 28 32 25	1 28 35 23	2 24 35 21	35 13 8 1	32 28 13 12	34 2 28 27	15 10 26	10 28 4 34	3 34 27 16	10 18	
10 Force (Intensity)	3 35 13 21	35 10 23 24	28 29 37 36	1 35 40 18	13 3 36 24	15 37 18 1	1 28 3 25	15 1 11	15 17 18 20	26 35 10 18	36 37 10 19	2 35	3 28 35 37
11 Stress or pressure	10 13 19 35	6 28 25	3 35	22 2 37	2 33 27 18	1 35 16	11	2	35	19 1 35	2 36 37	35 24	10 14 35 37
12 Shape	10 40 16	28 32 1	32 30 40	22 1 2 35	35 1	17 32 1 28	32 15 26	2 13 1	1 15 29	16 29 1 28	15 13 39	15 1 32	17 26 34 10
13 Stability of the object's composition		13	18	35 24 30 18	35 40 27 39	35 19	32 35 30	2 35 10 16	35 30 34 2	2 35 22 26	35 22 39 23	1 8 35	23 35 40 3
14 Strength	11 3	3 27 16	3 27	18 35 37 1	15 35 22 2	11 3 10 32	32 40 28 2	27 11 3	15 3 32	2 13 28	27 3 15 40	15	29 35 10 14
15 Duration of action by moving object	11 2 13	3	3 27 16 40	22 15 33 28	21 39 16 22	27 1 4	12 27	29 10 27	1 35 13	10 4 28 15	19 29 39 35	6 10	35 17 14 19
16 Duration of action by Stationary object	34 27 6 40	10 26 24		17 1 40 33	22	35 10	1	1	2		25 34 6 35	1	20 10 16 38
17 Temperature	19 35 3 10	32 19 24	24	22 33 35 2	22 35 2 24	26 27	26 27	4 10 16	2 18 27	2 17 16	3 27 35 31	26 2 19 16	15 28 35
18 Illumination Intensity		11 15 32	3 32	15 19	35 19 32 39	19 35 28 26	28 26 19	15 17 13 16	15 1 19	6 32 13	32 15	2 26 10	2 25 16
19 Use of energy by moving object	19 21 11 27	3 1 32		1 35 6 27	2 35 6	28 26 30	19 35	1 15 17 28	15 17 13 16	2 29 27 28	35 38	32 2	12 28 35
20 Use of energy by stationary object	10 36 23			10 2 22 37	19 22 18	1 4					19 35 16 25		1 6
21 Power	19 24 26 31	32 15 2	32 2	19 22 31 2	18 2 35	26 10 34	26 35 10	35 2 10 34	19 17 34	20 19 30 34	19 35 16	28 2 17	28 35 34
22 Loss of energy	11 10 35	32		21 22 35 2	21 35 2 22		35 32 1	2 19		7 23	35 3 15 23	2	28 10 29 35
23 Loss of Substance	10 29 39 35	16 34 31 28	35 10 24 31	33 22 30 40	10 1 34 29	15 34 33	32 28 2 24	2 35 34 27	15 10 2	35 10 28 24	35 18 10 13	35 10 18	28 35 10 23
24 Loss of Information	10 28 23			22 10 1	10 21 22	32	27 22				35 33	35	13 23 15
25 Loss of Time	10 30 4	24 34 28 32	24 26 28 18	35 18 34	35 22 18 39	35 28 34 4	4 28 10 34	32 10 1	35 28	6 29	18 28 32 10	24 28 35 30	
26 Quantity of Substance	18 3 28 40	3 2 28	33 30	35 33 29 31	3 35 40 39	29 1 35 27	35 29 25 10	2 32 10 25	15 3 29	3 13 27 10	3 27 29 18	8 35	13 29 3 27
27 Reliability		32 3 11 23	11 32 1	27 35 2 40	35 2 40 26		27 17 40	1 11	13 35 8 24	13 35 1	27 40 28	11 13 27	1 35 29 38
28 Measurement Accuracy	5 11 1 23			28 24 22 26	3 33 39 10	6 35 25 18	1 13 17 34	1 32 13 11	13 35 2	27 35 10 34	26 24 32 28	28 2 10 34	10 34 28 32
29 Manufacturing Precision	11 32 1			26 28 10 36	4 17 34 26		1 32 35 23	25 10		26 2 18		26 28 18 23	10 18 32 39
30 Object affected harmful factors	27 24 2 40	28 33 23 26	26 28 10 18			24 35 2	2 25 28 39	35 10 2	35 11 22 31	22 19 29 40	22 19 29 40	33 3 34	22 35 13 24
31 Object-generated harmful factors	24 2 40 39	3 33 26	4 17 34 26							19 1 31	2 21 27 1	2	22 35 18 39
32 Ease of manufacture		1 35 12 18		24 2			2 5 13 16	35 1 11 9	2 13 15	27 26 1	6 28 11 1	8 28 1	35 10 28 1
33 Convenience of Use	17 27 8 40	25 13 2 34	1 32 35 23	2 25 28 39		2 5 12		12 26 1 32	15 34 1 16	32 26 12 17		1 34 12 3	15 1 28
34 Ease of repair	11 10 1 16	10 2 13	25 10	35 10 2 16		1 35 11 10	1 12 26 15		7 1 4 16	35 1 13 11		34 35 7 13	1 32 10
35 Adaptability or versatility	35 13 8 24	35 5 1 10		35 11 32 31		1 13 31	15 34 1 16	1 16 7 4		15 29 37 28	1	27 34 35	35 28 6 37
36 Device complexity	13 35 1	2 26 10 34	26 24 32	22 19 29 40	19 1	27 26 1 13	27 9 26 24	1 13	29 15 28 37		15 10 37 28	15 1 24	12 17 28
37 Difficulty of detecting and measuring	27 40 28 8	26 24 32 28		22 19 29 28	2 21	5 28 11 29	2 5	12 26	1 15	15 10 37 28		34 21	35 18
38 Extent of automation	11 27 32	28 26 10 34	28 26 18 23	2 33	2	1 26 13	1 12 34 3	1 35 13	27 4 1 35	15 24 10	34 27 25		5 12 35 26
39 Productivity	1 35 10 38	1 10 34 28	18 10 32 1	22 35 13 24	35 22 18 39	35 28 2 24	1 28 7 19	1 32 10 25	1 35 28 37	12 17 28 24	35 18 27 2	5 12 35 26	

Contradiction Matrix for Solving **Technical Contradictions**

Glossary

TRIZ	Teoriya Resheniya Izobreatatelskikh Zadatch = the Theory of Inventive Problem Solving
39 Engineering Parameters	Fundamental descriptions used to define any engineering contradiction such as strength and weight
40 Inventive Principles	40 solutions to any contradiction
76 Standard Solutions	List of conceptual triggers that show you how to deal with harms and insufficiencies. They can be used with a Function Analysis diagram
Action	The change the subject makes to the object e.g. Needle *pierces* skin
Analogous Thinking	Locating the relevant and useful similarity at a conceptual level between systems that are dissimilar at the detail/application level (e.g the inventor of lawnmowers based the design on a small hand-held carpet trimmer). In TRIZ, analogous thinking can be achieved systematically by using the Prism of TRIZ
ARIZ	Algorithm of Inventive Problem Solving
Bad Solution Park	Call all spontaneous solution ideas 'bad solutions' and encourage their capture on stickies and then park them in a Bad Solution Park. They are the starting point of the TRIZ contradiction problem-solving process which is designed to keep the good in solutions but deal with anything bad
Benefits	An output we want or have from our system: it's definition does not contain a solution (e.g. We would say we want our car to be easy to park not small). We get our benefits from functions and features
Causes and Solutions Map	9-Boxes Map to locate the relationship between causes and effects of problems (and hazards) at different moments in time or different steps in a process, and different places in scale of a system, all its details and its context/environment
Component	An element of our system. In Function Analysis, this could be either the subject or the object
Concept Lists	Four lists of conceptual solution triggers based on patent analysis: the 40 Principles, the Trends of Technical Evolution, the76 Standard Solutions and the Effects Database
Conceptual Problem	A problem which describes only the problem functions with no detail of the type of system. Conceptual problems describe the functions which are needed or missing or excessive or harmful or insufficient

TRIZ for Engineers: Enabling Inventive Problem Solving, First Edition. Karen Gadd.

Conceptual Solution A solution which has no detail or ideas only its essential meaning. TRIZ conceptual solutions offer us the relevant knowledge which engineering teams can then expand to practical solutions by adding in relevant domain knowledge and experience

Constraints A reason for *why not* – a reason to stop, constrain, prevent, restrict, direct, confine an object or action. In engineering, constraints should be rigorously defined and tested to ensure that they really exist and are not assumed or imagined.

Context Map 9-Boxes Map which describes both all the essential and relevant time steps – the history, current position and future context of a problem or situation and its scale from the big picture of its environment/context through to the details and relevant small elements of the system

Contradiction A situation where we want conflicting requirements, e.g. we want something to be light but stiff

Contradiction Matrix 39 x 39 Matrix defined by the 39 Engineering Parameters which shows which of the 40 Inventive Principles other engineers have previously successfully used to solve contradictions similar to the ones being analysed.

Cost An input into our system, e.g. time, money, effort

Creativity Tools Simple and powerful thinking tools to overcome psychological inertia, and are good brain prompts to shift our focus and help us think fast and powerfully to both understand and solve problems

Features What we want from an object but may not be included in its functionality, e.g. wine glass attributes might be easy to make sparkle (tangible) and/or pleasing to look at (intangible)

Field Energy needed for interaction of two substances. In addition to the four fundamental fields – electromagnetic, gravitational, and nuclear fields of weak and strong interactions – TRIZ deals with engineering fields, such as mechanical, thermal, electric, magnetic, and chemical.

Function Analysis Aim of Function Analysis is to structure the whole model of the problem by restating it using simple language (no specialist terms, no acronyms) as a series of Subject-Action-Objects. It is a tool for both system understanding and problem solving.

Functionality Benefits you need from a system, e.g. the main function of a wine glass is to provide a vessel which is pleasant to drink from. Its Functionality is – hold liquid, enable to see liquid and smell liquid, do not heat up wine etc.

Functions Benefits are delivered by functions, which are delivered by features, e.g. I want a car which is easy to park (benefit), which could be delivered by being a small car. Functions are defined by Subject-Action-Objects

Harms Harms are outputs from our system which we don't want. This includes outputs which are not actively harmful but we don't need and are paying for, e.g. heat from a light bulb

Hazards Dangers – an unavoidable risk or chance of being injured or harmed; even when foreseeable. The scale or risk of a hazard can be estimated by the multiplying its likelihood with the downside or harm

Hazards Map Maps the relationship between causes and outcomes of hazards at different moments in time and different places in scale for a system and its environment

Ideal Outcome A TRIZ thinking tool. The Ideal Outcome allows us to understand what we really want, without thinking in constraints. We imagine what our Ideal Outcome would be if we had a magic wand

Ideality A measure of how good our system is: what are the benefits are system delivers, what are the costs and what are the harms

Ideality Audit We compare what we want (our Ideal Outcome) with what we have (our current processes) by working through each benefit in our Ideal Outcome, systematically, and ask whether our system delivers this benefit, and if it does, if it delivers enough of the benefit. We then assess the costs and harms of our system, and assess whether they are acceptable or not. Any insufficient, missing or excessive benefit, or any unacceptable cost or harm, is a problem

Ideality Equation How we understand our Ideality: Ideality Equation = Benefits/Costs + Harms. This helps us to understand that to improve our Ideality we can increase the benefits, decrease the costs and/or decrease the costs and harms

Ideality Plot A way of categorizing ideas according what's good about them (their Benefits) and what's bad about them (their Costs and Harms). We identify the level of Benefits (from high – which is good – to low – which is bad), and then their Costs and Harms (from high Costs and Harms – which is bad – to low Costs and Harms – which is good). This helps prioritize the implementation of ideas

Innovation Audit Shows exactly which solutions the engineering team have uncovered, demonstrates the TRIZ processes and tools which have been used, and records all the systematic innovation steps which have been followed. Also an innovation audit ensures that everyone understands that there is a systematic and complete problem understanding and solving process (which includes an Ideality Audit, Bad Solution Park and 9-Box TRIZ Solution Park) and that all the essential elements of this process are recorded not just the solutions. It shows that systematic innovation has been undertaken and provided results

Insufficiency A benefit we want but not enough of it

Needs What we want from our system

Object The component which is changed by the action, e.g. needle pierces *skin*

Parameter A feature of our system: contradictions are defined by contradictory features

Physical Contradiction When same system must satisfy opposite and mutually exclusive conditions, e.g. be hot/cold, present/absent, sharp/dull etc.

Primary Benefits The most important benefits of our system – must haves

Prime Output Prime purpose of our system, e.g. for an insulin pen this would be deliver insulin to user

Prism of TRIZ Every problem has been solved before, on a conceptual level. TRIZ has simple conceptual lists that help us access the world's knowledge of solved problems: we can use this to help solve the problem we have at hand

Psychological Inertia Mental habits which suggest imagined/false constraints or keep us stuck in the same old solution spaces and prevent innovation, clarity of understanding and thought. TRIZ tools break psychological inertia by subjugating constraints, moving us away from current solutions and images of problems, to re-construct and define the essential problem and see clever solutions in new and different ways

Requirements	Everything we really want, small and large. Accurate definitions are essential in problem solving. Systems exist to deliver needs/requirements/benefits. See also benefit/need
Resources	TRIZ is about getting what we want with the least harm and cost: we can do this by clever use of our resources. Before we can do this we must understand the functions that we want, and this clear thinking helps us look at what we have with new eyes, and see how to use our resources cleverly
Secondary Benefits	Other benefits we would like from our system but are less important than the Primary Benefits – nice to haves
Separate by Scale	Get opposite benefits by separating by system, e.g. a bicycle chain is rigid at subsystem level and flexible at system level
Separate in Space	Get opposite benefits in different places, e.g. a coffee cup is hot where it holds coffee and cold where we hold it
Separate in Time	Get opposite benefits at different times, e.g. umbrellas are big in use and small for storage
Separate on Condition	Get opposite benefits according to condition, e.g. a sieve is there for pasta (collects pasta) but not there for water (water flows through)
Separation Principles	Subset of the 40 Principles used to solve Physical Contradictions
Size-Time-Cost	Creativity tool for breaking psychological inertia. You imagine how you would solve this problem by expanding the parameters of size, time and cost to infinity and to zero, e.g. how would I solve this problem if I had an unlimited budget, or no budget?
Smart Little People	Creativity tool for understanding and solving problems where we imagine the elements of our system made up of many small, clever people
Solution Mode	Visualization of solutions. This is often simultaneous with problem understanding. More than anyone else engineers are more inclined to think about solutions rather than problems! Start describing a problem to engineers and after a few minutes they are listening with only a part of their brain, because they have thought of answers and are thinking hard about their own ideas and how to develop them (see Bad Solution Park)
Solution Trigger	Any idea or concept which helps us think up solutions
Stakeholder	Any person or institution who has involvement in our system, e.g. customers, suppliers, us
Subject	The component providing the action e.g. *Needle* pierces skin
Sub-system	Detail of our system, e.g. components
Super-system	Anything outside our system e.g. The product range, the company, the environment
System	The thing being studied: it can be a product, a process, or a level of analysis
Technical Contradiction	When two different parameters are in conflict so that when you improve one the other gets worse, e.g. strength vs. weight.

Thinking in Time and Scale	A thinking tool which models clever and creative thinking, and acts as a kind of lens onto our problem. We think about the different levels of our system, e.g. environment (supersystem), our system itself and the details or components (subsystem). We then think about how those levels change and interact over time, from the past to present to future. It can be used in a number of ways: to understand the context of the problem; we can map hazards and causes of problems; find solutions to problems; and identify resources
Trends of Technical Evolution	Based on patent analysis of how systems have develop over time. They show that systems develop in predictable ways, and if we can understand which trend a part of our system is following, we can predict its likely future development. The Trends are very useful for product development, developing next generation systems and for IP and patent work
Trimming	Trimming simplifies our system and gets rid of problems by removing components that involve harms, cost or complexity, while making sure we keep all their useful actions, by transferring those responsibilities to other parts of our system
Trimming Rules	Simple and systematic set of questions we follow to trim, giving us guidelines for the transferring of responsibilities for useful actions to other parts of our system. This allows us to remove components of our system without losing any of their useful actions. Can reduce part count and complexity. Can also lead to radical changes and a rapid, complete system redesign
TRIZ Effects Database	A list of over 2,500 physical effects, accessed by asking simple questions, e.g. how do I move a liquid?
X-Factor	A thinking tool that helps us understand what we want and see how to get it. With the X-Factor you pretend that a magical substance – called the X-Factor – will solve your problem for you, without worrying about practicalities. Once you have defined what benefit you want, you then look to see whether any resources currently in your system can deliver this benefit

Index

TRIZ for Engineers: Enabling Inventive Problem Solving, First Edition. Karen Gadd.

Lightning Source UK Ltd.
Milton Keynes UK
UKOW07f2200090516

273901UK00001B/1/P